STUDIES IN VIRAL ECOLOGY

STUDIES IN VIRAL ECOLOGY

Animal Host Systems

Volume 2

Edited By

CHRISTON J. HURST
Xavier University
Cincinnati, Ohio
USA

and

Universidad del Valle
Santiago de Cali, Valle
Colombia

WILEY-BLACKWELL

A John Wiley & Sons, Inc., Publication

Published by John Wiley & Sons, Inc., Hoboken, New Jersey
Published simultaneously in Canada

Library of Congress Cataloging-in-Publication Data:

Studies in viral ecology / edited by Christon J. Hurst.
v. cm.
Includes index.
Contents: v.1. Microbial and Botanical Host Systems (ISBN 978-0-470-62396-1)
– v.2. Animal Host Systems (ISBN 978-0-470-62429-6).
ISBN (set) 978-1-118-02458-4 (cloth)
1. Viruses–Ecology. I. Hurst, Christon J.
QR478.A1S78 2011
579.2–dc22

2010046370

Printed in Singapore.

oBook ISBN: 978-1-118-02571-0
ePDF ISBN: 978-1-118-02567-3
ePub ISBN: 978-1-118-02569-7

10 9 8 7 6 5 4 3 2 1

CONTENTS

VOLUME 1

VOLUME 2

DEDICATION

I dedicate these two volumes to the memory of my brother in spirit, Henry Hanssen. To me, he seemed a hero and I remember him most for his unfailing ability to present a sense of humanity in times of tragedy. We first met while studying together for our doctorates in Houston, Texas.

Henry was born in Colombia near Medellín and tragically orphaned as a young child after which he was lovingly raised by an aunt in Bogotá. Henry may have gained his tremendous sense of humanity from that experience. He had no biological children of his own but helped to raise two daughters. The first of those came into his life by a twist of luck while one day Henry was walking along a street in Colombia and heard what he thought might be a cat trapped inside of a garbage bin. Henry went over to free the cat and discovered instead a crying infant child in a plastic bag, presumably discarded there by a distraught mother. Henry took the baby to the police, and when no one stepped forward as a parent Henry adopted the child and eventually even helped to pay for her college tuition. The second daughter came through Henry's marriage to the love of his life.

When there arose need for representing humanity, Henry was undaunted by circumstance. His accomplishments included establishing an infant vaccination program against poliomyelitis in Angola at the personal request of Jonas Salk. Angola was in a state of civil war at that time and no one else was willing to undertake the necessary but frightening task. Henry showed equal humanitarianism to civilians and military on both sides of that conflict. Subsequently, Henry initiated a similar poliomyelitis vaccination program during a period of civil war in Central America and for his efforts was awarded honorary citizenship by one of the countries there. He then initiated a poliomyelitis vaccination program in his native Colombia, while that country's continuing civil war was in full strength.

I was proud to address Henry by the name of "brother" and always will think of him in that way. He addressed me by that same term of affection and he is lovingly remembered by everyone whom his life touched.

Henry Hanssen Villamizar (1945–2007)

PREFACE

Virology is a field of study which has grown and expanded greatly since the viruses as a group first received their name in 1898. Many of the people who presently are learning virology have come to perceive these acellular biological entities as being merely trinkets of nucleic acid to be cloned, probed, and spliced. However, the viruses are much more than merely trinkets to be played with in molecular biology laboratories. The viruses are indeed highly evolved biological entities with an organismal biology that is complex and interwoven with the biology of their hosting species. Ecology is defined as the branch of science which addresses the relationships between an organism of interest and the other organisms with which it interacts, the interactions between the organism of interest and its environment, and the geographical distribution of the organism of interest.

The purpose of this book is to help define and explain the ecology of viruses, i.e., to examine what life might seem like from a "virocentric" point of view, as opposed to our normal "anthropocentric" perspective. As we begin our examination of the virocentric life, it is important to realize that in nature both the viruses of macroorganisms and the viruses of microorganisms exist in cycles with their respective hosts. Under normal conditions, the impact of viruses upon their natural host populations may be barely apparent due to factors such as evolutionary coadaptation between the virus and those natural hosts. However, when viruses find access to new types of hosts and alternate transmission cycles, or when they encounter a concentrated population of susceptible genetically similar hosts such as occurs in densely populated human communities, communities of cultivated plants or animals, or algal blooms, then the impact of the virus upon its host population can appear catastrophic. The key to understanding these types of cycles lies in understanding the viruses and how their ecology relates to the ecology of their hosts, their alternate hosts, and any vectors which they utilize, as well as their relationship to the availability of suitable vehicles that can transport the different viral groups.

I hope that you will enjoy the information presented in this book set as much as I and the other authors have enjoyed presenting it to you. The written word is a marvelous thing, able to convey understanding and enthusiasm across unimaginable distances and through time.

CHRISTON J. HURST
Cincinnati, Ohio

CONTRIBUTORS

Michael J. Allen, Plymouth Marine Laboratory, Plymouth, UK

V. Gregory Chinchar, Department of Microbiology, University of Mississippi Medical Center, Jackson, MS

Ron A.M. Fouchier, Department of Virology, National Influenza Center, Erasmus Medical Center, Rotterdam, The Netherlands

Christon J. Hurst, Departments of Biology and Music, Xavier University, Cincinnati, OH; Engineering Faculty, Universidad del Valle, Ciudad Universitaria Meléndez, Santiago de Cali, Valle, Colombia

Laura D. Kramer, Wadsworth Center, New York State Department of Health, University of Albany, Albany, NY; Department of Biomedical Sciences, School of Public Health, University of Albany, Albany, NY

Vincent J. Munster, Laboratory of Virology, Rocky Mountain Laboratories, Division of Intramural Research, National Institute of Allergy and Infectious Diseases, National Institutes of Health, Hamilton, MT

Debi P. Nayak, Department of Microbiology and Immunology, David Geffen School of Medicine, University of California at Los Angeles, Los Angeles, CA

Audun Helge Nerland, The Gade Institute, University of Bergen, Bergen, Norway; Institute of Marine Research, University of Bergen, Bergen, Norway

Aina-Cathrine Øvergård, The Gade Institute, University of Bergen, Bergen, Norway; Institute of Marine Research, University of Bergen, Bergen, Norway

Leigh Owens, Discipline of Microbiology and Immunology, School of Veterinary and Biomedical Sciences, James Cook University, Townsville, Australia

Sonal Patel, Institute of Marine Research, University of Bergen, Bergen, Norway

Juan A. Raga, Marine Zoology Unit, Cavanilles Institute of Biodiversity and Evolutionary Biology, University of Valencia, Valencia, Spain

Tristan Renault, Laboratoire de Génétique et Pathologie, Ifremer La Tremblade, France

Jacques Robert, Department of Microbiology and Immunology, University of Rochester Medical Center, Rochester, NY

Declan C. Schroeder, Marine Biological Association of the UK, Plymouth, UK

Andrew T. Storfer, School of Biological Sciences, Washington State University, Pullman, WA

Norma P. Tavakoli, Wadsworth Center, New York State Department of Health, University of Albany, Albany, NY; Department of Biomedical Sciences, School of Public Health, University of Albany, Albany, NY

Marie-Françoise Van Bressem, Cetacean Conservation Medicine Group (CMED/CEPEC), Bogota, Colombia; Centro Peruano de Estudios Cetológicos (CEPEC), Museo de Delfines, Pucusana, Lima 20, Peru

Josanne H. Verhagen, Department of Virology, National Influenza Center, Erasmus Medical Center, Rotterdam, The Netherlands

William H. Wilson, Bigelow Laboratory for Ocean Sciences, West Boothbay Harbor, ME

ATTRIBUTION CREDITS FOR COVER AND SPINE ARTWORK

Cover credits

"Montage showing animal hosts", montage image used with permission of the artist, Christon J. Hurst. Those images incorporated into this montage were: Honeybees - Snapshot of a comb within a husbanded honeybee colony (source: image courtesy of S. J. Martin, provided by Declan C. Schroeder); Big eared townsend bat (*Corynorhinus townsendii*) - File:Big-eared-townsend-fledermaus.jpg (author unknown; public domain image, Bureau of Land Management, U. S. Federal Government); Laughing Kookaburra – Dacelo novaeguineae waterworks.jpg (author: Wikipedia user name Noodle snacks; Creative Commons Attribution-Share Alike 3.0 Unported license); Cotton Rat - File:Sigmodon hispidus1.jpg (author unknown; Centers for Disease Control and Prevention, U.S. Federal Government, public domain image); Heterocarpus shrimp - File: Heterocarpus ensifer.jpg (author unknown; National Oceanic and Atmospheric Administration, U.S. Federal Government, public domain image); Pillar coral - File:PillarCoral.jpg (author: Commander William Harrigan, NOAA Corps (ret.); National Oceanic and Atmospheric Administration, U.S. Federal Government, public domain image); Zebra striped Gorgonian wrapper, colonial anemone - File:Colonial anemone zebra.jpg (author: Nick Hobgood; Creative Commons Attribution-Share Alike 3.0 Unported license); Humpback whale - File:Humpback stellwagen edit.jpg (author: Whit Welles; Creative Commons Attribution 3.0 Unported license); Killer whales - File:Killerwhales jumping.jpg (author: Pittman; National Oceanic and Atmospheric Administration, U.S. Federal Government, public domain image); School of Goldband Fusilier, *Pterocaesio chrysozona* - File:School of Pterocaesio chrysozona in Papua New Guinea 1.jpg (author: Mila Zinkova; Creative Commons Attribution-Share Alike 3.0 Unported license); Abalone - File:Abalone OCA.jpg (author: Wikipedia user name Little Mountain 5; Creative Commons Attribution-Share Alike 3.0 Unported license.); and Giant Malaysian prawn *Macrobrachium rosenbergii* - File:Giant Malaysian Prawn.JPG (author: Wikipedia user name Syrist; Creative Commons Attribution-Share Alike 3.0 Unported license).

Spine credits

"Montage showing animal, botanical and microbial hosts", montage image used with permission of the artist, Christon J. Hurst. Those images incorporated into this montage were: Calliope Hummingbird - File:Calliope-nest.jpg (author: Wolfgang Wander; Creative Commons Attribution-Share Alike 3.0 Unported license); Cassava - File:Casava.jpg (author: Bob Walker; Creative Commons Attribution-ShareAlike 2.5 License); Tiger Salamander (Ambystoma tigrinum) - File: Salamandra Tigre.png (author: Carla Isabel Ribeiro; Creative Commons Attribution-Share Alike 3.0 Unported license); Volvox tertius (author: Matthew D. Herron; image supplied by and used with author's permission); Volvox aureus (author: Matthew D. Herron; image supplied by and used with author's permission); Molluscs (mostly bivalves) harvested from contaminated water in Zulia, Venezuela (author: Christon J. Hurst; image provided for use in this montage); and giant clam - File: Tridacna crocea.jpg (author: Nick Hobgood; Creative Commons Attribution-Share Alike 3.0 Unported license).

SECTION I

AN INTRODUCTION TO THE STRUCTURE AND BEHAVIOR OF VIRUSES

CHAPTER 1

DEFINING THE ECOLOGY OF VIRUSES*

CHRISTON J. HURST[1,2]

[1]Departments of Biology and Music, Xavier University, Cincinnati, OH

[2]Engineering Faculty, Universidad del Valle, Ciudad Universitaria Meléndez, Santiago de Cali, Valle, Colombia

CONTENTS

1.1 INTRODUCTION

The goal of virology is to understand the viruses and their behavior. Virology is an interesting subject and even has contributed to the concepts of what we consider to represent dieties and art. Sekhmet, an ancient Egyptian goddess, was for a time considered to be the source of both causation and cure for many of the diseases that we now know to be caused by viruses (Figure 1.1). Influenza, a viral-induced disease of vertebrates, was once assumed to be caused by the influence of the stars, and that is represented by the origin of it's name which is derived from Italian. The following was a

* This chapter represents a revision of "Defining the ecology of viruses", which appeared as chapter 1 of the book *Viral Ecology*, edited by Christon J. Hurst, published in 2000 by Academic Press. All of the artwork contained in this chapter appears courtesy of Christon J. Hurst.

Studies in Viral Ecology: Animal Host Systems: Volume 2, First Edition. Edited by Christon J. Hurst.

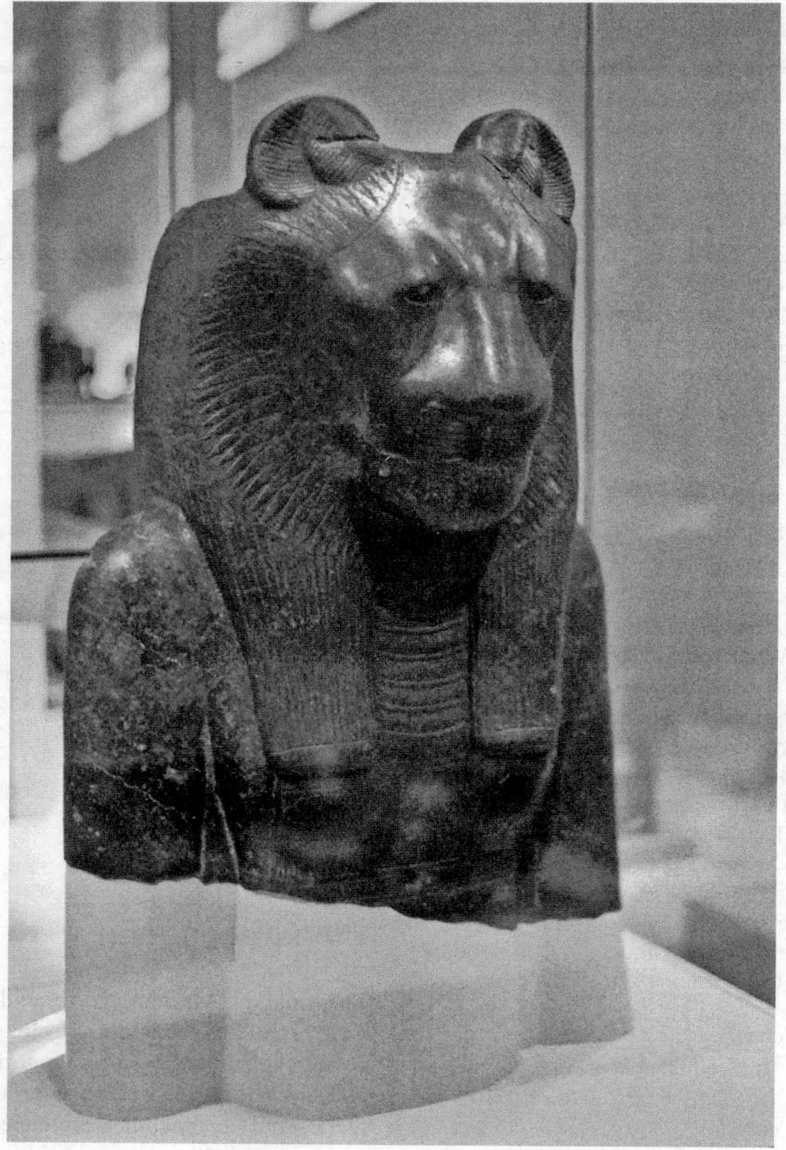

FIGURE 1.1 Image of Sekhmet, "Bust Fragment from a colossal statue of Sekhmet", Cincinnati Art Museum, John J. Emery Fund, Accession #1945.65 Cincinnati, Ohio. Originally the warrior goddess of Upper Egypt, Sekhmet was for a time believed to be the bringer of disease. She would inflict pestilence if not properly appeased, and if appeased could cure such illness.

rhyme which children in the United Sates sang while skipping rope during the influenza pandemic of 1918–1919:

I had a little bird
It's name was Enza
I opened a window
And in-flew-Enza.

(Source: The flu of 1918, by Eileen A Lynch, The Pennsylvania Gazette November/December 1998 (http://www.upenn.edu/gazette/1198/lynch.html).

And a bit more recently an interesting poem was written about viruses (Source: Michael Newman, 1984):

"The Virus"

Observe this virus: think how small
Its arsenal, and yet how loud its call;
It took my cell, now takes your cell,
And when it leaves will take our genes as well.
Genes that are master keys to growth
That turn it on, or turn it off, or both;
  Should it return to me or you
  It will own the skeleton keys to do
  A number on our tumblers; stage a coup.

But would you kill the us in it,
The sequence that it carries, bit by bit?
The virus was the first to live,
Or lean in that direction; now we give
Attention to its way with locks,
And how its tickings influence our clocks;
  Its gears fit in our clockworking,
  Its habits of expression have a ring
  That makes our carburetors start to ping.

This happens when cells start to choke
As red cells must in monoxic smoke,
When membranes get the guest list wrong
And single-file becomes a teeming throng,
And growth exists for its own sake;
Then soon enough the healthy genes must break;
  If we permit this with our cells,
  With molecules abet the clanging bells;
  Lend our particular tone to our death knells.

The purpose of this book is to define the ecology of viruses and, in so doing, try to approach the question of what life is like from a "virocentric" (as opposed to our normal anthropocentric) point of view. Ecology is defined as the branch of science which addresses the relationships between an organism of interest and the other organisms with which it interacts, the interactions between the organism of interest and its environment, and the geographic distribution of the organism of interest. The objective of this chapter is to introduce the main concepts of viral ecology. The remaining chapters of this book set, Studies in Viral Ecology volumes 1 and 2, will then address those concepts in greater detail and illustrate the way in which those concepts apply to various host systems.

1.1.1 What is a Virus?

Viruses are biological entities which possess a genome composed of either ribonucleic acid (RNA) or deoxyribonucleic acid (DNA). Viruses are infectious agents which do not possess a cellular structure of their own, and hence are "acellular infectious agents". Furthermore, the viruses are obligate intracellular parasites, meaning that they live (if that can be said of viruses) and replicate within living host cells at the expense of those host cells. Viruses accomplish their replication by usurping control of the host cell's biomolecular machinery. Those which are termed "classical viruses" will form a physical structure termed a "virion" that consists of their RNA or DNA genome surrounded by a layer of proteins (termed "capsid proteins") which form a shell or "capsid" that protects the genomic material. Together, this capsid structure and its enclosed genomic material are often referred to as being a "nucleocapsid". The genetic coding for the capsid proteins generally is carried by the viral genome. Most of the presently known virus types code for their own capsid proteins. However, there are some viruses which are termed as being "satellite viruses". The satellite viruses encapsidate with proteins that are coded for by the genome of another virus which coinfects (simultaneously infects) that same host cell. That virus which loans its help by giving its capsid proteins to the satellite virus is termed as being a "helper virus". The capsid or nucleocapsid is, in the case of some groups of viruses, surrounded in turn by one or more concentric lipid bilayer membranes which are obtained from the host cell. There exist many other types of acellular infectious agents which have commonalities with the classical viruses in terms of their ecology. Two of these other types of acellular infectious agents, the viroids and prions, are included in this book set and are addressed within their own respective chapters (Volume 1, chapters 10 and 12). Viroids are biological entities akin to the classical viruses and likewise can replicate only within host cells. The viroids possess RNA genomes but lack capsid proteins. The agents which we refer to as prions were once considered to be nonclassical viruses. However, we now know that the prions appear to be aberrant cellular protein products which, at least in the case of those afflicting mammals, have acquired the potential to be environmentally transmitted. The natural environmental acquisition of a prion infection occurs when a susceptible host mammal ingests the bodily material of an infected host mammal. The reproduction of prions is not a replication, but rather seems to result from a conversion of a normal host protein into an abnormal form (Volume 1, chapter 10). The Acidianus two-tailed virus, currently the sole member of the viral family Bicaudaviridae, undergoes a morphological maturation following its release from host cells and this is unique among all of the biological entities now considered to be viruses suggesting that this species may represent the initial discovery of an entirely new category of biological entities.

1.1.2 What is Viral Ecology?

Ecology is the study of the relationships between organisms and their surroundings. Viral ecology is, therefore, the relationship between viruses, other organisms, and the environments which a virus must face as it attempts to comply with the basic biological imperatives of genetic survival and replication. As shown in Figure 1.2, interactions between species and their constituent individual organisms (biological entities) occur in the areas where there exist overlaps in the temporal, physical, and biomolecular (or biochemical) aspects of the ecological zones of those different species. Many types of interactions can develop between species as they share an environment. One of the possible types of interactions is predation. When a microorganism is the predator, that predator is referred to as being a pathogen and the prey is referred to as being a host.

When we study viral ecology we can view the two genetic imperatives that every biological entity must face, namely, that it survive and that it reproduce, in the perspective of a biological life cycle. A generalized biological life cycle is presented in Figure 1.3. This type of cycle exists, in its most basic form, at the level of the individual virus or individual cellular being. However, it must be understood that in the case of a multicellular being this biological life cycle exists not only at the level of each individual cell, but also at the tissue or tissue system level, and at the organ level. This biological life cycle likewise exists on even larger scales, where it operates at levels which describe the existence of each species as a whole, at the biological genus level, and also seems to operate further upward to at least the biological family level. Ecologically, the life cycles of those different individuals and respective species which affect one another will become interconnected both temporaly, geographically, and biologically. Thus, there will occur an evolution of the entire biological assemblage and, in turn, this process of biotic evolution will be obliged to adapt to any abiotic changes that occur in the environment which those organisms share. While a species physiologic capacities establish the potential limits of the niche which it could occupy within this shared environment, the actual operational boundaries of it's niche are more restricted and defined by it's interspecies connections and biological competitions.

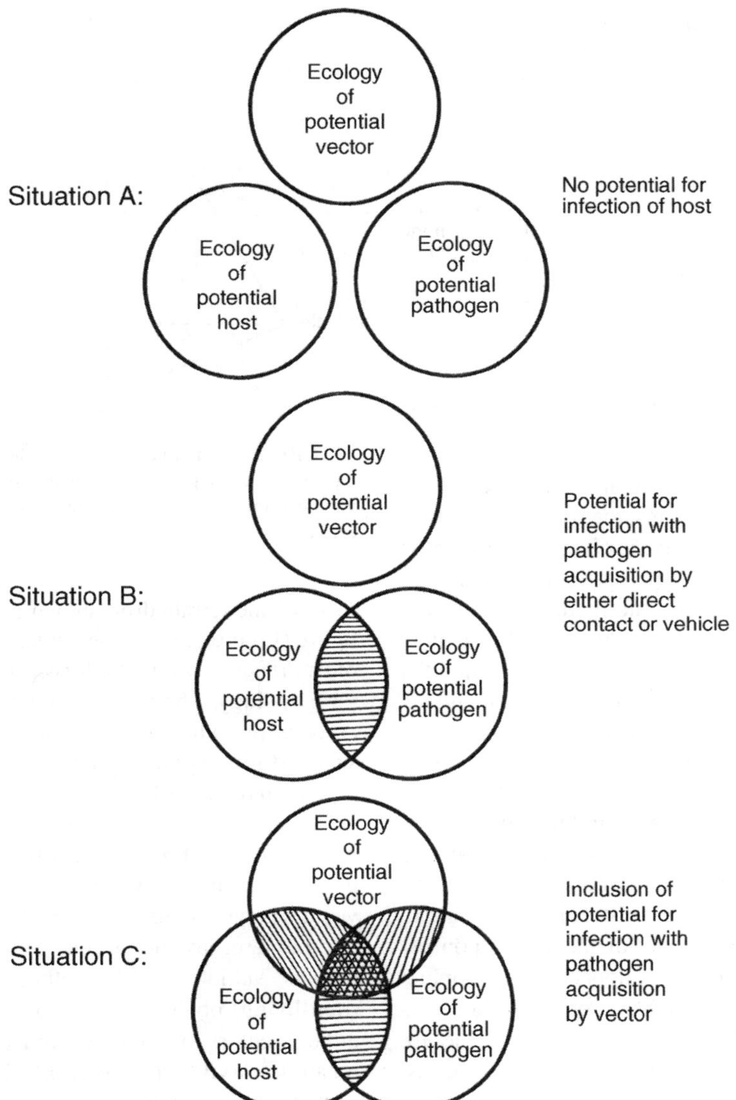

FIGURE 1.2 Interactions between organisms (biological entities) occur in the areas where the physical and chemical ecologies of the involved organisms overlap. Infectious disease is a type of interaction in which a microorganism acts as a parasitic predator. The microorganism is referred to as a pathogen in these instances.

1.1.3 Why Study Viral Ecology?

The interplay which occurs between a virus and the living organisms which surround it, while all simultaneously pursue their own biological drive to achieve genetic survival and replication, creates an interest for studying the ecology of viruses (Doyle, 1985; Fuller, 1974; Kuiken et al., 2006; Larson, 1998; Morell, 1997; Zinkernagel, 1996). While examining this topic, we improve our understanding of the behavioral nature of viruses as predatory biological entities. It is important to realize that in nature both the viruses of macroorganisms and the viruses of microorganisms normally

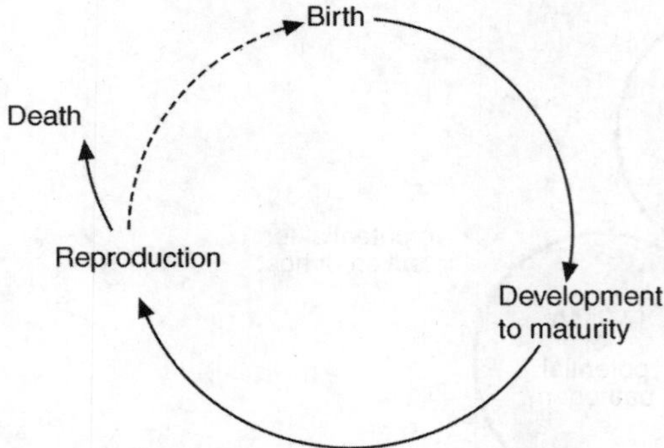

FIGURE 1.3 Generalized biological life cycle. Ecologically, the life cycles of different organisms which affect one another are temporally interconnected.

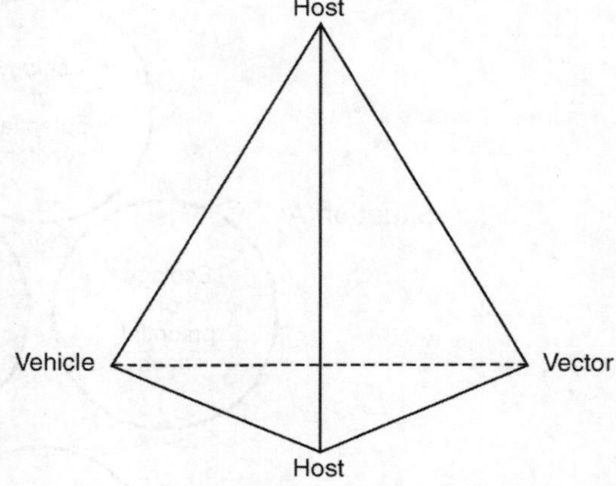

FIGURE 1.4 The lines connecting the four vertices of this tetrahedron represent the possible routes by which a virus can move from one host organism to another host organism.

exist in a cycle with their respective hosts. Under normal conditions, the impact of viruses upon their natural hosts may be barely apparent due to factors such as evolutionary coadaptation between the virus and its host (evolutionary coadaptation is the process by which species try to achieve a mutually acceptable coexistence by evolving in ways which enable them to adapt to one another). However, when viruses find access to new types of hosts and alternate transmission cycles, or when they encounter a concentrated population of susceptible genetically similar hosts such as occurs in densely populated human communities, communities of cultivated plants or animals, or algal blooms, then the impact of the virus upon its host population can appear catastrophic (Nathanson, 1997; Subbarao et al., 1998).

As we study viral ecology we come to understand not only those interconnections which exist between the entities of virus and host, but also the interconnections between these two entities and any vectors or vehicles which the virus may utilize. As shown in Figure 1.4, this interplay can be represented by the four vertices of a tetrahedron. The possible routes by which a virus may move from one host organism to another host organism can be illustrated as the interconnecting lines between those vertices which represent two hosts (present and proximate) plus one vertice apiece representing the concepts of vector and vehicle. Figure 1.5, which represents a flattened form of the tetrahedron shown in the previous figure (Figure 1.4) can be considered our point of reference as we move forward in examining viral ecology. The virus must survive when in association with the present host and then successfully move from that (infected) host organism (center of Figure 1.5) to another host organism. This movement, or transmission, may occur via direct contact between the two host organisms or via routes which involve vectors and vehicles (Hurst and Murphy, 1996). Vectors are, by definition, animate (living) objects. Vehicles are, by definition, inanimate (non-living) objects. Any virus which utilizes either vectors or vehicles must possess the means to survive when in association with those vectors and vehicles in order to sustain its cycle of transmission within a population of host organisms. If a virus replicates enough to increase its population while in association with a vector, then that vector is termed to be "biological" in nature. If the virus population does not increase while in association with a vector, then that vector is termed to be "mechanical" in nature. Because viruses are obligate intracellular parasites, and vehicles are by definition non-living, then we must assume that the virus cannot increase its population while in association with a vehicle.

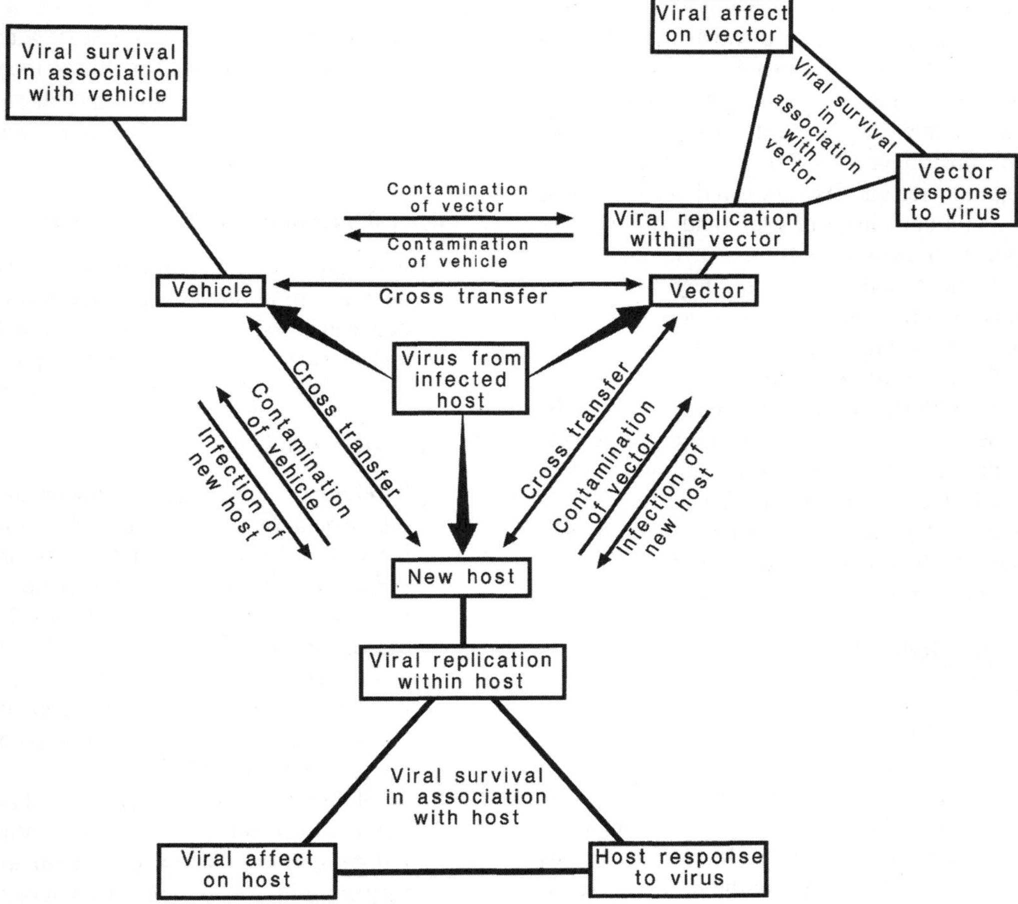

FIGURE 1.5 Viral ecology can be represented by this diagram, which represents a flattened form of the tetrahedron shown in the previous figure (Figure 1.4). The virus must successfully move from an infected host organism (center of figure) to another host organism. This movement, or transmission, may occur via direct transfer or via routes which involve vehicles and vectors. In order to sustain this cycle of transmission within a population of host organisms, the virus must survive when in association with the subsequently encountered hosts, vehicles and vectors.

Environmentally, there are several organizational levels at which a virus must function. The first and most basic of those levels is the individual host cell. That one cell may comprise the entire host organism. Elsewise, that host cell may be part of a tissue. If within a tissue, then the tissue will be contained within a larger structure termed either a tissue system (plant terminology) or an organ (plant and animal terminology). That tissue system or organ will be contained within an organism. The host organism is exposed to the open (ambient) environment, where it is but one part of a population of other organisms belonging to its same species. The members of that host species will be surrounded by populations of other types of organisms. Those populations of other types of organisms will be serving as hosts and vectors for either the same or other viruses. Each one of these organizational levels represents a different environment which the virus must successfully confront. A virus' affects upon it's hosts and vectors will draw responses against which the virus must defend itself if the virus is to survive. Also, the virus must always be ready to do battle with it's potential biological competitors. Contrariwise, the virus must be open to considering newly encountered (or reencountered) species as possible hosts or vectors. Because of their acellular nature, when viruses are viewed in the ambiental environments (air, soil and water) they appear to exist in a form that essentially is biologically inert. However, they have a very actively involved behavior when viewed in these many other organismal environments.

Considering the fact that viruses are obligate intracellular parasites, their ecology must be presented in terms which also include aspects of the ecology of their hosts and any vectors which they may utilize. Those factors or aspects of viral ecology which we study, and thus which will be considered in this book set, include the following:

Host Related Issues

1. what are the principal and alternate hosts for the viruses;
2. what types of replication strategies do the viruses employ on a host cellular level, host tissue or tissue system level, host organ level, the level of the host as a whole being, and the host population level;
3. what types of survival strategies have the viruses evolved that protect them as they confront and biologically interact with the environments internal to their host (many of those internal environments are actively hostile, as the hosts have developed many powerful defensive mechanisms);
4. what direct effects does a virus in question have upon its hosts, i.e. do the hosts get sick and, if the hosts get sick, then how severe is the disease and does that disease directly threaten the life of the host;
5. what indirect effects does the virus have upon its hosts, i.e., if the virus does not directly cause the death of the hosts or if viral-induced death occurs in a temporaly delayed manner as is the case with slow or inapparent viral infections, then how might that virus affect the fitness of the host to compete for food resources or to avoid the host's predators;

General Transmission-Related Issue

6. what types of transmission strategies do the viruses employ as they move between hosts, including their principal and alternate transmission routes which may include vehicles and vectors; and

Vector-Related Issues

7. in reference to biological vectors (during association with a biological vector the virus will replicate and usually is carried within the body of the vector), what types of replication strategies do the viruses employ on a vector cellular level, vector tissue or tissue system level, vector organ level, the level of the vector as a whole being, and also on a vector population level;
8. in reference to biological vectors, what types of survival strategies have the viruses evolved that protect them as they confront and biologically interact with the environments internal to their vectors (those internal environments may be actively hostile, as vectors have developed many powerful defensive mechanisms);
9. in reference to biological vectors, what direct effects does a virus in question have upon its vectors, i.e. do the vectors get sick and, if the vectors get sick, then how severe is the disease and does that disease directly threaten the lives of the vectors;
10. in reference to biological vectors, what indirect effects does the virus have upon its vectors, i.e., if the virus does not directly cause the death of the vectors or if viral-induced death occurs in a temporaly delayed manner as is the case

with slow or inapparent viral infections, then how might that virus affect the fitness of the vectors to compete for food resources or to avoid the vector's predators;

11. in reference to mechanical vectors, what types of survival strategies have been evolved by those viruses which are transmitted by (and during that event usually carried on the external surfaces of) mechanical vectors, since while in association with a mechanical vector the virus must successfully confront any compounds naturally present on the body surface of the vector plus confront the passively hostile ambiental environments of either air, water or soil through which the vector will be moving; and

Vehicle-Related Issue

12. what types of survival strategies have been evolved by those viruses which are transmitted by way of vehicles and which thereby must successfully confront the passively hostile ambiental environments of either air, water or soil as the virus itself is transferred through those environments.

If biological curiosity alone were not a sufficient reason for studying viral ecology, then perhaps we would study the viruses out of a desire to both understand them as predators and to contemplate the ways in which we might enlist their aid as ecological tools.

1.2 SURVIVING THE GAME: THE VIRUS AND IT'S HOST

Remember that: *so long as the virus finds a new host, whether or not the current host survives is unimportant.* Although it may be beneficial to not kill a current host until that host has reproduced to help provide a new generation of potential host organisms, if the host to virus ratio is large enough, then even this latter point may be unimportant.

This section presents in general terms the relationship between a virus and host. The generalities of relationships between viruses, vectors, and vehicles will be discussed in section 1.3 of this chapter. The specific subject of the practical limits to viral virulence in association with hosts and vectors will be addressed in section 1.4 of this chapter.

While in association with a host, the virus has only one principle goal. This goal is for the virus to replicate itself to a sufficient level that it can achieve transmission to another host. This goal can be attained by one of two basic strategies. The first of these strategies would be a productive infection, for which five basic patterns can be defined. The second strategy would be a non-productive infection. The goal of a productive infection is for the virus to produce infectious viral particles (those capable of infecting cells) which are termed "virions", during the virus' association with the current host. Subsequent spread of the infection to the next host occurs by transfer of these produced virions. Contrastingly, some of those agents which exhibit a non-productive pattern may either seldom or never produce actual virions. Thus, the usual goal of a non-productive strategy of infection is to pass the infection to the next host by directly transferring only the viral genomic sequences (van der Kuyl et al., 1995). The patterns of productive infection are:

"Short term - initial" in which viral production has only a short term initial course, after which the viral infection ends and there no longer is a presence of that virus within the body of the host individual although subsequent reinfection can occur, the outcome from this pattern of infection depends upon the virus type and historical exposure to that type within the host population, the situation being that in otherwise healthy members of a multicellular host population with which the virus has coevolved, these infections are usually mild and by themselves normally associated with a fairly low incidence of mortality;

"Recurrent" in which repeated episodes of viral production occur, this pattern often has a very pronounced initial period of viral production, after which the virus persists in a latent state within the body of the host with periodic reinitiations of viral production that usually are not life threatening;

"Increasing to end-stage" in which viral infection is normally associated with a slow, almost inocuous start followed by a gradual progression associated with an increasing level of viral production and eventual death of the host, in these instances death of the host may relate to destruction of the host's immunological defense systems which then results in death by secondary infections;

"Persistent-episodic" is a pattern that represents a prolonged nonfatal infection which may persist for the remainder of the hosts natural lifetime associated with a continuous production of virions within the host, but interestingly the infection only episodically results in symptoms, the viral genome does not become quiescent, the host remains infectious throughout the course of this associative interaction, and very notably some members of the family Picobirnaviridae often produce this pattern of productive infection;

"Persistent but inapparent" is a pattern that represents a prolonged nonfatal infection which seemingly never results in overt symptoms of illness attributable to that particular virus, the viral genome never becomes quiescent and viral infections that follow this pattern are persistently productive with the host often remaining infectious for the remainder of their natural lifetime, with notable examples of viruses which produce this pattern being members of the family Anelloviridae, and it also occurs in certain rare instances of infection by Human immunodeficiency virus 2 which is a member of the genus *Lentivirus* of the family Retroviridae.

There are two options to the "short term - initial" pattern. The first option is a very rapid, highly virulent approach which is termed "fulminate" (seemingly explosive) and usually results in the rapid death of the host organism. This first option usually represents the product of an encounter between a virus and a host with which the virus has not coevolved. The second option is for the virus to be less virulent, causing an infection which often progresses more slowly, and appears more benign to the host. The "recurrent" and "increasing to end-stage" patterns incorporate latency into their scheme. Latency is the establishment of a condition in which the virus remains forever associated with that individual host organism and generally shows a slow and possibly only sporadic replication rate that, for some combinations of virus and host, may never be life threatening to the host. The strategy of achieving a non-productive, or virtually non-productive, pattern of infection involves achieving an endogenous state (Terzian et al., 2001). Endogeny implies that the genome of the virus is passed through the host's germ cells to all offspring of the infected host (van der Kuyl et al., 1995; Villareal, 1997).

The product of interspecies encounters between a virus and it's natural host will usually lead to a relatively benign (mild, or not directly fatal), statistically predictable, outcome that results from adaptive coevolution between the two species. Still, these normal relationships do not represent a static coexistance between the virus and the natural host, but rather a tenuous equilibrium. Both the virus species and it's evolved host species will be struggling to get the upper hand during each of their encounters (Moineau et al., 1994. The result will normally be some morbidity and even some mortality among the host population as a result of infection by that virus. Yet, because the virus as a species may not be able to survive without this natural host species

(Alexander, 1981), excessive viral-related mortality in the host population is not in the long term best interest of the virus. Some endogenous viruses have evolved to offer a survival-related benefit to their natural host, and this can give an added measure of stability to their mutual relationship. Two examples of this type of relationship are the hypovirulence element associated with some strains of the Chestnut blight fungus, and the endogenous retroviruses of placental mammals. The hypovirulence (reduced virulence) which the virus-derived genetic elements afford to the fungi that cause Chestnut blight disease reduce the virulence of those fungi (Volume 1, chapter 9). This reduced virulence allows the host tree, and in turn the fungus, to survive. Placental mammals, including humans, permanently have incorporated species of endogenous retroviruses into the chromosomes of their genomes. It has been hypothesized that the incorporation of these viruses has allowed the evolution of the placental mammals by suppressing maternal immunity during pregnancy (Villareal, 1997).

However, the impact of a virus upon what either is, or could become, a natural host population can sometimes appear catastrophic. The most disastrous, from the host's perspective, are the biological invasions which occur when that host population encounters a virus which appears new to the host (Kuiken et al., 2006). Three categories of events can lead to biological invasions of a virus into a host population. These categories are: first, that this virus species and host species (or sub-population of the host species) may never have previously encountered one another (examples of this occurring in human populations would be the introduction of measles into the Pacific islands and the current introduction of HIV); second, if there have been previous encounters, the virus may have since changed to the point that antigenically it appears new to the host population (an example of this occurring in humans would be the influenza pandemic of 1918–1919); and third, that even if the two species may have had previous encounters, this subpopulation of the host species subsequently may have been geographically isolated for such a length of time that most of the current host population represents a completely new generation of susceptible individuals (examples in humans are outbreaks of viral gastroenteritis found in remotely isolated comunities on small islands as related to the occasional arrival of ill passengers by aircraft or watercraft). Sadly, the biological invasion of the HIV viruses into human populations seems to be successful (Caldwell and Caldwell, 1996), and the extreme host death rate associated with this invasion can be assumed to indicate that the two species have not had time to coevolve with one another. The sporadic, but limited, outbreaks in human populations of viruses such as those which cause the hemmorrhagic fevers known as Ebola and Lassa represent examples of unsuccessful biological invasions. The limited chain of transmission for these latter two illnesses (for Lassa, see: Fuller, 1974), with their serial transfers often being limited to only two or three hosts in succession, represents what will occur when a virus species appears genetically unable to establish a stabile relationship with a host species. The observation of extremely virulent and fulminate symptomatology, as associated with infections by Lassa and Ebola in humans, can generally be assumed to indicate either that the host in which these drastic symptoms are observed is not the natural host for those viruses or, at the very least, that these two species have not had time to coevolve. In fact, the extreme symptomatology and mortality which result in humans from Ebola and Lassa fevers seems to represent an overblown immune response on the part of the host (Spear, 1998). While having the death of a host individual occur as the product of an encounter with a pathogen may seem like a dire outcome, this outcome represents a mechanism of defense operating at the level of the host population. If a particular infectious agent is something against which members of the host population could not easily defend themselves, then it may be better to have that particular host individual die (and die very quickly!) to reduce the possible spread

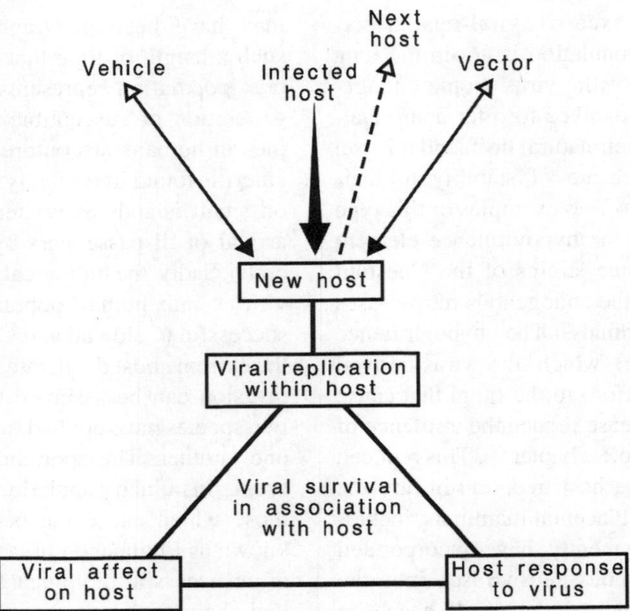

FIGURE 1.6 Viruses can arrive at their new host (filled arrows) either directly from the previously infected host, via an intermediate vehicle, or via an intermediate vector. Viral survival in association with that new host depends upon: viral replication within that new host, the effects which the virus has upon that host, and the response of that host to the virus. Successful viral survival in association with this new host will allow a possible subsequent transfer of the virus (open arrows) to its next host either directly, via a vehicle, or via a vector. This represents a segment from Figure 1.5.

of the contagion to the other members of the host population.

1.2.1 Cell Sweet Cell, and Struggles at Home

As diagramed in Figure 1.6, viruses can arrive at their new host (solid arrows) either directly from the previously infected host, via an intermediate vehicle, or via an intermediate vector. Viral survival in association with the new host will first depend upon the virus finding it's appropriate receptor molecules on the host cell's surface (Spear, 1998). After this initial location, the virus must be capable of entering and modifying the host cell so that the virus can reproduce within that cell. If the host is multicellular, then the virus may first have to successfully navigate within the body of the host until it finds the particular host tissue which contains it's correct host cells.

Within a multicellular host, the virus may face anatomically associated barriers including membranous tissues in animals. The virus also may face non-specific, non-immune biological defenses (Moffat, 1994), including such chemical factors as the enzymes found in both tears and saliva, and the acid found in gastic secretions. The types of anatomical and non-specific, non-immune defenses encountered can vary depending upon the viral transmission route and the portal by which the virus gains entry into the host's body. After a virus finds it's initial host cell and succeeds in beginning it's replication, the effects which the virus has upon the host can then draw a defensive biological response. The category of non-specific non-immune responses which a virus may encounter at this stage include even such things as changes in host body temperature for mammals. As if in a game of spy versus spy, the virus most importantly must survive the host's specific immune defenses

(Beck and Habicht, 1996; Gauntt, 1997; Levin et al., 1999; Litman, 1996; Ploegh, 1998; Zinkernagel, 1996).

The listing and adequate explanation of antiviral defense techniques would by itself be enough to nearly fill a library. But, I will attempt to summarize some of them here and help the reader to track those through this book set.

Molecular antiviral defenses begin at the most basic level which would be non-specific mechanisms. These conceptually include DNA restriction and modification systems (volume 1, chapter 5), progressing upward with greater complexity to the use of post transcriptional processing (Russev, 2007). Countering these defenses is done by such techniques as using virally-encoded restriction-like systems to chop-up the DNA genome of their host cells to provide a ready source of nucleic acids for the production of progeny viral genomes. There also are viruses which try to shut down the the post-transcriptional defenses, most clearly noticed among some viruses infective of plants. Plants in fact heavily rely upon molecular defenses such as post-transcriptional control, (volume 1, chapter 11) and beyond that technique the plants try to wall off an infection, essentially trying to live their lives despite presence of the infectious agent and hoping not to pass the infection along to their offspring through viral contamination of their germ cells.

Antimicrobial peptides are a defensive mechanism found in all classes of life, and represent a main part of the insect defensive system (volume 2, chapter 10). Higher on the scale of defensive responses are things which we term to be immunological in nature (Danilova, 2006). Some of these we term to be innate, others we call adaptive. A good starting point for this discussion of immunological responses is the capacity for distinguishing self versus non-self, accompanied by the capability for biochemically destroying cells that are determined to be non-self. This approach exists from at least the level of fungi (volume 1, chapter 9) upwards for the non-animals, and among the animals this approach begins with at least the corals (volume 2, chapter 5). Determining and acting upon the distinction of self versus non-self likely may have developed as a system that helps to support successful competition for growth in a crowded habitat, but it serves well against pathogenic organisms. As a health issue, this process sadly plays a role in autoimmune diseases and we try to suppress it when hoping to use organ and tissue transplantation to save human lives.

Apoptosis, the targeting of individual cells within the body of the host for selective destruction by the host, commonly exists across the animal kingdom. This mechanism is used by many invertebrates (volume 2, chapters 6 and 7) as wells as vertebrates to destroy any virally infected cells which may be present within their bodies. However, apoptosis is a weapon that can be used by both of the combatants. Using apoptosis to destroy virally-infected cells before the virus contained within those cells can assemble progeny virions is an effective approach when used carefully by the host. As might be expected, some viruses therefore defensively try either to shut-down the process of apoptosis, or at least to shut-down that process until the virus is ready to use apoptosis as a mechanism for assisting in the liberation of assembled virions from the infected host cell.

Vertebrates, and some of the invertebrates, have more complex body plans and can use them with good effectiveness in combating infections. With the evolutionary development of more complex body plans, comes the possibility of dedicating cells and even organs to the task of fighting pathogenic invaders. Those invertebrates with more complex body plans are represented in the anti-viral fight by their use of lymphoid organs to actively collect and either sequester or actively assault and destroy the microbial offenders. Some of the aquatic crustaceans (volume 2, chapter 7) tend to rely upon sequestering an infection and must hope to breed a new generation of their own progeny before they, themselves, are killed by the infection which they have sequestered within their body. At the same time, the infected parents must hope not to pass along the sequestered infection to their offspring through contamination of their eggs and sperm. Such collection and sequestration techniques are found upward through the evolutionary line and likewise used by the vertebrates. Many viruses have found ways around these issues, as is the case with endogenous viruses and retrotransposons that insert and maintain themselves in the genome of their host, passing directly through the germ cell line. Some viruses infect and replicated within the immune cells! Some viruses are shed along with the eggs of inertebrates and thus are ready to await the hatching of those offpsring. Still other viruses, as in the case of viviparous mammals, simply cross the placenta to infect the fetus.

Interferons and their homologues are protein systems which vertebrates have developed and use effectively against some viruses, and correspondingly many viral groups contain mechanisms for suppressing interferon production (Muñoz-Jordán and Fredericksen, 2010). Although the "walling-off" of a pathogen still occurs in vertebrates, with an example being the development of tubercules in some mycobacterial infections, active mechanisms for hunting down and destroying pathogens and pathogen-infected cells within their bodies is highly developed. With vertebrates, the end goal can be percieved as ridding the body of the pathogen even if that end goal is not always achieved. The jawed vertebrates possess immune systems which are termed adaptive, and these produce protein antibodies that can be highly specific (volume 2, chapters 8, 9, 11–14).

Options for surviving the immune defenses of the host can include such techniques as:

"You don't know me" (a virus infecting an accidental host, in which case a very rapid proliferation may occur, an example being Lassa fever in humans);

"Being very, very quiet" (forming a pattern of latency in association with the virus' persistence within that host, an example being herpesviruses);

"Virus of a thousand faces" (antigen shifting, an example being the lentiviruses);

"Keep to his left, that's his blind spot" (maintaining low antigenicity, an approach used by viroids and prions);

"Committing the perfect crime" (infecting the immune system, an approach taken by many retroviruses and herpesviruses); and

"Finding a permanent home" (taking up permanent genetic residency within the host and therefore automatically being transmitted to the host's progeny, an approach taken by viroids, endogenous retroviruses, and LTR retrotransposons).

Each virus must successfully confront it's host's responses while the virus tries to replicate to sufficient numbers that it has a realistic chance of being transmitted to another candidate host. Failure to successfully confront the host's responses will result in genetic termination of the virus and, on a broader scale, such failure may eventually result in extinction for that viral species.

1.2.2 I Want a Niche, Just Like the Niche, That Nurtured Dear Old Mom and Dad

The initial tissue type in which a virus replicates may be linked inextricably with the initial transmission mode and portal (or site) of entry into the body of the host. For example, those viruses of mammals which are acquired by fecal - oral transmission tend to initiate their replication either in the nasopharyngial tissues or else in the gastrointestinal tissues. There then are subsequent host tissue and organ types affected, some of which may be related to the virus' efforts at trying to reach it's proper portal of exit. Others of the host tissues affected by the virus may be unrelated to interhost viral transmission, although the affect upon those other tissues may play a strong role in the severity of illness which is associated with that viral infection. An example of the latter would be the encephalitic infection of brain neurons in association with echoviral conjunctivitis, an infection which initially would be acquired from fomites as part of a fecal-oral transmission pattern. In this case, the encephalitis causes nearly all of the associated morbidity but does not seem to benefit transmission of the virus (personal observation by author C. J. Hurst).

1.2.3 Being Societal

Successful viral survival in association with this new host will allow a possible subsequent transfer of the virus (Figure 1.6, open arrows) to its next host either directly, via a vehicle, or via a vector. The movement of a viral infection through a population of host organisms can be examined and mathematically modeled. An epidemic transmission pattern, characterized by a short term, higher than normal rate of infection within a host population is represented by the compartmental model shown in Figure 1.7 (Hurst and Murphy, 1996). An endemic transmission pattern, characterized by a long term, relatively constant incidence rate of infection within a host population is represented by the compartmental model shown in Figure 1.8 (Hurst and Murphy, 1996).

1.3 STEPPIN' OUT AND TAKING THE A TRAIN: REACHING OUT AND TOUCHING SOMEONE BY VECTOR OR VEHICLE

Remember that: *host-vector choices, cycles and vehicle utilizations as they exist today may (and probably do!) reflect evolutionary progression from prior species interactions and ecological relationships.*

After a virus has successfully replicated within the body of it's current (present) host, it must seek successful transmission to it's next (proximate) host. The resulting chain of transmission usually is the end-all of viral reproduction. These are three basic approaches by which this can be attained: transmission by direct contact between the present and proximate hosts, transmission mediated by a vector (Brogdon and McAllister, 1998; Hurst and Murphy, 1996; Mills and Childs, 1998), and transmission mediated by a vehicle (Hurst and Murphy, 1996). While considering these approaches, it is important to keep in mind that the chains of transmission originate by random chance followed by evolution.

1.3.1 "Down and Dirty" (Just Between Us Hosts)

This heading is one which can be used to describe host to host transmission (transmission by host to host contact). While this is one of the most notorious, it is not the most common route of viral transmission between animals. This route only serves to a limited extent in microbes. Even worse, this route essentially does not seem to function in vascular plants due to the relative immobility of those hosts.

1.3.2 "The Hitchhiker" (Finding a Vector)

Transmission by vectors may be the most prevalent route by which the viruses of plants are spread among their hosts. This route clearly also exists for some viruses of animals. However, this route has not yet been defined in terms of viruses which infect microbes. Vectors are, by definition, animate objects, and more specifically they are live organisms. Being a vector implies, although by definition does not require, that the entity serving as vector has self-mobility. Thus, plants could serve by definition as vectors, although when we consider the topic of viral vectors we usually tend to think in terms of the vectors as being invertebrate animals. Vertebrate animals can also serve as vectors, as likewise can some cellular microbes.

There are two categories of vectors: biological and mechanical. As was stated earlier, if the virus increases it's numbers while in association with a vector, then that vector is termed

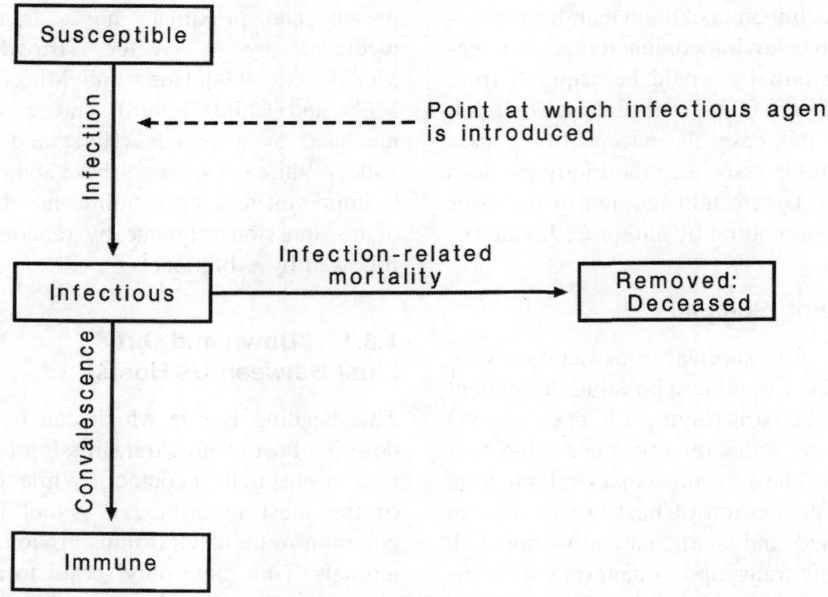

FIGURE 1.7 Epidemic transmission of a virus within a host population is represented by this type of compartment model (Hurst and Murphy, 1996). Each of the boxes, referred to as compartments, represents a decimal fraction of the host population with the sum of those decimal fractions equaling 1.0. The compartments which represent actively included members of the host population are those labeled susceptible, infectious, and immune. This model incorporates only a single category of removed individuals, representing those whose demise was due to infection related mortality. The solid arrows represent the rates at which individual members of the host species move between the different compartments during the course of an epidemic. Those rates of movement are often expressed in terms of individuals per day as described by Hurst and Murphy (1996). Used with permission of the author and Cambridge University Press.

as being biological. Conversely, the vector is termed to be mechanical if the virus does not increase it's numbers while in association with that vector. Beyond this there lie some deeper differences between mechanical and biological vectors. These differences include the fact that the acquisition of a virus by a biological vector usually involves a feeding process. Phagic habits of the biological vector result in the virus being acquired from an infected host when the vector ingests virally contaminated host body materials acquired through a bite or sting. Subsequent transfer of the infection from the contaminated biological vector to the virus' next host occurs when the biological vector wounds and feeds upon the next host. Actual transference of the virus to that next host occurs incidentally when the vector contaminates the wound by discharging viruses contained either in the vector's saliva, regurgitated stomach or intestinal contents, or else discharged feces and urine. Essentially any animal is capable of serving as a potential biological vector provided that the wound which it inflicts while feeding upon a host plant or animal will not result in the death of that new host until the virus would have had the chance to replicate within and subsequently be transmitted onward from that new host. There are many issues surrounding the question of what makes a good biological vector. These issues include: physical contact between the virus' host and the potential vector during a feeding event, viral reproduction within that potential vector, and that the infected vector be able to survive long enough to transmit the virus to a new host. It also helps if there is some factor driving the vector to pass along the infection, such as the

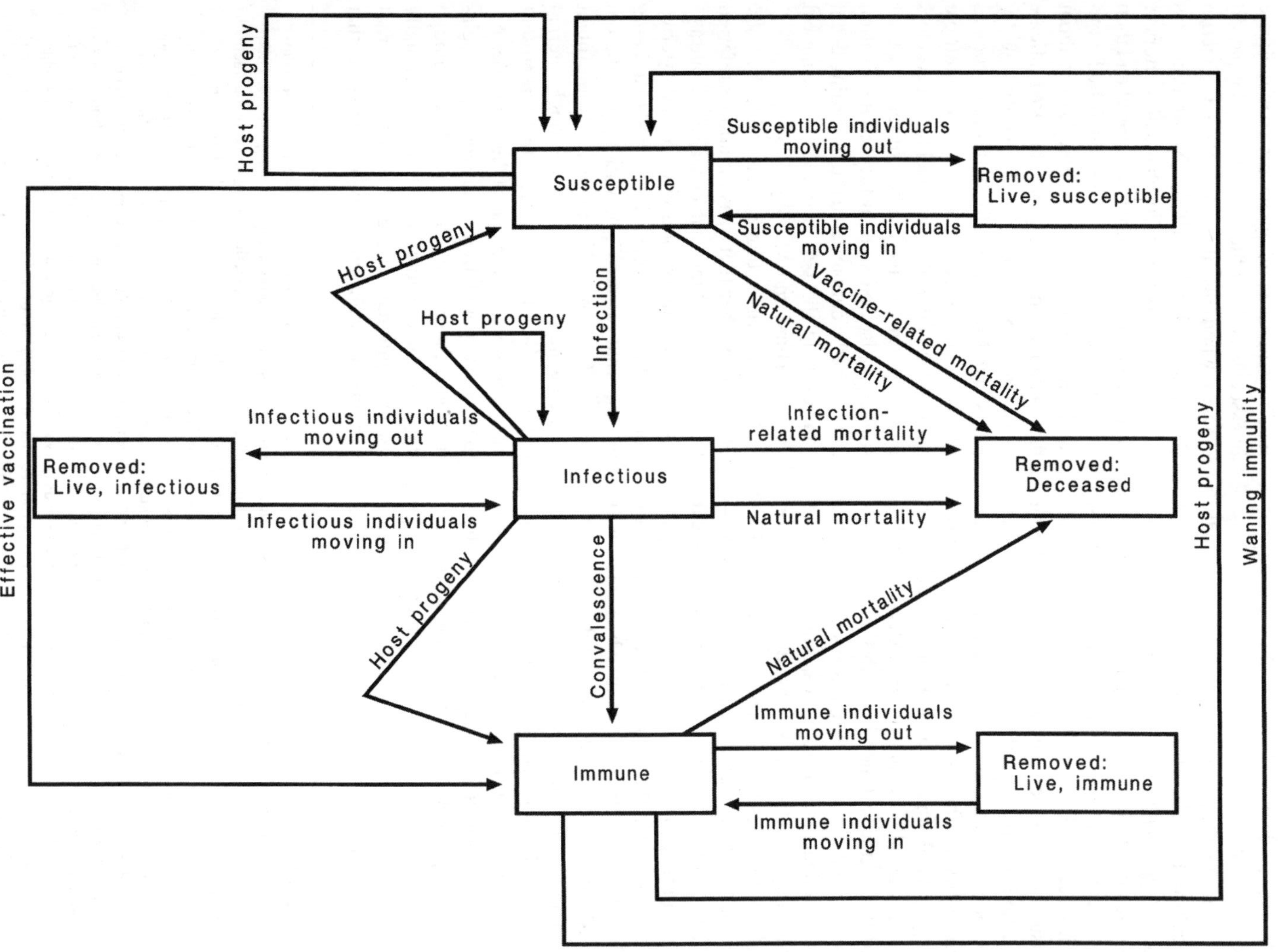

Host progeny
Susceptible
Susceptible individuals moving out
Removed: Live, susceptible
Susceptible individuals moving in
Host progeny
Infection
Vaccine-related mortality
Natural mortality
Host progeny
Infectious individuals moving out
Infection-related mortality
Effective vaccination
Removed: Live, infectious
Infectious
Removed: Deceased
Host progeny
Waning immunity
Infectious individuals moving in
Natural mortality
Host progeny
Convalescence
Natural mortality
Immune individuals moving out
Immune
Removed: Live, immune
Immune individuals moving in

virus finding it's way into the vector's saliva, or the virus increasing the physical aggressiveness of the vector.

The fact that biological vectors usually acquire the viral contaminant while wounding and ingesting tissues from an infected host brings us to another distinguishing difference between biological and mechanical vectors: viral contamination of a biological vector usually is associated with the virus being carried internal to the body of the vector. Replication of the virus then occurs within the body of the biological vector. Contrastingly, viral contamination of a mechanical vector usually occurs on the external surface of the vector and the virus subsequently tends to remain on the external surface of the mechanical vector. One possible example of mechanical vectoring would be the acquisition of plant viruses by pollinating animals such as bees and bats during their feeding process. These pollinators can serve as mechanical vectors if subsequently they are able to passively transfer the virus from their body surface to the next plant from which they will feed. In the case of these pollinators, the acquired virus presumably is carried external to the pollinator's body. Conversely, it is possible that a plant being visited by a pollinator might become contaminated by viruses afflicting that pollinator, and the plant could then passsively serve as a mechanical vector if subsequent pollinators should become infected when they visit that plant. Biting flies can serve as biological vectors if, during feeding, they ingest a pathogen which can replicate in association with that fly and then be passed onward when the fly bites it's next victim (Hurst and Murphy, 1996). Non-biting flies can passively serve as mechanical vectors if they feed upon contaminated material and then subsequently transmit those microbial contaminants to the food of a new host without that pathogen having been able to replicate while in association with the non-biting fly (Hurst and Murphy, 1996). Arthropods such as wasps, which repeatedly can sting multiple animals, could serve as mechanical vectors by transporting viruses on the surfaces of their stingers. Also, passive surface contamination of pets that occurs unrelated to a feeding event can result in the pets serving as mechanical vectors (Hurst and Murphy, 1996).

When a virus is transported inside the body of the vector, then that transportation is referred to as being an "internal carriage". Contrastingly, transportation of a virus on the external body surfaces of a vector is referred to as being an "external carriage". As will be described in volume 1, chapter 11, there are some plant viruses which are transported through internal carriage by invertebrates that represent mechanical vectors (because the virus does not increase its population level when in association with those invertebrates). Thus, although the biological vectoring of a virus usually involves internal carriage, the fact of internal carriage does not alone always indicate that

FIGURE 1.8 Endemic transmission of a virus within a host population is represented by this type of compartment model (Hurst and Murphy, 1996). This model is essentially an extension of the model presented in Figure 1.7. This model contains the same three compartments (susceptible, infectious, and immune) representing actively included individuals and the category of individuals removed by infection related mortality as were described for Figure 1.7. This model differs in that it must also consider the various possible categories of live removed individuals which can move into and out from the compartments of actively included individuals. Their removal represents the fact that they do not interact with the actively included individuals in such a way that the virus can reach them, often due to spatial isolation. This model also includes the fact that the immune status of individuals can naturally wane or diminish with time such that immune individuals return to the compartment labeled susceptible; production of host progeny, representing reproductive success of the members of the host species; natural mortality, as a means of removing members of the population; and the possible use of vaccination to circumvent the infectious process plus the associated vaccine - related mortality. Please notice that the progeny of infectious individuals may be susceptible, infectious, or immune at the time of their birth depending upon the type of virus which is involved and whether or not that viral infection is passed to the progeny. Used with permission of the author and Cambridge University Press.

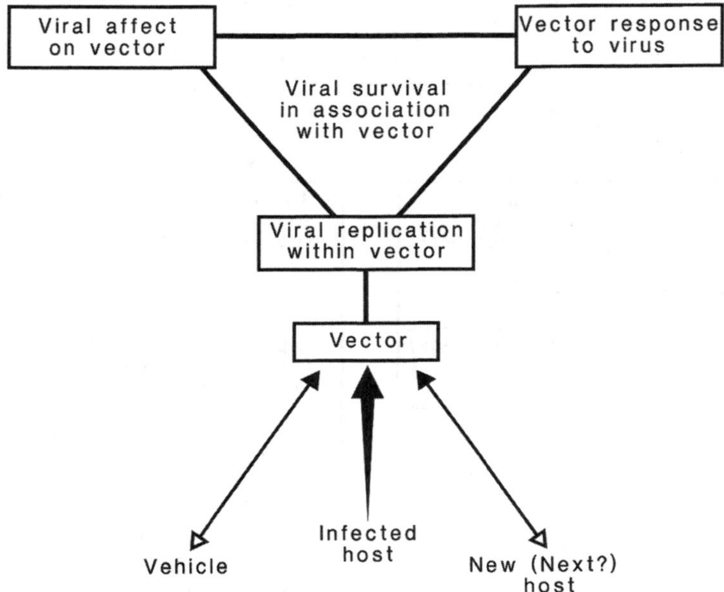

FIGURE 1.9 This figure addresses viral association with a biological vector and represents a segment from Figure 1.5. Vectors are, by definition, animate objects and are categorized either as 'biological', meaning that the virus increases in number during association with that vector, or 'mechanical', meaning that the virus does not increase in number during association with that vector. Biological vectors seem to have far greater importance than do mechanical vectors in terms of the spread of viral infections. Viruses can arrive at the biological vector (filled arrows) either directly from an infected host or via an intermediate vehicle. Transmission of the virus, via this vector, to a new host (or perhaps more accurately the 'next' host since, in the case of viruses, biological vectors may be considered as alternate hosts) requires that the virus both survive and replicate while in association with that biological vector. Thus, examining viral survival in association with a biological vector also involves considering the effects which viral replication has upon that vector and the response of that vector to the virus. Successful viral survival in association with the vector will allow a possible subsequent transfer of the virus to its next host either directly or via a vehicle (open arrows).

the vectoring is biological. Humans, interestingly, can serve as mechanical vectors via internal carriage for plant viruses that would be consumed with food and later excreted in feces (Zhang et al., 2006).

Because a virus must (by definition!) replicate in association with the biological vector; we can view the viral - vector association (Figure 1.9) in the same manner as was done for that of a virus and it's host (Figure 1.6). Indeed, it often is difficult to know which species is actually the viral host and which is actually the viral vector; to distinguish which is the victim and which serves as the messenger. Traditionally, we have often taken the view that humans are a high form of life and that there is a decreasing heirarchy down to the microbes. From this traditional, and sadly very anthropocentric, viewpoint we might assume that any living thing that transmits a virus between humans must be the vector as humans surely must be in the respectible position of serving as the host. Another version of this philosophy would consider a vertebrate to be the host and any invertebrate to be the vector. Still a third version has been based upon relative size, with the largest creature considered as the host and the smaller considered as the vector. Since we stated earlier that this chapter is intended to consider life from a virocentric perspective, we could easily accept the virocentric view which finds that there may be no clear distinction

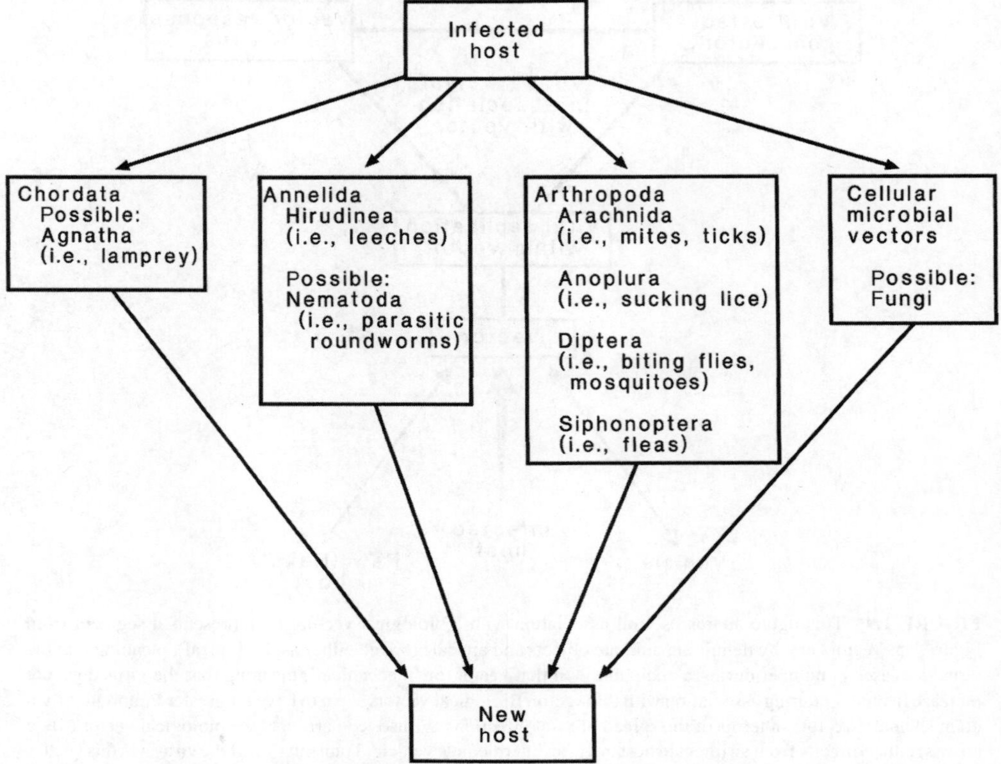

FIGURE 1.10 The transmission of a virus via a biological vector can be represented by this diagram. The virus is acquired as the biological vector feeds upon natural bodily fluids or else enzymatically liquified bodily components of the infected host. Subsequent transmission of the virus to a new host results when the vector releases contaminated excretions or secretions while feeding upon that new host.

between host and vector. Rather, any biological vector can likewise be viewed as a host. The argument as to which one, the traditional host or traditional vector, really serves as the host would then become moot.

Because many types of viruses are capable of infecting more than a single species of host, we are also left to ponder about determining which is the principle host versus those which serve as alternate hosts. Settlement of the distinction asked by this latter question is usually done by examining the comparative virulence of the virus in the different types of hosting species. That species for which the virus seems less virulent is assumed to be the more natural, most coevolved, host. It then is assumed that the species for which the virus seems to have greater virulence are alternate hosts. While trying to appreciate this conundrum, it must be understood that from a virocentric perspective both the principle and alternate hosts, as well as any biological vectors utilized by a virus, will all represent hosting species, and thus we may never be able to sort out the answers. Any further discussion of this particular issue is best left to only the most insistent of philosophers! Perhaps the only things left to be said of this issue are that examples of the transmission of a virus by a biological vector are represented in Figure 1.10, and that ecological interactions between a virus and it's principle hosts, alternate hosts, and biological vectors can be represented by the example shown in Figure 1.11.

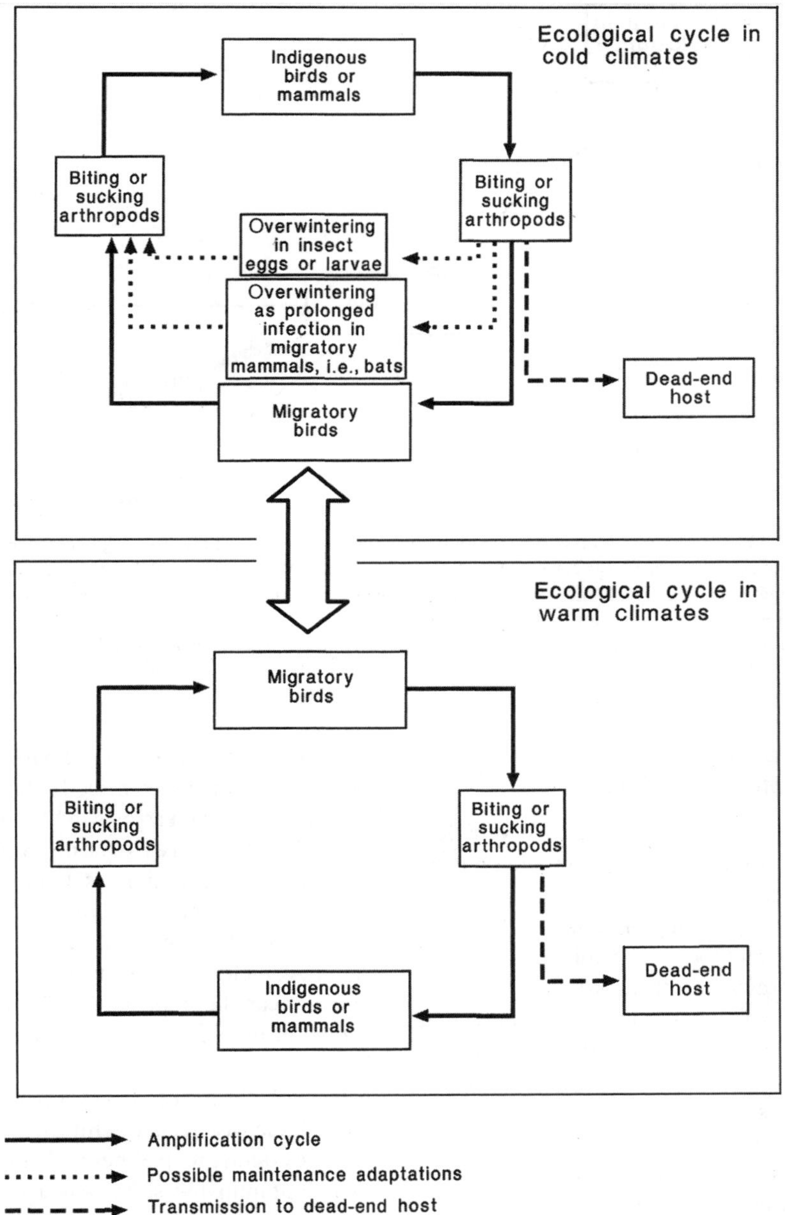

FIGURE 1.11 This figure represents a generalization of the ecological interactions which lead to insect-transmitted viral encephalitids. These infections generally are either enzootic or epizootic, meaning that their natural hosts are animals. Humans normally represent dead - end hosts for these viruses, meaning that the virus is not efficiently transmitted from infected humans to other hosts. The example shown in this figure is of a virus which has evolved ecological cycles both in warm, tropical climates and in cold, temperate climates. The cycle that has evolved in the warm climates can utilize arthropod vectors which do not have to go through the process of overwintering, thus allowing for an active year-round transmission cycle. Migratory birds, which may travel thousands of miles during their seasonal migrations, can shuttle the virus infection to the temperate zones. In the temperate zones, the virus' ecological cycle may need to include strategies for overwintering in insect eggs or larva and the possibility of survival as a prolonged infection in animals which may migrate lesser distances, such as bats.

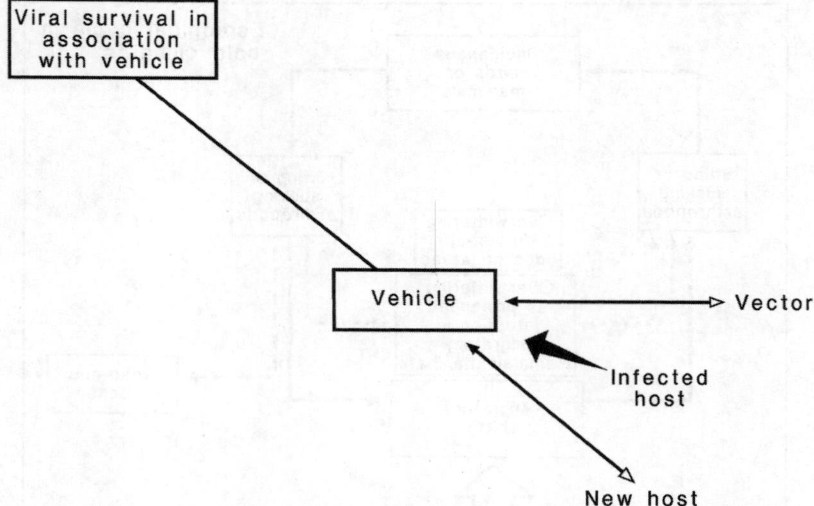

FIGURE 1.12 This figure addresses viral association with a vehicle and represents a segment from Figure 1.5. Viral transmission between hosts can occur by means of a vehicle. Vehicles are by definition inanimate objects. Viral contaminants can reach the vehicle (filled arrows) either directly from an infected host or via an intermediate vector. Transmission of the virus, via this vehicle, to a new host requires that the virus survive in association with the vehicle. Transference of the virus to its next host can occur either directly or via a vector (open arrows).

1.3.3 "In a Dirty Glass" (Going There by Vehicle)

Viruses also can be transmitted by vehicles. Vehicles are, by definition, inanimate objects. More specifically, the term vehicle applies to all objects other than living organisms. There are four general categories of vehicles and these are: foods, water, fomites (pronounced fo mi tez, defined as contaminated environmental surfaces which can serve in the transmission of pathogens), and aerosols. Figure 1.12 represents viral association with a vehicle. Transmission of the virus, via a vehicle, to a new host first requires contamination of that vehicle (shown by the filled arrows in Figure 1.12). The virus must then survive while in association with the vehicle. Because viruses are by definition obligate intracellular parasites, and by definition vehicles are non-living, then a virus neither can replicate on nor within a vehicle. Likewise, because vehicles are by definition non-living, we do not expect that any specific antiviral response will be produced by the vehicle. Transference of the virus to its next host can occur either directly or via a vector (shown as the open arrows in Figure 1.12). One possible indication as to the difference between a vector and a vehicle is that, while a live mosquito can serve as a biological vector, after it's death that same mosquito instead represents a vehicle. The transmission of a virus via a vehicle can be represented by the diagram shown in Figure 1.13. Acquisition of the virus by the next host or vector from that contaminated vehicle results from either ingestion of the vehicle (associated with foods and water), surface contact with either contaminated water or a contaminated solid object (a fomite), or inhalation (aerosols). Although, from a human perspective, we might tend to associate waterborne transmission with animals and in particular human diseases (volume 2 chapter 13); the waterborne approach will play a major role in viral transmission for viruses that infect cyanobacteria (volume 1 chapter 6), algae (volume 1 chapter 7) and seaweeds (volume 1

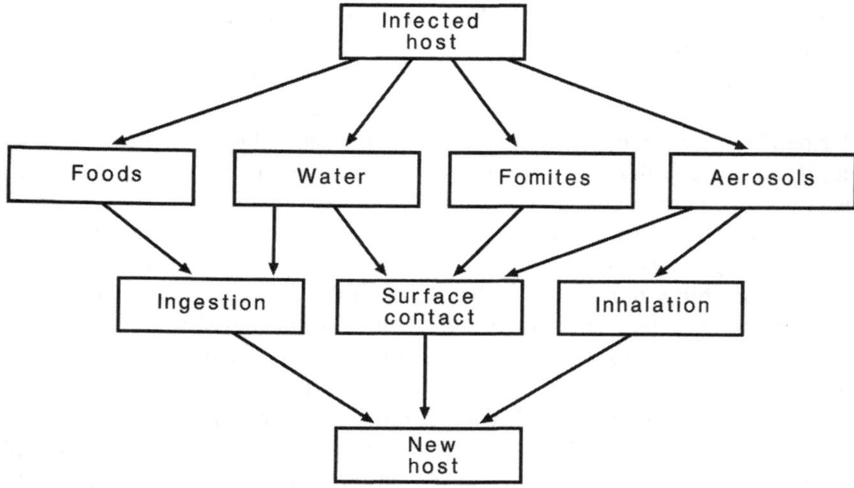

FIGURE 1.13 The transmission of a virus via a vehicle can be represented by this diagram. Food items can be contaminated by the action of an infected host. Alternatively, the food in question may actually be the body of an infected host that subsequently is consumed by a susceptible, predatory new host. Viral contaminants present in water can be acquired by a new host either directly, as the result of external or internal exposure to the contaminated water including ingestion of the water; or indirectly, following contact between the new host and an environmental surface (serving as a secondary, intermediate vehicle) that has been contaminated by that water. Fomites are solid environmental (non-food) objects whose surfaces may be involved in the transfer of infectious agents. Viral aerosols may result in the infection of a new host either directly through inhalation of the aerosol, or indirectly following contact between the new host and some other vehicle (either food, water, or a fomite) contaminated by that aerosol.

chapter 8). The are even viruses of terrestrial plants, including some carmoviruses of the viral family Tombusviridae, which seem as though they might be transmitted by water. The list of vehicles associated with viral transmission even includes agricultural tools and other work implements. The topic of vehicle-associated transmission of pathogens is discussed at length in the reference by Hurst and Murphy (1996).

1.3.4 Bringing Concepts Together

Biological entities exist over a spectrum of complexities, ranging from the viruses, viroids and prions (yes, even the prions are biological entities!) to multicellular organisms. The process of maintaining the viability of even the largest of organisms is, and perhaps must, be organized at small levels. Biologically, this has been achieved by a highly evolved process of internal compartmentalization of functions with a systemic coordination. If we consider for a moment one of the most enormous of the currently living multicelled organisms, the blue whale (*Balaenoptera musculus*), we notice that this kind of compartmentalization and coordination begins all of the way down at the level of the subcellular structures and organelles within each individual cell. The compartmentalization and coordination then continue upward through a number of levels including the various individual types of cells, the tissues into which those cells are organized, the organs which the tissues comprise, and finally the total internal coordination of all of these through nerve signaling and hormonal regulation. At every one of these biological levels there is a "taking from" and a "leaving behind" exchange of material with respect to the immediate surrounding environment. This results in the existence of dramatic environmental differences at all levels, even down to the many microenvironments which exist within the organizational regions of a single cell.

Every virus must try to comply with the basic biological imperatives of genetic survival and replication. While complying with these imperatives the viruses must, as obligate intracellular parasites, not only face but also survive within and successfully be transported through the various environments which are internal to the host. Those viruses which are transmitted by biological vectors must also have evolved the capability to survive and be transported through internal environments faced within the vector. Viruses which are transmitted by mechanical vectors generally must possess an additional evolved ability to survive on the surface of that vector. Likewise, both those viruses transmitted by mechanical vectors and viruses transmitted by vehicles must possess the ability to survive exposure to natural ambiental environments encountered either in the atmosphere, hydrosphere or lithosphere. These numerous environments are summarized in Figure 1.14. Conditions confronted at the interface zones, as indicated by the dashed lines in Figure 1.14, represent areas of still additional environmental complexity. While viruses appear biologically inert when viewed in the ambiental environments, they display their biology and interact with their surroundings when they reach the environments internal to their hosts and biological vectors.

The adaptability of a species in terms of its biological cycle and biological needs will determine that species' potential distribution range. This potential distribution range is limited in actuality to a smaller range based upon interspecies relationships and competitions. Ourselves being large multicellular creatures, we humans normally think of a distribution range as being geographical in nature. As microbiologists, many of us have come to understand the concept of distribution range in finer detail; an example being the depth within a body of water where a particular species of microorganism normally will be found. At the level of viral ecology, the concept of species distribution range encompasses everything from tissue and organ tropisms (those tissues and organs which a virus seems to attack preferentially) upwards to the geographical availability of host species, vector species, and the prevailing directional flow of appropriate vehicles such as air and water. The larger, geographical end of this scale is represented in Figure 1.15.

While considering the factors addressed in Figure 1.15, it is important to keep in mind that albeit the virus' election of hosts, vectors, and routes of transmission would all originate by random chance, the attainment of reliable continued viral success would require that such random selection events be followed and strengthened by evolution. This explains the reason why viruses do not appear suddenly to develop the ability to use a different vehicle. Indeed, it is perhaps likely that in order to use a vehicle such as air or water, the virus must have preadapted itself to the conditions which it will encounter in association with that vehicle. Nearly each individual species of virus which achieves transmission by vehicles, seems invariably to use only one type of vehicle. This trait likewise seems to hold true for all species belonging to any given viral genus. Furthermore, this identification seems to nearly always hold true at the level of viral family. In fact, this is one of the defining characteristics of the ecology of a viral group. The only virus which seems to have evolved the ability to utilize more than a single vehicle is the Hepatitis A virus (Hurst and Murphy, 1996), which has evolved a most remarkable ability to be effectively transmitted both by water and on fomites. Perhaps accordingly, the Hepatitis A virus currently exists in a genus (*Hepatovirus*) of its own. We should not be surprised if we eventually would discover other members of that viral genus, and subsequently discern those other members to likewise use these same two vehicles. It is for these reasons, that fears expressed in the public press that viruses such as Ebola will suddenly take flight and be transmitted over large distances via aerosol transmission amount to nothing more than frightening speculation. Why is it just speculation? Because that route of transmission is not a part of the

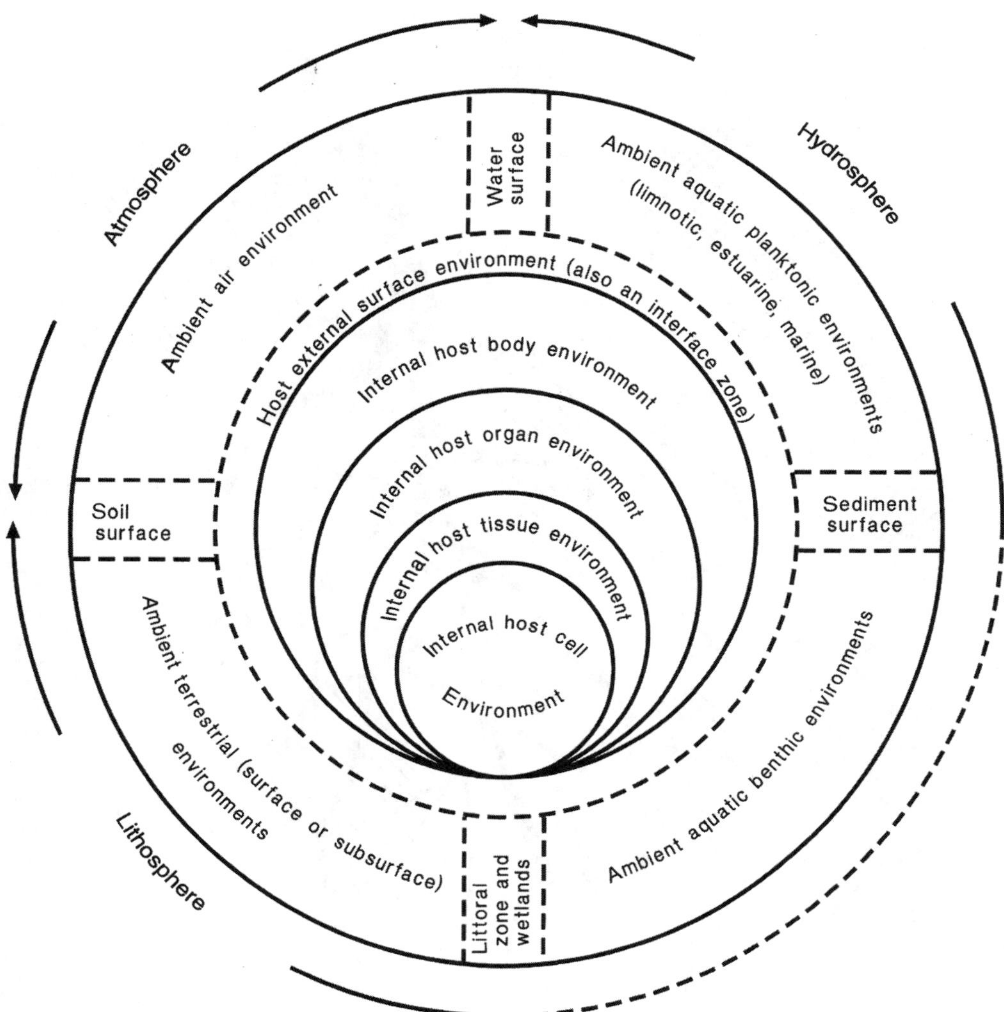

FIGURE 1.14 This figure integrates the concepts of host, vehicle and biological vector by representing the environments potentially faced by a virus. As obligate intracellular parasites, the viruses must face, survive within, and successfully be transported through environments which are internal to the host. Those viruses which are transmitted by biological vectors must also have evolved the capability to survive and be transported through internal environments faced within the vector. Viruses which are transmitted by vehicles and mechanical vectors must additionally possess an evolved ability to survive in natural ambiental environments (atmosphere, hydrosphere and lithosphere). Conditions confronted at the interface zones, as indicated by dashed lines, represent areas of additional environmental complexity.

virus' ecology. Invasive medical devices such as syringes, endoscopes and other surgical implements, plus transplanted animal tissues including transfused blood and blood products, and grafted plant material, represent exceptions to this rule. These devices and transplanted tissues represent unnatural vehicles which, by their nature, allow the virus an abnormal access to the interior of a new host (Hurst and Murphy, 1996). Any virus which would naturally be transmissible by direct contact with either an infected host or any type of

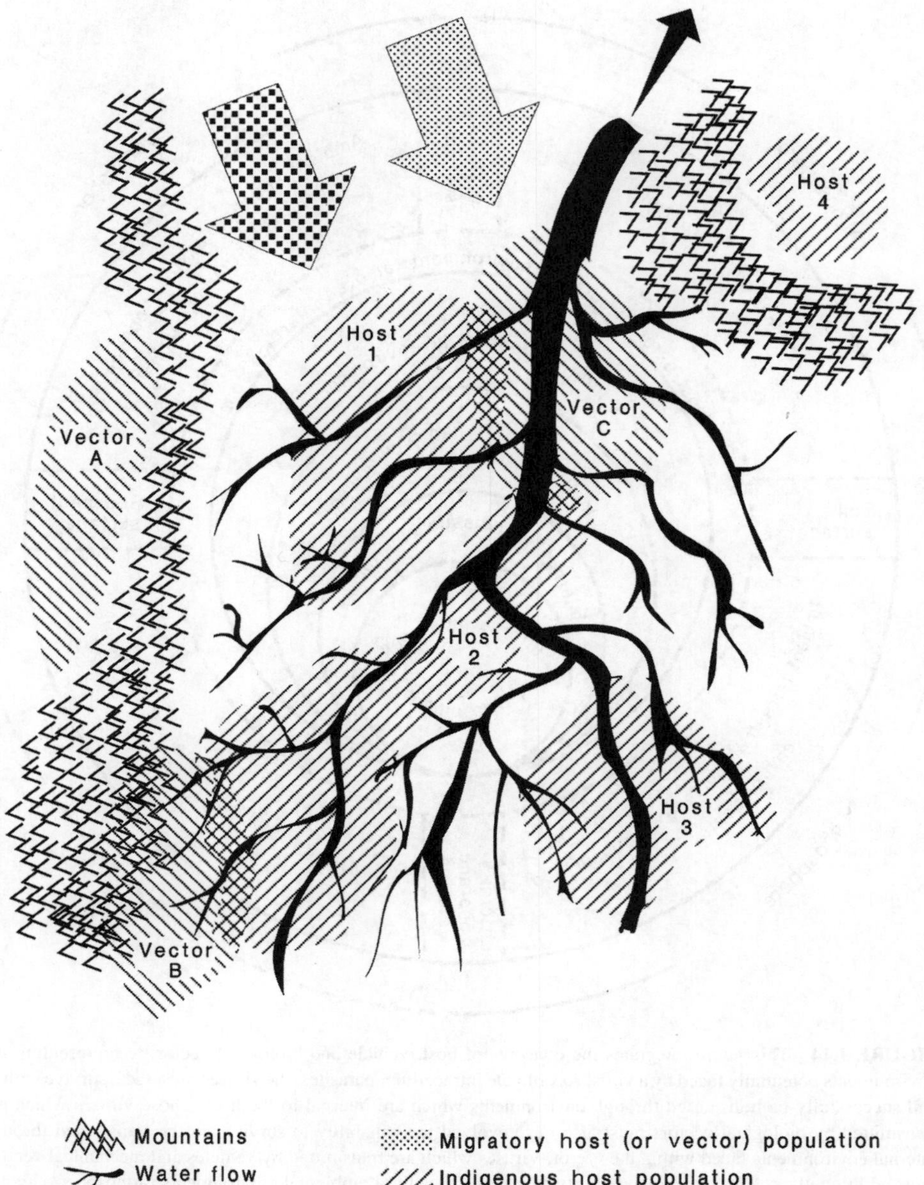

FIGURE 1.15 This figure presents a hypothetical example of the way in which the ecology of a virus is delineated by the spatial relationships between its potential hosts, vectors, and vehicles. The figure represents a viral infection existing in a watershed basin whose area covers tens of millions of hectares. An assumption is made that the four potential indigenous host populations and three potential indigenous vector populations are terrestrial organisms whose ecological areas are delineated and that these organisms do not migrate outside of their own respective ecological areas. Indigenous host populations 1, 2, and 3 reside in riverine ecological areas within the basin. Indigenous vector population B has a highland ecology, while vector population C has a lowland ecology, and both of these vector populations reside within the basin. Indigenous vector population A and indigenous host population 4 are excluded from participation in the viral infection cycle due to their geographical isolation and, because of their

vector can also be transmitted by one of these unnatural vehicular routes.

Viruses occasionally will appear in association with "apparently new" (unexpected) hosts and biological vectors. These latter occurrences with unexpected hosts or vectors represent the identification of sporadic events which occur when geographical boundaries are breached by the movement of those potential hosts and vectors for which the virus in question already has a preevolved disposition. These preevolved dispositions may represent, at some basic level, the renewal of old acquaintances between a virus, vector, and host. Alternatively, if these particular viral, host, and vector species truely never have met before, then an important aspect which can factor into these encounters is the biological relatedness between these "apparently new" hosts or vectors and those other hosts or vectors which the virus more normally would use.

1.3.5 Is There no Hope?

Many host-related factors do play a role in the transmission of viral-induced illnesses. These include:

"Finding the wrong host"– the "oops" or accidental occurrence factor wherein viruses occasionally will encounter and successfully infect living beings other than their natural hosting species, an event which represents a mistake not only for the host (which often will be fated to die for want of having inherited an evolved capability to mount an effective defense against that virus) but also is a mistake for the virus (which often will not be able to subsequently find one of its natural hosts and hence also loses it's existence);

"Only the good die young"– culling the herd for communal protection can have some advantage for the host population as a whole if those individuals that demonstrate a lesser ability to resist the virus are weakened enough by the infection that they then are more easily killed by predators (this is an act that both reduces the likelihood that other members of the host population will become infected by that virus strain and also may improve the gene pool of the host species by selectively eliminating it's most susceptible members);

"Being your own worst enemy"– behavioral opportunities for disease transmission do exist, and ethnic or social customs often play a role in disease transmission (including the probable reality that a lack of male circumcision has spelled disaster for the human population of Africa by facilitating the heterosexual transmission of HIV) (Caldwell and Caldwell, 1996), and in fact most of those vector borne diseases that aflict humans can be avoided by changes in host behavior.

If we view this situation from the human perspective, there does exist a basis for hope in terms of the health of hosts. Our most important

geographical exclusion from the basin, we do not need to be concerned with the nature of their ecological zones. Vector population B is capable of interacting in a cycle of transmission involving host population 2. Vector population C is capable of interacting in a cycle of transmission invloving host populations 1 and 2. None of the indigenous vector populations is capable of interacting in a cycle of transmission involving host population 3. A virus capable of being transmitted by surface waters could move from host population 3 to host population 2, since host population 2 is located downstream of host population 3. That same surface waterborne route could not spread the virus to host population 1, because host population 1 is not situated downstream of either host populations 2 or 3. Likewise, neither could the surface waterborne route spread the virus in a upstream direction from host population 1 to host population 2, nor from host population 2 to host population 3. Alternatively, a migratory host or vector population could carry the virus from host population 1 to host populations 2 and 3, as likewise could air flow if the virus is capable of being transmitted as an aerosol.

TABLE 1.1 Categories of Physical Barriers

Thermal
Acoustic (usually ultrasonic)
Pressure
  barometric
  hydrostatic
  osmotic
Radiation
  electronic
  neutronic
  photonic
  protonic
Impaction (includes gravitational)
Adhesion (adsorption)
  electrostatic
  van der Waals
Filtration (size exclusion)
Geographic features
Atmospheric factors (includes such meterological aspects as humidity, precipitation, and prevailing winds)

advantage lies in the use of barriers, which represent a very effective means by which we can reduce the transmission of all types of infectious agents. Barriers can be classified by their nature as physical (Table 1.1), chemical (Table 1.2), and biological (Table 1.3). In many cases, these barriers already exist in nature. Natural examples of barriers include both high and low temperatures (thermal, a physical barrier), sunlight (radiation, a physical barrier), the natural salinity of water (both osmotic, a physical barrier and also dessicant, a chemical barrier), and ecological competition (competitive, a biological barrier). The intentional use of barriers can involve both individual and combined applications. One example of a combined barrier application is the retorting of canned products, a process which employs a combination of elevated temperature and hydrostatic pressure to achieve either disinfection or sterilization (this process is similar to autoclaving). Many of these barrier concepts, such as filtration acting as a physical barrier, can be applied at different levels. For example: some particle exclusion filtration devices have pore sizes small enough that they can act as a filtration barrier against virus particles themselves; natural latex condoms and disposable gloves act as filtration barriers against a liquid vehicle (they contain pores which are larger than the virus particles yet smaller than the droplets of liquid in which the virus is contained); window screens and mosquito netting act as filtration barriers against flying vectors; and walls, fences, doors and gates can act as filtration barriers against infected hosts. The ingestion of food and water is associated with digestive treatments such as pH changes and secreted enzymes, both of which represent chemical barriers. When viewed from the virocentric perspective, the use of barrier techniques for preventing viral transmission would represent cause for despair instead of hope. There is, however, a notable exception represented by the idea of some viruses such as the polyhedrin- forming members of the viral families Baculoviridae and Reoviridae seem to require digestive treatment as an aid to their infectivity for their insect hosts.

TABLE 1.2 Categories of Chemical Barriers

Ionic (includes pH and salinity)
Surfactant
Oxidant
Alkylant
Desiccant
Denaturant

TABLE 1.3 Categories of Biological Barriers

Immunological (includes specific as well as nonspecific)
  naturally induced (intrinsic response)
  naturally transferred (lacteal, transovarian, transplacental, etc.)
  artificially transferred (includes injection with antiserum and tissue transfers such as transfusion and grafting)
Biomolecular resistance (not immune-related)
  lack of receptor molecules
  molecular attack mechanisms (includes nucleotide-based restrictions)
  antibiotic compounds (metabolic inhibitors, either intrinsic or artificially supplied)
Competitive (other species in ecological competition with either the virus, its vectors, or its hosts)

1.4 WHY THINGS ARE THE WAY THEY ARE

The ability of a virus to pass on it's genetic content is the key consideration of the virus. We now understand how this gets done on a molecular level. What still remains to be understood are how this thing gets done and has come about at the species level.

1.4.1 To Kill or Not to Kill - A Question of Virulence

One of the nagging questions which a virus must face is what should be the extent of it's virulence, i.e., whether or not it should kill it's hosts and biological vectors as a consequence of their encounters (Ewald, 1993; Lederberg, 1997). When considered in purely evolutionary terms, virulence is the ability of the disease agent to reduce reproductive fitness of that host. The relative virulence of a virus with respect to one of it's hosts or biological vectors is generally presumed to be a marker of co-evolution. More specifically stated, it seems that the less virulent is the virus for one of it's hosts or vectors, the more greatly coevolved is the relationship. Why should this be so? It should clearly be the case that, were a virus to infect an individual member of a host or biological vector population prior to that individual having reached reproductive age, it would be in the virus' best interest to not kill that host or vector. Contrariwise, in a very strict sense, death of that host or biological vector should not matter to the virus if that individual host or biological vector has passed the end of the normal reproductive lifespan. The reason for this latter philosophy is that, even if this particular host were to survive, it would not produce more susceptible offspring. Additionally, within each species of potential host or biological vector, there would be a strong genetic drive to enable their infants to mount sufficient immunological defense so as to reach the age of reproductive maturity. That same genetic drive does not, by definition, act upon the preservation of individuals who have passed their reproductive years. One example of the result from interaction of these forces is the fact that infections caused by the Hepatitis A virus can go nearly unnoticed in human infants, yet Hepatitis A virus infections can be disastrous in human adults.

Figure 1.16 represents the question of how the success of a virus relates to its' virulence. The virus will not be successful if the result of viral infection is too deleterious in terms of affecting the ability of the present host or biological vector to survive before that virus has been able to achieve transmission to it's next host or biological vector.

1.4.2 Genetic Equilibrium (versus Disequilibrium)

One of the hallmarks of relationships between virus species and their host species is their apparent goal of reaching a mutually acceptable genetically-based equilibrium (Dennehy et al., 2006; Lederberg, 1997; Zinkernagel, 1996). Some viruses also seem to have interchanged genetic material with their hosts while striving to evolutionarily reach a level of mutual coexistence.

There are many considerations associated with an apparent genetic equilibrium. In most instances of endemic viral infection in populations of a coevolved host or biological vector, the infections appear relatively unnoticed or relatively innocuous. This may change when the virus encounters a concentrated population of genetically similar susceptible hosts or biological vectors concentrated within a small radius, perhaps resulting in an epidemic. It also may change when the virus invades a population of novel hosts or vectors (hosts or vectors to which that virus appears to be new); this is termed a "biological invasion". Excessive virulence may represent reduced genetic fitness with respect to the virus, host, or biological vector. Limited virulence on the part of the virus seems to represent a state of coevolution but with some remaining flux in the virus-host

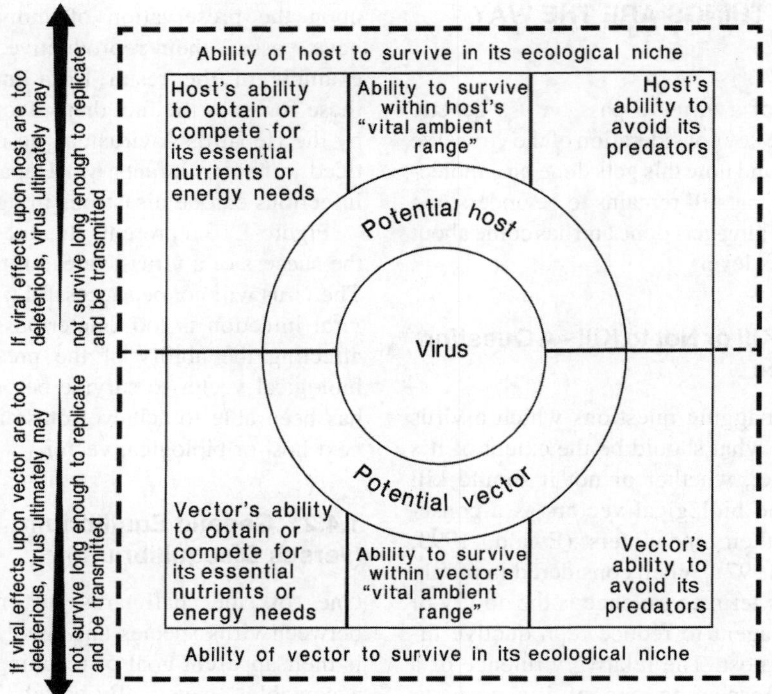

FIGURE 1.16 This figure represents the question of how the success of a virus relates to its' virulence. Success requires that the virus replicate within the bodies of its hosts and any biological vectors to concentrations which are high enough that the virus has a reasonable chance of being passed onward to infect either its next host or its next biological vector. The virus will not be successful if, within this period of replication, the result of viral infection is too deleterious in terms of affecting the ability of the present host or biological vector to survive within its own respective ecological niche. The survival requirements of those potential hosts and biological vectors include: the respective ability of those hosts and biological vectors to compete for their essential needs; their ability to survive within their own vital ambiental range as defined by factors such as temperature, plus either humidity and altitude (if terrestrial) or depth and salinity (if aquatic); and their ability to avoid being consumed by predatory individuals.

interaction. This state may have a beneficial effect by acting as a genetic screening upon both the host species and the viral species. In contrast, avirulence may represent a far more evolved steady state, although evolutionarily it may not be the final state, between the viral and host populations. Avirulence is normally acquired by repeated successive passage of the virus through members of a host or biological vector population.

What are the considerations associated with an apparent genetic disequilibrium? If the virus seems to make all of the members of a species extremely sick, then presumably it normally may not be hosted or vectored by that species. If a virus causes a reduction in the genetic fitness of the host (ability of the host to pass on it's genetic heritage) then the virus is viewed as being in disequilibrium with the host. Incompatible genetic differences may both fuel the fires of virulence and allow a constant state of genetic disequilibrium to exist. Genetic equilibria need time to establish. Constant disequilibrium may be viewed as a competitive strategy effected via "Evolutionary Cheating" (included in loving memory of Dr. Alex Frasier who taught me to understand evolution). Evolutionary cheating involves finding ways to change the rules of fair competition and thereby tilt the playing field in favor of your species. One good example of evolutionary cheating would be to eat your competing species. Viruses tend to steal genes from their hosts (Balter, 1998), and this would represent another example of evolutionary cheating.

1.4.3 Uniqueness versus Commonality (There Are Hussies and Floozies in the Virus World)

1.4.3.1 Numbers of Major Viral Groups (Viral Families and Floating Genera) Affecting Different Host Categories: From examining the list of approved viral taxonomic groups published by the ICTV (International Committee on Taxonomy of Viruses, Master Species list for November 2009, which is available as 2009_5F00_v3 on their website http://www.ictvdb.org/) is was possible to determine the host ranges of the 100 major viral groups (88 families plus 12 unassigned or 'floating' genera). These groups are listed alphabetically in Table 1 of Chapter 2 (if you are curious, searching each of those 2,289 viral species on the internet took 8 days of diligence). From that knowledge, the relative specificity of those major viral groups can be ascertained with regard to the host categories for which they are infective. Each of the major viral groups was associated only with either prokaryotic hosts or eukaryotic hosts. As such, none of the major viral groups crossed the imposing biochemical divide between prokaryotes and eukaryotes.

1.4.3.1.1 Prokaryotic Host Categories There are 18 known major viral groups that are associated with prokaryotic hosts, and summarizing these by category of host the results are:

Archaea - a total of 10 major viral groups contain member species which infect archaea, with 8 of those viral groups being unique to only this host category, and the other 2 viral groups being common which means that they include viral species infective of additional host categories;

Bacteria - a total of 10 major viral groups contain member species which infect bacteria, with 7 of those viral groups being unique to only this host category, and the other 3 viral groups being common which means that they include viral species infective of additional host categories;

Cyanobacteria - a total of 2 major viral groups contain member species which infect cyanobacteria with none of those viral groups being unique to only this host category.

Among those major viral groups associated with prokaryotes, we can assess which groups have commonality as expressed in terms of their possesing a general capacity for association with more than one host category (the hussies!), and those are:

1 viral group is common to Archaea + Bacteria

1 viral group is common to Archaea + Bacteria + Cyanobacteria

1 viral group is common to Bacteria + Cyanobacteria

1.4.3.1.2 Eukaryotic Host Categories There are 82 known major viral groups that are associated with eukaryotic hosts, and summarizing these by category of host the results are:

Algae – a total of 4 major viral groups contain member species which infect algae, with 1 of those viral groups being unique to only this host category, and the other 3 viral groups being common which means that they include viral species infective of additional host categories;

Fungi – a total of 14 major viral groups contain member species which infect fungi, with 6 of those viral groups being unique to only this host category, and the other 8 viral groups being common which means that they include viral species infective of additional host categories;

Invertebrates – a total of 22 major viral groups contain member species which infect invertebrates, with 9 of those viral groups being unique to only this host category, and the other 13 viral groups being common which means that they include viral species infective of additional host categories;

Plants – a total of 33 major viral groups contain member species which infect plants, with 25 of those viral groups being unique to only this host category, and the other 8 viral groups being common which means that they include viral species infective of additional host categories;

Protozoa – a total of 3 major viral groups contain member species which infect protozoans, with 1 of those viral groups being unique to only this host category, and the other 2 viral groups being common which means that they include viral species infective of additional host categories; and

Vertebrates – a total of 33 major viral groups contain member species which infect vertebrates, with 22 of those viral groups being unique to only this host category, and the other 11 viral groups being common which means that they include viral species infective of additional host categories.

Among those major viral groups associated with eukaryotes, we can assess which groups have commonality as expressed in terms of the general capacity for association with more than one host category (the hussies!), and those are:

Viruses Infecting only Microbial or Botanical Hosts

1 viral group is common to Algae + Protozoa

1 viral group is common to Fungi + Protozoa

3 viral groups are common to Plants + Fungi

Viruses Infecting Invertebrate Animal Hosts

1 viral group is common to Invertebrates + Fungi + Plants

1 viral group is common to Invertebrates + Fungi + Plants + Algae

Viruses Infecting Vertebrate Animal Hosts

7 viral groups are common to Invertebrates + Vertebrates

1 viral group is common to Invertebrates + Vertebrates + Fungi

2 viral groups are common to Invertebrates + Vertebrates + Plants

1 viral group is common to Invertebrates + Vertebrates + Fungi + Plants + Algae

The absolute floozies were the Reoviridae, a viral family that produces infectious virions and presently is known to have representation in five host categories of eukaryotes excepting only the protozoa; and the Metaviridae and Pseudoviridae which are the two viral families that represent LTR (long terminal repeat) retrotransposons and are known to each be associated with four host categories of eukaryotes.

Table 1.4 gives an asessment of relative specificity in terms of the percentage of major viral groups that were determined associated with (unique to) only a single host category, plus those major viral groups that were associated with only one additional host category. The absolute numbers of viral groups associated with each host category differed, with the greatest numbers of viral groups being known for vertebrates and plants. This relative wealth of information may be an absolute indication that in fact some host categories are more fertile ground for the evolution of new viral groups, but there also is an important associated truth which is that this difference in numbers of identified viral groups likely reflects the far greater amount of time and money that have been spent on researching viruses of vertebrates and plants. Among the eukaryotic host categories, those major viral groups infective for plants and vertebrates tended to be more

TABLE 1.4 Relative Specificity of the Viral Taxonomic Groups as Compared by Category of Host

Host Category	Viral groups unique to that host category	Viral groups common to one additional host category	Summary of viral groups either unique to that host category or common to just one additional host category
Eukaryotes			
Algae	25% (1 of 4)	25% (1 of 4)	50% (2 of 4)
Fungi	43% (6 of 14)	29% (4 of 14)	71% (10 of 14)
Invertebrates	41% (9 of 22)	32% (7 of 22)	73% (16 of 22)
Plants	76% (25 of 33)	9% (3 of 33)	85% (28 of 33)
Protozoa	33% (1 of 3)	67% (2 of 3)	100% (3 of 3)
Vertebrates	67% (22 of 33)	21% (7 of 33)	88% (29 of 33)
Prokarytoes			
Archaea	80% (8 of 10)	10% (1 of 10)	90% (9 of 10)
Bacteria	70% (7 of 10)	20% (2 of 10)	90% (9 of 10)
Cyanobacteria	0% (0 of 2)	50% (1 of 2)	50% (1 of 2)

The viral taxonomic groups represented in this table are the 88 families and 12 floating genera currently listed by the ICTV (International Committee on Taxonomy of Viruses, Master species list of November 2009 (2009_5F00_v3) which is available on the website http://www.ictvdb.org/) and those groups are listed along with their host ranges in Table 1 of Chapter 2).

unique, ranging from 67–76%, with the extent of uniqueness being either 43% or less for viral groups associated with the other categories of eukaryotes. Among the prokaryotic host categories, those major viral groups infective for archaea and bacteria tended to be more unique, ranging from 70–80%, while the extent of uniqueness was zero for viral groups associated with the only other category of prokaryotes, which was the cyanobacteria. The vast majority of the major viral groups either were unique to a single host category or common to only one additional host category (71–100%) except for the viruses of algae and cyanobacteria (50%).

1.4.3.2 What Might be Reflected When We Look at the Concept of Uniqueness versus Commonality for the Major Viral Groups? Figure 1.17 gives a visual representation for this concept of assessing uniqueness versus commonality. The most obvious separation was observed to be an apparently absolute distinction between those major viral groups associated with eukaryotic host categories (Figure 1.17a) versus prokaryotic host categories (Figure 1.17b). The second most obvious separation is not quite as absolute, but nevertheless represents a clear distinction between viral groups associated with animals versus non-animals. Among those major viral groups associated with animals, the majority of commonalities were limited to the host categories of vertebrates and invertebrates, with only a relatively small percentage of those viral groups extending between the animals and non-animals. Among those major viral groups associated with non-animals, the majority of commonalities were between the host categories of fungi and plants. Half of those viral groups which were common to fungi and plants were able to cross the divide into invertebrates.

Invertebrates often serve as biological vectors for viruses, and this accounts for many, but not all, of the viral group associations which exist between the host categories of invertebrates and either vertebrates, fungi, or plants. It also is very possible that the apparent separations or 'divides' visualized as we examine Figures 1.17a. and 1.17b. can give us clues as to when the presently known major viral groups evolved, i.e., that all presently known viral groups may have arisen since the separation of prokaryotes and eukaryotes, with there being a second major point representing the development of animals.

1.4.4 Evolution

As we look at the relationships between viruses and their hosts and vectors, we might ask ourselves that age-old question of "Which came first, the virus or the cell?" (Koonin et al., 2006). It is perhaps more likely that the viruses and cells arose simultaneously. Presumably they have been struggling to come to terms for a long time, (Claverie, 2006; Forterre and Prangishvili, 2009). We do not

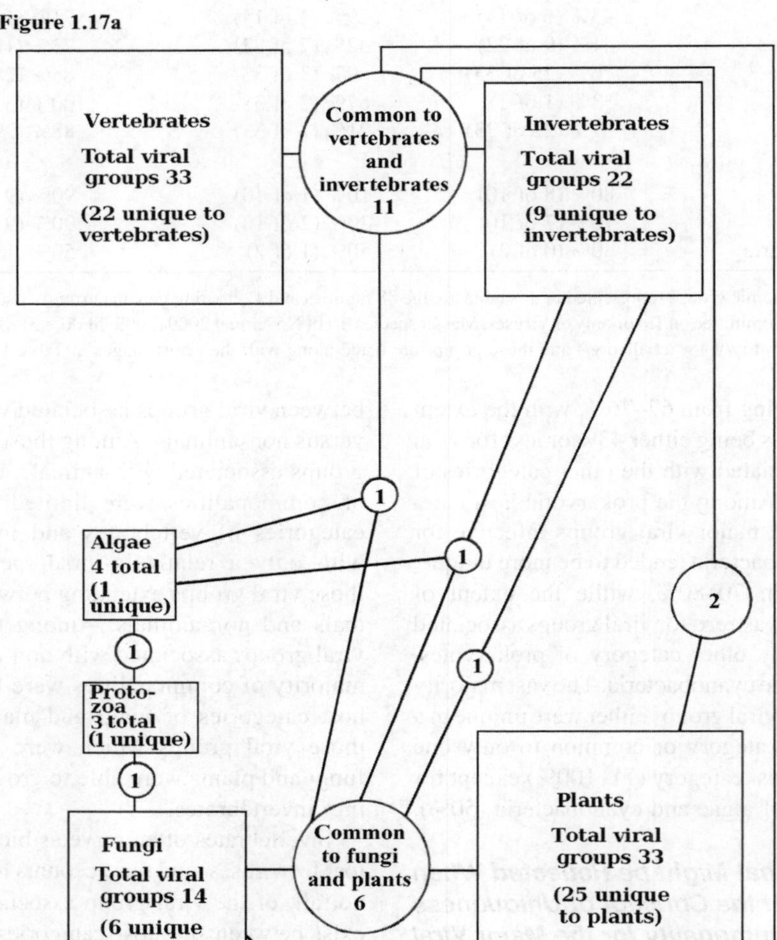

FIGURE 1.17 This figure represents the number of major viral groups, those having the taxonomic classification level of either family or unassigned "floating" genus, know to be associated with eukaryotic host categories (Figure 1.17a) and prokaryotic host categories (Figure 1.17b). The boxes represent host categories. The circles represent interconnections, which are zones that illustrate the fact that many of the major virus groups overlap and are common to more than a single host category. The areas within the boxes and circles are in relative proportion to the numbers of viral families and floating genera being represented, thus giving a visual presentation of viral diversity. The connecting lines represent possibilities for viral-mediated gene flow between host categories. To date, there are no viral families or floating genera known capable of crossing the boundary between eukaryotic and prokaryotic host categories. The names of the virus families and floating genera are listed in Table 1 of chapter 2.

Figure 1.17b

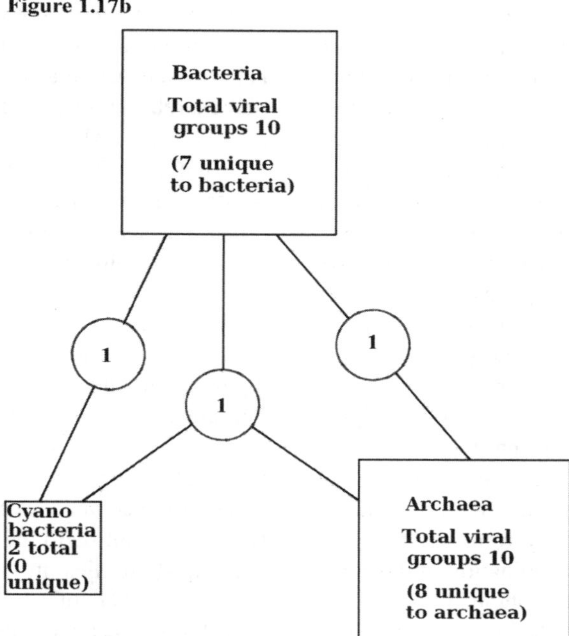

FIGURE 1.17 *(Continued).*

know either to what, or to where, the viruses are leading. Although in a true biological sense it is not necessary for the viruses to "lead" anywhere. From a virocentric view, a perfectly organized virus reproducing from host to host (perhaps with a few vectors included for spice) and transmitting its genetic information over time is a sufficient trend. In considering the evolution of viruses, we must remember the wisdom of Niles Eldredge (1991), that no existing biological entity can be said to represent an end product of evolution. Rather, it is only the extinct biological life forms that clearly can be said to have represented end products of evolution. Likewise, we do not and perhaps never may know if viruses arose only once or else have arisen at many times, with their evolutionary arisal bounded only by the practical limits of some definable adaptive zone. Understanding this comes from the realization that thus far, sabre-toothed cats have evolved at three different times during history and that they evolved from different lineages (Eldredge, 1991). Their evolution at each time would have corresponded to the opening of the appropriate niche, and each of their extinctions would have corresponded to the closing of that niche. For just as it is true that the availability of a niche can drive evolution, so too can the closure of a niche drive extinction.

Although the lack of viral fossils restricts our efforts at following the evolution of viruses, we can draw hypotheses by looking at parallels between a few of the virus groups and their hosts. To begin this process, we have seen that some of the presently existing viral families (we know nothing about those viral families that may be extinct) seem restricted to different host groups. It is likely that as time has gone by, these viruses and their hosts have coevolved and perhaps even undergone phylogenation (the evolution of phylogenetic groupings) in parallel. For example, those viruses which we know as the Myoviridae seem restricted to infecting prokaryotic cells. This could suggest either that the ancestors of the Myoviridae are relatively new or else relatively ancient. Members of the Siphoviridae, which also infect prokaryotes, have developed a relatively stabile mechanism of endogeny (in their case referred to as lysogeny), which may be suggestive of these viruses having had a long period of coevolutionary adaptation with their host cells. We can see that the viroids of plants, which genetically bear a link to the viruses (chapter 2 addresses viral taxonomy, and prions are specifically addressed in chapter 12 of volume 1) seemingly have developed such a highly evolved endogenous state that they never produce anything resembling a virion and indeed may not use or even need a natural route of transmission because they remain internal to their host. Additional examination of the existing viral groups, and the establishment of parallels between these and the known evolution of animal phyla, reveals that virus groups such as the Iridoviridae, which do produce virions, seem restricted to invertebrates and poikilothermic vertebrates. This latter examination could lead to the suggestion that ancestors of the iridoviruses followed the animal phylogenation pathway upward to a point just short of the evolution of euthermia. The retroviruses have gone onward to infect euthermic animals, and it has been hypothesized that at least some retroviruses have coevolved with their hosts to the extent that they allowed development of the placental mammals (Villareal, 1997).

Why are the viruses still around? The viruses might serve as an evolutionary benefit to the cellular organisms by gradually transferring genetic information between different sources and serving as a source for genomic development (de Lima Fávaro et al., 2005; Piskurek and Okada, 2007; Todorovska, 2007; Williams, 1996). Perhaps this is the reason why their hosting species continue allowing the viruses to exist. Perhaps the pure beauty of a virus, when viewed as an evolutionary element, is that it can break free from one host to enter another host. Gradually, that virus could coevolve until at last it might settle upon a permanent home as some endogenous genetic element within a single hosting species. Alternatively, the virus may play the role of eternally being a rebel in search of a cause. Oh, to be so free as a species!

What will the viruses become with time? As stated above, in a strictly evolutionary sense it is not necessary for the viruses to be leading to anywhere. However, if we can draw parallels and make the assumption that the relationship between virus and host moves with time towards avirulence and an eventual genetic equilibrium, then we can make hypotheses. Perhaps some of the viruses will indeed continue the way of being predatory outsiders. Others, however, seem destined for symbiosis and thus to become a part of us. We see at least two clues pointing to the latter type of destiny. One of these lies in Villareal's hypothesis (Villareal, 1997) that by evolving to have the same biological agenda as their placental mammalian hosts, the endogenous retroviruses have symbiotically joined with their hosts to create a single species. The hypovirulence elements of the fungi which cause Chestnut blight disease are another clue (Volume 1, Chapter 9), these elements apparently evolved from a virus and seemingly have achieved symbiosis. The hypovirulence elements sustain their existence by reducing the virulence of their host fungi, so that in turn the host fungus does not kill the tree upon which the fungus feeds, enabling all to survive.

Alas, it might also be true that the evolution of viruses represents a question which we cannot yet even try to answer.

1.5 SUMMARY (CAN THERE BE CONCLUSIONS?)

The ecology of a virus primarily consists of it's interactions with the organisms that serve as it's hosting species (principle hosts, alternate hosts, and vectors. The routes by which viruses achieve transmission between these other organisms represent a second aspect of the ecology of viruses. Furthermore, an examination of the interactions between a virus and it's hosts and biological vectors brings up many questions. Principle among these questions is the reason why the outcome of viral infections sometimes appears to be so disastrous, and yet at other times appears unnoticeable.

One of the founding principles in biology is that natural selection serves as the basis for the population dynamics which produce the many different outcomes that we observe as scientists. When we use this principle as the lense through which to examine interactions between viruses and their host and vector species, we notice that many possible strategies exist, more than can be explained. The strategies which we do find in evidence began at random and exist because selection has not done away with them. While we do not know how the viruses have arisen, or what will be their destiny, we can assume that there may be viruses for as long as there are cells.

ACKNOWLEDGEMENT

I am thankful to my friend and former colleague Dr. H. D. Alan Lindquist for his assistance with the preparation of a much older version of this chapter. My hopes and best wishes go out to him.

REFERENCES

Alexander, M. (1981). Why microbial predators and parasites do not eliminate their prey and hosts. *Ann. Rev. Microbiol.* 35, 113–133.

Balter, M. (1998). Viruses have many ways to be unwelcome guests. *Science* 280, 204–205.

Beck, G. and Habicht, G. S. (1996). Immunity and the invertebrates. *Scientific American* 275(5), 60–66.

Brogdon, W. G. and McAllister, J. C. (1998). Insecticide resistance and vector control. *Emerging Infectious Diseases* 4, 605–613.

Caldwell, J. C. and Caldwell, P. (1996). The African AIDS epidemic. *Scientific American* 274(3), 62–68.

Claverie, J.-M. (2006). Viruses take center stage in cellular evolution. *Genome Biology* 7(6), 110 (5 pages).

Danilova, N. (2006). The Evolution of Immune Mechanisms. *Journal of Experimental Zoology* 306B, 496–520.

de Lima Fávaro, L. C., de Araújo, W. L., de Azevedo, J. L., and Paccola-Meirelles, L. D. (2005). The biology and potential for genetic research of transposable elements in filamentous fungi. *Genetics and Molecular Biology* 28(4), 804–813.

Dennehy, J. J., Friedenberg, N. A., Holt, R. D., and Turner, P. E. (2006). Viral Ecology and the Maintenance of Novel Host Use. *American Naturalist* 167, 429–439.

Doyle, J. (1985). *"Altered Harvest"*. Viking, New York.

Eldredge, N. (1991). *"Fossils: the Evolution and Extinction of Species"*, pp. 4–30. Harry N. Abrams, New York.

Ewald, P. W. (1993). The evolution of virulence. *Scientific American* 268(4), 86–93.

Forterre, P. and Prangishvili, D. (2009). The great billion-year war between ribosome- and capsid-encoding organisms (cells and viruses) as the major source of evolutionary novelties. *Ann. N. Y. Acad. Sci.* 1178, 65–77.

Fuller, J. G. (1974). *"Fever!: the Hunt for a New Killer Virus"*. Reader's Digest Press, New York.

Gauntt, C. J. (1997). Nutrients often influence viral diseases. *ASM News* 63, 133–135.

Hurst, C. J. and Murphy, P. A. (1996). The transmission and prevention of infectious disease. *In "Modeling Disease Transmission and It's Prevention by Disinfection"* (C. J. Hurst, ed.), pp. 3–54. Cambridge University Press, Cambridge.

Koonin, E. V., Senkevich, T. G., and Dolja, V. V. (2006). The ancient virus world and evolution of cells. *Biology Direct* 1, 29 (27 pages).

Kuiken, T., Holmes, E. C., McCauley, J., Rimmelzwaan, G. F., Williams, C. S., and Grenfell, B. T. (2006). Host Species Barriers to Influenza Virus Infections. *Science* 312, 394–397.

Larson, G. (1998). *"There's a Hair in my Dirt! a Worm's Story"*. HarperCollins, New York.

Lederberg, J. (1997). Infectious disease as an evolutionary paradigm. *Emerging Infectious Diseases* 3, 417–423.

Levin, B. R., Lipsitch, M., and Bonhoeffer, S. (1999). Population biology, evolution, and infectious disease: convergence and synthesis. *Science* 283, 806–809.

Litman, G. W. (1996). Sharks and the origins of vertebrate immunity. *Scientific American* 275(5), 67–71.

Mills, J. N. and Childs, J. E. (1998). Ecologic studies of rodent reservoirs: their relevance for human health. *Emerging Infectious Diseases* 4, 529–537.

Moffat, A. S. (1994). Mapping the sequence of disease resistance. *Science* 265, 1804–1805.

Moineau, S., Pandian, S., and Klaenhammer, T. R. (1994). Evolution of a lytic bacteriophage via DNA acquisition from the *Lactococcus lactis* chromosome. *Appl. Environ. Microbiol.* 60, 1832–1841.

Morell, V. (1997). Return of the forest. *Science* 278, 2059.

Muñoz-Jordán, J. L. and Fredericksen, B. L. (2010). How Flaviviruses Activate and Suppress the Interferon Response. *Viruses* 2, 676–691.

Nathanson, N. (1997). The emergence of infectious diseases: societal causes and consequences. *ASM News* 63, 83–88.

Newman, M. (1984). The Virus. *Focus* 6(3), 14. July 1984, Bethesda Research Laboratories, Gaithersburg, Maryland.

Piskurek, O. and Okada, N. (2007). Poxviruses as possible vectors for horizontal transfer of retroposons from reptiles to mammals. *Proc. Natl. Acad. Sci.* 104(29), 12046–12051.

Ploegh, H. L. (1998). Viral strategies of immune evasion. *Science* 280, 248–253.

Russev, G. (2007). RNA Interference. *Biotechnology & Biotechnological Equipment* 21(3), 283–285.

Spear, P. G. (1998). A welcome mat for leprosy and lassa fever. *Science* 282, 1999–2000.

Subbarao, K., Klimov, A., Katz, J., Regnery, H., Lim, W., Hall, H., Perdue, M., Swayne, D., Bender, C., Huang, J., Hemphill, M., Rowe, T., Shaw, M., Xu, X., Fukuda, K. and Cox, N. (1998). Characterization of an avian influenza A (H5N1) virus isolated from a child with a fatal respiratory illness. *Science* 279, 393–396.

Terzian, C., Pélisson, A., and Bucheton, A. (2001). Evolution and phylogeny of insect endogenous retroviruses. *BioMed Central Evol. Biol.* 1, 3 (8 pages).

Todorovska, E. (2007). Retrotransposons and their role in plant-genome evolution. *Biotechnology & Biotechnological Equipment* 21(3), 294–305.

van der Kuyl, A. C., Dekker, J. T., and Goudsmit, J. (1995). Distribution of baboon endogenous virus among species of african monkeys suggests multiple ancient cross-species transmissions in shared habitats. *J. Virol.* 69, 7877–7887.

Villareal, L. P. (1997). On viruses, sex, and motherhood. *J. Virol.* 71, 859–865.

Williams, N. (1996). Phage transfer: a new player turns up in cholera infection. *Science* 272, 1869–1870.

Zhang, T., Breitbart, M., Lee, W.-H., Run, J.-Q., Wei, C.L., Soh, S. W. L., Hibberd, M. L., Liu, E. T., Rohwer, F., and Ruan, Y. (2006). RNA viral community in human feces: prevalence of plant pathogenic viruses. *PloS Biol* 4(1), e3.

Zinkernagel, R. M. (1996). Immunology taught by viruses. *Science* 271, 173–178.

CHAPTER 2

AN INTRODUCTION TO VIRAL TAXONOMY WITH EMPHASIS ON ANIMAL HOSTS AND THE PROPOSAL OF AKAMARA, A POTENTIAL DOMAIN FOR THE GENOMIC ACELLULAR AGENTS*

CHRISTON J. HURST[1,2]

[1]Departments of Biology and Music, Xavier University, Cincinnati, OH

[2]Engineering Faculty, Universidad del Valle, Ciudad Universitaria Meléndez, Santiago de Cali, Valle, Colombia

CONTENTS

2.1 INTRODUCTION

Taxonomy is literally the naming of taxons (in plural also termed taxa), which by definition are groupings of items based upon identifiable similarities. The viruses are a group of biological entities which have in common the fact that they possess a nucleic acid genome that is composed either of DNA or RNA. That nucleic acid genome is surrounded and protected by a shell of proteins which is termed either to be a nucleocapsid or, more simply, a capsid. The nucleocapsids of some viruses are, in turn, surrounded by a lipid membrane. The viruses have been grouped by many different methods. Those viral taxonomic groupings which presently are recognized by the ICTV (International Committee on Taxonomy of Viruses) divide these biological entities into families, genera and species. There also currently exist a few recognized viral order groups.

This chapter also introduces the idea that the taxonomy of the viruses and their biological relatives could be extended to the domain level. There currently exist three biological domains, Archaea, Bacteria and Eukarya, which consist only of cellular organisms. The establishment of these three existing domains and the

* This chapter represents a revision of "An introduction to viral taxonomy and the proposal of Akamara, a potential domain for the genomic acellular agents", which appeared as chapter 2 of the book *Viral Ecology*, edited by Christon J. Hurst, published in 2000 by Academic Press. All of the artwork contained in this chapter appears courtesy of Christon J. Hurst.

Studies in Viral Ecology: Animal Host Systems: Volume 2, First Edition. Edited by Christon J. Hurst.

taxonomic placement of biological entities within them largely is based upon the ribosomal RNA nucleotide sequence of those constituent organisms. This article proposes the creation of an additional biological domain that would represent the acellular infectious agents which possess nucleic acid genomes (termed 'genomic acellular infectious agents' for the purpose of this proposal). The proposed constituents of this domain are the agents commonly termed to be either viruses, satellite viruses, virusoids or viroids. The proposed domain title is Akamara (ακαμαρα), whose derivation from Greek would translate as meaning 'without chamber' or 'without vault', and is suggested as describing the fact that these agents lack a cellular structure of their own. A possible organizational structure within this proposed new domain is also suggested, with its occupants shown as being divided into two kingdoms, plus phyla and classes premised upon basic characteristics of the organisms' genomic biochemistry. The kingdom Euviria (true viruses) is suggested as containing the 'conventional' viruses plus those viral-like agents which likewise possess genomes that code for their own structural 'shell' or 'capsid' proteins. The kingdom Viroidia would contain the genus Deltavirus plus the viroids and virusoids, whose members are RNA agents that have in common the trait that their genomic structure has endowed them with the capacity for evolutionary survival even though their genomes do not code for such structural proteins. The members of the kingdom Euviria are suggested as being subdivided into two phyla based upon whether their genome is of RNA (phylum Ribovira) or DNA (phylum Deoxyribovira), and these are further subdivided into classes based upon whether their genomes are 'negative sense' single stranded RNA versus 'plus sense' single stranded RNA, double stranded RNA, single stranded DNA or double stranded DNA. The kingdom Viroidia is zsuggested to contain one phylum, Viroida, encompassing the viroids, virusoids, and genus Deltavirus, all of which possess RNA genomes.

2.2 THE EXISTING VIRAL FAMILIES

The viruses recognized by the ICTV have been assigned into genera, and nearly all of these genera have been grouped into families. Those genera which have not been incorporated into families are considered to be "floating genera." This concept is very fluid (pun intended! never accept taxonomy as though it were written into stone tablets either by some God or biological committee). The viral family groupings and floating genera which existed at the time when the ICTV published its Master Species List of November 2009 (2009_5F00_v3), which is available on the website of the International Committee on Taxonomy of Viruses http://www.ictvdb.org/, are listed in Table 2.1. Some of these viral families have been placed into higher taxonomic levels up to that of order (Table 2.2). Those viral families and floating genera which affect animal hosts are depicted, along with their basic morphological characteristics, in Figures 2.1–2.5. An examination of the drawings of the virus groups presented in those first five figures reveals that most of the known viral capsid structures can be categorized as being either helical or icosahedral in form. The basic form of a helical capsid structure is represented in Figure 2.6. A sculpture which interestingly and unintentionally resembles the membrane envelope with its associated proteins that encloses the helical capsid of a filovirus is presented in Figure 2.7. The icosahedral capsid structure is represented in Figures 2.8 and 2.9.

2.3 THE PROPOSED DOMAIN AKAMARA

It has been suggested (Morell, 1996) that with the sequencing of an archaeon microbe (Bult et al., 1996), the last of life's three domains has been elucidated, and that these domains are the Archaea, Bacteria and Eukarya. That assessment leaves out something very important, namely, the viruses and the other acellular infectious agents. Indeed, it must be

TABLE 2.1 Listing of Viral Taxonomic Groups - Families and Floating Genera

Viral Group (Name)	Taxonomic Level (Family vs. Unassociated or "Floating" Genus)	Nature of Genome	Host Range as Presently Known	Refer to Figure Number
Adenoviridae	Family	DNA, double-stranded	Vertebrates	2.1
Alloherpesviridae	Family	DNA, double-stranded	Vertebrates	2.1
Alphaflexiviridae	Family	RNA, single-stranded (+ sense)[a]	Fungi, plants	Vol.1[c]
Ampullaviridae	Family	DNA, double-stranded	Archaea	Vol.1
Anelloviridae	Family	DNA, single-stranded	Vertebrates	2.2
Arenaviridae	Family	RNA, single-stranded (± sense)[b]	Vertebrates	2.4
Arteriviridae	Family	RNA, single-stranded (+ sense)	Vertebrates	2.5
Ascoviridae	Family	DNA, double-stranded	Invertebrates	2.1
Asfarviridae	Family	DNA, double-stranded	Vertebrates	2.1
Astroviridae	Family	RNA, single-stranded (+ sense)	Vertebrates	2.5
Avsunviroidae	Family	RNA, single-stranded (viroid)	Plants	Vol.1
Baculoviridae	Family	DNA, double-stranded	Invertebrates	2.1
Barnaviridae	Family	RNA, single-stranded (+ sense)	Fungi	Vol.1
Benyvirus	Floating genus	RNA, single-stranded (+ sense)	Plants	Vol.1
Betaflexiviridae	Family	RNA, single-stranded (+ sense)	Plants	Vol.1
Bicaudaviridae	Family	DNA, double-stranded	Archaea	Vol.1
Birnaviridae	Family	RNA, double-stranded	Invertebrates, vertebrates	2.3
Bornaviridae	Family	RNA, single-stranded (– sense)[d]	Vertebrates	2.4
Bromoviridae	Family	RNA, single-stranded (+ sense)	Plants	Vol.1
Bunyaviridae	Family	RNA, single-stranded (± sense)	Invertebrates, plants, vertebrates	2.4
Caliciviridae	Family	RNA, single-stranded (+ sense)	Vertebrates	2.5
Caulimoviridae	Family	DNA, double-stranded	Plants	Vol.1
Chrysoviridae	Family	RNA, double-stranded	Fungi	Vol.1
Cilevirus	Floating genus	RNA, single-stranded (+ sense)	Plants	Vol.1
Circoviridae	Family	DNA, single-stranded	Vertebrates	2.2
Closteroviridae	Family	RNA, single-stranded (+ sense)	Plants	Vol.1
Coronaviridae	Family	RNA, single-stranded (+ sense)	Vertebrates	2.5
Corticoviridae	Family	DNA, double-stranded	Bacteria	Vol.1

(continued)

TABLE 2.1 (*Continued*)

Viral Group (Name)	Taxonomic Level (Family vs. Unassociated or "Floating" Genus)	Nature of Genome	Host Range as Presently Known	Refer to Figure Number
Cystoviridae	Family	RNA, double-stranded	Bacteria	Vol.1
Deltavirus	Floating genus	RNA, single-stranded (– sense)	Vertebrates	2.4
Dicistroviridae	Family	RNA, single-stranded (+ sense)	Invertebrates	2.5
Emaravirus	Floating genus	RNA, single-stranded (± sense)	Plants	Vol.1
Endornaviridae	Family	RNA, double-stranded	Fungi, plants	Vol.1
Filoviridae	Family	RNA, single-stranded (– sense)	Vertebrates	2.4
Flaviviridae	Family	RNA, single-stranded (+ sense)	Invertebrates, vertebrates	2.5
Fuselloviridae	Family	DNA, double-stranded	Archaea	Vol.1
Gammaflexiviridae	Family	RNA, single-stranded (+ sense)	Fungi	Vol.1
Geminiviridae	Family	DNA, single-stranded	Plants	Vol.1
Globuloviridae	Family	DNA, double-stranded	Archaea	Vol.1
Guttaviridae	Family	DNA, double-stranded	Archaea	Vol.1
Hepadnaviridae	Family	DNA, partially double-stranded	Vertebrates	2.1
Hepeviridae	Family	RNA, single stranded (+ sense)	Vertebrates	2.5
Herpesviridae	Family	DNA, double-stranded	Vertebrates	2.1
Hypoviridae	Family	RNA, double-stranded	Fungi	Vol.1
Idaeovirus	Floating genus	RNA, single-stranded (+ sense)	Plants	Vol.1
Iflaviridae	Family	RNA, single-stranded (+ sense)	Invertebrates	2.5
Inoviridae	Family	DNA, single-stranded	Bacteria	Vol.1
Iridoviridae	Family	DNA, double-stranded	Invertebrates, vertebrates	2.1
Leviviridae	Family	RNA, single-stranded (+ sense)	Bacteria	Vol.1
Lipothrixviridae	Family	DNA, double-stranded	Archaea	Vol.1
Luteoviridae	Family	RNA, single-stranded (+ sense)	Plants	Vol.1
Malacoherpesviridae	Family	DNA, double-stranded	Invertebrates	2.1
Marnaviridae	Family	RNA, single-stranded (+ sense)	Algae	Vol.1
Metaviridae	Family	RNA, single-stranded (+ sense?)	Fungi, invertebrates, plants, vertebrates	2.5
Microviridae	Family	DNA, single-stranded	Bacteria	Vol.1
Mimiviridae	Family	DNA, double-stranded	Protozoa	Vol.1

TABLE 2.1 *(Continued)*

Viral Group (Name)	Taxonomic Level (Family vs. Unassociated or "Floating" Genus)	Nature of Genome	Host Range as Presently Known	Refer to Figure Number
Myoviridae	Family	DNA, double-stranded	Archaea, bacteria, cyanobacteria	Vol.1
Nanoviridae	Family	DNA, single-stranded	Plants	Vol.1
Narnaviridae	Family	RNA, single-stranded (+ sense)	Fungi	Vol.1
Nimaviridae	Family	DNA, double-stranded	Invertebrates	2.1
Nodaviridae	Family	RNA, single-stranded (+ sense)	Fungi (experimentally), invertebrates, vertebrates	2.5
Ophioviridae	Family	RNA, single-stranded (– sense)	Plants	Vol.1
Orthomyxoviridae	Family	RNA, single-stranded (– sense)	Invertebrates, vertebrates	2.4
Ourmiavirus	Floating genus	RNA, single-stranded (+ sense)	Plants	Vol.1
Papillomaviridae	Family	DNA, partially double-stranded	Vertebrates	2.1
Paramyxoviridae	Family	RNA, single-stranded (– sense)	Vertebrates	2.4
Partitiviridae	Family	RNA, double-stranded	Fungi, plants	Vol.1
Parvoviridae	Family	DNA, single-stranded	Invertebrates, vertebrates	2.2
Phycodnaviridae	Family	DNA, double-stranded	Algae, protozoa	Vol.1
Picobirnaviridae	Family	RNA, double-stranded	Vertebrates	2.3
Picornaviridae	Family	RNA, single-stranded (+ sense)	Vertebrates	2.5
Plasmaviridae	Family	DNA, double-stranded	Bacteria	Vol.1
Podoviridae	Family	DNA, double-stranded	Bacteria, cyanobacteria	Vol.1
Polemovirus	Floating genus	RNA, single-stranded (+ sense)	Plants	Vol.1
Polydnaviridae	Family	DNA, double-stranded	Invertebrates	2.1
Polyomaviridae	Family	DNA, partially double-stranded	Vertebrates	2.1
Pospiviroidae	Family	RNA, single-stranded (viroid)	Plants	Vol.1
Potyviridae	Family	RNA, single-stranded (+ sense)	Plants	Vol.1
Poxviridae	Family	DNA, double-stranded	Invertebrates, vertebrates	2.1
Pseudoviridae	Family	RNA, single-stranded (+ sense)	Algae, fungi, invertebrates, plants	2.5

(continued)

TABLE 2.1 (*Continued*)

Viral Group (Name)	Taxonomic Level (Family vs. Unassociated or "Floating" Genus)	Nature of Genome	Host Range as Presently Known	Refer to Figure Number
Reoviridae	Family	RNA, double stranded	Algae, fungi, invertebrates, plants, vertebrates	2.3
Retroviridae	Family	RNA, single-stranded (+ sense)	Vertebrates	2.5
Rhabdoviridae	Family	RNA, single-stranded (– sense)	Invertebrates, plants, vertebrates	2.4
Rhizidiovirus	Floating genus	DNA, double-stranded	Fungi	Vol.1
Roniviridae	Family	RNA, single-stranded (+ sense)	Invertebrates	2.5
Rudiviridae	Family	DNA, double-stranded	Archaea	Vol.1
Salterprovirus	Floating genus	DNA, double-stranded	Archaea	Vol.1
Secoviridae	Family	RNA, single-stranded (+ sense)	Plants	Vol.1
Siphoviridae	Family	DNA, double-stranded	Archaea, bacteria	Vol.1
Sobemovirus	Floating genus	RNA, single-stranded (+ sense)	Plants	Vol.1
Tectiviridae	Family	DNA, double-stranded	Bacteria	Vol.1
Tenuivirus	Floating genus	RNA, single-stranded (± sense)	Plants	Vol.1
Tetraviridae	Family	RNA, single-stranded (+ sense)	Invertebrates	2.5
Togaviridae	Family	RNA, single-stranded (+ sense)	Invertebrates, vertebrates	2.5
Tombusviridae	Family	RNA, single-stranded (+ sense)	Plants	Vol.1
Totiviridae	Family	RNA, double-stranded	Fungi, protozoa	Vol.1
Tymoviridae	Family	RNA, single-stranded (+ sense)	Plants	Vol.1
Umbravirus	Floating genus	RNA, single-stranded (+ sense)	Plants	Vol.1
Varicosavirus	Floating genus	RNA, single-stranded (– sense)	Plants	Vol.1
Virgaviridae	Family	RNA, single-stranded (+ sense)	Plants	Vol.1

This information has been summarized from the ICTV Master Species List of November 2009 (ICTV Master Species List 2009_5F00_v3 available on the website of the International Committee on Taxonomy of Viruses http://www.ictvdb.org/).

[a]The RNA genome has the same sense as messenger RNA and can be translated directly.

[b]The RNA genome is considered to be ambisense, meaning that it has some sections which are of the same sense as messenger RNA, but other sections of the genome must be copied to produce an opposite strand which in turn can be translated.

[c]The members of this group are presented in Volume 1.

[d]The RNA genome must be copied to produce an opposite strand which in turn can be translated.

TABLE 2.2 Viruses Which have been Assigned to Taxonomic Orders

Order	Family	Subfamily	Genus
Caudovirales	Myoviridae	Undesignated	*"T4-like viruses"*
	Podoviridae	Autographivirinae	*"PhiKMV-like viruses"*
			"SP6-like viruses"
			"T7-like viruses"
		Picovirinae	*"AHJD-like viruses"*
			"Phi29-like viruses"
		Undesignated	*"BPP-1-like viruses"*
			"Epsilon15-like viruses"
			"LUZ24-like viruses"
			"N4-like viruses"
			"P22-like viruses"
			"Phieco32-like viruses"
	Siphoviridae	Undesignated	*"c2-like viruses"*
			"L5-like viruses"
			"Lambda-like viruses"
			"N15-like viruses"
			"PhiC31-like viruses"
			"PsiM1-like viruses"
			"SPbeta-like viruses"
			"T1-like viruses"
			"T5-like viruses"
Herpesvirales	Alloherpesviridae	Undesignated	*Batrachovirus*
			Cyprinivirus
			Ictalurivirus
			Salmonivirus
	Herpesviridae	Alphaherpesvirinae	*Iltovirus*
			Mardivirus
			Simplexvirus
			Varicellovirus
		Betaherpesvirinae	*Cytomegalovirus*
			Muromegalovirus
			Proboscivirus
			Roseolovirus
		Gammaherpesvirinae	*Lymphocryptovirus*
			Macavirus
			Percavirus
			Rhadinovirus
	Malacoherpesviridae	Undesignated	*Ostreavirus*
Mononegavirales	Bornaviridae	Undesignated	*Bornavirus*
	Filoviridae	Undesignated	*Ebolavirus*
	Paramyxoviridae	Paramyxovirinae	*Avulavirus*
			Henipavirus
			Morbillivirus
			Respirovirus
			Rubulavirus
		Pneumovirinae	*Metapneumovirus*
			Pneumovirus

(continued)

TABLE 2.2 (*Continued*)

Order	Family	Subfamily	Genus
	Rhabdoviridae	Undesignated	*Cytorhabdovirus*
			Ephemerovirus
			Lyssavirus
			Novirhabdovirus
			Nucleorhabdovirus
			Vesiculovirus
Nidovirales	Arteriviridae	Undesignated	*Arterivirus*
	Coronaviridae	Coronavirinae	*Alphacoronavirus*
			Betacoronavirus
			Gammacoronavirus
		Torovirinae	*Bafinivirus*
			Torovirus
	Roniviridae	Undesignated	*Okavirus*
Picornavirales	Dicistroviridae	Undesignated	*Cripavirus*
	Iflaviridae	Undesignated	*Iflavirus*
	Marnaviridae	Undesignated	*Marnavirus*
	Picornaviridae	Undesignated	*Aphthovirus*
			Avihepatovirus
			Cardiovirus
			Enterovirus
			Erbovirus
			Hepatovirus
			Kobuvirus
			Parechovirus
			Sapelovirus
			Senecavirus
			Teschovirus
			Tremovirus
	Secoviridae	Comovirinae	*Comovirus*
			Fabavirus
			Nepovirus
		Undesignated	*Cheravirus*
			Sadwavirus
			Sequivirus
			Torradovirus
			Waikavirus
Tymovirales	Alphaflexiviridae	Undesignated	*Allexivirus*
			Botrexvirus
			Lolavirus
			Mandarivirus
			Potexvirus
			Sclerodarnavirus
	Betaflexiviridae	Undesignated	*Capillovirus*
			Carlavirus
			Citrivirus
			Foveavirus
			Trichovirus
			Vitivirus

TABLE 2.2 (*Continued*)

Order	Family	Subfamily	Genus
	Gammaflexiviridae	Undesignated	*Mycoflexivirus*
	Tymoviridae	Undesignated	*Maculavirus*
			Marafivirus
			Tymovirus

This information has been summarized from the ICTV Master Species List of November 2009 (ICTV Master Species List 2009_5F00_v3 available on the website of the International Committee on Taxonomy of Viruses http://www.ictvdb.org/).

remembered that the first life forms whose genomes were sequenced in entirety were not cellular in nature, but rather the viruses MS2 (having an RNA genome) and SV40 (having a DNA genome), and we have had knowledge of their full genomes for more than 30 years (Fiers et al. 1976, 1978).

Perhaps the time has come to suggest a fourth biological domain to give a higher taxonomic home to the viruses and their genomic relatives. One logical suggestion would be to include as a group the 'conventional' viruses plus those satellite viruses whose genomes likewise code for their own structural 'shell' or 'capsid' proteins, and which are also commonly defined as 'viruses'. A second group within this domain might consist of the viroids, virusoids, and the viral genus Deltavirus, which share strong commonalities with respect to their RNA genomic structure and the fact that they do not code for such structural proteins. These infectious agents are excluded from the three existing domains for two reasons: first, their genomes do not code for ribosomal RNA which is the defining characteristic for membership in the three domains and second, they lack a cellular structure of their own, a fact which also kept them officially excluded from the older kingdom classifications.

The conventional viruses are very heterogenous with respect to their genomic structure and vary widely in the extent of genetic coding that they carry. Some of them such as T4, a member of the family Myoviridae, carry a major amount of the genetic coding which is necessary to replicate themselves. Other viruses, such as the human polioviruses which belong to the family Picornaviridae, carry just barely more than the limited amount of genome needed to code for their structural proteins. In comparison, the RNA genomes of the viroids and virusoids as a group are more homogenous and uniquely seem to evidence an evolutionary stability as infectious agents despite the fact that their genomes do not code for any such structural proteins. The variety of agents known as virusoids "borrow" encapsulating proteins from a helper virus. The viroids either have done away with the need for encapsulating proteins or perhaps never possessed them. The genome of the Hepatitis D virus, which is the constituent species of the floating genus Deltavirus, represents what seems to be an interesting evolutionary anomaly. This agent of humans is essentially identical to that of the viroids which are plant pathogens, with exception of the fact that the Hepatitis D virus' genome carrys the genetic coding for a protein that it apparently has picked-up from a cellular host (Brazas and Ganem, 1996; Robertson, 1996). Despite its very limited coding capacity, as with the virusoids, the hepatitis D virus needs to "borrow" enveloping structural proteins which are coded for by a helper virus.

The assignment of taxonomic levels for cellular organisms was initially based upon their similarities at the level of physical traits and aided by a trail of fossilized remains. This approach has since been superseded by the suggestion that such assignments could be based upon molecular chemistry, specifically the nucleotide sequence of the organism's ribosomal RNA. These assignments based upon RNA sequence are assumed to represent

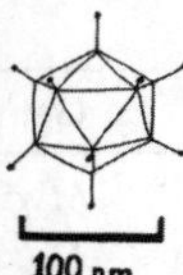

Plate 2.1.1 Family: Adenoviridae. Nucleic acid: DNA. Genome: Double stranded, 1 linear segment (26–45 Kbp). Morphology: Non-enveloped. Virion: Icosahedral (Mw = 1.5–1.8 × 10^8). Nucleocapsid: Icosahedral.

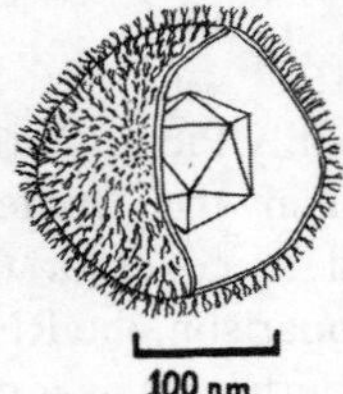

Plate 2.1.2 Family: Alloherpesviridae. Nucleic acid: DNA. Genome: Double-stranded, 1 linear segment (134 Kbp). Morphology: Enveloped. Virion: Quasi-spherical [(Mw = (approximate) 4.6 × 10^8)]. Nucleocapsid: Icosahedral.

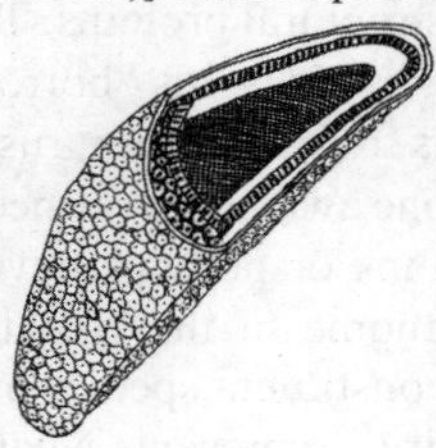

Plate 2.1.3 Family: Ascoviridae. Nucleic acid: DNA. Genome: Double-stranded, 1 circular segment (100–180 Kbp). Morphology: Enveloped. Virion: Oblong convex disc (Mw not specified). Nucleocapsid: Complex Distinguishing feature: Virion looks like a narrow contact lens! Contains a nucleoprotein core successively enclosed by an internal membrane, a surrounding layer of protein subunits in a generally hexagonal pattern, and the external lipid envelope.

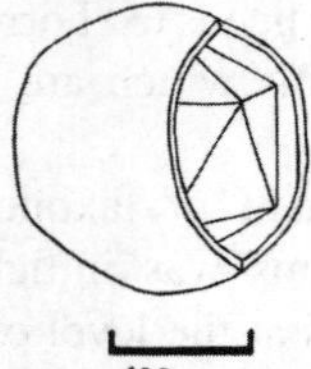

Plate 2.1.4 Family: Asfarviridae (was: genus African swine fever-like viruses). Nucleic acid: DNA. Genome: Double stranded, 1 linear segment (170–190 Kbp) Morphology: Enveloped. Virion: Spherical (Mw not specified) Nucleocapsid: Icosahedral.

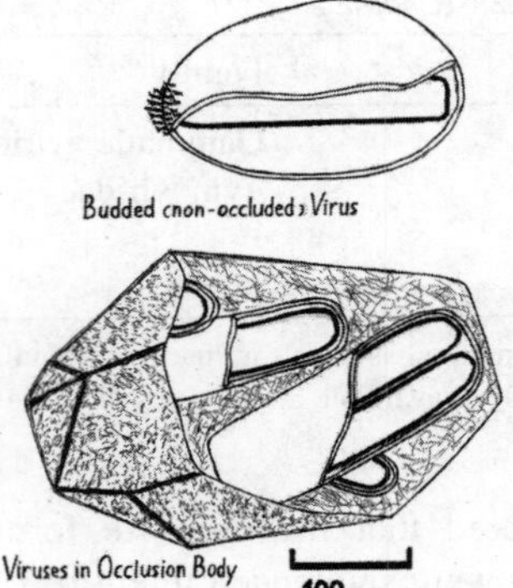

Plate 2.1.5 Family: Baculoviridae. Nucleic acid: DNA. Genome: Double stranded, 1 circular segment (80–180 Kbp). Morphology: Enveloped. Virion: Bacilliform (Mw not specified). Nucleocapsid: Bacilliform. Distinguishing feature: In one phenotype the virions are found occluded inside a crystalline protein "polyhedra" or "occlusion body". Virions of the other phenotype are referred to as being "budded" or "non-occluded".

Plate 2.1.6 Family: Hepadnaviridae. Nucleic acid: DNA. Genome: Single-stranded (partially double stranded), 1 circular molecule (approx. 3.2 Kb); formed by the partial co-hybridization of 2 unequal length segments the longer of which is Negative sense and the shorter is Positive sense. Morphology: Enveloped. Virion: Spherical (pleomorphic) (Mw = 1.6–1.8 × 10^6). Nucleocapsid: Icosahedral.

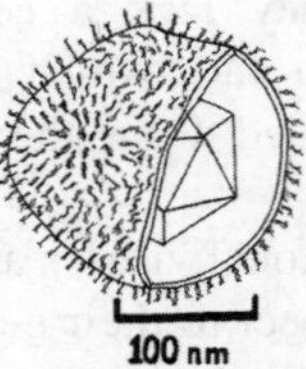

Plate 2.1.7 Family: Herpesviridae. Nucleic acid: DNA. Genome: Double stranded, 1 linear segment (125–240 Kbp). Morphology: Enveloped. Virion: Quasi-spherical [(Mw = (approximate) 4.6 × 10^8)]. Nucleocapsid: Icosahedral.

FIGURE 2.1 Relative sizes and basic information for those viruses which possess double stranded DNA genomes.

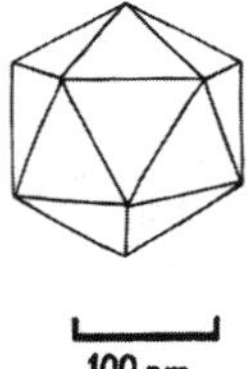

Plate 2.1.8 Family: Iridoviridae. Nucleic acid: DNA. Genome: Double-stranded, 1 linear segment (150 – 280 Kb). Morphology: Enveloped. Virion: Icosahedral (Mw not specified). Nucleocapsid: Icosahedral.

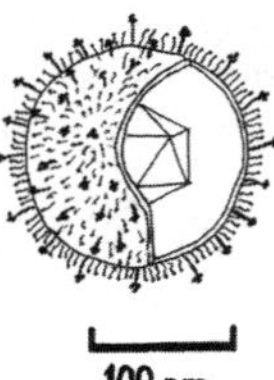

Plate 2.1.9 Family: Malacoherpesviridae. Nucleic acid: DNA. Genome: Double-stranded, 1 linear segment (125–240 Kbp). Morphology: Enveloped. Virion: Quasi-spherical [(Mw = (approximate) 4.6×10^8)]. Nucleocapsid: Icosahedral.

Plate 2.1.10 Family: Nimaviridae. Nucleic acid: DNA. Genome: Double-stranded, 1 circular segment (293–307 Kbp). Morphology: Enveloped (trilaminar). Virion: Ovoidal, (Mw not specified). Nucleocapsid: Helical. Distinguishing feature: Thread-like polar extension.

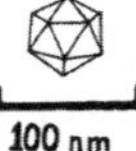

Plate 2.1.11 Family: Papillomaviridae (was part of Family Papovaviridae). Nucleic acid: DNA. Genome: Single-stranded (partially double-stranded), Negative sense, 1 circular segment (8 Kbp). Morphology: Non-enveloped. Virion: Icosahedral (Mw = $2.5–4.7 \times 10^7$). Nucleocapsid: Icosahedral.

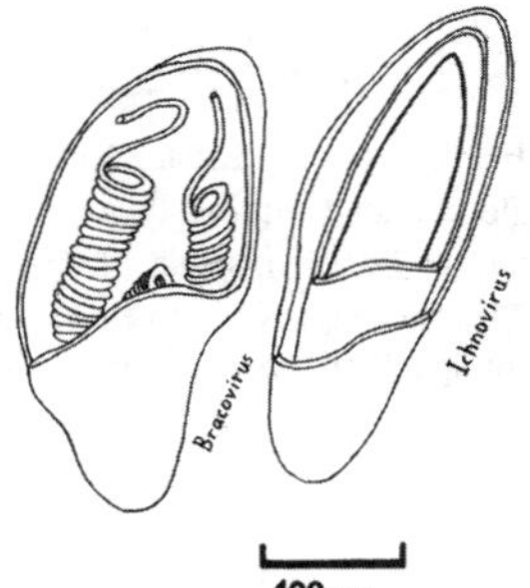

Plate 2.1.12 Family: Polydnaviridae. Nucleic acid: DNA. Genome: Double-stranded, Multiple circular segments (number of segments may vary by species) (150–250 Kb total). Morphology: Enveloped. Virion: Ellipsoidal (Mw not specified). Nucleocapsid: Helical. Distinguishing features: The two genera grouped within this family differ greatly in morphology. Members of the genus Bracovirus contain only a single unit membrane envelope. Members of the genus Ichnovirus uniquely have a double envelope consisting of two concentric unit membranes surrounding the nucleocapsid.

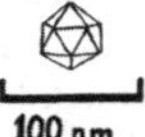

Plate 2.1.13 Family: Polyomaviridae (was part of: Family Papovaviridae). Nucleic acid: DNA. Genome: Single-stranded (partially double-stranded), Negative sense, 1 circular segment (5 Kbp). Morphology: Non-enveloped. Virion: Icosahedral (Mw = $2.5–4.7 \times 10^7$). Nucleocapsid: Icosahedral.

Plate 2.1.14 Family: Poxviridae. Nucleic acid: DNA. Genome: Double-stranded, 1 linear segment (130–375 Kb). Morphology: Enveloped. Virion: Ovoidal (Mw = 1.2×10^8). Nucleocapsid: Cylindrical or biconcave (Genus specific). Distinguishing feature: Virions contain both an external envelope plus internal surface and core membranes.

FIGURE 2.1 *(Continued)*

Plate 2.2.1 Family: Anelloviridae. Nucleic acid: DNA. Genome: Single-stranded, variously identified as having either Negative or Positive sense, 1 circular segment (3.4–3.9 Kb). Morphology: Non-enveloped Virion: Icosahedral, (Mw not specified). Nucleocapsid: Icosahedral.

Plate 2.2.2 Family: Circoviridae. Nucleic acid: DNA. Genome: Single-stranded, apparently contains both Positive and Negative sense coding, 1 circular segment (approx. 1.7 – 2.2 Kb). Morphology: Non-enveloped. Virion: Icosahedral (Mw not specified). Nucleocapsid: Icosahedral.

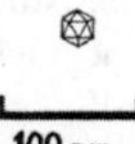

Plate 2.2.3 Family: Parvoviridae. Nucleic acid: DNA. Genome: Single stranded, strands of either sense can be encapsidated, 1 linear segment (4.7–5 Kb). Morphology: Non-enveloped. Virion: Icosahedral ($Mw = 5.5–6.2 \times 10^6$). Nucleocapsid: Icosahedral.

FIGURE 2.2 Relative sizes and basic information for those viruses which possess single stranded DNA genomes.

Plate 2.3.1 Family: Birnaviridae. Nucleic acid: RNA. Genome: Double stranded, 2 linear segments (Approx. 6 Kb total). Morphology: Non-enveloped. Virion: Icosahedral ($Mw = 5.5 \times 10^7$). Nucleocapsid: Icosahedral.

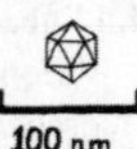

Plate 2.3.2 Family: Picobirnaviridae. Nucleic acid: RNA. Genome: Double-stranded, 1–3 presumably linear segments (4.5–6.2 Kbp total). Morphology: Non-enveloped. Virion: Icosahedral (Mw not specified). Nucleocapsid: Icosahedral.

Plate 2.3.3 Family: Reoviridae. Nucleic acid: RNA. Genome: Double stranded, 10–12 linear segments (1–3.9 Kbp per segment). Morphology: Non-enveloped. Virion: Icosahedral ($Mw = 1.2 \times 10^8$). Nucleocapsid: Icosahedral. Distinguishing feature: Nucleocapsid contains several concentric protein layers. Members of the genus *Cypovirus*, which infect insects, have capsids with only a single layer and form protein polyhedra.

FIGURE 2.3 Relative sizes and basic information for those viruses which possess double stranded RNA genomes.

the phylogenetic origin and evolutionary history of the organisms and they have largely confirmed the preexisting eukaryote classifications that had been based upon physical traits. Similarly, defining the genetic relatedness of the viruses on the taxonomic levels of order, family, genus and species, as elaborated by the ICTV was initially based upon morphologic and antigenic characteristics of the viruses. These older viral classifications have subsequently been refined and largely confirmed based upon the nucleotide sequence and organizational structure of the viral genomes. The proposed taxonomic structure for the genomic acellular infectious agents (Figure 2.10) suggests a logical placement of the existing ICTV taxonomic classifications into a higher-level schematic by progressing upwards using successively more basic attributes of the viral genomes. The suggested domain name is Akamara (ακαμαρα), whose derivation from Greek [α (without) + καμαρα (vault, chamber)] could represent the fact that these life forms do not possess a cellular structure of their own.

All of the groups of infectious agents shown in Figure 2.10 are depicted as belonging to a common domain as it would seem perhaps improbable to premise an accurate grouping of these agents based upon which of the three commonly suggested evolutionary sources represented their respective origins. In examining this point we should remember the three possible theories about how viruses began. These are: that the viruses may be remnants of the primordial soup, might represent degenerated cellular organisms, or be regulatory cellular elements that have gone awry.

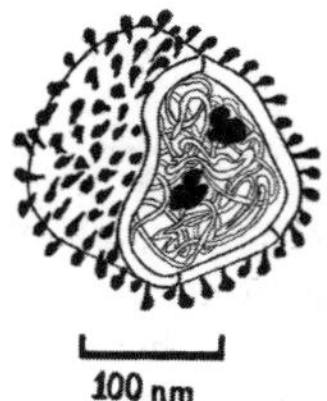

Plate 2.4.1 Family: Arenaviridae. Nucleic acid: RNA. Genome: Single-stranded, 2 circular segments of which at least one is Ambisense and the other either Negative sense or Ambisense, (11 Kb total). Morphology: Enveloped. Virion: Spherical (pleomorphic), (Mw not specified). Nucleocapsid: Helical. Distinguishing feature: Virions contain ribosomes from host cell.

Plate 2.4.2 Family: Bornaviridae. Nucleic acid: RNA. Genome: Single-stranded, Negative sense, 1 linear segment (8.9 Kb). Morphology: Enveloped. Virion: Spherical, (Mw not specified). Nucleocapsid: Helical.

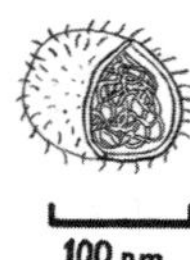

Plate 2.4.3 Family: Bunyaviridae. Nucleic acid: RNA. Genome: Single-stranded, 3 linear or pseudo-circular segments which differ by genus with regard to their sense, often at least one is Ambisense and the others variously are either Negative sense or Ambisense, (10.5–22.8 Kb). Morphology: Enveloped. Virion: Spherical (pleomorphic), (Mw $= 3.0–4.0 \times 10^8$). Nucleocapsid: Helical.

Plate 2.4.4 Floating genus: *Deltavirus*. Nucleic acid: RNA. Genome: Single-stranded, Negative sense, 1 circular segment (1.7 Kb). Morphology: Enveloped. Virion: Spherical (Mw not specified). Nucleocapsid: Presumably icosahedral. Important note: Deltavirus depends upon a coinfecting helper virus to both allow its replication and provide its capsid proteins. The capsid of the Hepatitis delta virus is provided by Hepadnavirus Hepatitis B virus, which otherwise is capable of infecting and replicating independently.

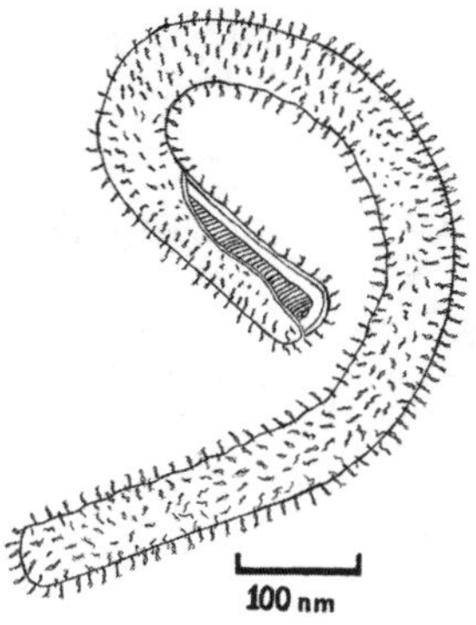

Plate 2.4.5 Family: Filoviridae. Nucleic acid: RNA. Genome: Single-stranded, Negative sense, 1 linear segment (19 Kb). Morphology: Enveloped. Virion: Filamental (pleomorphic), (Mw $= 4.2 \times 10^6$). Nucleocapsid: Helical.

Plate 2.4.6 Family: Orthomyxoviridae. Nucleic acid: RNA. Genome: Single-stranded, Negative sense, 6 to 8 linear segments (12–15 Kb total). Morphology: Enveloped. Virion: Spherical (pleomorphic) (Mw $= 2.5 \times 10^8$). Nucleocapsid: Helical.

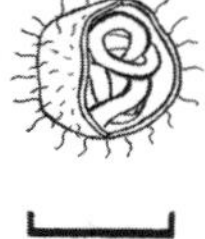

Plate 2.4.7 Family: Paramyxoviridae. Nucleic acid: RNA. Genome: Single-stranded, Negative sense, 1 linear segment (15–19 Kb). Morphology: Enveloped. Virion: Spherical (pleomorphic) (Mw $= 5.0 \times 10^8$ and upwards). Nucleocapsid: Helical.

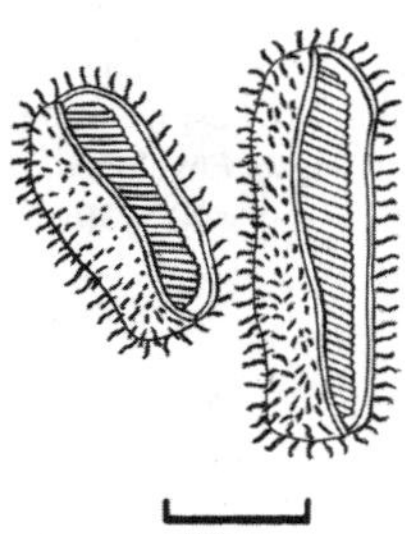

Plate 2.4.8 Family: Rhabdoviridae. Nucleic acid: RNA. Genome: Single-stranded, Negative sense, 1 linear segment (11–15 Kb). Morphology: Enveloped. Virion: Bullet as well as bacilliform (Mw $= 0.3–1.0 \times 10^9$). Nucleocapsid: Helical.

FIGURE 2.4 Relative sizes and basic information for those viruses which possess single stranded RNA genomes having either negative sense or ambisense coding.

Plate 2.5.1 Family: Arteriviridae (was: genus Arterivirus). Nucleic acid: RNA. Genome: Single stranded, Positive sense, 1 linear segment (Approx. 15 Kbp). Morphology: Enveloped. Virion: Spherical (Mw not specified). Nucleocapsid: Icosahedral.

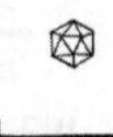

Plate 2.5.2 Family: Astroviridae. Nucleic acid: RNA. Genome: Single stranded, Positive sense, 1 linear segment (6.8–7.9 Kb). Morphology: Non-enveloped. Virion: Icosahedral (Mw = 8.0×10^6). Nucleocapsid: Icosahedral.

Plate 2.5.3 Family: Caliciviridae. Nucleic acid: RNA. Genome: Single-stranded, Positive sense, 1 linear segment (7.4–7.7 Kb). Morphology: non-enveloped. Virion: Icosahedral (Mw = 1.5×10^7). Nucleocapsid: Icosahedral.

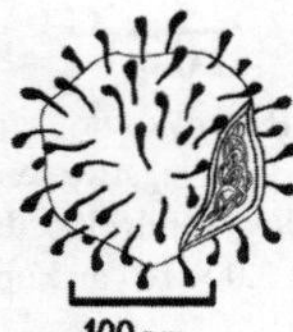

Plate 2.5.4 Family: Coronaviridae. Nucleic acid: RNA. Genome: Single-stranded, Positive sense, 1 linear segment (21–37 Kb). Morphology: Enveloped. Virion: Spherical (pleomorphic), (Mw = 4.0×10^8). Nucleocapsid: Helical.

Plate 2.5.5 Family: Dicistroviridae. Nucleic acid: RNA. Genome: Single-stranded, Positive sense, 1 linear segment (Approx. 9.3 Kb). Morphology: Non-enveloped. Virion: Icosahedral (Mw not specified). Nucleocapsid: Icosahedral.

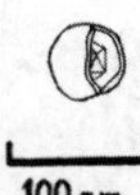

Plate 2.5.6 Family: Flaviviridae. Nucleic acid: RNA. Genome: Single-stranded, Positive sense, 1 linear segment (9.6–12.3 Kb). Morphology: Enveloped. Virion: Spherical (Mw = 6.0×10^7). Nucleocapsid: Icosahedral.

Plate 2.5.7 Family: Hepeviridae. Nucleic acid: RNA. Genome: Single-stranded, Positive sense, 1 linear segment (7.2 Kb). Morphology: Non-enveloped. Virion: Icosahedral (Mw not specified). Nucleocapsid: Icosahedral.

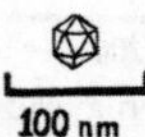

Plate 2.5.8 Family: Iflaviridae. Nucleic acid: RNA. Genome: Single-stranded, Positive sense, 1 linear segment (Approx. 10.1 Kb). Morphology: Non-enveloped. Virion: Icosahedral (Mw not specified). Nucleocapsid: Icosahedral.

Plate 2.5.9 Family: Metaviridae. Nucleic acid: RNA. Genome: Single-stranded, Positive sense 1 linear segment (4–11 Kb). Morphology: Enveloped. Virion: None. Nucleocapsid: These are retrotransposons for which no capsid has been identified although they do form intracellular aggregations of irregularly ovoid enveloped particles presumably possessing nucleoprotein cores. These are termed "viral-like particles" because they are not considered infectious, and thus by definition are not virions.

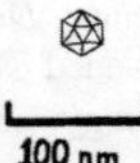

Plate 2.5.10 Family: Nodaviridae. Nucleic acid: RNA. Genome: Single stranded, Positive sense, 2 linear segments (4.5 Kb total). Morphology: Non-enveloped. Virion: Icosahedral (Mw = 8.0×10^6). Nucleocapsid: Icosahedral.

FIGURE 2.5 Relative sizes and basic information for those viruses which possess single stranded RNA genomes having positive sense coding.

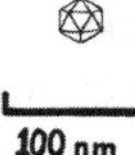

Plate 2.5.11 Family: Picornaviridae. Nucleic acid: RNA. Genome: Single-stranded, Positive sense, 1 linear segment (7.2–9 Kb). Morphology: Non-enveloped. Virion: Icosahedral (Mw = $8.0–9.0 \times 10^6$). Nucleocapsid: Icosahedral.

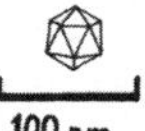

Plate 2.5.12 Family: Pseudoviridae. Nucleic acid: RNA. Genome: Single-stranded, Positive sense, 1 linear segment (4.2–12 Kb). Morphology: Non-enveloped. Virion: None. Nucleocapsid: Icosahedral. Distinguishing feature: These are retrotransposons and form intracellular aggregations of particles with icosahedral capsids termed "viral-like particles" but those are not considered infectious and therefore by definition are not virions.

Plate 2.5.13 Family: Retroviridae. Nucleic acid: RNA. Genome: Single-stranded, Positive sense, 2 identical dimers of a linear segment (7–10 Kb per dimer). Morphology: Enveloped. Virion: Spherical (Mw not specified). Nucleocapsid: Icosahedral.

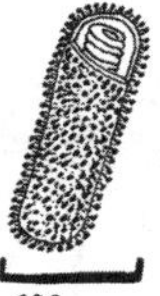

Plate 2.5.14 Family: Roniviridae. Nucleic acid: RNA. Genome: Single-stranded, Positive sense, 1 linear segment (26 Kb). Morphology: Enveloped. Virion: Bacilliform (Mw not specified). Nucleocapsid: Helical.

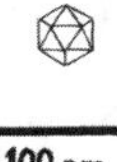

Plate 2.5.15 Family: Tetraviridae. Nucleic acid: RNA. Genome: Single-stranded, Positive sense, 1 to 2 linear segments (6.6–7.8 Kb total). Morphology: Non-enveloped. Virion: Icosahedral (Mw = 1.6×10^7). Nucleocapsid: Icosahedral.

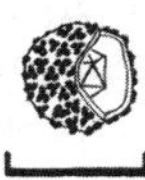

Plate 2.5.16 Family: Togaviridae. Nucleic acid: RNA. Genome: Single-stranded, Positive sense, 1 linear segment (10–12 Kb). Morphology: Enveloped. Virion: Spherical (Mw = 5.2×10^7). Nucleocapsid: Icosahedral.

FIGURE 2.5 *(Continued)*

The suggested subdominal classifications as shown in Figure 2.10 would group together the numerous agents whose genomes code for their own structural 'shell' or 'capsid' proteins as one kingdom (Euviria, signifying 'true' viruses). The viroids, a group of agents which share a unique and very homogenous single-stranded RNA genomic organization which somehow has enabled them to evolutionarily persist despite the fact that they do not code for proteins, are suggested as constituting a second kingdom (Viroidia) along with other groups of related agents whose genomes likewise do not code for their structural 'shell' or 'capsid' proteins. This may be perceived as a key biological difference, since all of the cellular organisms as well as the Euviria completely code for their own structural proteins. The kingdom Euviria is suggested as being divided into two phyla, which separate the Euviria with

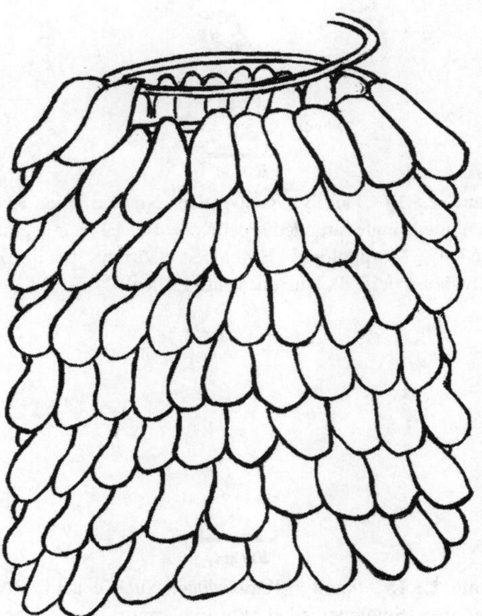

FIGURE 2.6 Drawing of a helical capsid structure showing how the capsid proteins attach to the helical coil of the viral nucleic acid genome. Presumably all of the capsid proteins are identical to one another in a helical structure.

respect to whether their genomes are composed of RNA (Ribovira) or DNA (Deoxyribovira). It is suggested in turn, that these two phyla of the Euviria could logically be subdivided into classes based upon whether their genomes are double stranded or single stranded. Those Euviria which possess single stranded RNA genomes could logically be divided as to whether their genomes are "plus" sense, meaning that they can be directly translated by ribosomes, or are "negative" sense. This assignment of phyla and classes is premised upon principal biochemical differences in the agents' genomes and also follows basic commonalities in terms of the molecular strategies of these infectious agents. As noted above, the viroids, virusoids, and floating genus Deltavirus are shown as being assigned a separate kingdom level, named Viroidia, and the suggested schematic carries their placement intact to the phylum level as all of the agents grouped into that category possess RNA genomes. The current ICTV nomenclature structure groups these genomic acellular infectious agents at the levels of order through species and could be adopted directly into this proposed new domain. An example of that existing ICTV structure is shown within the box that is inside of Figure 2.10. That ICTV nomenclature structure is now considered to be based upon viral nucleotide sequence commonalities at the lowest levels, progressing through commonalities in genome organization according to a 'bottom up' philosophy. The levels of taxonomy proposed here can be seen as logically continuing that trend by progressing upward to the trait of strandedness at the class level, where a distinction of single versus double stranded genome is used and aided by the designation of plus versus negative sense in the case of those viruses whose strategies are based upon single stranded RNA genomes, and to the still more basic distinction of RNA (Ribovira) versus DNA (Deoxyribovira) genome at the phylum level. Simultaneously, this proposed structure could also be perceived as progressing from the top downward to meet the current ICTV structure. Perhaps someday, the ICTV will allow me to present this concept in the open journal literature!

Perhaps it is necessary to ask if these genomic acellular infectious agents do indeed deserve to be considered among the living and therefore assigned into a taxonomic structure. In examining this point, we should remember the three possible theories about how viruses began, id est, that these agents as we know them represent the evolutional product from either primordial remnants, degenerated cells, or rogue cellular elements. It is indeed possible and perhaps likely that the origin and evolutionary course which the various groups of genomic acellular infectious agents have followed to attain their present forms does in fact reflect a combination of contributions from all three sources. Many of the large viruses certainly have replicative traits that seem to mimic some of the molecular complexity which we associate with cellularity. In contrast, the idea

FIGURE 2.7 "Vivaldi" by Shigeo Kawashima, 2008; Collection of the Franklin Park Conservatory, Columbus, Ohio, USA. A sculpture made of bamboo, plastic zip ties and yarn, that has an interesting similarity of appearance to the filoviruses.

that these groups of acellular infectious agents began as sub-cellular elements that gained an independence is certainly suggested by some of the smaller viruses, which scarcely carry any more genetic coding than that which is required for their few capsid proteins, and by the viroids and virusoids which do not code for capsid proteins. However, since these groups of acellular infectious agents seem to have evolutionarily taken on an identity of their own, then perhaps we should recognize that identity as a life form. When considering the possibility that these groups of acellular infectious agents might in fact have begun as cellular organisms that have since lost biochemical complexity, if we should consider them now to be non-living based upon the application of complex definitions of living creatures, then

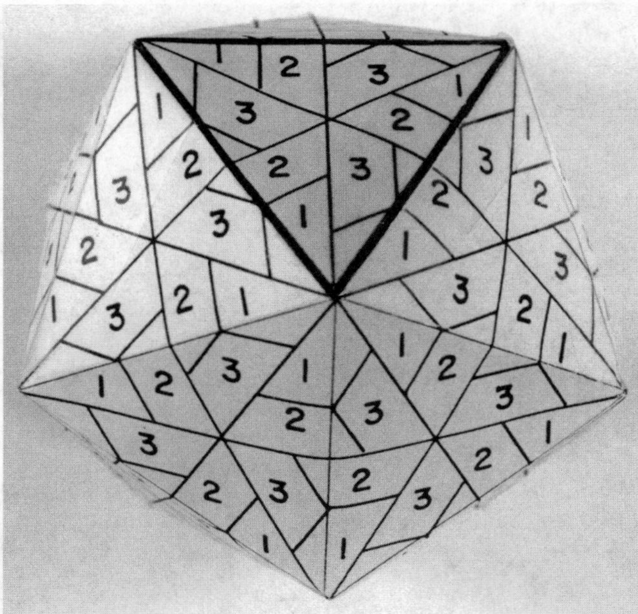

FIGURE 2.8 Photograph of the assembled model published by Hurst et al. (1987) showing the protein arrangement in an icosahedral capsid structure. This particular structure is a representation of the viral family Picornaviridae. The members of this viral family produce capsids that contain multiple copies of three major (larger sized, numbered 1, 2, and 3) capsid proteins and one minor (smaller sized) capsid protein. The relative positions of the three major capsid proteins are shown as the trapezoids numbered 1, 2, and 3. The trapezoidal shape is used for illustrative purposes, as the true shapes of these proteins is more complex and not truely trapezoidal. The darkly outlined triangle represents one of the twenty sides of the viral capsid. Although these sides are often referred to as "faces", the word icosahedron literally interprets from the greek as meaning that this structure has twenty surfaces upon which it could sit.

we would be faced with deciding at which exact point in the process of evolutionary simplification the term 'life' would cease to be applicable. If, alternatively, we could more simply define life by indicating that living things are naturally existing organic entities which are capable of catalyzing their biochemical self replication, then yes, these

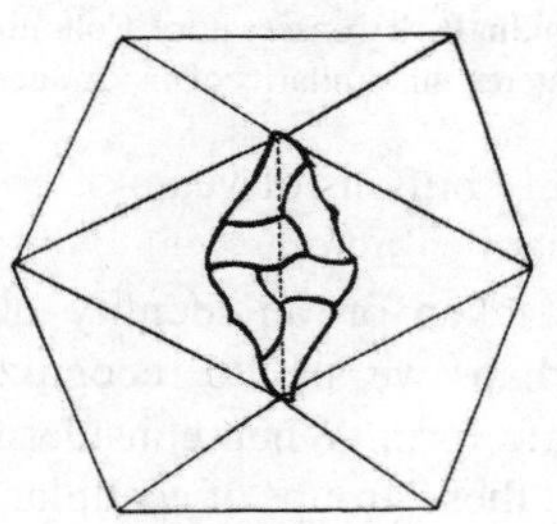

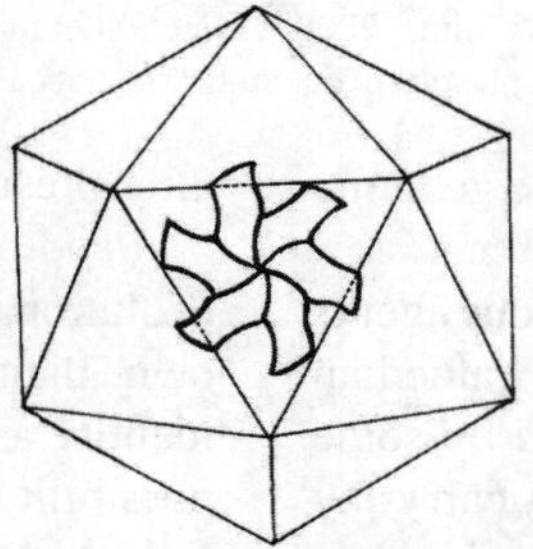

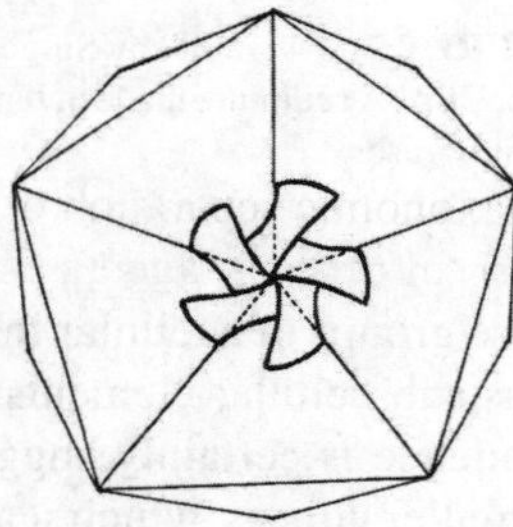

FIGURE 2.9 Drawing of an icosahedral capsid structure showing what would be a mirror image of the shape of the capsid proteins for the viral family Bromoviridae. Unlike the picornaviral model, the bromoviral capsid seems to contain multiple copies of only one type of capsid protein. Presumably, those copies of the same protein would be rotated into different relative positions such that they can arrange into an icosahedron. This drawing shows how those capsid proteins combine to produce the two-fold (left image), three-fold (center image) and five-fold (right image) axes of symmetry which define an icosahedral structure.

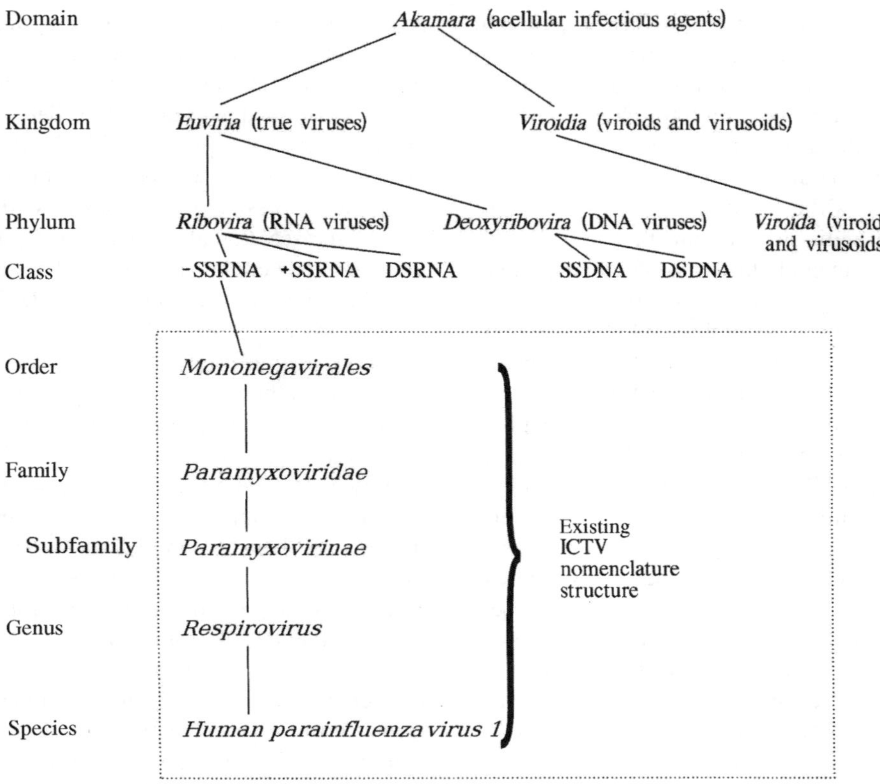

FIGURE 2.10 Proposed new domain, Akamara, plus proposed taxonomic structure at the levels of kingdom, phylum, and class. The abbreviated designations at the class level represent: -SSRNA, 'negative' sense single stranded RNA genome; +SSRNA, 'positive' sense single stranded RNA genome; DSRNA, double stranded RNA genome; SSDNA, single stranded DNA genome; and DSDNA, double stranded DNA genome. The existing ICTV nomenclature structure for viruses, an example of which is shown within the inset box, covers only taxonomic levels from order through species and could be adopted directly into this domain.

genomic acellular infectious agents are a form of life.

It is not necessarily implied that the suggested taxonomic separation of these genomic acellular infectious agents into kingdoms, phyla and classes represents a strict phylogeny. This is a departure from the current philosophy that the existing ICTV taxonomic nomenclature, which goes no higher than the level of order, does carry strong evidence for common phylogeny. Taxonomic systems change with the evolution of our scientific philosophy, and the proposed categories presented in this chapter are only a suggestion. However, it might be necessary to admit that we may never be able to use strict phylogeny to base an exact classification of these agents at the very highest taxonomic levels. This is due to the idea that after presumably billions of years of coevolution and biochemical interactions within their host cells, the physical appearance of these modern descendants may be very different from that of their evolutionary ancestors and they have left no identifiable fossilized remains to guide us. Instead, basic biochemistry has been used as the basis for this proposed division of the genomic acellular infectious agents into kingdoms, phyla and classes. If there is a fault to be found with this proposal, it is perhaps that this organizational structure could be seen as relating to a cellular origin for the genomic acellular infectious agents. However, this proposal is not intended to imply that cellularity was an initiating condition for the evolution of these acellular agents. Rather, as indicated earlier in this article, we do not and perhaps cannot know the conditions under which these life forms initially began, and the same present-day endpoint could have resulted from gradual evolution regardless of whether either the viruses or other genomic acellular infectious agents began as primordial components, cellular organisms, or rogue cellular elements. Likewise, it would not be possible to place these organisms into the three domains described by Woese et al. (1990), since those domains are defined by the nucleotide sequences of their constituent organism's genes that code for ribosomal RNA whereas the genomes of the viruses, satellite viruses, virusoids and viroids do not code for ribosomal RNA. There is a suggestion that single point evolutionary connections exist at the branch junctures where the three domains that are used to describe the cellular organisms separate from one another. However, assigning a single point of evolutionary connection between the life forms contained in this newly proposed domain Akamara and the domains of Woese et al. (1990) may not be possible as that might necessitate stating that these acellular organisms either evolved into or from the cellular organisms which are represented by the three existing domains.

2.4 CONCLUSIONS

The purpose of this chapter is to help us take stock of what we as virologists now have available in terms of taxonomy for the viruses and their relatives. As we now have passed the century mark for use of the name virus (Beijerinck, 1898) and official biological recognition of the viruses by a scientific commission (Loeffler and Frosch, 1898), it would seem time that the scientific community consider extending the existing viral taxonomy by recognizing the viruses on a domain level. The taxonomic schematic proposed in this chapter is, of course, only one possible suggestion and no taxonomic scheme can be considered infallable. However, this proposal is logically based and represents an assessment which relies upon the successive generations of sound biochemical research that has been conducted and published by tens of thousands of virologists world wide. What does this proposal leave out? Classification of those RNA viruses which utilize reverse transcription to produce a DNA equivalent of their genome during the course of their replication could be perceived as requiring a separate class or even phylum within the proposed kingdom Euviria. Those viruses which possess single stranded RNA genomes that are positive sense imitate messenger RNA molecules. Those viruses that possess single stranded RNA genomes which are ambisense, meaning that their genomes are partially negative sense and partially positive sense, may be seen as representing a significant departure from those viruses which possess single stranded RNA genomes that strictly are positive sense. The reason why this represents a molecular departure is because the negative sense regions of those viruses' RNA genomes must be copied to form a matching positive sense strand before they can be translated into protein. Perhaps the ambisense single stranded RNA viruses could be grouped with the negative sense single stranded RNA viruses because the members of this latter viral group also have incorporated the same molecular departure into their biochemistry. This commonality is the reason why the ambisense and negative sense single stranded RNA viruses were combined together into Figure 2.4.

What else is left in the field of biological entities? There are reasons for biologists to potentially consider at least some plasmids to represent a type of infectious agent, and perhaps therefore to be a form of life and worthy of an eventual home within some taxonomic scheme. A noteworthy example of this might be the conjugational plasmids which carry coding for specific protein structures that are then expressed by their cellular host organisms and are important in facilitating the transmission of that plasmid to a new cellular host. We certainly allow the LTR (long terminal repeat) retrotransposons to be considered viruses and presume that these may have had an initial viral origin, the ICTV has grouped those into the viral families Metaviridae and Pseudoviridae. The viral species called Acidianus two-tailed virus, which currently is the sole member of the viral family Bicaudaviridae, has the unique trait of undergoing a dramatic morphological rearrangement which is considered to be a maturation following its release from the host cell (Prangishvili et al, 2006) and this is a strong departure from any of the other viruses. This difference seems sufficient to potentially suggest that at some future time the Acidianus two-tailed virus will become the first identified member of a new non-viral category of biological entities. The prions of mammals are considered infectious agents and in that sense might eventually be thought to also represent a form of life despite the fact that they apparently do not carry any genomic coding of their own (and thus would be defined as agenomic or non-genomic) as they move from one host to another. Indeed, we must consider that perhaps an infectious agent would not need to carry any genomic material with it if its new host cells already possessed all of the coding necessary for replicating that agent. In fact, while two of the traits which prions possess, a measure of resistance to acidic conditions and to proteolytic enzymes (Volume 1, Chapter 10), would seem to contribute to the pathogenic process associated with the prions, these same two traits could also be seen as representing an evolutionary adaptation to the acidic conditions and enzymatic milieu encountered in the digestive tract of their host animals during the course of these prions' natural transmission as an enterically acquired infection. These remaining questions represent fodder for future thought.

REFERENCES

Beijerinck, M. W. (1898). Ueber ein Contagium vivum fluidum als Ursache der Fleckenkrankheit der Tabaksblatter. *Verhandelingen der Koninklijke Akademie Wetenschappen te Amsterdam, II* 6(5), 1–21.

Braza, R. and Ganem, D. (1996). A cellular homolog of hepatitis delta antigen: implications for viral replication and evolution. *Science* 274, 90–94.

Bult, C. J., White, O., Olsen, G. J., Zhou, L., Fleischmann, R. D., Sutton, G. G., Blake, J. A., FitzGerald, L. M., Clayton, R. A., Gocayne, J. D., Kerlavage, A. R., Dougherty, B. A., Tomb, J.-F., Adams, M. D., Reich, C. I., Overbeek, R., Kirkness, E. F., Weinstock, K. G., Merrick, J. M., Glodek, A., Scott, J. L., Geoghagen, N. S. M., Weidman, J. F., Fuhrmann, J. L., Nguyen, D., Utterback, T. R., Kelley, J. M., Peterson, J. D., Sadow, P. W., Hanna, M. C., Cotton, M. D., Roberts, K. M., Hurst, M. A., Kaine, B. P., Borodovsky, M., Klenk, H.-P., Fraser, C. M., Smith, H. O., Woese, C. R., and Venter, J. C. (1996). Complete genome sequence of the methanogenic archaeon, Methanococcus jannaschii. *Science* 273, 1058–1073.

Fiers, W., Contreras, R., Duerinck. F., Haegeman, G., Iserentant. D., Merregaert. J., Min. Jou. W., Molemans. F., Raeymaekers. A., Van den, Berghe, A., Volckaert, G., and Ysebaert, M., (1976). Complete nucleotide sequence of bacteriophage MS2 RNA: primary and secondary structure of the replicase gene. *Nature* 260, 500–507.

Fiers, W., Contreras, R., Haegeman, G., Rogiers, R., Van de Voorde, A., Van Heuverswyn, H., Van Herreweghe, J., Volckaert, G., and Ysebaert, M. (1978). Complete nucleotide sequence of SV40 DNA. *Nature* 273, 113–120.

Hurst, C. J., Benton, W. H., and Enneking, J. M. (1987). Three-dimensional model of human rhinovirus type 14. *Trends in Biochemical Sciences* 12, 460.

Loeffler and Frosch (1898). Berichte der Kommission zur Erforschung der Maulund Klauenseuche bei dem Institut fur Infektionskrankheiten in Berlin. *Zentralblatt fur Bakteriologie, Parasitenkunde und Infektionskrankheiten. 1 Abt. Medizinsch-hygienische Bakteriologie und tierische Parasitenkunde* 23, 371–391.

Morell, V. (1996). Life's last domain. *Science* 273, 1043–1045.

Prangishvili, D., Vestergaard, G., Häring, M., Aramayo, R., Basta, T., Reinhard, R., and Garrett, R. A. (2006). Structural and genomic properties of the hyperthermophilic archaeal virus ATV with an extracellular stage of the reproductive cycle. *J. Molecular Bio.* 359: 1203–1216.

Robertson, H. D. (1996). How did replicating and coding RNAs first get together? *Science* 274, 66–67.

Woese, C. R., Kandler, O., and Wheelis, M. L. (1990). Towards a natural system of organisms: proposal for the domains Archaea, Bacteria, and Eucarya. *Proc. Natl. Acad. Sci USA* 87, 4576–4579.

CHAPTER 3

VIRUS MORPHOLOGY, REPLICATION, AND ASSEMBLY*

DEBI P. NAYAK
Department of Microbiology and Immunology, David Geffen School of Medicine, University of California at Los Angeles, Los Angeles, CA

CONTENTS

* This is a revised version of the chapter that appeared in *Viral Ecology*, edited by Christon J. Hurst, published in 2000 by Academic Press.

3.1 INTRODUCTION

Viruses are unique life forms different from all other living organisms, either eukaryotes or prokaryotes, for three fundamental reasons: (1) the nature of environment in which they grow and multiply, (2) the nature of their genome, and (3) the mode of their multiplication. First, they are obligate intracellular parasites, that is, can function and multiply only inside another living organism that may be a prokaryotic or eukaryotic cell depending on the virus. Viruses are acellular and metabolically inert outside the host cell. Although there are other examples of obligatory parasites among the eukaryotes and prokaryotes, the nature of the intimate relationship between viruses and

Studies in Viral Ecology: Animal Host Systems: Volume 2, First Edition. Edited by Christon J. Hurst.

their host (i.e., environment) is much different. For example, some viruses extend their parasitic behavior to another level of mutual coexistence with their host; that is, they not only exist intracellularly but can also, and do in some cases, integrate their genome into the genome of their host and thus tie their fate to the fate of the host. In fact, under these conditions, the integrated viral genome behaves as a host gene(s), undergoing similar regulatory control in transcription and replication and similar evolutionary changes as do the host gene(s). Second, whereas all other living forms can use only DNA (and not RNA) as their genetic material (genome) for information transmission from parent to progeny, viruses can use either DNA or RNA as their genome; that is, some viruses can use only RNA (and not DNA) as their genetic material. Therefore, these classes of RNA viruses have developed new sets of enzymes for replicating and transcribing RNA from an RNA template, as such enzymes (RNA-dependent RNA polymerase (RDRP)) are not normally found in eukaryotic or prokaryotic cells. Finally, all eukaryotic and prokaryotic cells divide and multiply as a whole unit, that is, $1 \rightarrow 2 \rightarrow 4 \rightarrow 8$ and so on. However, viruses do not multiply as a unit. In fact, they have developed a much more efficient way to multiply just as complex machines are made in a modern factory. Different viral components are made separately from independent templates, and then these components are assembled into the whole and infectious units, also called virus particles (virions), just as the complex machines are efficiently assembled from individual components. Similarly, disassembly of the virus components occurs during the infection process, leading to genome replication, transcription of messenger RNAs (mRNAs), translation of viral proteins, and assembly of the virus particles and their release from the infected cell into the environment for continuing the next infectious cycle. In this chapter, aspects of viral morphology, mode of viral replication, and viral morphogenesis including budding and release are discussed.

Although all viruses exhibit this common mode of replication or infectious cycle, viruses are a heterogeneous group of microorganisms that vary with respect to size, morphology, and chemical composition. The size of virions ranges from 20 nm (parvovirus, family Parvoviridae) to ~300 nm (poxvirus, family Poxviridae) in diameter, compared to the size of *Escherichia coli*, which is about 1000 nm in length. However, some filamentous viruses such as filoviruses (family Filoviridae) may be 800 nm or even longer. In addition to size, the shape of viruses also varies. Some viruses are round (spherical, spheroidal), others filamentous, and still others pleomorphic. Usually, naked (nonenveloped) viruses have specific shapes and sizes (Figure 3.1), whereas some enveloped viruses (particularly those possessing helical nucleocapsids) are highly pleomorphic (e.g., orthomyxoviruses), with shapes varying from spherical to filamentous (Table 3.1, Figures 3.1–3.5). Viruses are different from viroids and prions. Viroids are small, circular, single-stranded infectious RNA molecules without a protein coat or capsid and cause a number of plant diseases including potato spindle tuber disease, cucumber pale fruit disease, citrus exocortis disease, and cadang-cadang (coconuts) disease, and so on. On the other hand, prions are infectious protein molecules without any DNA or RNA and thought to cause transmissible and/or inherited neurodegenerative diseases known as transmissible spongiform encephalopathies. These include Creutzfeldt–Jakob disease, kuru, and Gerstmann–Straussler syndrome in humans, as well as scrapie in sheep and goats and mad cow disease in cattle. The infectious prion proteins are modified forms of normal proteins encoded by a host gene. The normal prion protein that has alpha helices in its secondary structure is converted into beta sheets for the secondary structure in diseased animals.

3.2 CHEMICAL COMPOSITION

The chemical composition of a virus depends on the nature of that virus, that is, the nature of

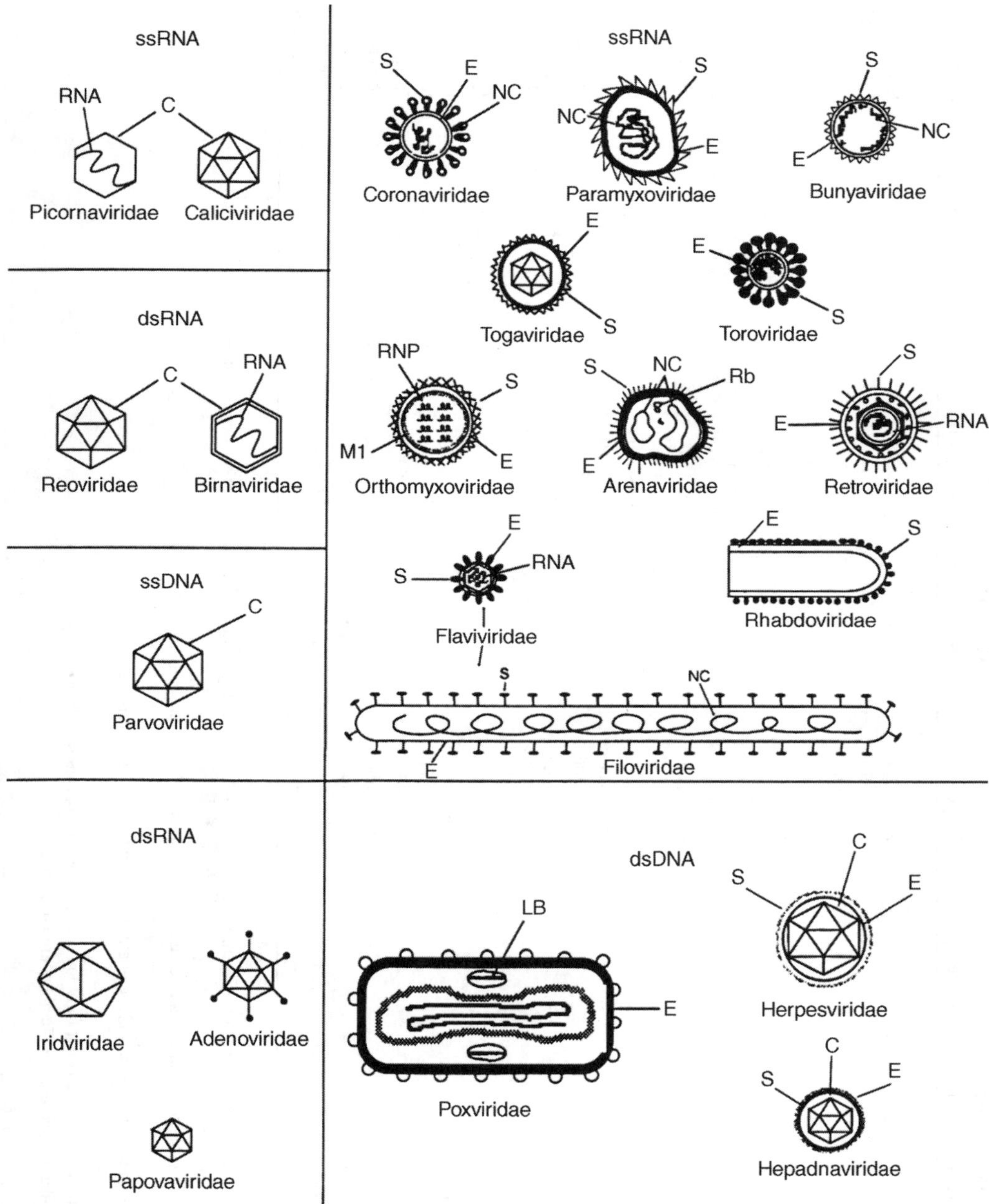

FIGURE 3.1 Schematic presentation of different forms of viral structures. C, capsid; S, spike on viral envelope; E, viral lipid envelope; NC, nucleocapsid (i.e., capsid proteins in association with RNA or DNA); M1, matrix protein of influenza virus; LB, lateral bodies present in poxviruses; ss, single-stranded, ds, double-stranded RNA or DNA.

the viral genome (RNA or DNA), the composition of the protein shell called the viral "nucleocapsid" surrounding the genome, and the presence or absence of viral membrane depending on whether the virus is enveloped or naked. All viruses have nucleocapsids and therefore contain nucleic acids and proteins. The nucleic acid is the genome that contains the

TABLE 3.1 Properties of the Virions of the Major Genera of DNA and RNA Animal Viruses

Viruses	Genome Nature of Nucleocapsid	Envelope	Shape	Genome Polarity	Size (nm)	Transcriptase in Virion	Symmetry
RNA viruses							
Enterovirus	S,[a] 1[b]	−	Icosahedral	+	~20–30	−	Icosahedral
Rhinovirus	S, 1	−	Icosahedral	+	20–30	−	Icosahedral
Calicivirus	S, 1	−	Icosahedral	+	20–30	−	Icosahedral
Alphavinis	S, 1	+	Spheroidal	+	50–60	−	Icosahedral
Flavivirus	S, 1	+	Spheroidal	+	40–50	−	Icosahedral
Orthomyxovirus	S, 8	+	Spheroidal[c]	−	80–120	+	Helical
Paramyxovirus	S, 1	+	Spheroidal	−	100–150	+	Helical
Coronavirus	S, 1	+	Spheroidal	+	80–220	−	Helical
Arenavirus	S, 2	+	Spheroidal	±[d]	85–120	+	Helical[e]
Bunyavirus	S, 3	+	Spheroidal	±[d]	90–100	+	Helical[e]
Retrovirus	S, 1[f]	+	Spheroidal	+	100–120	+[g]	Icosahedral[h]
Rhabdovirus	S, 1	+	Bullet shaped	−	175 × 70	+	Helical
Reovirus	D, 10	−	Icosahedral	±	70–80	+	Icosahedral
Orbivirus	D, 10	−	Icosahedral	±	50–60	+	Icosahedral
Filovirus	S, 1	+	Filamentous	−	≥80 × 800	+	Helical
DNA viruses							
Papillomavirus	D, circular	−	Icosahedral	±	55	−	Icosahedral
Polyomavirus	D, circular	−	Icosahedral	±	45	−	Icosahedral
Adenovirus	D, linear	−	icosahedral	±	70–80	−	Icosahedral
Hepadnavirus	D, circular partiae	+	Spheroidal	±	40–50	+[l]	Icosahedral
Herpesvirus	D, linear	+	Spheroidal	±	150	−	Icosahedral
Iridovirus[i]	D, linear	+[i]	Spheroidal	+	125 × 300	+	Icosahedral
Poxvirus	D, linear	+	Brick shaped	+	300 × 240 × 140[j]	+	Complex
Parvovirus	S, linear	−	Icosahedral	+,−[k]	20	−	Icosahedral

[a]D = double-stranded; S = single-stranded.
[b]Genome, the number indicates the segments of RNA present in the virus particle. All RNA genome is haploid except retrovirus (diploid).
[c]Pleomorphic including filamentous forms.
[d]Ambisense (contains coding for protein on both genomic and complementary RNA strands).
[e]Circular helical nucleocapsid.
[f]Diploid, two molecules of the same RNA (+ strand) segment are present in one virus particle.
[g]Reverse transcriptase (RT).
[h]The capsid structure of mature retroviruses is not fully known, although it appears icosahedral.
[i]Insect iridoviruses have no envelope; vertebrate members are enveloped.
[j]Length × width × thickness.
[k]Some virus particles contain plus-strand and others contain minus-strand DNAs.
[l]P protein functions as reverse transcriptase.

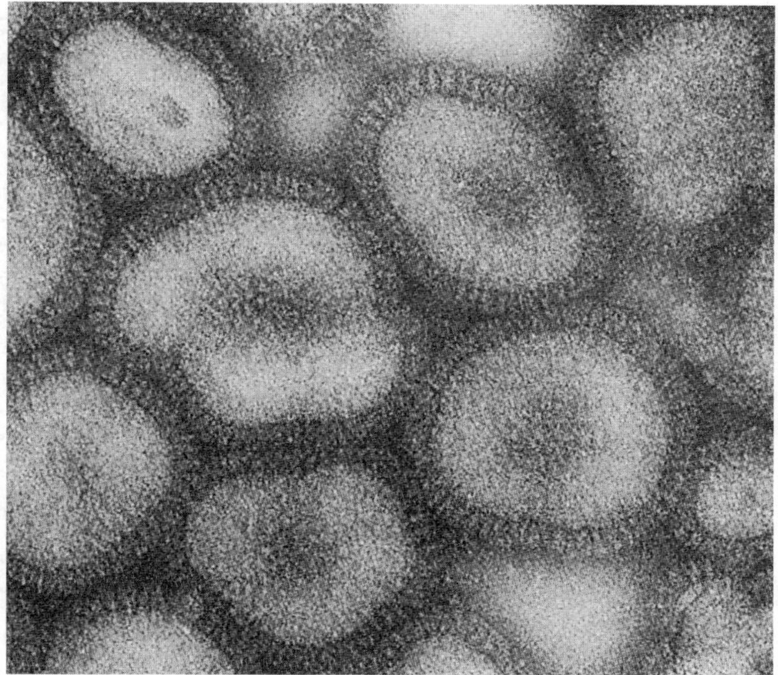

FIGURE 3.2 Influenza virus morphology. Transmission electron micrographs of *Influenzavirus A*. Courtesy of K. G. Murti of St. Jude Children's Research Hospital of Memphis, Tennessee.

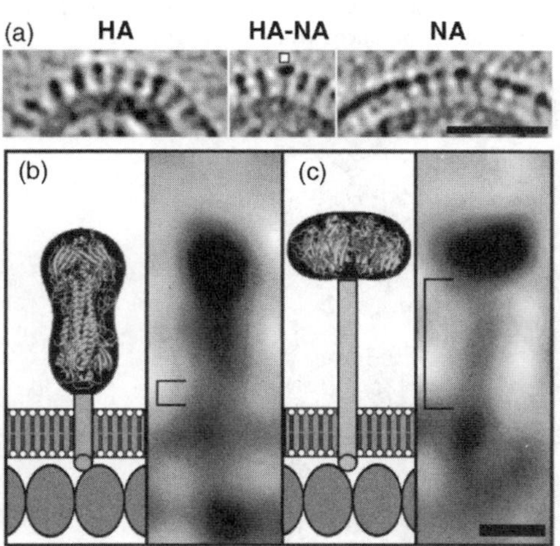

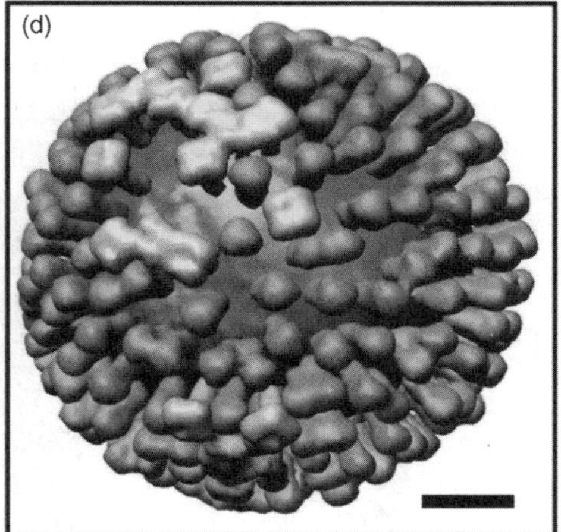

FIGURE 3.3 Model virus with HA and NA spikes by cryo-ET analysis. (a) HA cluster (left), (b) single NA (marked) in a cluster of HA (middle), and (c) cluster of mainly NA spikes (right). (b and c) The stem length of HA and NA (square brackets in (b) and (c), respectively). The structures of the stem, transmembrane domain, and ectodomain are shown schematically. Molecules in the matrix layer are inferred to be packed in a monolayer (scale bar 5 nm). (d) Model of distribution of HA (green), NA (gold), and lipid bilayers (blue) in a single virion (scale 20 nm). Reproduced from Harris et al. (2006) with permission. (*See the color version of this figure in Color Plate section.*)

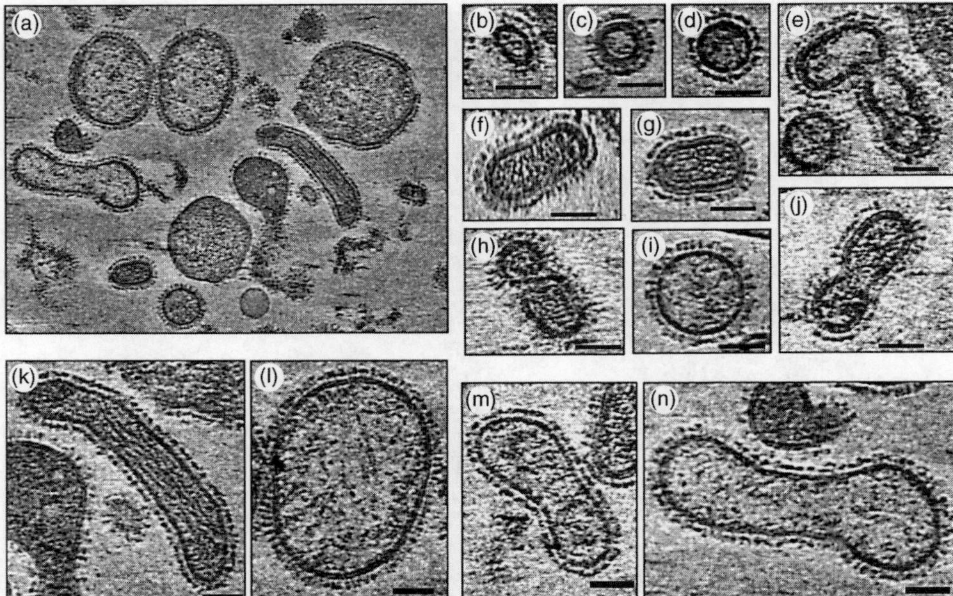

FIGURE 3.4 Cryoelectron tomography of A/PR8 (an antigenic variant of influenza A virus, a member of the genus *Influenzavirus A*) showing highly pleomorphic virion architecture. (a) A density slice from a 3D cryoelectron tomography reconstruction of influenza A virus strain PR8. PR8 virus was grown in MDCK cells at 0.001 MOI. The tilt series spanning –70° to 70° sample tilt were recorded in a TF20 cryoelectron microscope using the Batch Tomography program (FEI Company) and then reconstructed using the Inspect3D (FEI Company) and refined by Protomo program (Winkler and Taylor, 2006). (b–n) Comparison of central slices of viral particles extracted from different cryoelectron tomograms. Different virus particles were picked at random. No attempt was made to determine the percentage of each virus form in the population. Each virus particle contained electron dense spots (RNP) inside and spikes outside. Both HA and NA spikes, as identified based on morphology described in Harris et al. (2006), were visible on the outer membrane (scale bar 50 nm). Reproduced from Nayak et al. (2009).

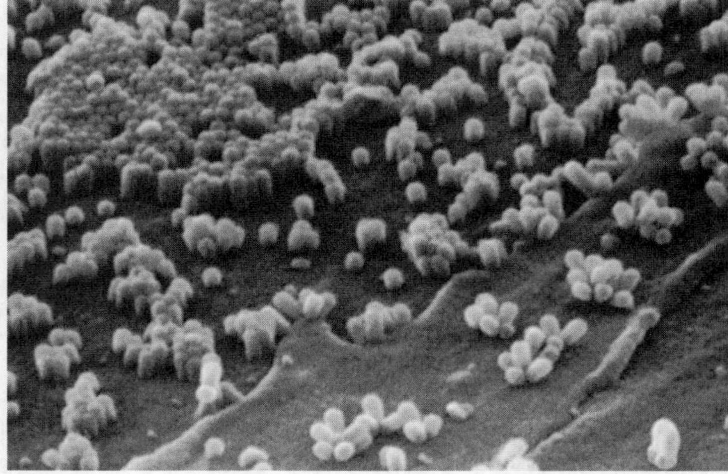

FIGURE 3.5 Scanning electron micrographs of influenza viruses budding from infected cells. Spherical virus particles nearly complete are seen budding from infected cells (×40,000). These micrographs were provided by David Hockley of the National Institute for Biological Standards and Control at Hertfordshire, UK and reproduced from Nayak et al. (2009).

information necessary for viral function and multiplication, and this information is passed from the parent to progeny viruses. Some viruses contain extragenomic nucleic acid, for example, tRNA in retroviruses (family Retroviridae) and ribosomal RNA in arenaviruses (family Arenaviridae). Viral proteins have three primary functions: (1) they provide the shell to protect the nucleic acid from degradation by environmental nucleases, (2) facilitate transfer of the genome from virus to host and from one host to another, and (3) provide many of the enzymatic and regulatory functions needed for transcription and replication so that viruses can survive, multiply, and perpetuate. In addition to the capsid shell, many viruses also possess an envelope (or viral membrane) around the nucleocapsid. The envelope in these viruses is critical for their transmission from one host to another. The naked nucleocapsids of enveloped viruses are noninfectious or poorly infectious because they lack the viral receptor binding protein for attachment to the host receptor. The viral envelope contains lipids and carbohydrates in addition to "envelope- or membrane-associated" viral proteins. The viral genome codes for most, if not all, of the proteins associated with the viral envelope. Lipids of the viral membrane, on the other hand, are synthesized by the host cell and derived from it. Therefore, viral lipid composition varies depending on the host cell in which the virus grows and also on the type of the cellular membrane (e.g., ER, Golgi, plasma, or nuclear membrane) from which the particular type of virus buds. However, lipid composition of viral membrane does not completely mimic that of the cellular membrane but rather is selectively enriched in specific host lipid components. Enveloped viruses in most cases are assembled and bud from specialized membrane microdomains called lipid rafts that are enriched in long saturated fatty acids, cholesterol, and sphingolipids (Nayak and Hui, 2004; Nayak et al., 2009). The carbohydrate content of the viral envelope is usually determined by the nature of glycosylation (*N*-glycosylation, *O*-glycosylation, complex versus simple sugar addition) of the viral envelope proteins, which may in turn undergo other modifications, such as myristoylation, palmitoylation, sulfation, and phosphorylation.

3.2.1 Viral Nucleic Acid (Genome)

Genomes of different viruses are widely diverse in size and complexity. Some are composed of DNA, while others of RNA. As mentioned earlier, only in viruses is RNA known to function as a genome. Viral DNA genomes vary in complexity ranging from 5 kb containing 5–6 genes (parvoviruses, members of the viral family Parvoviridae; SV40 (Simian virus 40), family Polyomaviridae, genus *Polyomavirus*) to over 300 kb (avipoxviruses, family Poxviridae, genus *Avipoxvirus*) containing more than 200 genes and having complex organization. Some DNA genomes are double-stranded (SV40), some are partially double-stranded (hepatitis B virus (HBV), family Hepadnaviridae, genus *Orthohepadnavirus*), and still others are single-stranded (parvoviruses) (Tables 3.1 and 3.2). The single-stranded viral DNAs can be of plus or minus polarity. Some DNA genomes are circular (and supercoiled), while others are linear. Some linear DNA genomes become circular intermediates during replication. Many viral DNA genomes are terminally redundant in their nucleotide sequences.

RNA genomes of viruses also vary in length and complexity but not as widely as do DNA genomes. For the RNA viruses known to date, the range of variation is from $\sim$7 kb for rhinoviruses, which are divided into the species of human rhinovirus A, B, and C, all belonging to genus *Enterovirus* of the family Picornaviridae, to $\sim$30 kb for coronaviruses (family Coronaviridae). Coronavirus RNA represents the largest stable single-stranded RNA found in nature. Viral RNA can be single- or double-stranded (Tables 3.1 and 3.3).

The viral RNA genome may be nonsegmented, consisting of a single RNA molecule, or segmented, consisting of multiple segments. Usually, viral genomes are haploid, but some are diploid (e.g., retroviruses; Figure 3.6). Some viral RNA genomes may be linear, whereas others have partial terminal complementarity

TABLE 3.2 Replication of DNA Viruses

Virus	Form of DNA	Polymerase	Activity	Presence in Virion	Replication Site in Cell
Papovaviruses	ds[a]	Host	DNA *pol*	–	Nucleus
Adenoviruses	Ds	Viral	*DNA pol*	–	Nucleus
Herpesviruses	Ds	Viral	DNA *pol*	–	Nucleus
Poxviruses	Ds	Viral	DNA *pol*	–[b]	Cytoplasm
Parvoviruses	Ss	Host	DNA *pol*	–	Nucleus
Hepadnaviruses	Partially Ds	Viral	Reverse transcriptase	+	Nucleus/cytoplasm

[a]ds, double-stranded; ss, single-stranded.

[b]Virions contain DNA-dependent RNA transcriptase and many other enzymes, but not DNA-dependent DNA polymerase.

assuming panhandle structures (e.g., orthomyxoviruses, family Orthomyxoviridae). Some of the single-stranded RNA genomes are of plus or "positive" polarity, meaning that they can be translated directly into proteins, and others are of minus or "negative" polarity, meaning that they cannot function as mRNAs and as such cannot be directly translated into proteins. Therefore, these negative-strand viral RNAs should be used as a template to synthesize a translatable complementary strand (mRNA), and still other viral RNAs are ambisense (Table 3.1). The plus-polarity naked viral genomes (except for retroviruses), completely free from all viral proteins, are infectious when introduced into a permissive cell, whereas minus-polarity naked genomes are noninfectious. Viruses possessing the minus-polarity genome therefore must carry an enzyme, RDRP, inside the virus particle to initiate the infectious cycle.

TABLE 3.3 Replication of RNA Viruses

	Virus	Form of RNA	Source of Nucleic Polymerase	Nature of Polymerase Activity	Presence of Polymerase in Virion	Viral Replication Site Within Host Cell
A	Paramyxovirus, rhabdovirus	ss[a] (–), unsegmented	Viral	RDRP	+	Cytoplasm
B	Bunyavirus, arenavirus	ss[b] (±), segmented	Viral	RDRP	+	Cytoplasm
C	Orthomyxovirus (influenza virus)	ss (–), segmented	Viral	RDRP	+	Nucleus
D	Rotavirus, reovirus, and orbivirus	ds[c] (±), segmented	Viral	RDRP	+	Cytoplasm
E	Picornavirus (poliovirus, hepatitis A), togavirus (Sindbis virus) Coronavirus	ss (+), unsegmented	Viral	RDRP	–	Cytoplasm
F	Retrovirus (HIV)	ss (+), unsegmented, diploid	Viral	Reverse transcriptase	+	Nucleus

[a]ss = single-stranded.

[b]± = ambisense genome.

[c]ds = double-stranded; (+) or (–) indicate positive or negative polarity, respectively.

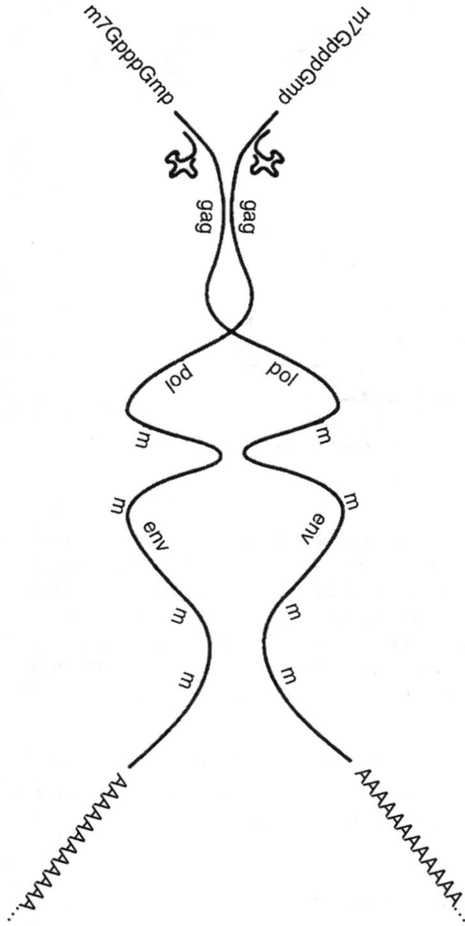

FIGURE 3.6 Features of the retrovirus (family Retroviridae) genome. The diploid RNA genome includes the following from 5′ to 3′: the m^{7}Gppp capping group, the primer tRNA, the coding regions, the M_6A residues (m), and the 3′ poly(AAAAA) sequence. Reprinted with permission from Fields and Knipe (1990).

Similarly, retroviruses must possess reverse transcriptase (RT, RNA-dependent DNA polymerase) in virus particles to initiate the infectious cycle inside host cells. However, using reverse genetics (RNA → DNA → RNA), many of the RNA genomes of both plus and minus polarity can be converted into infectious double-stranded DNA, thus permitting artificially induced mutational changes and genetic analysis of the viral genome, as well as use in DNA vaccination and as vectors in gene therapy. Some of the DNA viral genomes (adenoviruses, family Adenoviridae; hepadnaviruses, family Hepadnaviridae) and RNA (polioviruses, antigenic variants of human enterovirus C, family Picornaviridae, genus *Enterovirus*) viral genomes possess a covalently linked terminal protein at the 5′-end of a genomic nucleic acid strand, which provides critical functions for initiating DNA or RNA replication. Some positive-strand RNA viral genomes are also capped at the 5′-end and polyadenylated at the 3′-end (togaviruses, family Togaviridae), while others are not capped at the 5′-end (polioviruses) but possess polyadenylation (poly(A)) at the 3′-end. The minus-strand RNA genomes do not possess the cap at the 5′-end or the poly(A) at the 3′-end. Usually, the 5′ and 3′ ends of the minus-strand RNA genome are partially complementary, often forming panhandles by intrastrand hybridization and functioning as their own promoters for transcription and replication.

Organization of genes in the RNA genome varies between different groups of viruses. For positive-strand naked RNA viruses (e.g., polioviruses), which are translated into a single large polyprotein, the 5′-end of the genome is not capped but is rather covalently linked to a small protein VPg (Figure 3.7). The 5′-end of these viral genomes contains an untranslated region possessing a highly ordered secondary structure for internal ribosome entry, followed next in sequence by the genes of capsid proteins (VP4, VP2, VP3, VP1). The genes for nonstructural proteins including proteases and viral replicase (an RNA-dependent RNA polymerase) are located in the 3′-half of the genome. However, for the plus-strand enveloped RNA viruses (e.g., Sindbis virus, Family Togaviridae, genus *Alphavirus*), the genes for the nonstructural proteins are present at the 5′-end and structural proteins including capsid and envelope proteins are present in the 3′-half of the genome. Structural genes of this latter type of viruses are translated from a separate subgenomic mRNA, whereas their nonstructural proteins are translated from the genomic RNA. The large plus-strand coronavirus RNA

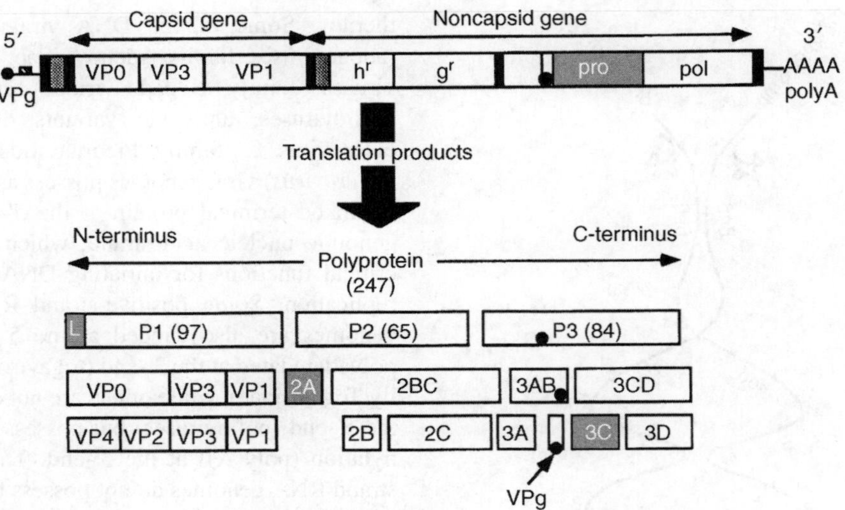

FIGURE 3.7 Organization of picornaviral (family Picornaviridae) genome (plus-strand RNA) and its translation products. The virus RNA has VPg protein attached to its 5′-end and poly(A) at its 3′-end. The order and the position of virally encoded proteins are shown. P1, P2, and P3 indicate three intermediate precursor proteins cleaved from the polyprotein. These precursor proteins are further cleaved by virus-encoded proteases into mature functional proteins. Numbers in parentheses indicate molecular weights in thousands. h^r and g^r indicate host range and guanidine resistance determinants, respectively. 2A and 3C are proteinases involved in cleavage of the polyprotein and precursor proteins into mature viral proteins. VPg, VPO, and so on indicate specific viral proteins.

genome possesses the nonstructural genes in the 5′-half and structural genes in the 3′-half of the genome. The gene for the highly abundant nucleoprotein (N protein) of coronaviruses is present at the 3′-end of the genome.

For unsegmented minus-strand RNA genomes, the order of genes for both vesiculoviruses (family Rhabdoviridae) and paramyxoviruses (family Paramyxoviridae) are similar. Structural genes for capsid (N and P proteins) and envelope proteins are at the 3′-half, and the large polymerase (L) gene occupies the entire 5′-half of the minus-strand RNA genome (Figure 3.8). The 3′-end of the template (minus-strand) RNA is transcribed into a leader (ℓ) sequence not present in the mRNA, and the region between two genes is separated by an element called the EIS. It consists of an "E" (end) sequence for transcription termination and polyadenylation of a gene, an "I" (intergenic) sequence that allows the viral transcriptase to escape (therefore the "I" sequence is not represented in the mRNA), and "S" (the start) sequences that denote the start of the next gene. EIS sequences in the genome vary for different viruses in these groups.

3.2.2 Viral Proteins

Proteins are major constituents of the viral structure, and their main functions, as indicated previously, are to protect the nucleic acid from nucleases and provide receptor binding site(s) for virus attachment, which is required for efficient transmission of virus from one host to another. Viral proteins can be classified as nonstructural or structural. Nonstructural proteins are those proteins that are encoded by the virion genome and expressed inside the virus-infected host cells, but not found in the virion particles. These nonstructural proteins usually have regulatory or catalytic functions that are involved in viral replication or transcription processes, as well as in modifying host functions. Structural proteins are broadly defined as proteins found in virus particles. The majority of these structural proteins constitutes the viral

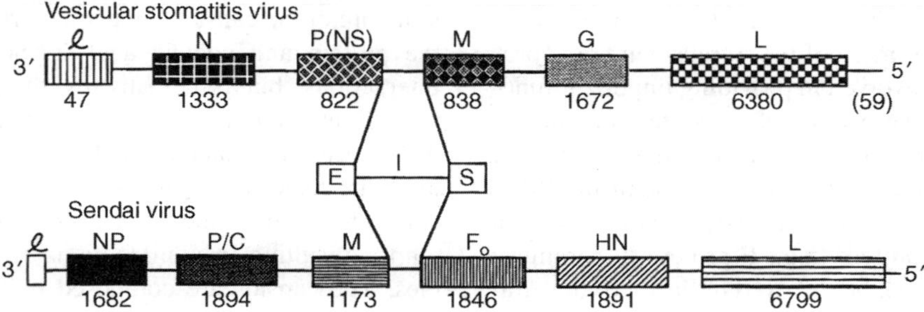

FIGURE 3.8 Genome of unsegmented negative-strand RNA viruses (vesicular stomatitis virus (VSV, refers to three species belonging to the family Rhabdoviridae and genus *Vesiculovirus*)) and Sendai virus, a species of the family Paramyxoviridae, genus *Respirovirus*. Numbers underneath rectangles represent the number of nucleotides in each gene (shown above the line), ℓ, leader sequence; E, end (or transcription termination) sequence; I, intergenic sequence (not transcribed); S, start sequence of mRNA of the next gene; N, NP = nucleoproteins; P/C, P (NS) = phosphoprotein; M = matrix protein; G, F, HN = glycoproteins; L = polymerase protein.

capsid or core and are intimately associated with the viral genome to form the nucleocapsid. The cores of some viruses also contain regulatory or catalytic proteins as minor structural proteins (e.g., proteins with enzymatic functions, such as transcriptase (RDRP) or reverse transcriptase) (Tables 3.2 and 3.3). In addition, some viruses include host proteins such as histones associated with the viral genome in virus particles (e.g., SV40 minichromosome; Simian virus 40, family Polyomaviridae, genus *Polyomavirus*) or ribosomes, as is the case with arenaviruses. Although these minor virus- and host-coded proteins are critically involved in virus replication and infectivity, they are not essential for formation of viral capsids.

In addition to having viral capsids, the enveloped viruses possess membranes (or envelopes) surrounding the viral capsids. These viral membranes, as noted earlier, contain lipids derived from the host membrane and proteins specified by the viral genome. Two types of proteins are found in the viral membrane: transmembrane (TM) proteins and matrix proteins.

3.2.2.1 Transmembrane Proteins

Transmembrane proteins can be type I (such as influenza virus hemagglutinin (HA), family Orthomyxoviridae; VSV G protein, family Rhabdoviridae, genus *Vesiculovirus*), type II (such as influenza virus neuraminidase (NA)), and type III (such as influenza virus M2), depending on their molecular orientation, or complex proteins, containing multiple transmembrane domains (TMDs) (such as E1 glycoprotein of coronaviruses). Enveloped viruses may contain only one (as in the G protein of VSV (vesicular stomatitis virus, refers to three species belonging to the family Rhabdoviridae, genus *Vesiculovirus*)), two (as in the HN and F proteins in paramyxoviruses), or multiple transmembrane proteins (as in the influenza viruses, herpesviruses, family Herpesviridae; poxviruses, family Poxviridae, etc.) on their envelope. Again, viruses containing multiple transmembrane proteins may have proteins of different orientations such as type I, type II, and type III (e.g., influenza viruses, family Orthomyxoviridae). These transmembrane proteins are often glycosylated via *N*- or *O*-glycosidic bonds and their carbohydrate moieties can be composed of simple sugars, usually consisting of mannose molecules, complex sugars, including galactose, glucosamine, galactosamine, fucose, and mannose, and sialic acid residues. Proper glycosylation of viral proteins is often important to provide the necessary molecular stability, solubility, oligomer formation, and intracellular transport of viral proteins, as well as for modulating the host immune response, including epitope masking and unmasking. These glycans may also play an important role in apical sorting of proteins within the polarized epithelial cells. It often is the case that one or more of these transmembrane proteins are involved in providing important functions in the processes of virus life cycle such as receptor binding (e.g., HA in influenza viruses, HN in paramyxoviruses, G protein in VSV). The same protein (influenza viruses, VSV) or a different protein (e.g., F protein in paramyxoviruses) can be involved in fusion of the viral envelope with cellular membranes, uncoating, and entry of the viral genome inside the cell. In addition, some other viral membrane protein can aid in releasing mature viruses from the infected cells and spreading of viruses from cell to cell (e.g., function of the neuraminidase protein in releasing influenza viruses after budding). These envelope proteins are important not only in virus infectious cycle but also for host defense, where they elicit both neutralizing antibodies and CTL (cytotoxic T lymphocyte) responses against the virus infection in infected hosts and therefore play a critical role in vaccination and protection against viral infections.

3.2.2.2 Matrix Proteins In addition to the transmembrane proteins, the majority of these enveloped viruses also contain another type of membrane protein called a matrix protein (e.g., M1 protein of influenza viruses) that forms a shell underneath the membrane enclosing the capsid (Figure 3.3). The matrix proteins are therefore likely to interact with the lipid bilayer and transmembrane proteins of the viral envelope on the outer side and with the nucleocapsid on the inner side. Matrix proteins are also usually the most abundant proteins in enveloped virus particles and are critical for the budding of enveloped viruses. Some enveloped viruses containing icosahedral capsids do not possess typical matrix proteins around the nucleocapsids underneath the membrane (e.g., togaviruses).

Viruses vary greatly in size and shape. They can be spherical, cylindrical (rod shaped), or even pleomorphic (Figures 3.1 and 3.4). Primarily, the virus structure is determined by the nature of the capsid and whether the capsid is naked or surrounded by an envelope. The structure of the capsid is in part determined by the protein and nucleic acid (nucleocapsid) interactions, but principally by the protein–protein interactions of the capsid protein(s). In most cases, the nucleic acid is incorporated after the majority of the protein shell of the capsid has been formed, or capsids can remain empty, resulting in noninfectious virus particles. The capsids are composed of repeating protein subunits called capsomeres. Capsomeres are composed of multimeric units of a single protein, or often heteromeric units of more than one protein.

3.2.3 Lipids

In addition to nucleic acids (DNA or RNA) and proteins, enveloped viruses contain lipids in their membrane. These lipids constitute integral components of viruses and are critically involved in many aspects of virus life cycle including entry, fusion, uncoating, and delivery of viral genome into the host cell for initiating the infectious cycle; transport and assembly of viral components; and budding and release of virus particles. Although viruses bud from host membranes and all viral lipids are acquired from the host membranes, some of the viral lipids do not match quantitatively with that of the host membranes. A number of factors are involved in the selection of viral lipids, including the budding site of enveloped viruses in the infected cells and the type of cells in which viruses are grown, as well as the type of organelles such as plasma membrane, Golgi complex, nucleus, and so on from which the virus buds. Since different cell types and subcellular organelles possess varying lipid composition, viruses budding from membranes of different organelles will have different lipid composition. For example, herpesvirus, a complex DNA virus, buds from the inner nuclear membrane. However, fully mature infectious herpesvirus exits from the basal layer of infected epithelial cells. Hepadna, rota, and spuma viruses bud from the endoplasmic reticulum (ER). Coronaviruses (family Coronaviridae) and vaccinia virus (family Poxviridae, genus

Orthopoxvirus) acquire their envelope from the intermediate pre-Golgi compartment (IC). Vaccinia virus is further surrounded by membrane envelope, which is a part of complex maturation process, before being released from the plasma membrane. Bunyaviruses (family Bunyaviridae) and togaviruses (family Togaviridae, genera *Alphavirus* and *Rubivirus*) acquire their envelope on the Golgi complex. However, whereas Sindbis virus (SIN, family Togaviridae, genus *Alphavirus*) exits from the apical membrane, Semliki Forest virus (SFV, also family Togaviridae, genus *Alphavirus*) buds from the basolateral membrane. The assembly and budding of some viruses such as orthomyxo-, paramyxo-, filo-, retro-, and rhabdoviruses occur only at the plasma membrane, although orthomyxoviruses and paramyxoviruses bud from the apical plasma membrane whereas the filo-, retro-, and rhabdoviruses bud from the basolateral plasma membrane. Viruses budding from the different domains of the same membrane will have different lipid composition. Furthermore, both cellular and viral membranes are mosaic in nature and contain different lipid microdomains. Among these, lipid rafts are known to play many important functions in both cellular and viral biology and often function as the budding site for many viruses.

3.2.3.1 Lipid Rafts Lipid rafts are operationally defined as cholesterol-dependent microdomains resistant to solubilization by nonionic detergents such as TX-100 at low temperature. Lipid rafts consist of sphingolipid–cholesterol clusters, usually varying in size and are present in the plasma membrane, apical transport vesicles, and Golgi and *trans*-Golgi membranes. Lipid rafts vary in size, ~50 nm in diameter (Pralle et al., 2000) and smaller than the caveolae that also exhibits TX-100 insolubility similar to the lipid rafts. Lipid raft microdomains are formed by lateral organization and phase separation of lipids between l_o phase and l_d or l_a phases (Nayak and Hui, 2004). l_o and l_d or l_a phases refer to ordered and disordered phases of lipid in the membrane, respectively. l_o phase separation also leads to asymmetric distribution of different lipids in the exoplasmic versus cytoplasmic lipid leaflets. These lipid microdomains containing l_o phase have been variously called by different names, such as detergent-insoluble GSL (glycosphingolipid)-enriched domains (DIGs), GSL-enriched membranes (GEMs) or microdomains, detergent-resistant membranes (DRMs), Triton-insoluble membranes (TIMs), GSL/sphingolipid–cholesterol rafts, lipid rafts, or simply rafts.

Lipid raft microdomains contain glycerophospholipids, (glyco)-sphingolipids, GPI lipids bearing predominantly saturated fatty acids, cholesterols, and gangliosides such as GMI and GM2. These lipids form tight packing and cholesterol contributes to tight packing by filling the interstitial space between the long saturated acyl chains and sphingolipids resulting in the formation of l_o state of lipids in lipid raft microdomains. The tight lateral packing of sphingolipids and cholesterol leads to TX-100 insolubility at low temperature. Lipid rafts exclude most of the membrane proteins including the TM proteins except for proteins with GPI anchor, and palmitoylation, prenylation, acylation, myristoylation are partitioned in these microdomains. However, some TM proteins such as influenza virus HA and NA without acyl modification are included in the lipid raft microdomain of influenza virus envelope. Protein–lipid and protein–protein interactions may contribute to coalescence, growth, and stability of lipid rafts.

3.3 MORPHOLOGY

Viruses vary greatly in size and shape. They can be spherical, cylindrical (Figure 3.1), or even pleomorphic (Figure 3.4). Primarily, the virus morphology is determined by the nature of the capsid structure and whether the capsid is naked or surrounded by envelope. The structure of the capsid is in part determined by the protein and nucleic acid (nucleocapsid) interactions but principally by the protein–protein interactions of capsid proteins. In most cases, nucleic acid is incorporated after the majority of the protein shell of the capsid has been formed, or capsids can remain empty, resulting in production of noninfectious virus particles. The capsids are composed of repeating protein subunits called capsomeres. Capsomeres are composed of multimeric units of single or often heteromeric units composed of more than one protein. Formation of the viral capsid and its shape is primarily determined by three-dimensional (3D) structure of the capsid proteins, which in turn is determined by the specific amino acid sequence encoded by the viral nucleic acid. The amino acid sequence is considered the primary structure of the protein, whose three-dimensional structure is composed of secondary structures such as α helices, β sheets, and random coils. These secondary structures interact with each other, forming the tertiary and quaternary structures, which are usually stabilized by noncovalent interactions (sometimes by covalent disulfide linkages) and represent folding of the proteins into relatively stable structures of microdomains (e.g., globular heads). In addition, extended and flexible regions of proteins, called hinges, are also present, and these hinges become important for interaction with other members of the protein subunits that form the capsomeres. In most viruses, contacts between capsomeres are repeated, exhibiting symmetry. This is a process of self-assembly driven by the stability of interaction among the protein subunits forming the capsomeres and the capsomeres forming the capsid. Viral capsids have a helical (spring-like) or icosahedral-based (cuboidal or spherical) symmetry.

Until recently, the morphology of viruses was based on transmission electron microscopy (TEM) images of negatively stained viral particles or thin sections of virus-infected cells. However, staining and sectioning procedures often introduce artifacts in the shape, size, and morphology of virus particles during sample processing owing to the use of heavy metal stain at nonphysiological pH and sample drying. Viruses such as influenza viruses are particularly sensitive to these procedures owing to the flexible, pH-sensitive viral envelope. Recently, electron tomography (ET) has been used to reconstruct the 3D structure of viral particles in thin sections by combining different tilt views of the same sample. In addition, the size and shape of virus particles vary with different virus isolates and laboratory strains. Recently, cryoelectron microscopy (cryo-EM) and cryoelectron tomography (cryo-ET) have been used to examine the structure of these viruses in their natural state without fixing and staining (Calder et al., 2010). Furthermore, cryo-ET can be used to determine the 3D structure of each viral particle by combining different tilt views of the same viral particles (Baumeister, 2002). The 3D structures can then be computationally sliced to reveal the structural arrangement of proteins, nucleic acid, and lipids and their possible interactions in their native state within the virus particles. Some examples of morphology of different viruses are shown in Figures 3.3–3.5 and 3.9.

3.3.1 Helical Capsids

Helical capsids are usually flexible and rod like. The length of the helical capsid is usually determined by the length of the nucleic acids; that is, some defective interfering (DI) viruses having shorter nucleic acids will have a shorter helical nucleocapsid (e.g., DI RNA of VSV). Helical capsids can be naked, that is, without an envelope (e.g., tobacco mosaic virus, family Virgaviridae, genus *Tobamovirus*). However, there is no known example of an animal virus with a naked helical nucleocapsid. All animal viruses with helical capsids found to date are enveloped. However, such helical capsids when enclosed in an envelope can exhibit various morphologies, including filamentous (filoviruses), rod shaped (e.g., rhabdoviruses) or spherical, spheroidal or elongated (e.g., orthomyxo- or paramyxoviruses), and even pleomorphic (Figure 3.4), indicating that the helical capsid in these viruses is flexible. Some helical capsids can be further folded, forming supercoiled nucleocapsids (e.g., orthomyxoviruses). Helical capsids can package only

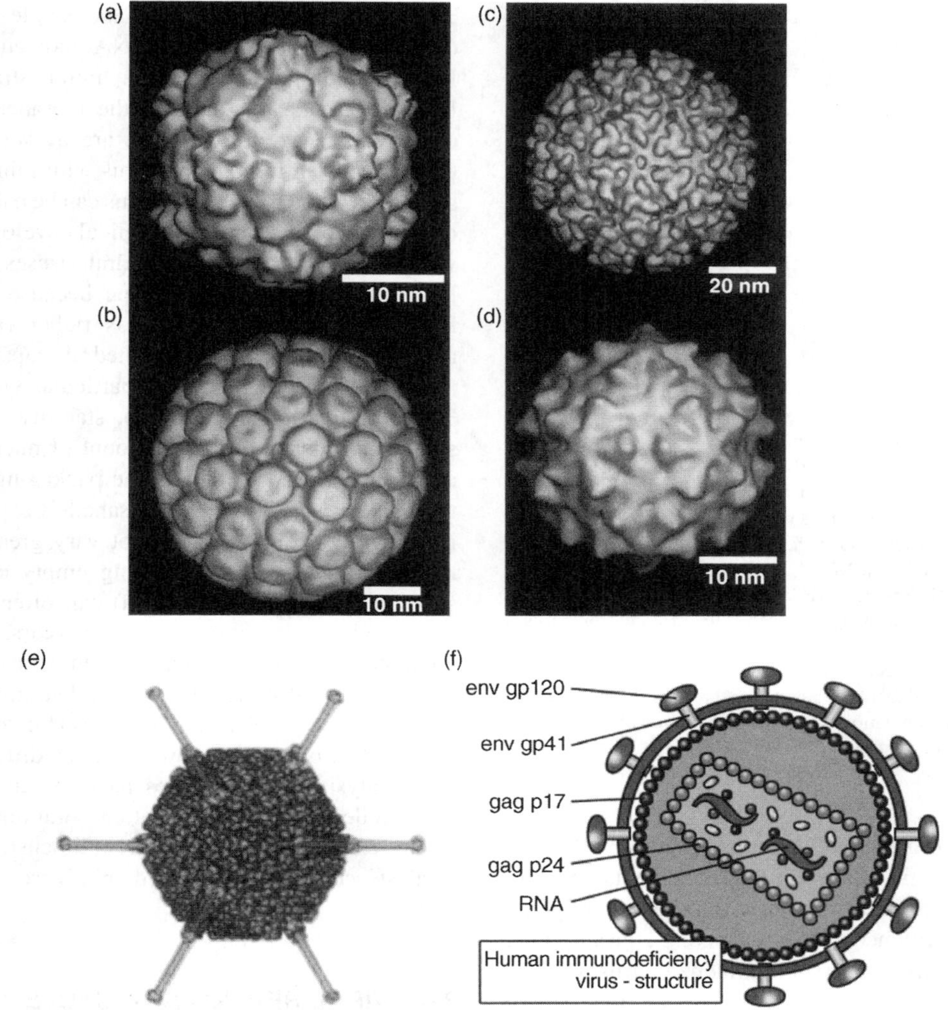

FIGURE 3.9 Structure of representative RNA and DNA viruses as determined by cryoelectron microscopy. (a) human rhinovirus 14, a ssRNA virus that is an antigenic variant of human rhinovirus B, family Picornaviridae, genus *Enterovirus*; (b) SV40, Simian virus 40, a dsDNA virus of the family Polyomaviridae, genus *Polyomavirus*; (c) Sindbis virus capsid, an ssRNA virus of the family Togaviridae, genus *Alphavirus*; (d) flock house virus, a positive-strand bipartite ssRNA insect virus of the family Nodaviridae, genus *Alphanodavirus*; (e) adenovirus, a dsDNA virus of the family Adenoviridae. The micrographs of human rhinovirus 14, SV40, Sindbis virus, and flock house virus were provided by and are reprinted with permission from Norm Olson and Jim Baker of Purdue University. The adenovirus micrograph was provided by and is reprinted with permission from Phoebe Stewart of UCLA. (f) Schematic presentation of HIV reproduced from Avert with permission.

single-stranded RNA, but not double-stranded DNA or RNA, possibly because of the rigidity of the double-stranded nucleic acids. However, some viruses with helical capsids may possess only one capsid containing one virion RNA (unsegmented) molecule (rhabdoviruses, paramyxoviruses) or multiple capsids containing multiple RNA segments (orthomyxoviruses) (Calder et al., 2010). Viruses containing multiple RNA segments can undergo reassortment

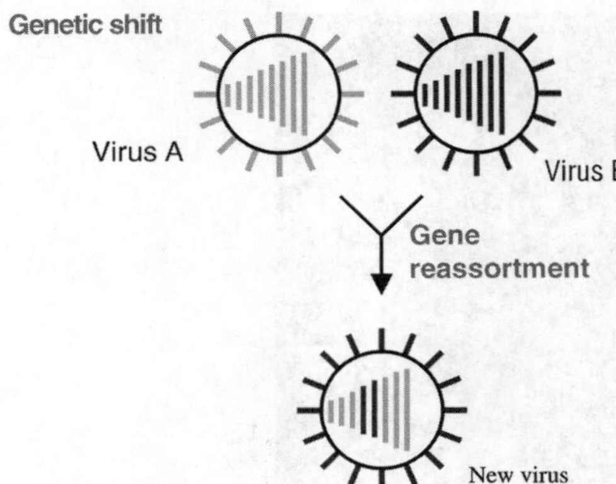

FIGURE 3.10 Reassortment of influenza virus RNA segments. When two influenza viruses (virus A orange and virus B blue color RNA segments) infect a single cell, progeny viruses from the infected cells will possess different combinations of vRNA segments. One progeny virus particle shown here contains six RNA segments (PB1, PB2, PA, NP, M, and NS RNA) from virus A and two RNA segments (HA and NA) from virus B. The new progeny virus will be antigenically different from virus A and will emerge as a potentially pandemic virus. In addition to gene reassortment, mutations in vRNA segments will facilitate adaptation, growth, and spread in the human population. (*See the color version of this figure in Color Plate section.*)

with other related viruses (Figure 3.10), thus exchanging different RNA segments and giving rise to new viruses with different antigenic and virulence determinants (e.g., the antigenic shift that occurs in influenza viruses). The genomic RNA is protected by the helical capsid in some viruses (e.g., paramyxo- and rhabdoviruses) but remains exposed in others (e.g., orthomyxoviruses). A single viral protein (e.g., NP protein of orthomyxoviruses) is usually involved in helical capsid formation.

3.3.2 Icosahedral Capsids

Viruses with icosahedral capsids possess a closed shell enclosing the nucleic acid inside (Figure 3.1). An icosahedron has 20 triangular faces, 30 edges, and 12 vertices and is characterized by a 5:3:2-fold rotational symmetry. Unlike helical nucleocapsids that package only single-stranded nucleic acid, icosahedral capsids can be used to package single- or double-stranded RNA and DNA molecules. However, although plus- or minus-strand DNA segments are found in the icosahedral capsids of parvoviruses, there are as yet no examples of an icosahedral virus with minus-strand RNA. An icosahedral virus can be naked or enveloped, but, unlike the helical enveloped viruses, the enveloped icosahedral viruses are less pleomorphic in their shape because the icosahedron capsid structure is rather rigid and, in addition, with icosahedral capsids, the overall size is fixed for a particular virus. The virus particle's formation, stability, and size do not depend on the amount of nucleic acid in the capsid. Although the packaging of the nucleic acid inside the icosahedral capsid is relatively fixed and does not vary greatly, noninfectious viruses containing empty capsids (i.e., without nucleic acid) can often be seen in virus populations. In recent years, the complete three-dimensional structures of several icosahedral viruses have been determined at the atomic level using the powerful tools of cryoelectron microscopy and X-ray diffraction analysis. Such analyses have led to the rational design of a number of antiviral drugs. Some examples of such three-dimensional viral structures are presented in Figures 3.3 and 3.9.

3.4 VIRAL REPLICATION CYCLE

To survive, viruses must multiply. Since viruses cannot multiply outside the host cell, they must infect host cells and use cellular machinery and energy supplies to replicate and produce the progeny viruses that must in turn infect other hosts and the cycle continues. Host–virus interaction at the cellular level is therefore obligatory for virus replication. Specific host cells can be susceptible (i.e., permissive) or nonsusceptible (i.e., resistant or nonpermissive) to a particular virus. Nonsusceptibility of cells can be at the attachment (e.g., lack of a suitable receptor for a virus at the cell surface) and entry/uncoating phases, at the intracellular phase (i.e., a block in synthesis of viral macromolecules), or at the assembly and exit phases. Furthermore, following infection, viruses can cause abortive (nonproductive) or productive infection. Only productive infection yields infectious progeny virus particles. Following abortive or productive infection, the host cell may survive or die, i.e., the cytopathic effect (CPE). CPE caused by a virus does not necessarily indicate the permissiveness of a cell to a virus leading to productive infection. The viral genome in abortive infection may be degraded, may become integrated into the host DNA, or exist as extrachromosomal (episomal) DNA in the surviving cell. The growth properties of such cells may be altered, including the possibility that they may become transformed and cancerous. Alternatively, cells containing the integrated viral DNA may behave normally, exhibiting little change in their normal properties. Malignant transformation of infected cells often depends on the site of viral genomic integration, leading to activation of cellular oncogenes, disruption or inhibition of tumor suppressor genes, or synthesis of viral oncogene products that are encoded by the virus in its genome. In the infected cells, the viral genome may remain dormant, resulting in a latent infection, and it can be activated later, producing infectious viruses, as occurs with herpesviruses. Alternatively, infected cells may yield virus at a low level without affecting cell survival, resulting in persistent infection, as occurs with LCMV (lymphocytic choriomeningitis virus, family Arenaviridae, genus *Arenavirus*). The effect of virus infection has been studied at both the cellular and organismic levels. At the organismic level, it is called "viral pathogenesis," while at the cellular level, it is called the CPE. Under these conditions, cells may undergo morphological changes, including rounding, detachment, cell death and cell lysis (either apoptotic or necrotic), and syncytium (giant multinucleated cell) formation, as well as inclusion body formation. Many of these changes are caused by the toxic effects of viral proteins affecting host macromolecular synthesis, including DNA replication, DNA fragmentation, mRNA transcription, translation, protein modification, and degradation, as well as other cellular synthetic and catalytic processes. Furthermore, since the same cellular machineries are directed toward viral macromolecular synthesis, the host is deprived of their functions. In addition to direct cell killing, virus infection can indirectly cause injury to tissues in a complex organism, as a result of both complex host–viral immune interactions (i.e., immunopathology) and by cytokine production causing inflammatory reactions. Usually, lytic viruses cause cell death and when a sufficient number of cells in a given tissue (e.g., lungs, liver, etc.) die, it leads to the loss of function of the tissue and the production of specific disease syndrome (pneumonia, hepatitis, etc.). Since infection of the host usually begins at a very low MOI (multiplicity of infection, expressed as virus:cell or virus:host ratio), the virus must be able to replicate efficiently and produce a large number of progeny viruses in a short period to infect and kill a sufficient number of cells to cause the disease syndrome. It is evident from the foregoing discussion that for successful replication, a virus must find susceptible host cells and it must be able to attach itself to and penetrate into the host cell and be uncoated, rendering the viral genome available for interaction of the viral and cellular machineries for transcription, translation, and replication of the viral genome. Finally, the newly synthesized viral components must be assembled into progeny viruses and released into the medium (i.e., outside environment) to infect other hosts. Whether with cultured cells in laboratory or the complex organisms in nature, the virus–host interaction always occurs at the level of single cells. Thus, the viral infectious cycle (also known as the viral growth cycle, replication, or multiplication cycle) can be divided into different phases, namely, (1) adsorption (attachment), penetration, and uncoating; (2) transcription, translation, and replication; and (3) assembly and release. The replication cycle of influenza (orthomyxo) viruses is diagrammatically shown in Figure 3.11.

Influenza virus

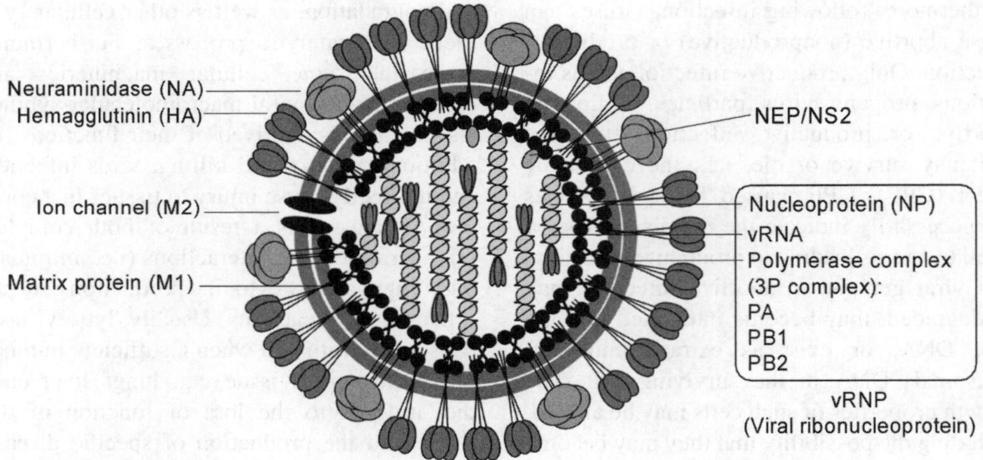

Family: Orthomyxoviridae
Genome: (–)ssRNA

Subtype: A (8 RNA segments) e.g., WSN
B (8 RNA segments)
C (7 RNA segments)

(a)

FIGURE 3.11 Schematic presentation of the infectious cycle of an influenza virus. (a) Schematic presentation of influenza virus structure. (b) Schematic presentation of influenza virus infection showing attachment, entry, and uncoating of a virus particle. The steps in the replication cycle are attachment mediated through HA and sialic acid receptor, entry into the cell via endosome, HA-mediated fusion of virus membrane with endosomal membrane at low pH, release of vRNP, transport of vRNP into the nucleus, and transcription (mRNA synthesis) and replication (cRNA and vRNA synthesis) of vRNP in the nucleus. (c) Schematic presentation of influenza virus infectious cycle showing export, assembly, and budding of a virus particle. The steps include export of vRNP from nucleus into cytoplasm, export of virus proteins, vRNP to the budding site, bud formation, and bud release by fusion and fission viral and cellular membranes. (*See the color version of this figure in Color Plate section.*)

3.4.1 Adsorption

Viral adsorption is defined as the specific binding of a virus to a cellular (host) receptor. It is the first step for the virus to enter into the cell. Viruses cannot cause disease if this first step is blocked. Vaccines and the resulting antibodies are designed primarily to block this step in the virus replication cycle. It is a receptor–ligand interaction in which viruses function as specific ligands and bind to the receptors present on the cell surface. Ligand functions of the virus are provided by the specific viral proteins present at the surface of the virus. For naked (i.e., nonenveloped) viruses, this function is performed by one of the capsid proteins and for enveloped viruses, one of the membrane proteins functions as the ligand (also variously known as the receptor binding protein, viral attachment protein, or antireceptor) for the host receptor. Usually, only one viral protein provides the receptor binding function, although one or more cellular proteins can function as receptor and coreceptor. For enveloped viruses, a classic example of a viral ligand (i.e., receptor

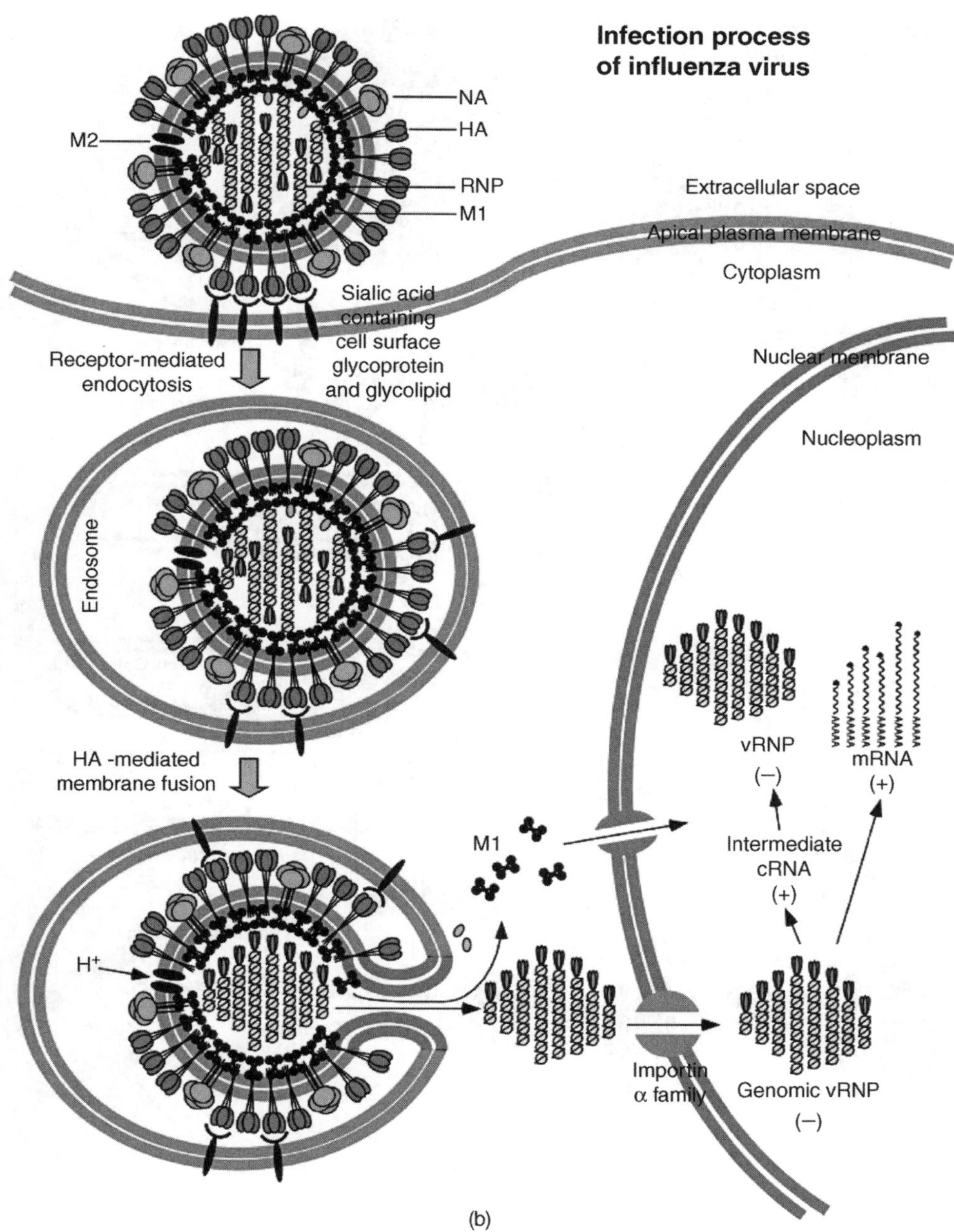

(b)

FIGURE 3.11 *(Continued)*

binding protein) is the influenza virus hemagglutinin and its receptor binding site is present on the globular head of the HA spike. Variation in the amino acid sequence of the receptor binding site of HA is a critical factor in species-specific susceptibility of influenza viruses (e.g., chicken versus human). For non-enveloped viruses, a classic example of a viral

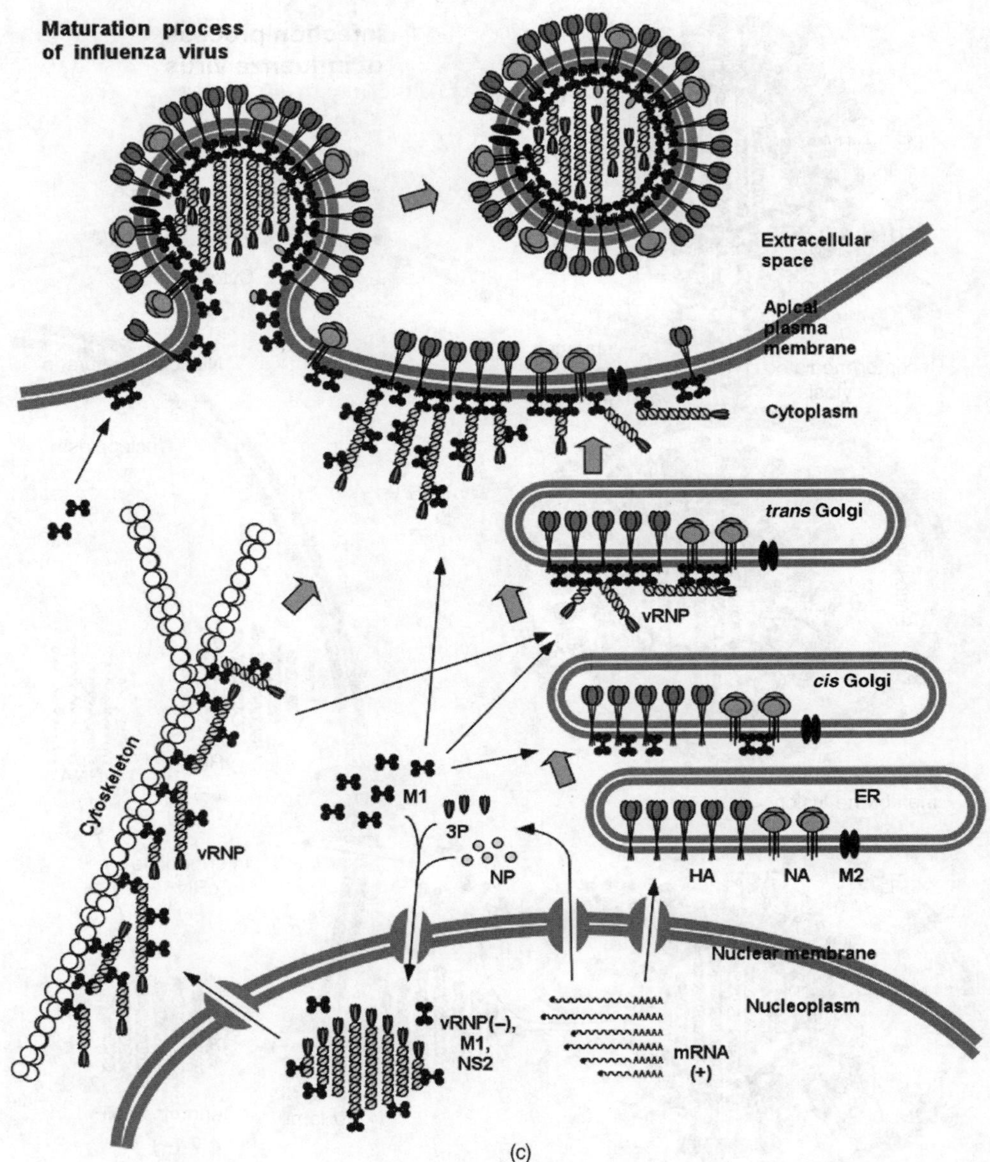

FIGURE 3.11 *(Continued)*

ligand is the VP1 of rhinoviruses. When five VP1 proteins are packed together within the viral capsid structure, the confluence of these grooves forms a depression called a canyon. The canyon is shown to be the site for interaction between human rhinovirus 14 (HRV14; an antigenic variant of human rhinovirus B, family Picornaviridae, genus *Enterovirus*) and the cellular molecule ICAM-1 (receptor for rhinovirus). The amino acids lining the floor of these canyons are highly conserved, but residues on the surface of the canyon are variable (Figure 3.12). Antibodies can bind to the surface epitopes in and around the proximity of

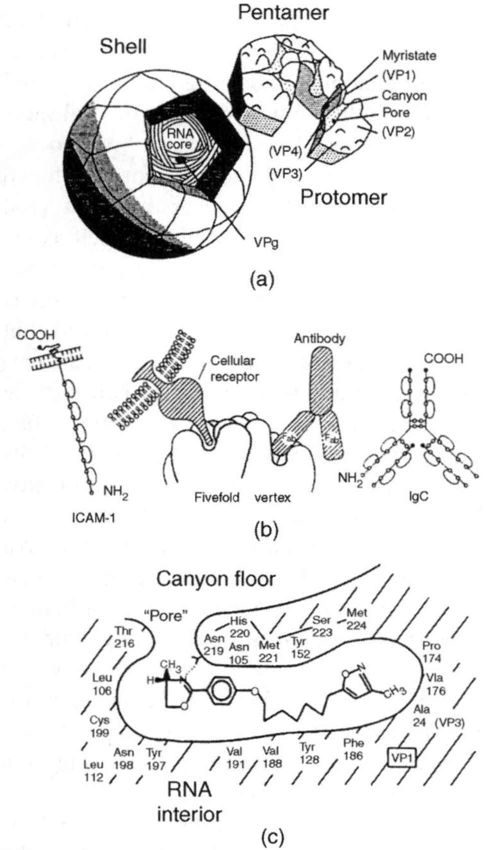

FIGURE 3.12 Key features in the function of cellular receptor interactions with an invading virus, similar to a typical picornavirus. (a) Exploded diagram showing internal location at the canyon-like center of the pentamer fivefold vertex with myristate residues on the NH_2 terminus of VP4. (b) Binding of cellular receptor (ICAM-1 molecule) to the floor of the canyon. Note that the binding site of the ICAM-1 molecule, identified as a major rhinovirus (species human rhinovirus A, B, and C) receptor, has a diameter roughly half that of an IgG antibody molecule. (c) Location of a drug binding site in VP1 of HRV14 (human rhinovirus 14) and identity of amino acid residues lining the wall. The drug shown here, WIN 52084, prevents attachment of HRV14 by deforming part of the canyon floor. The pentamer vertex lies to the right. Reprinted with permission from Fields and Knipe (1990).

the receptor binding site and thus interfere with virus attachment by steric hindrance. Viruses can accept mutations in these surface epitopes and thereby escape (and are thus known as escape mutants) neutralization by specific antibodies, but the receptor binding site usually does not undergo mutational changes because of its location inside the canyon and therefore remains conserved. This also appears to be the case with influenza virus hemagglutinin and other viral receptor binding sites that remain conserved despite the variation in the neutralizing epitopes of the same viral protein. Thus, the receptor binding site in the viral protein is usually a depression or canyon and is therefore protected from the mutational pressure of antibodies because antibodies do not have direct access to this region.

The cellular receptors of many viruses have recently been identified. Cellular receptors should be present at the cell surface and are carbohydrates, lipids, or proteins. Sialooligosaccharides present in glycoproteins or glycolipids function as receptors for orthomyxoviruses, paramyxoviruses, or polyomaviruses, as well as immunoglobulin superfamily molecules (CD4 for HIV, human immunodeficiency virus 1 and 2, of the family Retroviridae, genus *Lentivirus*, and many members of the family Picornaviridae, including ICAM-1 for both rhinoviruses (genus *Enterovirus*) and encephalomyocarditis virus (genus *Cardiovirus*), as well as Pvr for polioviruses (genus *Enterovirus*)). Hormone or neurotransmitter receptors function as receptors for a number of other viruses (e.g., epidermal growth factor for vaccinia virus, β-adrenergic receptor for reovirus (family Reoviridae), and acetylcholine receptor for rabies virus (family Rhabdoviridae, genus *Lyssavirus*)) and heparan sulfate for herpesviruses (family Herpesviridae). Some viruses have more than one receptor, one being the primary receptor and the other a coreceptor. A classic example of this is the case of CD4 and chemokine receptors (CXCR4, CCR5, etc.) functioning as the receptor and coreceptor for HIV, respectively. Both the receptor and the coreceptor are needed for productive HIV infection, although only one viral protein (gp120) provides the receptor binding sites for both receptor and coreceptors. Receptor–virus interaction is the major reason for the host and tissue tropism of viruses. It has been shown that lack of a specific coreceptor on surface of the cell provides resistance to HIV infection in some persons. Receptor–virus interactions are specific, and this is a noncovalent binding independent of energy or temperature. Thus, the kinetics of viral binding to cells can be determined at 4°C, which serves as a research aid since their interaction at this temperature prevents viral penetration and uncoating. Therefore, binding virus to cells at 4°C and subsequently raising the temperature to 37°C can be used to infect cells synchronously and to study the subsequent events such as uncoating and penetration of virus into host cells. The time course of viral adsorption follows first-order kinetics and is dependent on virus to cell concentration. Usually, susceptible cells contain a large number of virus receptors, in the range of 10^4–10^5 per cell.

3.4.2 Penetration and Uncoating

Following specific ligand to receptor interaction, the next steps in virus replication include entry/penetration of virus into the host cell and uncoating of the viral genome, which are energy-dependent processes and can be prevented experimentally in the laboratory by subjecting the virus–cell complex to low temperatures (4°C). Viruses attached to the cell surface can be detached by specific enzymatic treatment (e.g., neuraminidase treatment in the case of influenza virus). However, once the virus enters into the cell, it can be neither separated from the cell nor neutralized by antibodies. Penetration refers to the entry of the surface-bound virus particles inside the cell, where they exist free in the cytoplasm or inside the host cell vesicles (usually within endosomes). Quantitatively, penetration of virus particles is measured by the loss of the ability of antiviral antibodies to neutralize the cell-bound virus particles after adsorption, an effect that occurs because after the viral particles have entered the cell, they are protected and no longer accessible to antibodies outside the cell. Uncoating, on the contrary, refers to disruption of virus particles, causing partial or complete separation of nucleic acid from the capsid, and is needed for initiation of transcription and translation of the viral genome. Uncoating can be assessed by, among other things, changes in viral morphology or viral density, release of nucleocapsid and membrane proteins from enveloped virus particles, and the accessibility of viral genome to nucleases. For viruses such as orthomyxovirus and poliovirus, these processes are separated temporally (i.e., penetration is followed by uncoating in the cytoplasm), but for some viruses, both penetration and uncoating occur simultaneously at the cell surface (e.g., paramyxoviruses, HIV). Uncoating refers to the step in which the viral genome becomes functional transcriptionally or translationally. However, complete separation of nucleic acid from all capsid proteins is not required for most viruses. For naked viruses, uncoating is a post-penetration process that occurs in the endosome or nucleus. Viruses that undergo uncoating in the cytoplasm following endocytosis require low pH (~5) in the endosome for uncoating, whereas viruses that undergo fusion at the cell surface can undergo uncoating in a pH-independent manner.

Naked viruses such as the RNA-based picornaviruses enter into the cytoplasm of the infected cells via receptor-mediated endocytosis (Figure 3.13) or by phagocytosis (also called viropexis). In the endosome, the virus particle undergoes alteration in structural and antigenic properties and becomes acid labile and noninfectious. During uncoating, VP4 (a capsid protein) is released and the viral RNA is extruded from the capsid structure through the hole in the capsid caused by VP4 release into the cytoplasm. How the viral RNA gets through the endosomal membrane is not clear, but it is speculated that pore formation may occur by the interaction of the myristoylated NH_2 terminus of VP4 with the endosomal membrane (Flint et al., 1999). The viral RNA now becomes available for translation and replication (Figure 3.13, step 4a). However, only a small fraction of the viruses in the endosomes undergo successful uncoating. The majority of the virus particles in the endosomes, however,

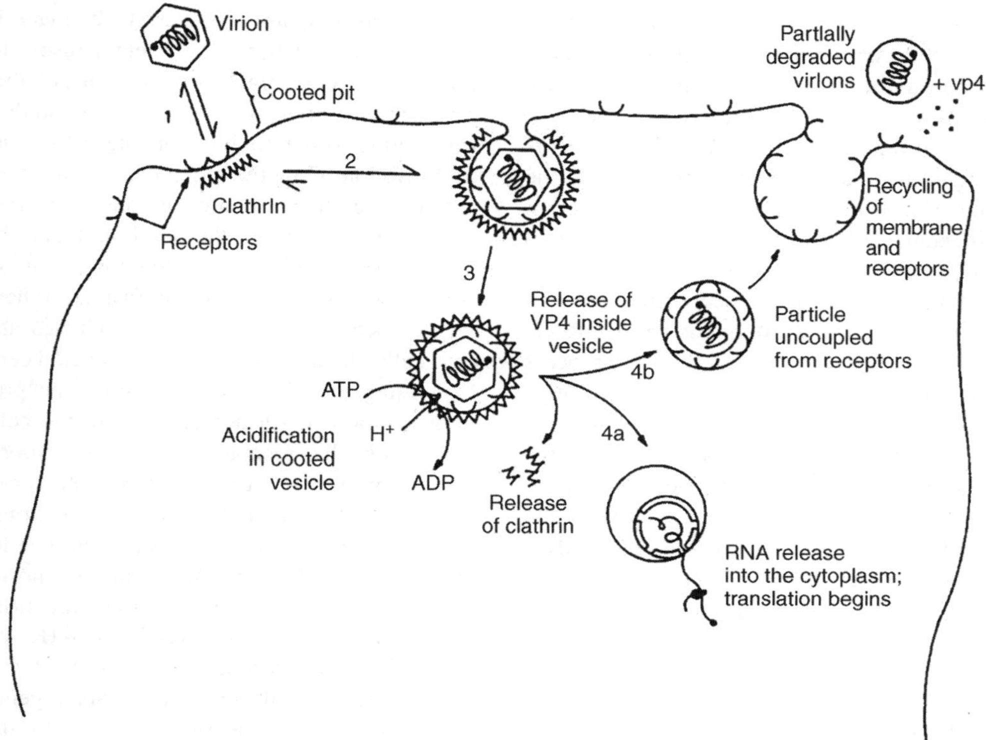

FIGURE 3.13 Receptor-mediated endocytosis of viruses such as polioviruses (steps 1 through 4a,b). The virus binds to cell surface receptors, usually glycoproteins, that undergo clustering at clathrin-coated pits (step 1) and is followed by invagination (step 2) and internalization (endocytosis) to form clathrin-coated vesicles (step 3). Acidification inside the coated vesicles, brought about by an energy-requiring ATPase-coupled proton pump, triggers the release of VP4 and unfolding of hydrophobic polypeptide patches previously buried inside the viral capsid. Fusion of the lipid bilayer with hydrophobic patches in the acid-unfolded capsid protein presumably triggers release and transfer of RNA from virion into the cytosol (step 4a), where ribosomes can begin translating the plus-strand viral genome. Fusion of uncoated vesicles with other kinds of intracellular lysosome-like vesicles may also be involved in the uncoating process. Some virus particles are not fully uncoated after acid-induced changes in the endosomes and are released into the extracellular medium via an abortive pathway (step 4b). These partially degraded extracellular virus particles are noninfectious. Reprinted with permission from Fields and Knipe (1990).

become noninfectious due to acid-induced structural alteration and are released outside the cell by the abortive pathway (Figure 3.13, step 4b). SV40 virus, a naked DNA virus, also enters into the cytoplasm via receptor-mediated endocytosis. Some alteration in the SV40 virion structure occurs in the endosome as VP3, a viral capsid protein, becomes exposed. However, in the case of SV40, the virus is extruded essentially intact from the endosome into the cytoplasm and targeted to the nucleus. Therefore, the uncoating of the SV40 genome occurs in the nucleus and not in the cytoplasm. In the nucleus, the viral minichromosome (viral DNA containing associated histone proteins) is released from the capsid and becomes available for transcription and replication. Therefore, although entry of SV40 virus into the cell likewise occurs via an endosome, its uncoating takes place within the nucleus in a pH-independent manner. However, how the SV40 virus is released from the endosome into the cytoplasm prior to nuclear entry remains unclear. Reovirus, a double-stranded naked RNA virus, uses the host proteolytic enzymes present in the lysosome to partially remove the outer capsid proteins and activate the core RNA transcriptase for initiation of viral mRNA synthesis.

For enveloped viruses, uncoating occurs through fusion of the viral membrane with the cellular membrane using pH-independent or pH-dependent pathways. As mentioned previously, in the pH-independent pathway, virus penetration and uncoating occur simultaneously at the cell surface after virus–host interaction. This is best illustrated by the entry process of paramyxoviruses and retroviruses (e.g., HIV). In both cases, viruses bind to the cell surface receptors (i.e., sialic acid present on the cell surface glycolipids or glycoproteins for paramyxoviruses and the receptor protein CD4 and coreceptors for HIV). Either one (gp160 for HIV) or two (F and HN for paramyxovirus) separate viral glycoproteins are involved in this binding and fusion process. Fusion-inducing proteins in the infecting virus must be cleaved for causing fusion to occur (e.g., gp160 → gp120 and gp41 for HIV and F → F1 and F2 for Sendai virus, the latter belonging to the family Paramyxoviridae, genus *Respirovirus*). For HIV, the gp120/gp41 complex undergoes conformational changes after binding to the cellular receptor and coreceptor, releasing the hydrophobic domain of gp41 that then functions as a fusion peptide and mediates fusion of the viral membrane with the plasma membrane, thereby releasing the nucleocapsid containing the viral RNA and reverse transcriptase into the cytoplasm. Subsequently, cyclophilin A, present in HIV particles, aids in the uncoating process by destabilizing the capsid and initiating reverse transcription of the viral RNA. For paramyxoviruses, HN protein binds to the sialic acid on the cell surface receptor and induces, in some way, conformational changes in the other viral envelope protein, known as the F1/F2 complex, thereby facilitating the fusion domain of F1 to cause fusion between the viral membrane and the plasma membrane and release of the viral nucleocapsid containing the RNA-dependant RNA transcriptase (RDRP) into the cytoplasm. For paramyxovirus, the entire viral replication process takes place in the cytoplasm, whereas for retroviruses the proviral DNA is formed in the cytoplasm after reverse transcription of the viral RNA and then transported to the nucleus for integration and transcription. How the receptor–protein interaction facilitates conformational changes leading to fusion of the viral and cellular membranes in a pH-independent manner is not fully understood. Furthermore, fusion for these viruses occurs not only between viruses and host cells but also between virus-infected cells expressing the cleaved viral membrane proteins on the cell surface and uninfected cells containing the receptors (and coreceptors) present on the cell surface. These cell to cell interactions lead to the formation of syncytium or multinucleated giant cells. Such multinucleated giant cells are important diagnostic markers for a number of viral infections (e.g., human respiratory syncytial virus (RSV), family Paramyxoviridae, genus *Pneumovirus*; mumps virus, family Paramyxoviridae, genus *Rubulavirus*; and measles viruses, family Paramyxoviridae, genus *Morbillivirus*). The process of fusion of HIV-infected cells with uninfected CD4+ T cells is implicated in the pathogenesis of AIDS, which causes depletion of CD4+ T cells in HIV-infected people.

For other enveloped viruses such as VSV and influenza viruses, penetration and uncoating are two separate events. Following receptor binding, these viruses enter the cytoplasm by receptor-mediated endocytosis, and fusion and uncoating occur within the endosome in a pH-dependent (low pH of ~5) manner. The fusion and uncoating of these viruses can be blocked by agents such as monensin, which increases endosomal pH. For VSV, the G protein binds to the receptor and becomes activated for fusion at low pH, even though it remains uncleaved. Although the VSV G protein contains a hydrophobic fusion region, the mechanism of its fusion process within the endosome is not well understood. The fusion and uncoating processes are best understood at the molecular level for influenza viruses. Again, for influenza viruses, although fusion and uncoating occur simultaneously, they are considered two separate events. Following binding to sialic acid on the cell surface receptor, influenza virus undergoes receptor-mediated endocytosis and the cleaved HA trimer (i.e., HA1/HA2 heterotrimer complex) present on the viral membrane undergoes conformational changes at the low pH (~5) of endosomes. Acidic pH specifically alters the structure of HA2, which attains the fusiogenic state (Figure 3.14). In conjunction with this process, HA1 is dissociated from the stem of the HA spike and the fusion peptide present at the NH_2 terminus of HA2, which normally remains buried in the protein interior of the HA trimer, is released and the polypeptide structural loop is transformed into a helix to form an extended coiled-coil structure that relocates and thrusts the boomerang-shaped hydrophobic fusion peptide toward and into the target (endosomal) membrane (Figure 3.14). This process first leads to hemifusion by mixing of the outer lipids of the bilayers of both viral and endosomal membranes and then to complete fusion of both lipid bilayers of the membranes, leading to the formation of a pore between the two compartments. Subsequently, the pore dilates leading to mixing of the cytosol and virion contents and delivery of the viral nucleocapsid into the cytoplasm (Figure 3.14). In addition to causing fusion, low pH also aids in the uncoating of the influenza virus nucleocapsid. Uncoating (Figure 3.11b) in this case is defined as the separation of a nucleocapsid from the virus matrix protein (M1). Therefore, with this type of virus, low pH (–5) is not only crucial to the outside of the virus particle (virion) for inducing conformational changes of HA1 and HA2 but is also needed inside the virus particle for separation of M1 from the nucleocapsid. Acidification of the virion interior is carried out by a viral protein called M2. A small number of M2 tetramers (16–20 per virus particle) are formed by the type III transmembrane M2 protein present on the viral membrane. These M2 tetramers constitute ion channels that remain closed at neutral pH and open at low pH (~5) to allow protons (H^+) to enter from the endosomes into the core of the virus particle. The resulting acidic pH inside virus particles causes dissociation of M1 from the viral RNP (also known as the vRNP or nucleocapsid) containing vRNA (minus strand), and so the M1-free viral RNP is released into the cytoplasm (Figure 3.11b). Both the opening of the M2 ion channel and the uncoating of some members of the species influenza A virus can be blocked by amantadine (or rimantadine), a drug currently used to treat influenza infection. The dissociation of M1 from the vRNP is important since the released vRNP can now be translocated into the host nucleus, where the transcription and replication of vRNA can occur. M1, on the other hand, interferes with the transport of vRNP into the nucleus and also inhibits the vRNP transcription. However, mutation(s) in the M2 channel can make the mutant virus resistant to amantadine (or rimantadine). Many of the epidemic/pandemic viruses have become resistant to amantadine (or rimantadine).

3.4.3 Targeting Viral Nucleocapsids to the Replication Site

Viral replication occurs either in the nucleus or in the cytoplasm of infected cells. For those viruses that replicate in the cytoplasm, which customarily are those with RNA genomes, except for the DNA-containing poxviruses, the uncoating process releases the viral nucleocapsid directly into the cytoplasm, which is the site of transcription and replication. For viruses that replicate within the nucleus, which tend to be the ones having DNA genomes with notable exceptions such as the RNA-containing influenza viruses and retroviruses, the nucleocapsids of these viruses, released in the cytoplasm after uncoating, must be targeted into the nucleus. Nuclear targeting requires that these viral nucleocapsids contain proteins possessing nuclear targeting signals (NTSs) or nuclear localizing signals (NLSs) that are recognized by the cellular nuclear targeting machinery and translocated into the cell nucleus via nuclear pores. However, the stage of uncoating at which nuclear targeting takes place varies with

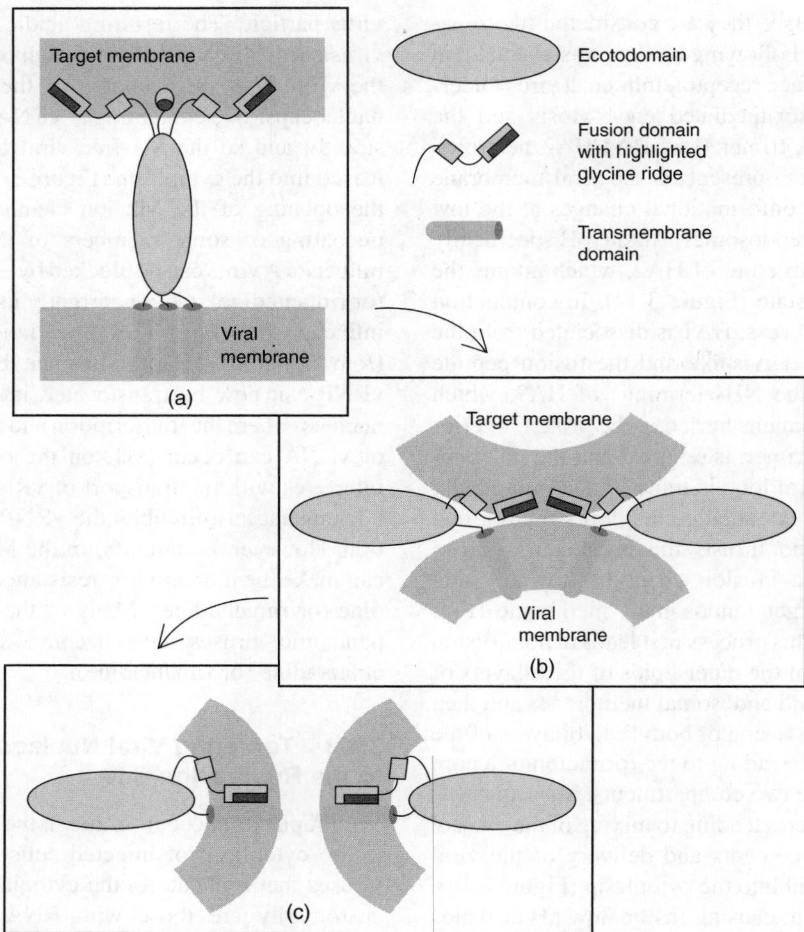

FIGURE 3.14 Boomerang model of influenza virus HA-mediated membrane fusion. (a) Cleaved influenza virus HA (HA0 into HA1 and HA2) undergoes pH-induced conformational change in the endosome and thrusts the boomerang-shaped fusion peptide toward the target cellular membrane where it inserts. (b) The ectodomain tilts to the plane of membranes. The boomerangs retrieve the target membrane and bring it to the close juxtaposition with the viral membrane such that lipid exchange forming hemifusion can occur. In this state, lipids of the outer leaflets, but not the inner leaflets, mix. At the point of hemifusion, the aqueous contents of the two vesicles still remain separated. (c) Eventually the fusion peptides and the transmembrane domains interact by virtue of the glycine edge fusion peptide, causing complete fusion of both lipid bilayers and leading to formation of the initial fusion pore opening. Multiple HA trimers are required for causing and opening the fusion pore. After opening of the initial narrow fusion pore, the pore dilates (not shown) releasing the viral nucleocapsid into the cytoplasm of the infected cell. Reprinted with permission from Tamm (2003). (*See the color version of this figure in Color Plate section.*)

viruses. For SV40, essentially the entire virus particle taken into the cytoplasm is transported into the nucleus, and it is only in the nucleus that uncoating of the capsid occurs concomitant with release of the viral minichromosome. For adenoviruses, uncoating occurs at the nuclear pore where the viral nucleocapsid docks and the viral DNA is delivered into the nucleus through the nuclear pore. For influenza viruses, uncoating occurs by dissociation of M1 from the vRNP during introduction of the nucleocapsid into the cytoplasm. This M1-free vRNP is then transported into the nucleus. For retroviruses, not only uncoating but also additional biosynthetic processes—including reverse transcription of the RNA genome and synthesis of the double-stranded proviral DNA—occur in the cytoplasm. Then the retroviral DNA along with integrase is translocated into the nucleus for integration of the proviral DNA into the host genome. Transcription of the retroviral genomic and subgenomic mRNAs occurs only from the integrated proviral DNA in the nucleus. For hepatitis B virus, the partially double-stranded DNA, the viral genome following uncoating in the cytoplasm, becomes fully double-stranded and circularized in the cytoplasm and then it is translocated into the nucleus for subsequent transcription of genomic and subgenomic mRNAs.

3.4.4 Postuncoating Events

The "immediate events" in the viral replication cycle, those that occur following uncoating, vary with the nature of the viral genome. For plus-strand RNA viruses except retroviruses, translation of the viral RNA follows immediately after uncoating. The viral RNA extruded from the capsid is then used by the host translation machinery for directing protein synthesis (Figure 3.13). For all other viruses, whether of DNA or RNA genome, the step immediately following uncoating is either transcription of the genome yielding functional mRNAs or reverse transcription of vRNA yielding proviral DNA (retroviruses).

3.4.5 Transcription of Viral Genes

From the transcription viewpoint, viruses can be classified into two major categories, that is, whether they possess a DNA genome or an RNA genome. Of the first group, the DNA genome of different viruses varies greatly in complexity between virus families, encoding from only 4–5 genes (polyomaviruses, family Polyomaviridae) to more than 200 genes (poxviruses) or open reading frames (ORFs). DNA viruses use DNA-dependent RNA polymerase that can be either virus specified (e.g., poxviral RNA polymerase) or host specified (e.g., RNA *pol* II) to generate their mRNAs. RNA viruses, however, must use RDRP, which is always virus specified and is therefore different and specific for each virus group.

3.4.5.1 Transcription of DNA Viruses

All DNA viruses except the poxviruses transcribe and replicate their genomic material in the host cell nucleus. Poxviruses transcribe and replicate in the cytoplasm. In addition, all DNA viruses except poxviruses use host *pol* II for transcription of their DNA into mRNAs. Poxviruses use virus-specific RNA polymerase for transcription of their genome. Viral DNA genomes as host DNA often possess the *cis*-acting elements, which are essential for successful transcription of their DNA. These DNA elements are called the viral promoter and enhancer. The promoter is the RNA polymerase binding site on viral DNA (e.g., TATA box, CAT box, GC box) localized in the vicinity (usually upstream) of the transcription initiation point. The enhancer element, which enhances transcription of the viral mRNA over the basal level, is found in the proximal or distal region of the promoter and may be located upstream or downstream of the promoter element. Transcription of the viral DNA genome can broadly be divided into the early and late phases. Early genes are usually catalytic and regulatory in nature, involved in regulating transcription of mRNAs and replication of viral DNA. Late genes usually produce mRNAs for structural viral proteins, which are the major components of the viral capsid or envelope. Early genes are usually transcribed prior to the initiation of viral DNA synthesis, and late genes are transcribed only after viral DNA synthesis is initiated. Thus, synthesis of the progeny viral DNA demarcates the dividing line between early and late gene transcription. However, for complex viruses such as HSV (human herpesvirus 1 and 2 of the family Herpesviridae, genus *Simplexvirus*), the different classes of regulatory genes, for example, immediate early (α), delayed early (β), and late

(γ1, γ2) are transcribed at different phases of the viral replication cycle, each having different regulatory functions for turning on or shutting off other viral genes. Viral genes can be transcribed from either of the two DNA strands, with the coding sequences thus running in a direction opposite to a duplex DNA. These viral genes usually possess the structural features of eukaryotic cellular genes, and the viral mRNAs similarly undergo posttranscriptional processing similar to cellular genes. The majority of viral mRNAs, like host mRNAs, are usually capped at the 5′-end, polyadenylated at the 3′-end, and may undergo posttranscriptional splicing prior to their exit from the host cell nucleus. However, poxviral mRNAs, which are also capped and polyadenylated, do not undergo splicing since they are made in the cytoplasm.

An example of transcription of a small double-stranded viral DNA genome (SV40) is shown in Figure 3.15. Transcription of the SV40 genome is carried out by the host cell's RNA polymerase II. Early mRNAs (large T and small T) are transcribed from the early promoter of the early DNA strand, whereas late mRNAs (i.e., the mRNAs for VP1, VP2, VP3, and agno proteins) are transcribed from the late promoter and the opposite DNA strand. Both early and late transcription in SV40 is initiated from the common control region in opposite directions at different phases of the replication cycle. This control region also regulates SV40 viral DNA replication. This region consists of a series of repeat elements with different functions: three 21-base repeats that together contain six copies of GC-containing hexamers serve as the promoter for early transcription. Downstream of these repeats is a TATA box and upstream are two 72-base repeats constituting the enhancer element (Figure 3.15, bottom). These three regulatory elements bind specific cellular factors and are important in regulating early transcription. Of these, the 21-bp repeats and the enhancer elements are also important in regulating late transcription. The switch from early to late transcription is brought about by binding of large T antigen to specific sites in the control region and a change in the replicative state of the viral DNA. Large T (LT) and small T (ST) antigens are two early proteins translated from two different mRNAs produced by differential splicing. The two late mRNAs have a common untranslated region and a common poly(A) addition site but are generated by differential splicing. Each of these late SV40 mRNAs is bicistronic, with alternative initiation codons. One of these late mRNAs is translated into VP2 and VP3, and the other into VP1 and the agno protein.

On entry into the cytoplasm of the infected cell, HBV, a partially double-stranded DNA virus, uses virus-specific reverse transcriptase (P) to synthesize the complete circular DNA that is then transported into the nucleus. Host cell *pol* II subsequently transcribes its genomic and subgenomic mRNAs from different initiation points (Figure 3.16). They are all capped at the 5′-end, unspliced, and have a common termination and poly(A) addition site at the 3′-end. Different classes of genomic-length (3.5 kb) RNAs, possessing different 5′ but common 3′ termini, function as a template for making cDNA or are translated into both the Pre-C and C proteins and the P protein. Subgenomic mRNAs are translated into the Pre-S1, Pre-S2, and S proteins as well as the X protein (Figure 3.16).

3.4.5.2 Transcription of RNA Viruses

Among the different families of RNA viruses, the RNA viral genome appears to be much less complex compared to the genomes of the highly complex DNA viruses. However, these RNA viruses use multiple strategies to encode different mRNAs and proteins. Unlike DNA viruses, the majority of the RNA viruses (except for retro-, orthomyxo-, and related viruses) replicate in the cytoplasm, so that their mRNAs cannot undergo RNA splicing. RNA viruses also possess genes both for regulatory and catalytic proteins and for structural proteins. However, transcription of mRNAs encoding these proteins is not as strictly demarcated with respect to the timing of their genomic nucleic acid replication as is found for DNA viruses. On the other hand, with RNA viruses, there is a

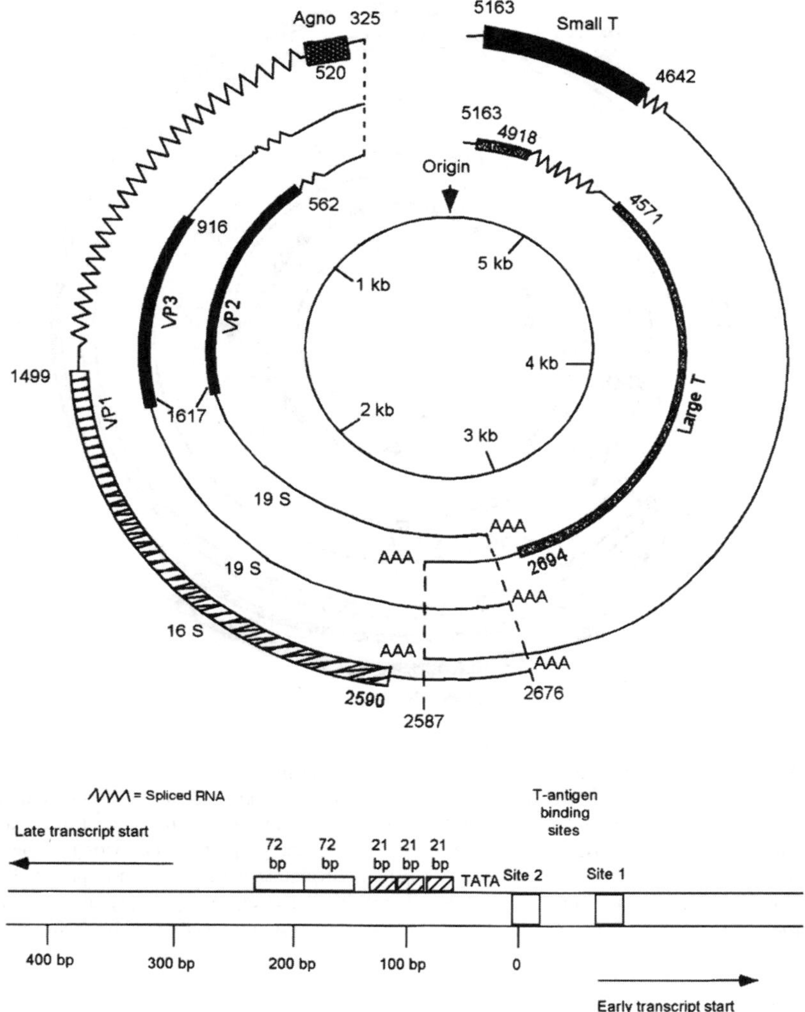

FIGURE 3.15 Genome and transcription map of SV40 (top). The origin of replication is shown at the top of the inner circle. The numbers indicate the nucleotide position in the SV40 DNA, while zigzag markings indicate spliced introns. Different shaded regions indicate different protein-coding sequences. The bottom drawing shows the details of the transcription regulatory elements in the proximity of the "origin" region and the direction of the early and late transcription.

great deal of variation at the level of transcription of different viral mRNAs. The mRNAs of the major structural proteins—such as the nucleoprotein (NP), matrix (M), and glycoproteins—are usually made in larger amounts compared to the lower amount of mRNAs synthesized for catalytic (e.g., polymerases) proteins. For nonsegmented negative-strand RNA viruses, the level of mRNA transcription is regulated by the promoter-proximal position of a gene (e.g., for VSV or paramyxoviruses, see Figure 3.8) or by an internal promoter that produces subgenomic mRNAs (e.g., togaviruses). The strategy used by different RNA viruses for mRNA transcription depends on the nature of the RNA genome (+ or − strand, segmented or nonsegmented) and whether the nucleocapsid is icosahedral or helical.

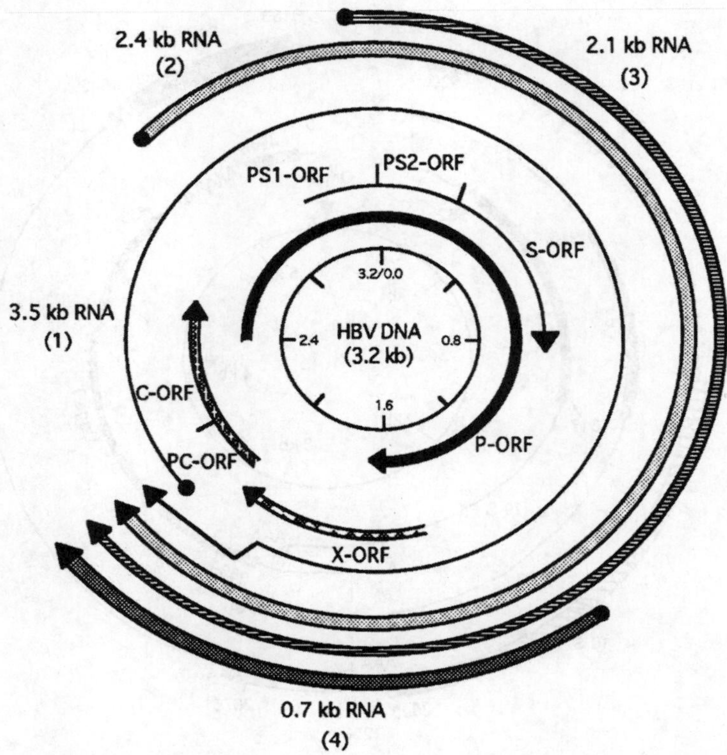

FIGURE 3.16 Replication, transcription, and translation of hepatitis B virus (family Hepadnaviridae, genus *Orthohepadnavirus*) DNA. Four RNA classes: 3.5 kb (1), 2.4 kb (2), 2.1 kb (3), and 0.7 kb (4) are transcribed. The 3.5 kb product (#1) is used for full-length DNA (minus-strand) synthesis. Different classes of 3.5 kb product also function as mRNAs whose translation products are HBcAg, the polypeptide consisting of the PC-ORF (precore), C-ORF (core) and P-ORF (P protein, also called polymerase or reverse transcriptase). The 2.4 kb mRNA (#2) makes a large protein consisting of the polypeptides PS1-ORF, PS2-ORF (presurface), and S-ORF (surface protein). The 2.1 kb mRNA (#3) makes the S-ORF (surface) protein, and the 0.7 kb mRNA (#4) encodes the X-ORF protein.

For plus-strand naked icosahedral RNA viruses (e.g., poliovirus), the entire viral genomic RNA functions as the only mRNA and is translated from one ORF into a large polyprotein that is then cleaved by specific proteases into different functional proteins representing the RNA polymerase and the capsid proteins (VP1, VP2, VP3, VP4) and so on (Figure 3.7).

For some enveloped plus-strand RNA viruses (e.g., togaviruses), the 5′-half of the viral genomic RNA encodes and is translated into nonstructural (catalytic) proteins involved in RNA transcription and replication, whereas a separate subgenomic 26S mRNA (+), made from an internal promoter on the minus-strand RNA template, encodes the structural proteins (i.e., capsid and envelope proteins). This 26S mRNA is synthesized in larger quantities than is the genomic-length RNA. However, another group (flaviviruses, family Flaviviridae) of enveloped plus-strand RNA viruses possesses one large ORF in its genomic RNA encoding a single large polyprotein that, as is the case with picornaviruses, is cleaved into specific proteins by a virus-encoded proteinase.

For coronaviruses, which contain a large plus-strand RNA genome of ~30 kb, multiple subgenomic mRNAs are found. However, each of these mRNAs possesses the same 5′-leader (i.e., leader-primed transcription) and the common 3′-end containing poly(A) sequences. These mRNAs therefore contain the nucleotide sequence of more than one ORF. Usually, however, only the first ORF at the 5′-end of mRNA is translated into protein.

Minus-strand RNA (–) viruses replicating in the cytoplasm may possess either one large genomic RNA molecule (nonsegmented) or two or more different subgenomic RNAs (segmented). For those viruses that possess one nonsegmented genomic RNA molecule (e.g., VSV), the viral genes are arranged sequentially in the genomic RNA (–) with stop, intergenic, and start (EIS) sequences (Figure 3.8). The viral RNA polymerase (RDRP) synthesizes the virus mRNAs by initiating transcription at the 3′-end (one entry) and then terminates at the stop sequence (E) of that gene, skips the intergenic sequence (I), and initiates at the start (S) sequence of the next gene, and so on. Therefore, the viral RNA polymerase sequentially transcribes the downstream genes and there is no independent internal entry of the RNA polymerase on the viral genome. Since RNA polymerase randomly falls off during transcription and cannot initiate *de novo* internally, the mRNA level (and, consequently, the protein level) is determined by the location of a particular gene in the viral genome. For example, mRNA of the capsid protein (N or NP) is present at the extreme 3′-end of the minus strand (i.e., proximal to the promoter) just after the leader (ℓ) sequence, and it is therefore made in the most abundant amount because it is the first gene to be transcribed by RDRP into mRNA. On the other hand, the L (polymerase) gene, encompassing nearly half of the genome, is located at the 5′-end of the viral RNA (distal to the promoter), so the L mRNAs and L proteins are made in the least amount (Figure 3.8). Each mRNA is capped at the 5′-end and polyadenylated at the 3′-end by the virally encoded RDRP.

Orthomyxoviruses, which are segmented minus-strand RNA viruses, possess 8 (genera *Influenzavirus* A and B) or 7 (genus *Influenzavirus* C) RNA segments that in total encode 10 mRNAs and 11 proteins for type A and B viruses. Orthomyxoviruses are transcribed and replicated in the nucleus. Orthomyxoviruses use a unique strategy to initiate transcription. They cannot initiate *de novo* mRNA transcription without a primer and must use the host's capped RNA as the primer at the 5′-end for mRNA transcription. One of the three proteins (PB2) of the viral polymerase complex (PB1/PB2/PA) recognizes the newly synthesized capped host RNA and PA possessing the endo nuclease activity cleaves it around 12–15 nucleotides from its 5′-end. Then another protein (PB1) of the polymerase complex uses the capped primer for viral mRNA initiation and chain elongation. Therefore, each influenza viral mRNA possesses at its 5′-end a capped nonviral RNA sequence acquired from the host nuclear RNA (Figure 3.19). In addition, two viral RNA segments (segments #7 and #8) generate both unspliced and spliced mRNAs, causing translational shift to a different reading frame. Furthermore, another small (87 amino acids) protein PB1-F2 is translated from an alternative reading frame of PB1 mRNA. In this process, eight influenza viral RNA segments of type A and B viruses give rise to 10 mRNAs and 11 proteins.

Segmented ambisense RNA viruses (e.g., arenaviruses) on infection produce a subgenomic mRNA using the 3′-end of the genomic RNA as the template, and later on in the infectious cycle use the antigenomic RNA as the template to generate the mRNA with the same polarity as the 5′-end of genomic RNA.

Viruses that possess double-stranded (ds) RNA viral genomes, such as reoviruses, are segmented and replicate in the cytoplasm. Their viral transcriptase, which is also present within the virus particles, synthesizes single monocistronic mRNAs from each dsRNA segment.

Retroviruses, although possessing a plus-strand RNA genome, contain reverse transcriptase in the virion. Transcription of retroviral mRNAs occurs in the nucleus from the integrated proviral DNA template by the host RNA *pol* II. Usually, both the unspliced genomic-length mRNA and the subgenomic mRNA, the latter being produced by splicing in the nucleus, function in protein translation.

3.4.6 Translation

Virions have evolved to become very efficient organisms that package a relatively small amount of genomic information as DNA or RNA in their capsids but use this information efficiently to generate the maximum number of functional proteins required to produce infectious progeny virions and cause the disease syndrome. For some viruses like VSV, all the viral proteins encoded by the genome and produced in the infected cells including the transcriptase are incorporated into the virion and become structural components of virus particles. For these viruses, there are by definition no nonstructural proteins; that is, there are no proteins that are encoded in the virion genome and produced in the infected cells but not incorporated into the virion. However, for the majority of viruses, one or more nonstructural proteins, either catalytic (enzymatic) or regulatory, are synthesized in virus-infected cells. These nonstructural proteins are required for the infectious cycle but are not incorporated into virion particles. Both structural and nonstructural proteins are translated from viral mRNAs, and the majority of viral mRNAs (except in the case of picornaviruses) possess structural features similar to that of the host mRNA (i.e., they possess a cap at the 5′-end, a translation initiation triplet (AUG) in the context of Kozak's rule, and translation termination triplets and poly(A) sequences at the 3′-end). These viral mRNAs undergo cap-dependent ribosome binding and ribosome scanning to locate the proper initiation triplet, a process that does not provide any advantage over the host mRNAs during translation. Therefore, after infection, the virus must overcome two major problems to achieve successful replication: (1) viruses must somehow overcome competition from host mRNAs for using translation machineries, and (2) viruses that possess only a limited amount of coding information must still be able to generate the considerable number of functional proteins needed for replication.

Viruses have developed a number of strategies to compete with host mRNAs for efficiently using the host translation machinery. These include the following: (a) Viral transcription machinery (especially in RNA viruses) are more efficient in generating high levels of mRNAs, so that they can outcompete host mRNAs in translation. (b) Some viral proteins target and interfere with the host transcription machinery, so that the host transcription level goes down or shuts off. Influenza viruses, however, use a novel system to their advantage. As mentioned previously, one of the influenza polymerase proteins, PB2, recognizes, binds to, and cleaves the newly synthesized capped host hnRNAs (heterogeneous nuclear RNAs) around 13–15 nucleotides, and the capped oligonucleotide is used as the primer for mRNA synthesis. The cleavage of host hnRNAs, in turn, prevents host mRNA synthesis and processing. In addition, this virus interferes with nuclear export of the host mRNAs. (c) Some viruses modify the host translation machinery to use that machinery for their advantage, while simultaneously shutting off host mRNA translation. This latter mechanistic approach is particularly evident for picornaviruses, which inactivate the cap binding protein and modify the host translational factors (e.g., eIF2, eIF3/4B) and thus shut off cap-dependent host mRNA translation. However, picornaviral mRNA can still be translated efficiently because it does not have a cap at the 5′-end but rather possesses a unique RNA secondary structure known as an internal ribosome entry site (IRES) and is independent of Kozak's rule (Kozak, 1986). Kozak's rule states that most eukaryotic mRNAs contain a short recognition sequence (ACCATGG) that facilitates binding of mRNA to the small subunit of the ribosome for initiation of protein translation (Kozak, 1986). The picornaviral mRNAs possessing an IRES can be translated efficiently in a cap-independent manner, while capped host mRNAs cannot be translated because of viral-mediated inactivation of some of the host translational factors.

Viruses have developed different strategies to produce a relatively large number of functional proteins from a small amount of genetic information using both transcriptional (or posttranscriptional) and translational (or posttranslational) processing.

3.4.6.1 Transcriptional (or Posttranscriptional) Generation of Different mRNAs Double-stranded DNA viruses can use both of their DNA strands to transcribe mRNAs, thereby increasing potential transfer of information into proteins. Some viruses that make mRNAs in the nucleus (RNA or DNA viruses) can generate different mRNAs from the same genomic strand by using unspliced mRNA or electing alternative splicing sites, thus even causing frameshifts in the subsequent translation. Influenza viral proteins M1, M2, NS1, NS2, and SV40 (such as VP1, VP2, and large T and small T antigens) are classic examples of generating different mRNAs and proteins through splicing. Some viruses use RNA editing (i.e., nontemplated nucleotide addition in the mRNA) to shift the translation frame. This latter technique is frequently used by paramyxoviruses to generate their V and C proteins. Hepatitis delta virus (family unassigned, genus *Deltavirus*) uses adenosine deaminase for RNA editing as part of the transcription process to generate its δ Ag-L antigen. Other viruses selectively use different promoters to generate genomic and subgenomic mRNAs (e.g., HBV, togaviruses). Also, as mentioned earlier, influenza virus PB1 mRNA often uses an alternative reading frame to produce PB1-F2 protein.

3.4.6.2 Translational (and Posttranslational) Generation of Different Viral Proteins The most common way to generate a number of functional proteins after translation is by proteolytic cleavage. These endoproteases, usually encoded by the virus, are sequence specific and can generate a number of functional proteins from one large viral polypeptide. Classic examples of this type of cleavage activity are found with poliovirus (picornavirus) and flavivirus proteins. Poliovirus RNA is translated into a large polypeptide that sequentially undergoes endoproteolytic cleavage by different poliovirus proteases at specific amino acid sites, generating 11 viral proteins (VP4, VP2, VP3, VP1, 2A, 2B, 2C, 3A, VPg, 3C, 3D) and other intermediate proteins (Figure 3.7). The importance of virus-specific proteases has been demonstrated in HIV infection, during which HIV protease inhibitors alone or in combination with RT inhibitors can be used in the treatment of AIDS to reduce virus load of the patient. Cleavage by host proteases is also sometimes critical to render viral proteins functional and viral particles infectious (e.g., conversion of influenza viral HA to HA1 and HA2, HIV gp160 to gp120 and gp41).

Different initiation codons are also used in bicistronic mRNAs to translate different proteins. Depending on the initiation codon used, either one or the other protein can be translated (e.g., NB protein and NA protein from the same mRNA in *Influenzavirus B* and PB1 and PB1-F2 in *Influenzavirus A*). Usually, one of the initiation codons is favored, thus regulating the levels of the two proteins produced from one bicistronic messenger RNA. Another strategy, often used by retroviruses, is translational frameshift or translational suppression of termination codons. Translational frameshift owing to ribosomal slippage causes generation of the *gag*-pro-*pol* fusion protein in avian leukosis virus (family Retroviridae, genus *Alpharetrovirus*). This protein is then cleaved by a virus-specific protease (usually aspartic proteases) to generate individual functional proteins. Similarly, some retroviruses use translational termination suppression to continue translation in the same reading frame. In the *gag*-UAG-*pol* sequence, translation is normally terminated after the *gag* protein at the UAG codon. Occasionally, termination at UAG can be suppressed by a minor host tRNA capable of inserting glutamine and thereby generating a *gag*–*pol* fusion protein that subsequently is cleaved by a viral protease to generate *gag* and *pol* proteins. Again, both frameshift and in-frame suppression produce only a minority of fusion proteins with *pol*, thus regulating the amount of *pol* protein needed in small amounts in virus-infected cells.

3.4.7 Replication of Viral Genome

The replication pathway of different viral genomes varies depending on the nature of the viral genome. The overall strategy of viral genome replication can be grouped into seven pathways depending on the nature of the genome (Figure 3.17). While all DNA viruses of eukaryotes except poxviruses replicate in the nucleus of their host cells, some use cellular DNA polymerase and others use DNA polymerase encoded by the virus genome (Table 3.2). Poxviruses replicate their genome in the

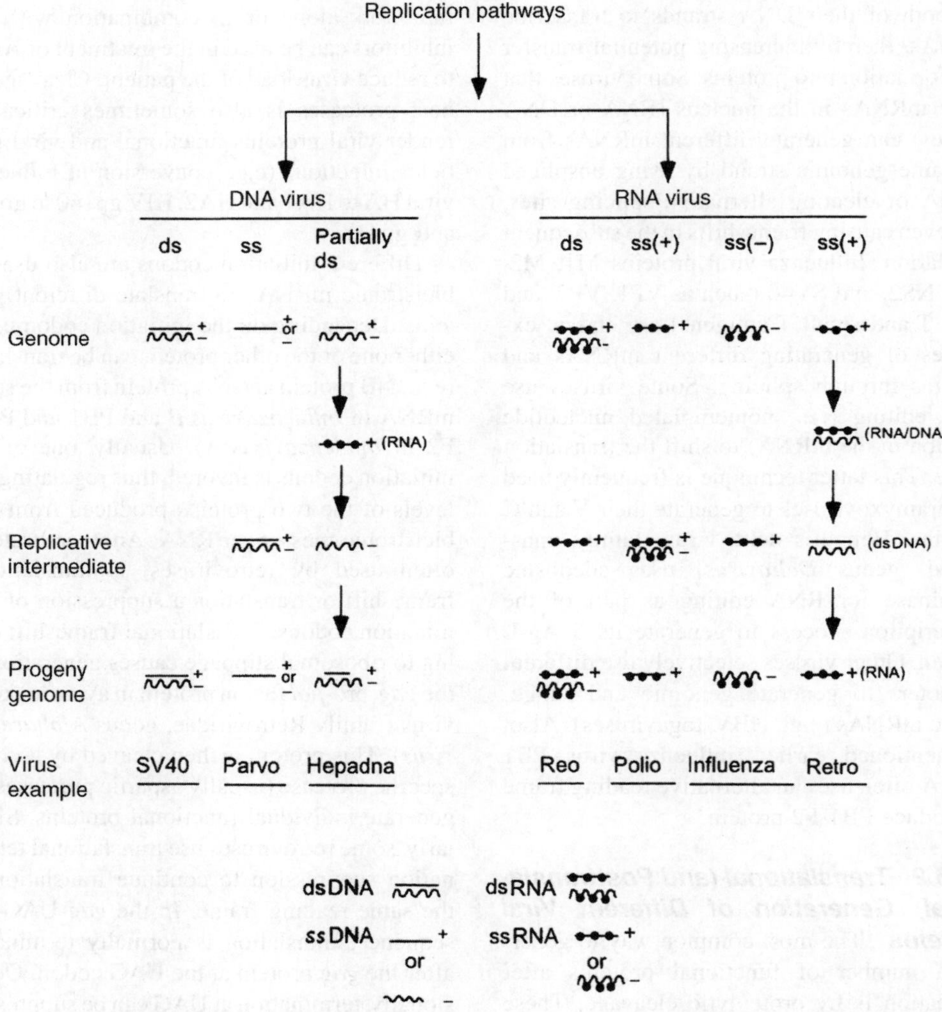

FIGURE 3.17 Seven replication pathways of the DNA and RNA genomes of viruses. Examples of different viruses with DNA or RNA genomes are indicated as ds, double-stranded; ss, single-stranded; and (+) and (–), positive and negative polarity, respectively. *Note*: the name "Polio" is now considered to represent antigenic variants of the species human enterovirus C belonging to the genus *Enterovirus*; the name "Parvo" refers to members of the family Parvoviridae and the name "Papova" refers to members of the families Papillomaviridae and Polyomaviridae.

cytoplasm and use polymerase encoded by the viral genome. All RNA viruses except retroviruses use an RDRP encoded by their own genome (Table 3.3). Some of these (minus-stranded RNA viruses) carry RDRP in the virus particle to initiate transcription/replication of viral RNA following their entry and uncoating inside the cell. Retroviruses require reverse transcriptase, an RNA-dependent DNA polymerase, in the virion particle to initiate replication. The majority of RNA viruses of eukaryotes replicate in the cytoplasm, except the orthomyxo- and related viruses and the retroviruses. Orthomyxoviruses require cellular capped 5′-RNAs as primers for mRNA transcription, and retroviruses require production of proviral DNA and its integration into the host DNA as a prelude to both transcription and replication of the viral genome.

3.4.7.1 Replication of DNA Genome

Smaller DNA viruses including the papillomaviruses, members of the family Papillomaviridae; polyomaviruses, members of the family Polyomaviridae; and parvoviruses, members of the family Parvoviridae, rely on the host cell DNA polymerase, whereas more complex DNA viruses use their own virus-encoded DNA polymerase (Table 3.2). The step for switching from transcription to replication of DNA viral genomes is primarily determined by the level of early viral proteins, which often are both regulatory and catalytic in nature. For SV40, when a sufficient amount of large T antigen is synthesized, binding of the LT antigen initiation to the transcription start site of early mRNAs (Figure 3.15) causes suppression of early mRNA transcription. The helicase activity of the virus LT antigen then unwinds the DNA molecule, creating a replication bubble, whereupon the host DNA primase–polymerase complex initiates DNA synthesis using an RNA primer, creating a replication fork. Synthesis of the SV40 DNA continues bidirectionally, creating circular intermediates (Figure 3.18c). Adenoviruses use asymmetric DNA replication, which initiates DNA synthesis at the 3′-end of one strand (template strand). At the 5′-end of that strand, a 55 kDa protein, covalently linked to the DNA, is needed for initiation of DNA replication. The new growing opposite DNA strand then displaces the preexisting opposite strand. The displaced strand forms a panhandle structure by pairing the inverted terminal repeats before its own replication begins (Figure 3.18a). In poxvirus DNA, two complementary forms are joined at the terminal repeat sections forming palindromes. During replication, concatamers of two genomic-length strands are formed. Unit length genomic molecules are then formed by separating the staggered ends and ligation (Figure 3.18e). Linear herpesvirus DNA becomes circularized inside the host cell nucleus and then replicates as a rolling circle, forming tandem concatamers. Finally, the unit length genomic DNA molecules are excised from concatamers (Figure 3.18b). Single-stranded parvoviral linear DNA has terminal palindromes that form hairpin structures. These hairpins then serve to covalently link the plus and minus DNA strands and self-prime the replication. The progeny viral DNA genomes are then made by strand displacement (Figure 3.18d).

Hepatitis B virus DNA uses reverse transcription for replication (Figure 3.18f). The partially dsDNA in the virion contains a complete minus and a partial plus strand. After infection of the cell, the virion-associated reverse transcriptase renders the partially double-stranded viral DNA into a circular duplex DNA in the cytoplasm that is then translocated into the nucleus and transcribed into a full-length plus-strand RNA by the host RNA *pol* II already present in the nucleus (Figure 3.16). This full-length plus-strand viral RNA is encapsidated, transported into cytoplasm, and reverse transcribed into a full-length minus-strand and a partial plus-strand DNA before being released as infectious virion.

3.4.7.2 Replication of RNA Genome

Viral RNA genomes can be single-stranded and composed of a plus or minus strand, or it can be double-stranded. Furthermore, while the genomes of some RNA viruses are segmented

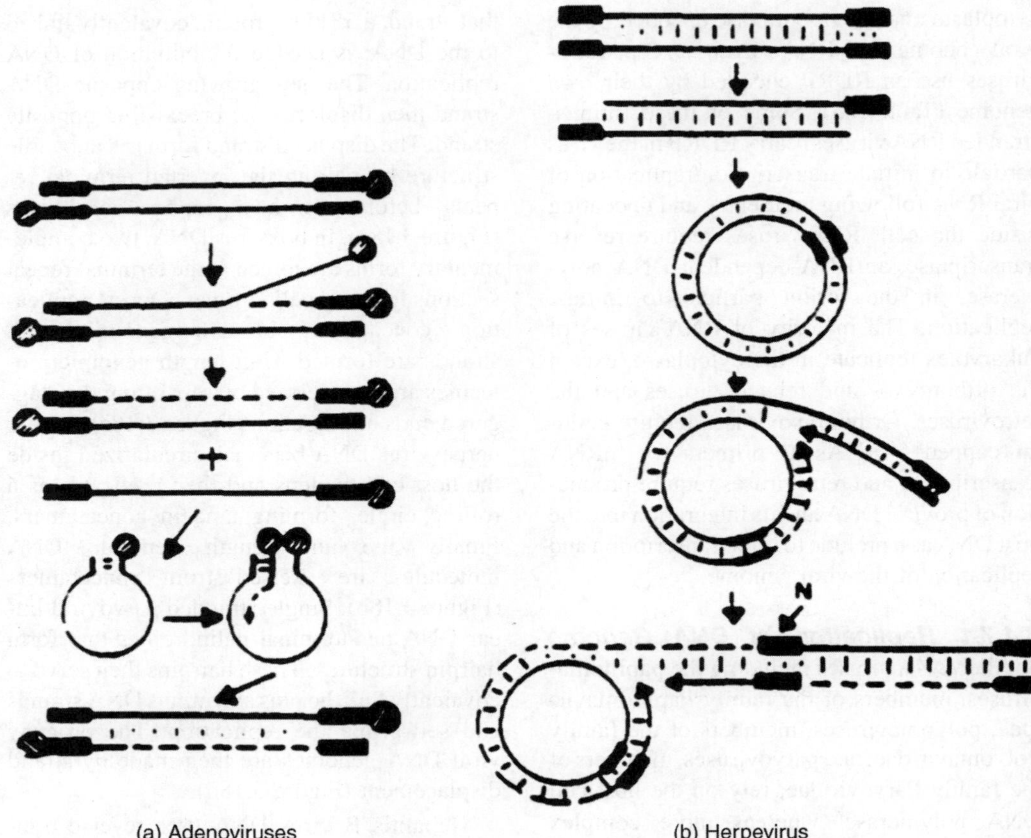

FIGURE 3.18 Replication pathways for viral DNA genomes. Dashed circles in (a) are terminal viral proteins attached to the 5′-end of the DNA strands. N in (b and d) represents endonuclease cleavage site. The heavy lines shown in (d) are palindromes and self-priming steps, with (+) and (–) representing strand polarity. The wavy lines in (f) represent RNA and the dashed lines represent DNA coding for direct repeats DR1 and DR2. Reprinted with permission from Davis et al. (1990).

(multiple RNA molecules), others are nonsegmented (i.e., one RNA molecule) (Tables 3.1 and 3.3). Switching from transcription to replication in the viral infectious cycle usually occurs after sufficient amounts of the capsid protein (e.g., nucleoprotein) are synthesized. The nucleoprotein functions as a regulator for switching from transcription to replication of the viral RNA genome.

Inhibition of nucleoprotein or protein synthesis will also inhibit vRNA (genomic RNA present in the virion) replication without necessarily interfering with mRNA synthesis. The same core enzyme (i.e., RDRP) is used for both transcription and replication, but the enzyme (and possibly the template RNA) becomes modified by viral nucleoprotein and cellular factors to effect the switch from transcription mode to replication mode. There are five different classes of RNA genomes, based on the different strategies that these viruses use for genome replication (Table 3.3).

Replication of Single-Stranded Viral RNA
Plus-strand (+) RNA viruses are copied into a complete minus-strand RNA that then serves as a template for synthesis of more plus strands via replicative RNA intermediates

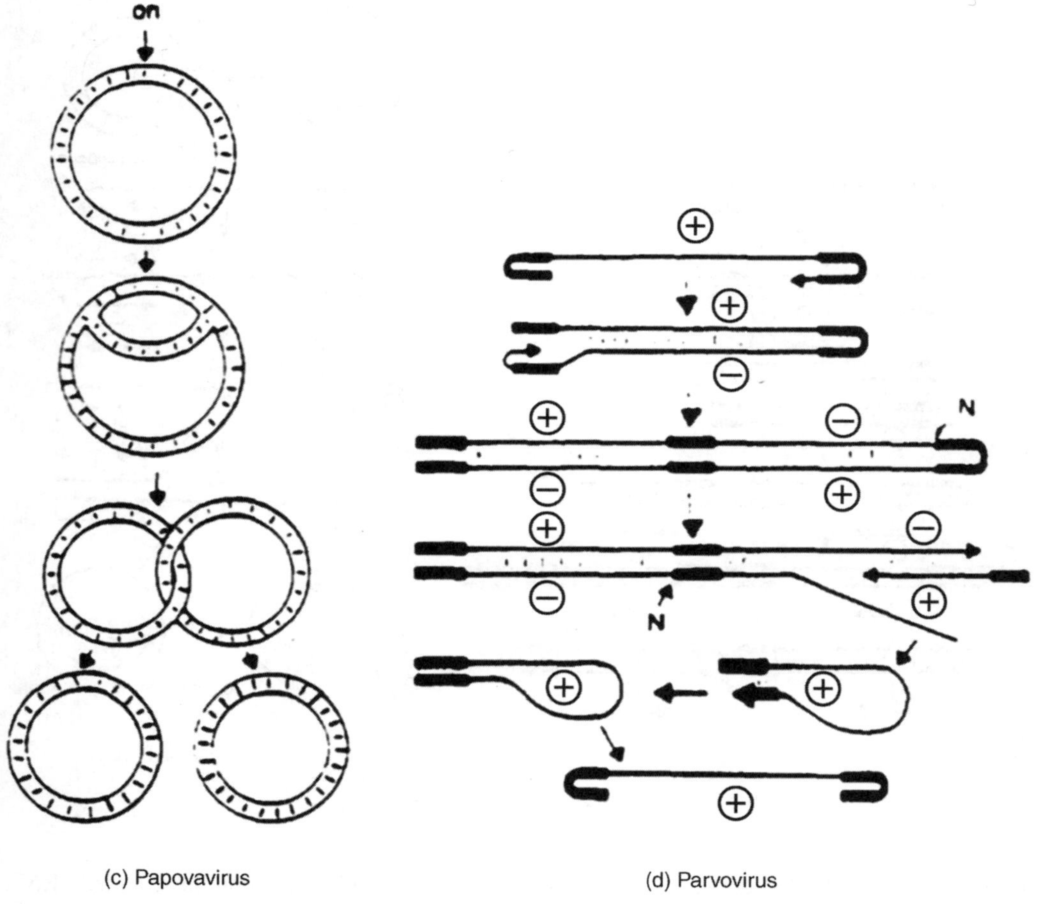

(c) Papovavirus

(d) Parvovirus

FIGURE 3.18 *(Continued)*

(Figure 3.17). Minus-strand nonsegmented RNA genomes are transcribed into two types of plus-strand RNAs: subgenomic mRNAs (plus sense), which represent specific portions of the genome and are translated into proteins, and full-length cRNA (plus sense), which represents a complete copy of the entire minus-strand genome and serves as the template for genomic RNA (minus sense) synthesis (Figure 3.19). The synthesis of cRNA is regulated by a switch from transcription to replication mode that occurs after sufficient amounts of capsid proteins (e.g., NPs in influenza viruses) are synthesized. The capsid proteins provide the antitermination factor required for full-length cRNA synthesis. The cRNA is then copied back into full-length minus-strand viral RNA, which is incorporated into virions. Orthomyxoviruses, which possess a segmented minus-strand RNA genome, likewise synthesize two classes of plus-strand RNA from the same minus-strand RNA template. However, the mRNAs and cRNAs of these viruses are different at their both 5′ and 3′ ends (Figure 3.19). As indicated earlier, the orthomyxoviral mRNAs possess nonviral sequences from capped host mRNAs at their 5′ ends and terminate 18–22 nucleotides prior to the 3′-end with the addition of poly(A) sequences. However, the orthomyxoviral template cRNAs are complete copies of vRNA from end to end without any nonviral sequence

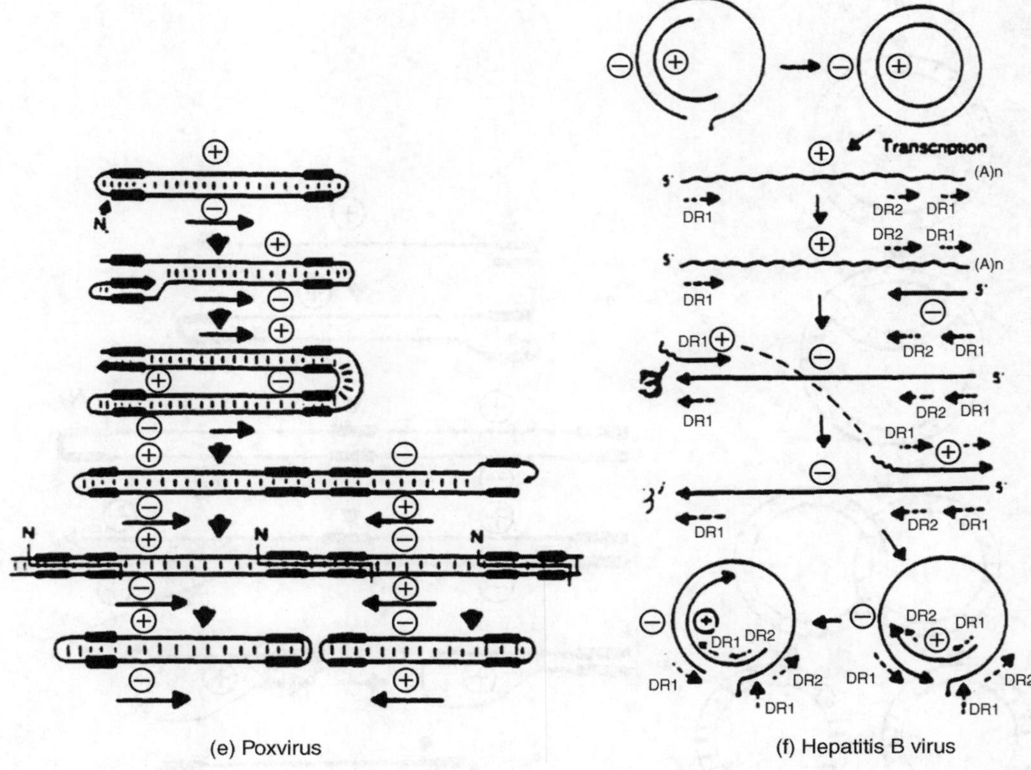

FIGURE 3.18 *(Continued)*

at the 5′-end or poly(A) sequences at the 3′-end. Therefore, for orthomyxoviral cRNA synthesis to occur, the viral polymerase must be able to initiate RNA synthesis without any host capped primer at the 5′-end and the RNA synthesis must not terminate until the complete 3′-end is reached, thus fully copying the entire template vRNA from the 3′-end to the 5′-end and without any polyadenylation. Such complete cRNAs then function as templates for vRNA (minus-strand) synthesis.

Replication of Double-Stranded Viral RNA
Each segment of double-stranded viral RNA genome is replicated independently. First, the genome is transcribed to generate plus-strand mRNAs within the incoming virion core by the virion-associated RDRP. Next, the mRNA is used as a template by RDRP to synthesize the minus RNA strand, and thereby mRNAs are converted into double-stranded RNA (Fig. 3.17) that is then packaged into progeny virion capsids.

Replication of RNA via a DNA Intermediate
Retroviruses contain a diploid genome consisting of two identical RNA molecules, a tRNA primer (Figure 3.6) and a reverse transcriptase, an RNA-dependent DNA polymerase that also possesses both RNase H and integrase activities. Conversion of the plus-strand viral RNA into double-stranded DNA is initiated by viral reverse transcriptase using the tRNA as a primer. The RNA-dependent DNA replication process is complex and requires strand switching twice (Figure 3.20). Eventually, a double-stranded proviral DNA is made in the cytoplasm that is translocated into the nucleus and integrated into the host genome. The integrated proviral DNA is then

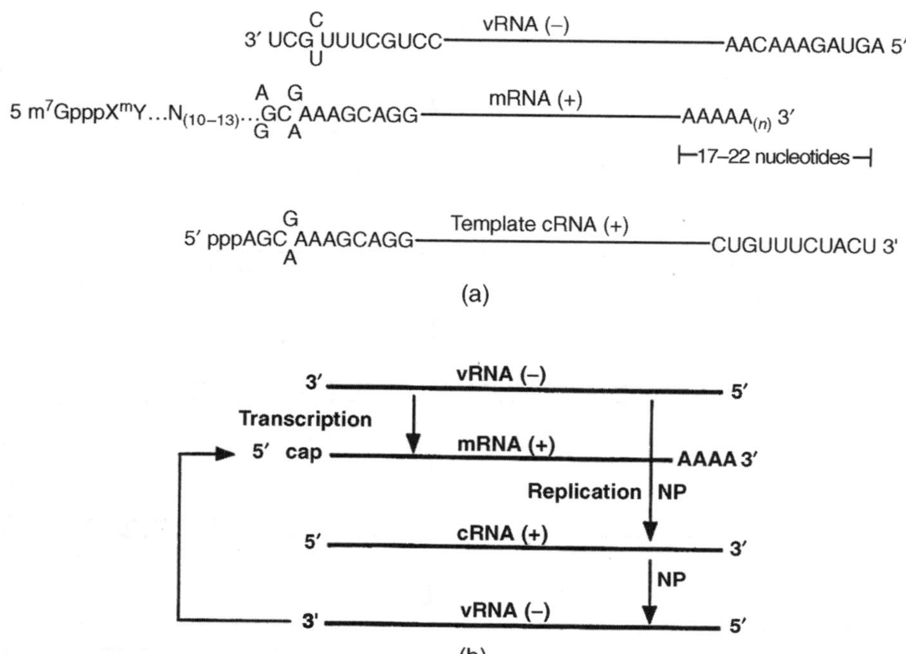

FIGURE 3.19 Transcription and replication of the influenza virus RNA (vRNA). (a) The three classes of influenza virus-specific RNAs found in the virus-infected cells, vRNA of minus (−) polarity; cRNA and mRNA, both of the latter two possessing (+) polarity. Note that the viral mRNA (+) possesses nonviral (host) capped sequences at the 5′-end and lacks sequences of 17–22 nucleotides from the 3′-end but contains poly(A) sequences. The template cRNA (+ strand), on the other hand, is an exact copy from end to end of the vRNA (−strand) and does not possess either cap at the 5′-end or poly(A) at the 3′-end. (b) The transcriptive and replicative processes of influenza viral RNA.

transcribed by the host RNA *pol* II into full-length plus-strand RNA and the full-length RNA is then transported into the cytoplasm and encapsidated into progeny virions.

3.5 ASSEMBLY AND MORPHOGENESIS OF VIRUS PARTICLES

As indicated earlier, compared to eukaryotes or prokaryotes, viruses use a unique multiplication strategy to produce their progeny. All cells, prokaryotic or eukaryotic, multiply as a whole unit from parent to progeny and in a geometric order, that is, 1, 2, 4, 8, and so duplicatively on. Viruses, however, do not multiply as units. Rather, they are assembled from component parts. Each component part of progeny virus particles is made separately, and they are often made in different amounts and at different locations and compartments within the host cell. These viral components are then put together to form the whole (infectious) virus particles (virions). In this assembly line type of process, all individual viral components need not be assembled at the same time, and in fact some components may be put together separately to form higher ordered structures, that is, subviral particles (e.g., capsid, nucleocapsid, RNPs), before they are assembled into a whole progeny virus particle. The number of steps involved and the complexity of the assembly process may vary greatly from one virus type to another. Some viruses, such as the polioviruses, have only a few components to assemble, and yet others, such as

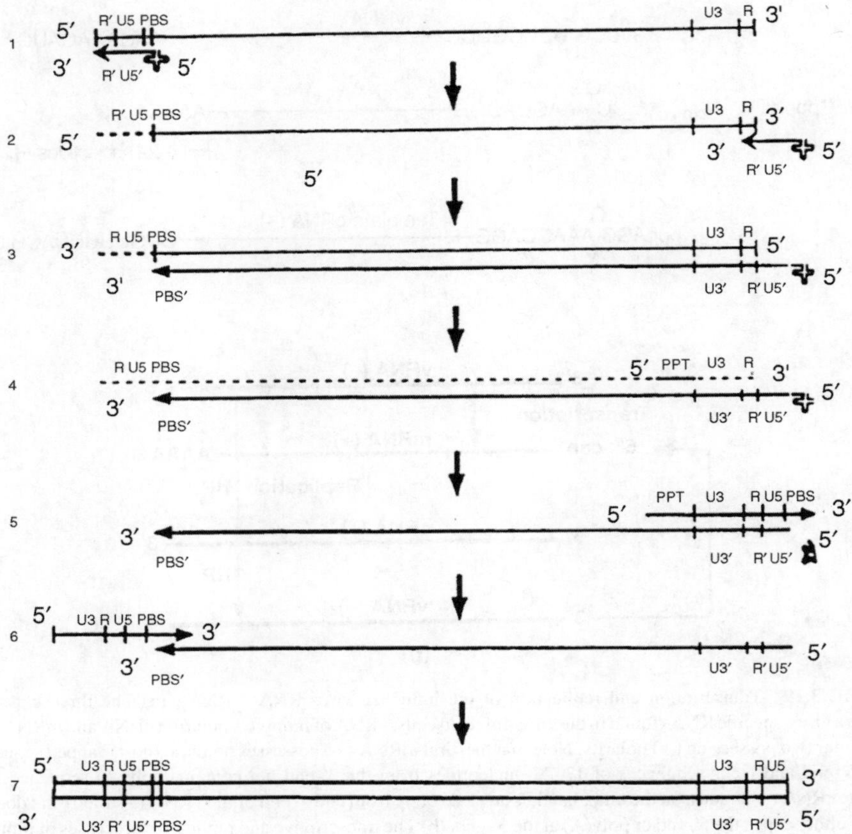

FIGURE 3.20 Reverse transcription of retroviral genomic RNA into double-stranded proviral DNA. Step 1: annealing of primer tRNA (shown as a cross-shaped symbol) to the primer binding site (PBS) and synthesis of minus-strand strong-stop DNA. The R and U5 RNA is degraded by the RNase H activity of the reverse transcriptase, and the strong-stop DNA is released. Step 2: The first strand switch (or transfer). The minus-strand strong-stop DNA is annealed to the 3′-terminus of the genomic RNA via R–R′ hybridization (first strand jump). Steps 3 and 4: Further synthesis of minus-strand DNA, during which the genomic RNA is further degraded by RNase H. However, a small piece of RNA (the polypurine tract (PPT)) remains undergraded and serves as a primer for synthesis of plus-strand strong-stop DNA (step 5). Step 5: Termination of plus-strand DNA synthesis at 18 nucleotides into the primer tRNA, thus generating the new primer binding site (PBS) sequence; the plus-strand DNA is then released from the minus-strand DNA. Step 6: The second jump (or the second transfer). The plus-strand strong-stop DNA is annealed to the 3′-terminus of the minus-strand DNA via PBS–PBS hybridization, completing the second jump. Step 7: Completion of the synthesis of the double-stranded proviral DNA. R, terminally redundant identical sequences at the 5′ and 3′ ends of viral RNA; U5, unique nucleotide sequences near the 5′-end of the viral genome between the R and PBS (primer binding site); U3, the region near the 3′-end of the viral RNA between the initiation site of the plus-strand DNA synthesis and R sequences; PPT, polypurine tract that escapes RNase H digestion and serves as a primer for the second strand DNA synthesis. Strong-stop DNA is the DNA copy of the region between the primer binding site (PBS) and the 5′-end of the viral RNA genome. Reprinted with permission from Mak and Kleiman (1997).

the poxviruses or herpesviruses, have many components to assemble and their assembly compared to polioviruses is a far more complex process involving multiple steps.

With respect to the assembly processes, viruses can be classified into two major subclasses: naked viruses and enveloped viruses. Naked viruses consist of only a nucleocapsid, that is, the capsid containing the genome (DNA or RNA) and no envelope. The assembly of the protein capsid and incorporation of genomic nucleic acid into the capsid to create this nucleocapsid will render the virus particle infectious. For these viruses, the virus receptor binding proteins are part of the capsid proteins. Enveloped viruses, however, are those in which the nucleocapsid is surrounded by a lipid membrane containing the transmembrane viral proteins. In enveloped viruses, one of the transmembrane viral proteins (and not the capsid protein) contains the receptor binding protein.

3.5.1 Assembly and Morphogenesis of Naked Viruses

The assembly of naked viruses occurs in the cytoplasm (most RNA viruses) or nucleus (DNA viruses). Nearly all cytoplasmic viruses with the exception of poxviruses are RNA viruses (e.g., the plus-sense RNA picornaviruses). The entire genomic RNA of these viruses is translated into a single giant polyprotein (Figure 3.7) that is cleaved by a virus-specific protease into P1 (a coat precursor protein); P1 is then further cleaved by the protease 3C into VP0, VP3, and VP1 (5S promoter). Five subunits (5S promoter)—each containing one molecule of VP0, VP3, and VP1—then assemble into pentamers (14S). Twelve pentamers form the 60-subunit protein shell (capsid) for the picornaviruses. The viral RNA genome is then incorporated into the capsid, forming what is called the "provirion." Subsequently, VP0 molecules in the provirion are cleaved into VP4 and VP2 (Figure 3.7), converting the provirion nucleocapsid into an infectious virion. This process of picornaviral capsid assembly is basically a self-assembly process whose rate depends on viral protein concentration.

For assembly of a naked virion to occur inside the nucleus, one of at least two distinct strategies can be used. The first of these would require that all capsid proteins, after their translation in the cytoplasm, must be transported into the nucleus either independently or cooperatively by forming a complex with other capsid proteins and that nucleocapsid assembly occurs around the viral genome in the host nucleus. This option is used by polyomaviruses whose DNA genomes, or minichromosomes, contain a single closed circular duplex DNA molecule complexed with cellular histone, which is organized into a nucleosome within the host nucleus. Polyomaviral capsid assembly then proceeds in a stepwise fashion around the viral minichromosome. The capsid of SV40, which is a member of this virus group, contains 360 copies of its major viral protein (VP1) assembled into 72 pentamers plus 30–60 copies of internal proteins VP2 and VP3. VP2 contains the full VP3 sequence plus 100 extra amino acids at the NH_2 terminus, which are critical for interacting with the SV40 minichromosome. The polyomaviral capsid proteins and minichromosomes first assemble into 200S structures called provirions that then mature into infectious virions. During this maturation, H1 histone protein is removed from the viral minichromosome and degraded.

Adenoviruses use a second type of strategy in which the capsid shell is first formed by the assembly of viral capsid proteins. Viral DNA, including core proteins, is then inserted into the empty capsid shells to form infectious virions. Both these nuclear DNA viruses and the cytoplasmic naked RNA viruses are primarily released to the extracellular environment by cell lysis.

3.5.2 Assembly, Morphogenesis, and Budding of Enveloped Viruses

The assembly and budding of enveloped viruses is much more complex than that of naked viruses. It involves not only assembly and formation of nucleocapsids but also envelopment of the nucleocapsid and budding of enveloped nucleocapsids from different cellular organelles and membranes specific for each group of viruses. Subsequently, buds are pinched off and the virus particles are released into the extracellular environment. The assembly and the budding site on the cellular membrane vary with different groups of viruses. Some viruses, such as poxviruses and rotaviruses, bud from the ER, while others, such as bunyaviruses, bud from the Golgi complex, and still others bud from the nuclear membrane, such as herpesviruses. Still other viruses (e.g., orthomyxo-, paramyxo-, rhabdo-, and retroviruses) use the plasma membrane (apical or basolateral) as the budding site.

Assembly, morphogenesis, and budding of enveloped viruses require multiple steps: (1) transport and assembly of viral components to the budding site and (2) the budding process including bud initiation, bud growth, and pinching off from the plasma membrane. Budding is a complex process and involves physical and structural as well as functional requirements of multiple biological components of both virus and host cell and the processes involved in budding are not fully understood in viral biology. Since the assembly and budding processes of orthomyxoviruses have been well studied, the steps involved in morphogenesis and budding of these viruses will be discussed in some detail. For comparison, rhabdovirus and retrovirus assembly will also be mentioned as needed. As noted earlier, orthomyxo- and paramyxoviruses are enveloped RNA viruses containing single-stranded RNA genomes of negative (minus) polarity, and they are assembled into nucleocapsids having helical symmetry (Table 3.1). Electron microscopic studies have demonstrated that these viruses bud from the plasma membrane into the outside environment and that complete virions are usually not found inside the cell during the productive infectious cycle.

For budding to occur, all viral components must be brought to budding site. With the majority of viruses enveloped or nonenveloped, assembly implies the formation of complete capsid, either helical or icosahedral including incorporation of the genome into the capsid. Furthermore, with the majority of enveloped viruses, capsid formation is a requirement for bud formation and bud release as is shown for retroviruses such as human immunodeficiency viruses (Ganser-Pornillos et al., 2008) and alphaviruses such as SFV (Garoff et al., 1994). Even for some enveloped viruses possessing helical nucleocapsids such as VSV, formation of nucleocapsid is critical for bud formation and bud release. In fact, the size of the nucleocapsid assessed by the size of vRNA determines the size and shape of the released VSV particles. For example, smaller DI virus particles contain shorter vRNA/RNP (nucleocapsid) compared to the bullet-shaped elongated virus particles of wild-type viruses containing the complete and longer vRNA/RNP (Pattnaik and Wertz, 1991). Therefore, with these viruses, capsid assembly is a critical requirement for virus budding. However, requirements of capsid assembly and budding are much more complex for viruses with segmented genome (e.g., influenza viruses) for a number of reasons. First, budding may occur in the absence of vRNPs (capsids) and/or with incomplete vRNPs. Furthermore, the viral genome consists of multiple segments of vRNAs/vRNPs. Therefore, budding of infectious virus particles requires that each segment of multiple vRNPs must be successfully incorporated into the bud. Second, all the components of the virus, that is, envelope containing the transmembrane proteins (HA, NA, and M2) and M1 and vRNPs, must be brought individually or as a complex to the budding site for bud initiation, bud growth, and finally release of infectious virus buds.

For elucidating the budding process of influenza viruses, the viral structure can be separated into three major subviral components, each of which must be brought to the assembly site for morphogenesis. These subviral components are (a) the viral nucleocapsid (or viral ribonucleoprotein (vRNP)) containing the vRNA, NP, and transcriptase/polymerase complex that together form the inner core of virus particle; (b) the matrix protein (M1), which forms an outer protein shell around the nucleocapsid and constitutes the bridge between the envelope and nucleocapsid; and (c) the envelope (or membrane), which forms the outermost barrier of these enveloped virus particles. The viral envelope contains virally coded transmembrane proteins and host cell lipids. Each of these subviral components must be brought to the budding site for interactions among themselves and for budding to occur. Depending on the virus groups, the budding site on the cell membrane varies. For example, viruses belonging to orthomyxo-, paramyxo-, rhabdo-, and retroviruses bud from the plasma membrane of infected cells. However, while the orthomyxo- and paramyxoviruses bud from the apical domain of plasma membrane in polarized epithelial cells, both *in vivo* (e.g., in bronchial epithelium) and in cultured polarized epithelial cells (e.g., Madin Darby canine kidney (MDCK) cells), the rhabdoviruses and retroviruses bud from the basolateral surface of polarized epithelial cells.

3.5.3 Assembly and Transport of Viral Components to the Budding Site

As mentioned above, for budding and release of influenza viruses to occur, all viral components must be brought to the budding site. Therefore, two questions arise: (1) How these subviral complexes are brought to the budding site? (2) What factors determine the selection of budding site? As mentioned above, there are three major subviral components within the influenza virus particle, namely, virus envelope, M1 (matrix protein), and the viral core (vRNP/nucleocapsid). The influenza virus envelope consists of a mosaic lipid bilayer and viral transmembrane proteins (HA, NA, and M2). Transport of the transmembrane envelope proteins (HA, NA, and M2) has been studied extensively (Nayak et al., 2004). As mentioned earlier, the viral membrane is a mosaic containing both raft- and nonraft-associated lipids. Both HA and NA proteins are inserted in the raft domains, whereas M2 is present in the nonraft lipid domains. These transmembrane proteins use cellular exocytic transport pathway for apical transport and possess the determinants for both lipid raft association and apical transport in their TMD. Lipid raft association of TMD is responsible for apical transport of both HA and NA (Lin et al., 1998; Barman and Nayak, 2000). However, as discussed later, transport of the envelope proteins is not the only or major determinant for the selection of the budding site of influenza viruses.

Next question is how the M1 protein, the most abundant viral protein, present underneath the lipid bilayer and forms the bridge between the envelope and viral core (vRNP) is brought to the budding site. M1 is not known to possess any apical determinant but possesses determinants for lipid binding, for RNA, RNP, or NP binding (Baudin et al., 2001; Noton et al., 2007; Watanabe et al., 1996; Ye et al., 1999), and for associating with HA and NA tails (Ali et al., 2000) and M2 tails (Chen et al., 2008). Therefore, it is likely that some M1 can be transported to the budding site of apical plasma membrane on the piggyback of HA and NA and also as a complex with vRNP.

Finally, how is the virus core (viral nucleocapsid) that consists of vRNP (minus-strand vRNA associated with NP), minor amounts of NEP (nuclear export protein), and 3P protein complex (three polymerase proteins PA, PB1, and PB2 forming a heterotrimeric complex) brought to the apical budding site. Helical nucleocapsid assembly occurs during the synthesis of minus-strand viral RNA for both segmented (influenza) and nonsegmented (VSV, Sendai) RNA viruses. In the absence of the capsid protein (N or NP), minus-strand vRNA synthesis does not occur. Each influenza viral nucleocapsid (vRNP) is a supercoiled helix with ribbon structure and a terminal loop, where the vRNA is coiled around NP monomer to form a hairpin structure and vRNA is exposed on the outer surface of NP (Elton et al., 2006). Therefore, RNP assembly involves the formation of these subviral complexes and their transport to the budding site, that is, the apical domain of the plasma membrane in polarized epithelial cells whether in cultured cells in laboratory or respiratory epithelium of infected animals. Furthermore, since influenza RNP is synthesized in the nucleus, it must be exported from the nucleus into cytoplasm before being transported to the apical plasma membrane. M1, a small protein possessing nuclear localization signal (NLS), can enter the nucleus, interact with both vRNP and NEP forming the daisy-chain complex of (Crm1 and RanGTP)–NEP–M1–RNP, and mediate nuclear export of v-RNP (Akarsu et al., 2003; Whittaker and Digard, 2006). M1–RNP complex has been demonstrated both in infected cells and in virions (Zhirnov, 1992; Ye et al., 1999). Interaction of M1 with RNP preventing transcription is critically required for the exit of vRNPs into cytoplasm and incorporation into virions (Nayak et al., 2004) since it is only transcriptionally inactive vRNPs with the polymerase complex present at the end of vRNP that are found in virus particles (Murti et al., 1988). Furthermore, lack of chain elongation of preexisting RNA molecules and requirements of capped RNA primers for *de novo in vitro* RNA transcription of vRNP molecules also support the presence of transcriptionally active vRNP within influenza virus particles.

Recent studies suggest that the viral NP or ribonucleoprotein (RNP) complex may possess an as yet undefined determinant for apical transport (Carrasco et al., 2004; Nayak et al., 2009). It was recently shown that influenza NP/RNP exits the nucleus from its apical side of nucleus and is transported to the apical plasma membrane of polarized MDCK cells. NP/RNP was also shown to interact with actin microfilaments (Avalos et al., 1997) and associate with lipid rafts (Carrasco et al., 2004). Therefore, it is likely that RNP along with the associated M1 can be directed to the apical budding site via its association with cortical actin microfilaments and lipid rafts. However, neither the apical determinant(s) of NP/RNP nor the cellular machinery involved in its apical transport has been identified.

Finally, since the genome of influenza virus is segmented, multiple vRNA/vRNP segments (eight separate segments for members of the genera *Influenzavirus A* and *B* versus seven segments for members of the genus *Influenzavirus C*) must be correctly assembled and incorporated into each infectious virus particle (see later). Each of these vRNA segments replicates independently in infected cells and can undergo reassortment during assembly and budding (Figure 3.10). When two or more viruses infect a single cell, released virus particles will have a set of different RNA segments arising from one or more viruses infecting the same cell (Figure 3.10). This is the major cause of antigenic shift and responsible for the emergence of pandemic influenza viruses. Depending on the set and combination of vRNAs, the virus particles will possess different antigenic epitopes and will have selective advantage of growth, virulence, and spreading.

However, although packaging of different RNP segments in the virus bud is critically important for infectivity of virus particle, assembly or incorporation of genomic segments does not appear to play a critical role in the budding of virions. However, M1 has been shown to play an important role in the assembly of virion components as it interacts with multiple components, such as viral RNA or viral RNP, and envelope proteins (HA, NA, and M2) and brings viral components together. M2 interacts with M1 via cytoplasmic tail and thereby plays an important role in virus assembly, genome packaging, and budding (Iwatsuki-Horimoto et al., 2006; McCown and Pekosz, 2005, 2006; Chen et al., 2008).

Although both orthomyxo- and paramyxoviruses bud from the apical domain of plasma membrane, there are two major differences: (1) as mentioned previously, since the viral genome of orthomyxoviruses is segmented, multiple RNA segments (eight separate RNA segments for members of the genera *Influenzavirus A* and *B* versus seven RNA segments for members of the genus *Influenzavirus C*) must be incorporated into infectious virions, whereas only one large RNA molecule is packaged in infectious paramyxovirus particles. (2) Since the transcription and replication of orthomyxovirus RNA and assembly of these viral nucleocapsids (vRNP) occur in the host nucleus, the viral nucleocapsids must be exported out of the nucleus into the cytoplasm for the final stages of viral assembly and for budding. In contrast, paramyxoviruses possess one single nonsegmented minus-strand vRNA, and all these steps, including assembly of viral nucleocapsids, take place in the cytoplasm. The processes involved in the assembly and transport of other nonsegmented minus-strand RNA viruses such as rhabdoviruses (VSV) are essentially similar to that of paramyxoviruses. However, VSV buds from the basolateral membrane, whereas paramyxoviruses bud from the apical domain of the plasma membrane.

3.5.4 Selection of Budding Site

As mentioned previously, different enveloped viruses bud from different membrane compartments of infected cells, and budding site plays an important role in the pathogenesis of specific viruses (Nayak,). Therefore, it becomes important to ask, how is the budding site of enveloped viruses (e.g., apical domain of plasma membrane in polarized epithelial cells for influenza viruses) selected? For the majority of the viruses, viral glycoproteins are thought to be important in the selection of the budding site since virus glycoproteins, even when expressed alone in the absence of other viral components, predominantly accumulate at the site of virus budding. For example, viruses such as hepatitis B virus, bunyaviruses, coronaviruses, and others that bud from the internal subcellular organelles possess intrinsic determinants for the same subcellular localization as the site of virus budding (Hobman, 1993). On the other hand, for viruses budding from the plasma membrane, the viral glycoproteins possess apical or basolateral sorting signals and are directed to the specific site where virus assembly and budding occur in polarized epithelial cells. The surface glycoproteins of viruses such as influenza virus (family Orthomyxoviridae) and human respiratory syncytial virus (family Paramyxovidae) budding from the apical plasma membrane possess apical sorting signal(s) and predominantly accumulate at the apical plasma membrane in polarized epithelial cells. Conversely, for viruses released from the basolateral membrane, their surface glycoproteins, possessing basolateral sorting signal, are transported basolaterally in polarized epithelial cells even when these proteins are expressed alone. VSV, SFV, vaccinia virus, and certain retroviruses including human immunodeficiency virus type 1 (HIV-1) exhibit basolateral budding. Furthermore, in different cells and tissues where some viruses bud from the opposite domains of the plasma membrane, their glycoproteins are distributed accordingly. For example, SFV buds apically from FRT (a Fisher rat thyroid-derived cell line) cells but basolaterally from CaCo-2 (human epithelial colorectal carcinoma) cells. Similarly, in the absence of any other viral protein, p62/E2, the envelope glycoproteins of SFV, is targeted apically in FRT cells but basolaterally in CaCo-2 cells (Zurzolo et al., 1992). However, there are examples of polarized virus budding occurring independently of the polarized envelope viral glycoprotein sorting. For example, although measles virus glycoproteins H and F are transported in a random fashion or to basolateral membrane, respectively, virus budding was observed to have occurred predominantly from the apical surface of polarized MDCK cells (Maisner et al., 1998). Similarly, the spike protein of coronavirus is not involved in the polarized budding of this virus (Rossen et al., 1998). Moreover, Lake Victoria marburgvirus (family Filoviridae, genus *Marburgvirus*) buds predominantly from the basolateral surface, while its glycoprotein is transported to the apical surface (Sanger et al., 2001).

However, accumulation of viral glycoproteins may not be the only or the major determinant in selecting the budding site. For example, using a mutant transfectant influenza virus (HAtyr) containing basolaterally targeted HA (Cys543 → Tyr543), it was shown that the basolateral targeting of HA did not significantly alter the apical budding of influenza virus (Barman et al., 2003; Mora et al., 2002). Over 99% of the virus particles containing the HAtyr were released from the apical side even though the majority of HAtyr viruses were directed to the basolateral side. However, when virus budding was examined by thin section transmission electron microscopy, empty virus-like structures (Barman et al., 2003) with the same size diameter as the virus particles at apical surface were often observed only in HAtyr-infected cells (Barman et al., 2003). Likely, these particles represent abortive virus buds containing HA and M1 but not vRNP, suggesting that vRNP may play a role in polarized budding of influenza virus. Furthermore, apical targeting of NP was also shown to be independent of M1 and NEP that did not accumulate at the apical membrane (Carrasco et al., 2004; Nayak et al., 2009). We also observed that NP/vRNP in VLP-infected polarized MDCK cells lacking the expression of viral envelope proteins accumulated at the apical plasma membrane similar to that observed in wt virus-infected cells (Nayak et al., 2009) and NP exited through apical side of the nucleus in both wild-type (wt) virus-infected and VLP-infected polarized MDCK cells. These results demonstrate that NP/vRNP can be transported independently to the apical plasma membrane of polarized epithelial cells in the absence of transmembrane viral proteins. Therefore, transmembrane proteins alone do not determine the site of virus budding and NP also plays an important role in apical budding of influenza A viruses. Both cortical actin microfilaments and lipid rafts may aid in apical transport since NP binds to both these host components.

Two steps are obligatory for virus assembly and morphogenesis to occur. First, as mentioned above, all viral components (or subviral particles) must be directed and brought to the assembly site, that is, the apical plasma membrane in polarized epithelial cells for assembly and budding of orthomyxoviruses and paramyxoviruses. Obviously, this step is the first obligatory requirement in virus assembly and morphogenesis, since if different viral components are misdirected to different locations or parts of the cell, virus assembly and morphogenesis cannot take place. Second, the viral components must interact with each other to form the proper virus structure during morphogenesis. It is possible that viral components may be directed to the assembly site but that defective interaction among these components will not yield infectious particles. However, although these two steps are obligatory, they alone may not be sufficient to form and release infectious virus particles. Therefore, virus components may be directed correctly to the assembly site and then they interact with each other to form virus particles, yet infectious viruses may not be released into the medium. For example, abortive virus morphogenesis in HeLa (human cervical carcinoma) cells infected with influenza viruses has been observed where virus particles are formed at the plasma membrane but not released (Gujuluva et al., 1994). Therefore, there are other factors regulating the pinching-off process causing bud release.

In addition, with influenza viruses, correct assembly of multiple vRNA/vRNP segments (eight separate segments for *Influenzavirus A and B* and seven segments for *Influenzavirus C*) will be required for incorporation into each infectious virus particle. Although packaging of different RNP segments in the virus bud is absolutely essential for infectivity of virus particle, assembly or incorporation of all eight RNA genomic segments is not critical for budding and bud release of virus particles. However, since infectivity of a virus particle depends on the correct incorporation of each vRNA segment, it becomes important to determine how all vRNA segments are selectively incorporated into infectious virions. Two models have been proposed for the incorporation of eight vRNA/vRNP segments into virions: "random packaging" and "specific packaging." The "random packaging" model predicts the presence of common structural elements in all vRNPs, causing them to be incorporated randomly into virions, and therefore incorporation of vRNPs will be concentration dependent. Support for this model comes from the observation that influenza A virions can possess more than eight vRNPs (9–11 vRNAs per virion) (Bancroft and Parslow, 2002; Enami et al., 1991), and at most 1 in 10 or more virus particles are infectious. On the other hand, the "specific packaging" model assumes that specific structural features are present in each vRNA/vRNP segment, enabling them to be selectively incorporated into virions. Evidence for this model is deduced mainly from the finding that the various vRNAs are equimolar within viral particles even though their concentrations in infected cells may vary (Smith and Hay, 1982). The selective packaging model has been favored by the earlier studies demonstrating that the small DI (also called von Magnus particles) vRNAs can competitively inhibit the packaging of their normal counterparts but not that of other vRNAs (Duhaut and McCauley, 1996; Nakajima et al., 1979; Nayak et al., 1985, 1989; Odagiri and Tobita, 1990). DI RNAs are smaller internally deleted viral RNA segments. They possess all the structural features of the wild-type viral RNA segments for replication and incorporation into virus particles. They selectively replace their progenitor viral RNA in virions. Recent studies have demonstrated the presence of segment-specific packaging signal (s) in 3′- and 5′-UTR as well as adjacent coding regions (varying with both specific RNA segment and 3′ or 5′ ends). Specific packaging signals are found for all eight RNA segments (Watanabe et al., 2003; Fujii et al., 2003, 2005; Liang et al., 2005; Muramoto et al., 2006; Ozawa et al., 2007) and incorporation of some specific RNA segments is critical for the incorporation of other RNA segments (Muramoto et al., 2006; Marsh et al., 2008).

ET studies of serially sectioned *Influenzavirus A* particles have shown that the RNPs of influenza A virus are organized in a distinct pattern (seven segments of different lengths surrounding a central segment). This finding argues against random incorporation of RNPs into virions and supports the "specific packaging" model (Noda et al., 2006). Such a model would require that specific vRNA–vRNA interaction among the specific eight vRNP segments should form multisegmental vRNP macromolecules prior to or during incorporation into virus particles and that these large vRNP complexes containing eight unique vRNPs should be stable. However, such intracytoplasmic multi-RNA/RNP complexes have not yet been demonstrated. More important, bud closure and virus release should not occur until such vRNP complexes containing eight specific vRNP segments are formed and incorporated in the bud. In support of this model, Fujii et al. (2003) demonstrated that the efficiency of infectious virion production correlated with the number of different vRNA segments. They observed that the higher the number of different vRNA segments, the higher was the efficiency of virion production. Recently, specific nucleotide residues in 3′ and 5′ ends (coding and noncoding) of PB1, PB2, and PA (Liang et al., 2008; Marsh et al., 2008), as well as in HA (Marsh et al., 2007), have been further shown to play a critical role for packaging of specific vRNA segment into progeny virions. The major weakness of this model is that bud closure and virus release do not appear to depend on the incorporation of eight specific RNA segments and particles with fewer RNP segments are found. Therefore, it is possible that segment-specific complex formation and incorporation of viral RNA may occur but may not affect bud closing and bud release.

3.5.5 Budding Process

For enveloped viruses, budding of infectious virus particles requires that all structural viral components be brought to the budding site for assembly and incorporation into the virus particles. For influenza viruses that selectively bud from the apical domain of polarized epithelial cells, all viral components must be brought to the budding site at the apical plasma membrane, and highly specific interactions are required both prior and subsequent to their arrival in order to achieve successful assembly. As mentioned earlier, transporting viral components to the budding site requires involvement of the exocytic pathway and its components. Similarly, during the assembly process, multiple cellular components including actin microfilaments and lipid rafts also play critical roles in concentrating the viral components and providing a favorable environment for their interaction and the progressive formation of higher order subviral complexes. The final budding process itself requires three major steps: bud initiation, bud growth, and bud completion, including releasing of the virus from the host cell membrane. Each of these steps involves interaction of multiple host and viral components.

3.5.5.1 Bud Initiation Bud initiation requires outward bending of the plasma membrane and involves transition of a more planar membrane structure to a curved structure at the budding site. Although the structural nature and biochemical properties as well as the physical forces at these sites responsible for membrane bending and bud initiation are unknown, it is likely that both lipid rafts and raft-associated proteins present at the budding site play an important role in causing membrane curvature and bud initiation. Lipid rafts producing asymmetry in lipid bilayers can cause intrinsic curvature of one lipid monolayer relative to the other monolayer leading to membrane bending (Nayak and Hui, 2004). Membrane deformation can be caused by selective transfer of lipids between the lipid bilayers, interaction of cholesterol with the budding leaflet, and hydrolytic cleavage of phosphocholine head groups of sphingomyelin by sphingomyelinase generating smaller head groups (Holopainen et al., 2000). In addition, BAR (Bin/amphiphysin/Rsv) domain is shown to cause membrane curvature (Peter et al., 2004) and is known to be present in a number of proteins involved in vesicle formation and recycling. (BAR domains are helical domains found in proteins involved in vesiculation processes including endocytosis, intracellular trafficking, budding, and so on that require membrane bending. BAR domains interact with endocytic and cytoskeletal machinery including GTPase, dynamin and possess dimerization motifs sensing and inducing membrane curvature. BAR domain containing proteins include endophilins, GTPase activating proteins, amphiphysin, arfaptin, and others.) However, the specific role of any of these host proteins in influenza virus budding largely remains unknown. In addition to lipid raft microdomains, accumulation of viral proteins including HA, NA, and M2 on the outer side of membrane and M1 proteins on the inner leaflet of the membrane plays a critical role in further facilitating the membrane bending at the budding site (Nayak and Hui, 2004). Among these, M1 interacting with the inner leaflet of lipid bilayers is likely to play a major role in bud initiation. Currently, we do know that clustering of M1 owing to M1/M1 interaction underneath the lipid bilayers can cause outward membrane bending and bud initiation.

3.5.5.2 Bud Growth Bud growth leading to bud maturation is the intermediate stage between the bud initiation and the bud release. Bud growth determines the size and the morphology of released virus particles. However, what factors or forces determine and regulate bud growth remains unclear. For most viruses, regardless of whether they contain icosahedral (e.g., SFV) or helical (e.g., VSV) nucleocapsids, the size of the nucleocapsids determines the size of the virions. However, there is room for variability in virion size. For example, influenza viruses are highly pleomorphic and the size of the released particles can vary from spheroidal to elongated and even filamentous (Figures 3.4 and 3.5), and the content of the nucleocapsids is not the major factor for bud growth. Influenza virus bud growth rather appears to depend on two forces, pulling and pushing. The pulling force is primarily provided by the transmembrane proteins along with M1 that pull nucleocapsids into the bud. On the other hand, the host cortical actin microfilaments that bind to viral RNPs provide the pushing force for incorporating the nucleocapsids and M1 into the bud. Electron tomography analysis of virus buds attached to the cell surface shows that helical nucleocapsids are oriented perpendicularly to the cell membrane while being incorporated into the buds and that buds essentially complete and still remaining attached to the cell membrane are of similar size (Figures 3.5 and 3.21).

As mentioned earlier, influenza virus particles are highly pleomorphic in shape and size

FIGURE 3.21 Virus buds at the cell surface by cryo-EM. At 12 hpi, WSN-infected MDCK cells were processed for thin section and examined by cryotomography. This picture represents one slice through inner core of the virus buds. One can see the parallel arrangement of the vRNPs inside the bud perpendicular to cell surface. The bud neck (⇒) shows gaps indicating possible absence of M1. HA and NA spikes are seen on the bud envelope. Reproduced from Nayak et al. (2009).

(Calder et al., 2010). Basically, there are two types of pleomorphism observed among influenza viruses: (1) strain specific, that is, strain to strain variation that may also vary depending on the host cell and (2) variation within the population of plaque purified virus in the same cell. Clearly, the genome of the virus strain is an important factor in determining the particle size and shape of a specific virus strain (e.g., Udorn versus WSN, both of which are antigenic variants of influenza A virus, belonging to the genus *Influenzavirus A*). Specific viral genes involved in determining filamentous versus spheroidal forms have been identified. Similarly, the roles of polarized epithelial cells and intact actin microfilaments were found to be critical in maintaining the filamentous form of Udorn virus. However, the cause of pleomorphism in plaque purified influenza viruses is not well understood. Whatever may be the viral and cellular factors involved in viral pleomorphism, these are likely to affect bud growth and closing and will eventually affect the shape and size of the virus particles. This is not to state that factors affecting bud closing will always affect bud size and bud shape; however, factors affecting bud shape and bud size will always affect bud closing. The viral and host factors affecting the size of virus particles will hinder or facilitate bud closing.

3.5.5.3 Bud Closing Bud closing is the final step for the scission of the bud and release of the virus particle into the outer environment. Bud closure would involve fusion of two ends of the apposing viral membranes as well as that of the apposing cell membranes leading to fission of the virus bud from the infected cell membrane (Figures 3.11c and 3.22). This would require bringing and holding the apposing membrane ends next to each other in close proximity, so that each end can find its counterpart causing fusion of corresponding lipid bilayers. Virus buds would then become separated from the membrane of the parent-infected cell. This model holds that two lipid bilayers are to be held in very close proximity for fusion to occur. Host and viral factors could have both positive and negative impact on bud release.

Some factors could interfere in bringing the apposing ends close to each other and therefore should be removed. Host factors such as actins and lipid rafts may interfere with bud release. Cellular actin microfilaments pushing the viral RNP into the bud may interfere with the final step of bud closing, and actin depolymerization is known to facilitate bud release. Other host factors could help in bringing and holding the membrane ends close to each other for fusion to occur and therefore should be brought to the pinching-off site. On the other hand, a number

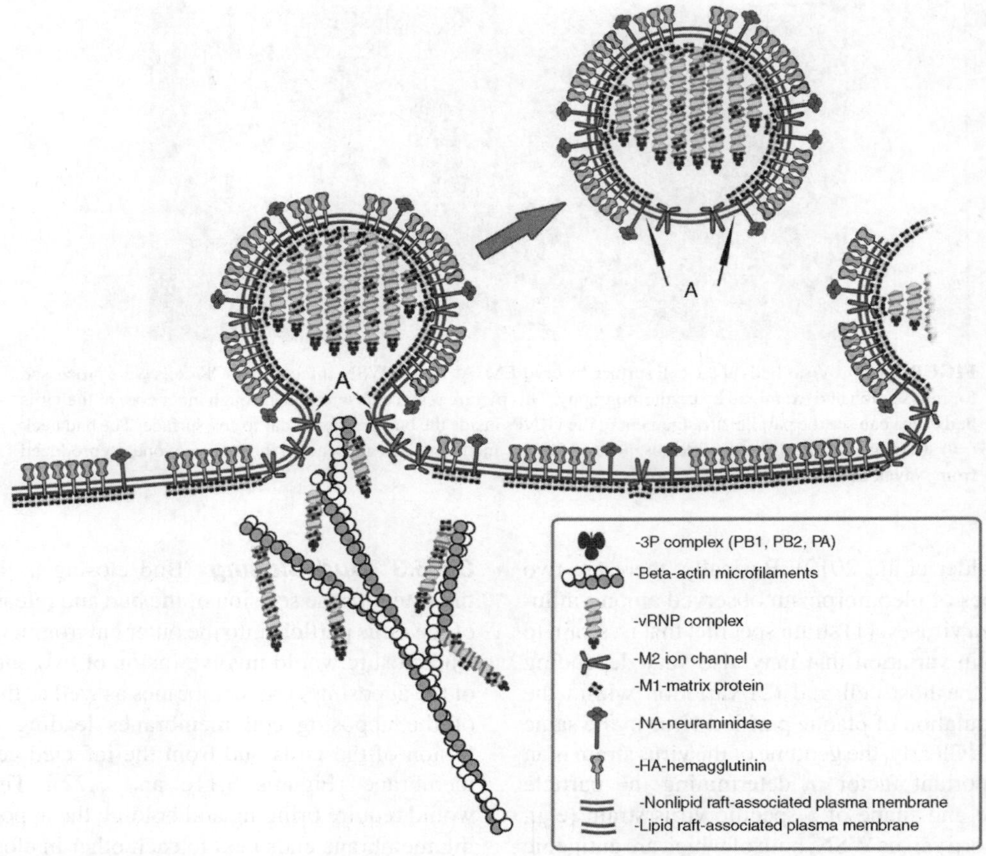

FIGURE 3.22 Schematic illustration of the pinching-off process of influenza virus bud. The pinching-off region (neck) is shown to be viral membrane devoid of lipid rafts (Barman and Nayak, 2007), devoid of HA and NA spikes outside and M1 inside the lipid bilayers (Harris et al., 2006), and may contain M2 (Schroeder et al., 2005). Reproduced from Nayak et al. (2009). (*See the color version of this figure in Color Plate section.*)

of host components have been shown to activate bud release. For example, VPS (vesicular protein sorting) components are shown to facilitate bud release of HIV and other enveloped viruses. Bud scission of these viruses depends on the interaction of their L domain(s) with the components(s) of VPS pathways involved in giving rise to multivesicular bodies (MVBs). Tsg101 and AIP1/Alix, the components of ESCRT (endosomal sorting complex required for transport), interact with L domains and require the function of AAA-ATPase of Vps4 for bud release of HIV (Fujii et al., 2007; Demirov and Freed, 2004). However, some other type of action must occur in the case of orthomyxoviruses, since influenza virus proteins do not contain any identifiable L domain (s) and influenza virus budding is not affected by dominant Vps4 (Chen et al., 2007) or proteasome inhibitors (Hui and Nayak, 2001; Khor et al., 2003).

Among the orthomyxoviral components, three viral proteins, namely, NA, M2, and M1, have been shown to play critical roles in both virus morphogenesis and bud release of influenza viruses. Specific NA mutants in TMD and CT (cytoplasmic tail) affected virus morphology, generating elongated virus buds (Jin et al., 1997; Barman et al., 2004). Similarly, some M1 mutants produced elongated virus particles indicating the involvement of M1 in the last step of bud release (Burleigh et al., 2005; Nayak et al., 2004). Complete or partial deletion of the WSN M2 tail was also shown to cause attenuation of virus growth and to produce elongated or even filamentous particles in some mutants, indicating an important role of the M2 tail in viral assembly and morphogenesis (Iwatsuki-Horimoto et al., 2006) and also suggesting that the M2 tail affected particle release in VLP (virus-like particle) assay and affected viral morphology (Chen et al., 2008). VLPs are virus-like particles possessing virus-like morphology, but are noninfectious because they do not contain any viral genetic material (DNA or RNA). VLPs are produced by expression and self-assembly of viral structural proteins in a variety of cell culture systems, including mammalian cell lines, insect cell lines, yeast, and plant cells.

As mentioned previously, lipid rafts and cortical actin microfilaments, though critical in many aspects of the budding process, are inhibitory in the final step of bud closing and therefore should be removed from the pinching-off site. Clearly, the role of specific host factors varies in the bud release of different enveloped viruses. Although some host factors critical for bud release of HIV and other enveloped viruses are identified, other host factors required for bringing and holding the apposing viral and cellular membranes next to each other for facilitating fusion and fission are yet to be identified. Among the better understood viral components, M2 may play a critical role in the pinching-off process of influenza viruses. M2 when present in the neck of the bud may aid in bud release (Schroeder et al., 2005) by bringing together lipid microdomains that are devoid of lipid rafts in this region (Figure 3.22). Absence of M1 protein underneath the lipid bilayers and absence of spikes on the outer surface may indicate the absence of lipid rafts (Figures 3.3 and 3.4). From CT analysis, such lipid microdomains are proposed to be the preferred sites for the bud pinching off (Harris et al., 2006).

As mentioned previously, bud closing for influenza viruses is very inefficient and only a small fraction of virus buds are released, while the majority of virus buds remain attached to the cell membrane even though they appear mature (Figures 3.5 and 3.21). Both host and virus factors may be contributing toward the rate-limiting step of the pinching-off process. Influenza virus budding also appears to be an active, energy-dependent process and metabolic inhibitors, such as antimycin A (AmA), carbonyl cyanide *m*-chlorophenylhydrazone (CCCP), carbonylcyanide *p*-trifluoromethoxy-phenylhydrazone (FCCP), and oligomycin that prevent the synthesis of ATP, and ATP analogues, such as ATPγS (adenosine 5′-*O*-(3-thiotriphosphate) and AMP-PNP (5′-adenylylimidodiphosphate), are shown to inhibit influenza virus budding (Hui and Nayak, 2001). Therefore, limited energy at the end of infectious cycle may be a factor for the inefficient release of virus particles. Among other host factors, actin microfilaments may interfere with bud closing, and, conversely, disassembly of cortical actin microfilaments may facilitate it. This notion is supported by several observations, including the release of virus particles in abortively infected HeLa cells (Gujuluva et al., 1994), conversion of filamentous Udorn (H3N2) virus to spherical virus, and enhanced release of WSN and PR8 (antigenic variants of influenza A virus) spherical particles in polarized MDCK cells (Roberts and Compans, 1998; Simpson-Holley et al., 2002) by microfilament-disrupting agents.

Scission of influenza virus buds from infected cells is the last step in the completion of virus life cycle. This step appears to be rate limiting and morphological analysis by thin section transmission microscopy (Barman and Nayak, unpublished), scanning electron microscopy (Figure 3.5), and cryotomography (CT) (Figure 3.21) shows that a large number of mature virus particles remain attached to the cell membrane and only a relatively small fraction of virus buds ($\sim$10%) are released. The kinetics of virus release relative to the presence of virus buds awaits further investigation in order to understand the cause(s)and mechanism(s) for such inefficient bud completion.

3.5.6 Role of Viral Budding in Viral Pathogenesis

Since the host is usually infected at very low MOI and the severity of the disease syndrome caused by lytic viruses largely depends on the number of the cells of the affected organs (or tissues) killed by the infecting virus, factors contributing to productive replication, release of infectious progeny virus, virus yield, and budding will have a major role in the development and severity of disease production. In addition, the site of budding can be an important contributory factor in viral pathogenesis, particularly for such respiratory viruses as influenza and Sendai viruses. The influenza and Sendai viruses bud from the apical surface of polarized epithelial cells (e.g., bronchial epithelial cells) into the lumen of the lungs and are therefore usually pneumotropic, that is, restricted to the lungs, and do not cause viremia or invade other internal organs. However, occasionally some influenza viruses such as the fowl plague viruses (H5 or H7, both being antigenic variants of influenza A virus, with the designation of H5 or H7 indicating the hemagglutinin subtype specificity) and WSN (H1N1) virus are not restricted to the lungs and produce viremia infecting other internal organs (pantropism) and cause a high degree of mortality in infected animals. The viruses restricted to lungs are called "pneumotropic," whereas the viruses that cause viremia and spread to other internal organs are called "pantropic." In humans, most of the influenza viruses are pneumotropic and do not spread to other internal organs. However, it is not clear whether the Spanish influenza of 1918, the most devastating influenza pandemic in recorded human history that killed 20–40 million people worldwide, particularly affecting young healthy adults, was only pneumotropic, that is, restricted in lungs or also pantropic and invaded other internal organs. During 1918, some people died due to influenza pandemic, in addition to pneumonia, showing evidence of massive pulmonary hemorrhage and edema (Taubenberger, 1998). The 1918 flu virus, like fowl plague virus, may have caused viremia and infected other organs. Therefore, it is possible that 1918 highly virulent viruses were not restricted to lungs in chicken or humans. In recent years H5N1, the Hong Kong chicken influenza virus, which is extremely virulent and pantropic for chicken, causing viremia and spreading to other internal organs, also caused high morbidity and mortality in infected humans. The majority of H5N1-infected people exhibited clinical pneumonia (or acute respiratory distress syndrome), gastrointestinal symptoms, and impaired hepatic and renal function and therefore exhibited pantropic characteristics in humans. However, since this virus did not spread from human to human, it did not emerge as a major pandemic. On the other hand, unlike the avian H5N1 virus, the recent H1N1 swine influenza virus spread efficiently among humans and developed as a pandemic. Luckily, H1N1 swine flu virus was mostly pneumotropic and caused only a moderate pandemic. Why some influenza viruses are pneumotropic and others are pantropic is an important question for predicting the outcome of a major influenza epidemic or pandemic.

The severity of viral pathogenesis depends on both viral and host factors, including host immunity and cytokine production. The virulence determinants of influenza viruses are complex and multigenic. However, one factor that is thought to be critical in viral growth and virulence is the cleavability of HA into HA1 and HA2. Influenza virus is normally restricted to the lungs because its HA can be cleaved by tryptase Clara, a serine protease restricted to the lungs. However, some HA variants containing multiple basic amino acids at the HA1–HA2 junction, found only in H5 and H7 avian subtypes, can be cleaved by furin and subtilisin-type enzymes that are present ubiquitously throughout the body, enabling such viruses to grow in other organs and possibly contributing to pantropism. In addition, the NA of some influenza viruses (e.g., WSN virus) binds to plasminogen and activates its conversion into plasmin in the vicinity of HA, and the activated plasmin cleaves HA into HA1 and HA2, rendering the virus infectious. This, therefore, enables viruses such as WSN, which lack multiple basic residues in its HA, to grow and multiply in tissues other than the lungs.

However, although the cleavage of HA into HA1 and HA2 is a major virulence factor, it is not the only factor contributing to the pantropism of a normally pneumotropic flu virus. For example, although WSN virus is pantropic and neurovirulent in the mouse, gene reassortment experiments demonstrated that the WSN NA gene responsible for the cleavage of HA was not sufficient for neurovirulence in chickens or mice. Other WSN genes, such as the M and NS genes, in addition to the NA gene, were required for neurovirulence and, therefore, likely affected pantropism. The function of M and NS genes in neurovirulence is not known. The M gene in Sendai virus has been shown to affect apical versus basolateral budding and contribute to the pantropism of F1-R Sendai virus mutant (Tashiro and Seto, 1997). Therefore, it is possible that, in addition to increased cleavability of HA, the pantropic virus causes alteration in apical budding, releasing more virus basolaterally. Since blood vessels are proximal to the basolateral surface of cells, basolateral budding would facilitate more viruses entering into the blood, causing viremia, and invading other internal organs. Therefore, pantropic influenza viruses such as WSN/33 virus or highly virulent Hong Kong H5N1 and H7N1 fowl plague viruses may also cause altered budding from apical and basolateral surfaces. Thus, altered budding may be considered an important trait for the virulence of a specific strain of influenza virus. However, the role of the altered budding in influenza virus pathogenesis remains to be determined.

Sendai virus, like influenza virus, is a pneumotropic mouse virus that buds apically. However, a Sendai virus mutant F1-R that exhibited pantropism possessed two potentially important characteristics (Tashiro and Seto, 1997): (1) ubiquitous cleavage mutation of F into F1 and F2 due to the presence of multiple basic residues and (2) altered budding from both apical and basolateral surfaces. Contrastingly, the Sendai virus mutants that exhibited only one of these two traits, either cleavage of F into F1 and F2 or altered budding, did not cause viremia or pantropism in the mouse. This would also support the argument that altered budding may be a factor that facilitates viremia and pantropism. Therefore, altered apical versus basolateral budding could be an important factor in the release of virus into the blood, invasion of internal organs, pantropism, and higher virulence of a specific virus strain.

3.6 CONCLUSIONS

The replication and morphogenesis processes of viruses are different from those of prokaryotic or eukaryotic organisms. In this chapter, some of the general steps involved in the viral infectious cycle, including entry, uncoating, transcription, translation, replication, and assembly processes and the possible role of budding in viral pathogenesis have been presented. Of these, viral morphogenesis is the most obscure phase in the virus life cycle. Yet knowledge of how the particles are formed during this morphogenetic stage is fundamental to understanding virus growth and multiplication and, therefore, is crucial in defining viral infectivity, transmission, virulence, tissue tropism, host specificity, and pathogenesis and contributes to an overall understanding of the disease process and progression of disease, including host morbidity and mortality. In addition, the site of budding can affect virus virulence and pathogenesis. Elucidation of the viral replication and assembly processes is critical in terms of enabling us to find ways to block these steps and thereby intervene in the viral life cycle and disease process. Much remains to be done to achieve these necessary research goals, particularly in terms of elucidating those stages of the viral assembly process that relate to how viral components are brought to the assembly site, how those components interact with each other at the assembly site, and how viral budding actually occurs. A better understanding of viral replication and morphogenesis may lead us to develop novel therapeutic agents capable of interfering with these critical steps in viral multiplication, pathogenesis, and virulence.

ACKNOWLEDGMENTS

Research in the author's laboratory was supported by grants from the National Institutes of Health. The author thanks Jonathan Rodger and Philip Postovoit for electronic reproduction of some of the figures. Author acknowledges the help of Sakar Shivakoti in manuscript preparation.

ABBREVIATIONS AND DEFINITIONS

Ambisense RNA	These RNAs are of partly positive-sense and partly negative-sense polarity.
ATPγS	(adenosine 5′-*O*-(3-thiotriphosphate) and AMP-PNP (5′-adenylylimidodiphosphate) are ATP analogues.
BAR (Bin–amphiphysin–Rvs) domains	BAR domains are helical domains found in proteins involved in vesiculation processes including endocytosis, intracellular trafficking, budding, and so on that require membrane bending. BAR domains interact with endocytic and cytoskeletal machinery including GTPase dynamin and possess dimerization motifs sensing and inducing membrane curvature. BAR domain proteins include endophilins, GTPase activating proteins, amphiphysin, Arfaptin, and so on.
Capsid coat or shell	The protein shell in contact with or directly surrounding the viral nucleic acid (genome).
CAP	The 5′-cap is found on the 5′-end of a eukaryotic mRNA molecule with the exception of some viral RNAs and consists of an altered guanine nucleotide connected to the mRNA via an unusual 5′- to 5′-triphosphate linkage. This guanosine is methylated on the 7 position and is referred to as a 7-methylguanosine cap, abbreviated m^7G. The 5′-cap of mRNA facilitates nuclear export, prevents mRNA degradation by exonucleases, and promotes ribosome binding and protein translation.
Capsomeres	These are morphological units that form capsids. Capsomeres consist of oligomers of one or more viral proteins.
CPE	Cytopathic effect; could be due to apoptosis, necrosis, or syncytium formation.
cRNA	Full-length plus-strand template RNA complementary to the minus-strand genomic RNA.
DI	Defective interfering viruses. DI virus particles contain a smaller viral genome, are noninfectious, and need the help of infec-

tious (wild-type) virus for replication but, in turn, interfere with replication of homologous infectious (standard) viruses.

Dominant negative — A mutation whose gene product interacts with the intracellular components as the wild-type gene product and thereby adversely affects the function of normal, wild-type gene product within the same cell.

Ectodomain — The portion of the transmembrane protein that remains exposed outside the cell or virus particle.

EIS — These are the *cis* elements of the unsegmented minus-strand RNA genome (e.g., VSV). "E" denotes the end of transcription termination and polyadenylation sequence; "I" stands for intragenic sequence not transcribed in messenger RNA (mRNA); "S" indicates the start sequence for the next mRNA.

Endocytic pathways (endocytosis) — The process of internalization of external macromolecules or viruses, which involves specific binding to cell surface receptors. Viruses use this mechanism to enter into host cells. In this process, clathrin-coated vesicles and subcellular organelles such as endosomes and lysosomes are involved.

Envelope — The viral membrane containing the lipid bilayer and associated proteins that surround the nucleocapsid of enveloped viruses and form the outermost barrier of the enveloped virus particle.

Enveloped viruses — Viruses that possess an envelope or membrane surrounding the nucleocapsid. For enveloped viruses, the naked nucleocapsid is not infectious.

Episomal, extrachromosomal — The state of existence of nucleic acid molecules that do not became integrated into host cell chromosomes. They exist and multiply independently within the cell nucleus or cell cytoplasm.

Escape mutants — Virus mutants that are not neutralized by antibodies. These viruses possess amino acid change in the epitope and therefore no longer bind to the neutralizing antibody.

Exocytic pathways (exocytosis) — The mechanism for transporting intracellular transmembrane or secretory proteins from intracellular compartments to the cell surface or extracellular environment. In this process, various subcellular compartments, such as endoplasmic reticulum and Golgi complexes, are involved in protein transport.

Genome — The complete genetic information (DNA or RNA) of an organism.

Glycosylation — In this process, one or several carbohydrate groups are attached to proteins during their transport through the exocytic pathways. Sugar residues are attached at specific sites to amino acids such as serine or threonine (for O-linked) or asparagine (for N-linked) carbohydrate moieties. These carbohydrate moieties are also called glycans.

HBV — Hepatitis B virus (family Hepadnaviridae, genus *Orthohepadnavirus*).

Helical capsids — These structures are spiral, spring-like, and flexible rods. The RNA genome in a helical capsid is exposed (influenza viruses) or enclosed (paramyxoviruses, rhabdoviruses) by the nucleoprotein molecules constituting the nucleocapsid.

H1N1 — Antigenic variant of influenza A virus (family Orthomyxoviridae, genus *Influenzavirus A*). H denotes hemagglutinin (H1–H15) and N stands for neuraminidase (N1–N9) subtypes.

HIV — Human immunodeficiency virus 1 and 2 (family Retroviridae, genus *Lentivirus*).

hnRNA — Heterogeneous nuclear RNA (hnRNA), also called pre-mRNA or immature mRNA, is an incompletely processed single strand of ribonucleic acid (RNA) present in the nucleus. It contains introns and exons and is processed by splicing to eliminate introns and become mature mRNA that can be translated into protein.

HSV — Human herpesvirus 1 and 2 (family Herpesviridae, genus *Simplexvirus*).

HRV14 — Human rhinovirus strain 14 (antigenic variant of human rhinovirus B, family Picornaviridae, genus *Enterovirus*)

ICAM-1 — Intracellular adhesion molecule-1, the receptor for rhinoviruses.

Icosahedron, icosadeltahedron, icosahedral symmetry, icosahedral capsids — Icosahedron is a structure with a two-, three-, and fivefold rotational symmetry. It is a polyhedron with 20 faces, 12 vertices, and 30 edges. Most icosahedral viruses have 60 (multiple of 60) subunits (e.g., polioviruses, togaviruses).

Inclusion bodies — Microscopic structures, produced in some virus-infected cells consisting of viral proteins, nucleic acids, and cellular elements (particularly cytoskeletal elements). Inclusion bodies can be intranuclear (herpesviruses) and intracytoplasmic (paramyxoviruses).

Kozak's rule — Most eukaryotic mRNAs contain a short recognition sequence (ACCATGG) that facilitates binding of mRNA to the small subunit of the ribosome and initiation of protein translation (Kozak, 1986).

LB — Lateral bodies found in poxviruses.

LCMV — Lymphocytic choriomeningitis virus (family Arenaviridae, genus *Arenavirus*).

Lipid rafts — These are lipid microdomains containing increased levels of glycosphingolipids, including sphingomyelins, cholesterol, and long saturated fatty acids but decreased phosphatidylcholines. These specialized membrane microdomains are relatively resistant to nonionic detergents, such as Triton X-100 or Brij-98, at low temperatures (e.g., 4°C). These specialized membrane microdomains play important roles in the assembly of signaling molecules, influence membrane fluidity and membrane protein trafficking, and regulate neurotransmission and receptor trafficking, virus assembly, and virus budding. Lipid rafts are more ordered and tightly packed compared to the surrounding nonraft lipid bilayer.

LT — Large T antigen of SV40.

MOI — Multiplicity of infection, that is, infectious units adsorbed per cell.

Naked or nonenveloped viruses — These viruses do not have any membrane and the nucleocapsids represent the infectious virus.

Nucleocapsid — The complete nucleic acid–protein complex of a virus particle. Sometimes the term viral ribonucleoprotein (vRNP) is used to indicate nucleocapsid (e.g., vRNP of influenza viruses).

Panhandle — A circular nucleic acid structure of single-stranded (ss) DNA or RNA with a double-stranded stem at the end produced by intrastrand hybridization due to partial complementarity of the nucleic acid sequences at both the 5′ and 3′ termini of ssRNA or DNA. The panhandle structures function as the promoter and are important for transcription and replication.

pfu — Plaque-forming unit.

Phagocytosis, viropexis — Uptake of particles by cells not totally dependent on receptor-mediated endocytosis. The particle on the surface is engulfed by the cell membrane into a phagocytic vesicle. These phagocytic vesicles then undergo similar changes as the endosome. Poxviruses enter cells by phagocytosis.

Poly(A) — Polyadenylation at the 3′-end of an RNA molecule.

Prions — These are infectious protein molecules without any DNA or RNA and thought to cause transmissible and/or inherited neurodegenerative diseases known as transmissible spongiform encephalopathies. These include Creutzfeldt–Jakob disease, kuru, and Gerstmann–Straussler syndrome in humans, as well as scrapie in sheep and goats and mad cow disease in cattle. The infectious prion proteins are modified forms of normal proteins encoded by a host gene. The normal prion protein having alpha helices in its secondary structure is converted into beta sheets for the secondary structure in diseased animals.

Protomer — The term often used to indicate a structural unit containing one or more nonidentical protein subunits. Promoters are used as a building block for virus capsid assembly.

RDRP — RNA-dependent RNA polymerase, also called RNA transcriptase and RNA replicase.

RNA of positive and negative polarity — The RNA strand of the same polarity as the mRNA-encoding proteins and is called positive-, plus-, or (+) strand RNA. When the RNA is of polarity opposite to the mRNA (i.e., cannot code for a protein), it is called negative-, minus-, or (–) strand RNA.

RNP — Ribonucleoprotein. Viral nucleoprotein binding to vRNA is called vRNP.

RT — Reverse transcriptase, RNA-dependent DNA polymerase.

ST — Small T antigen of SV40.

Structural and nonstructural proteins — Structural proteins are those proteins that are found in virions as components of capsid or envelope. Nonstructural proteins are those virally encoded proteins that are produced in the infected cells but not found in virions. Nonstructural proteins are usually catalytic and regulatory in nature and are also involved in modifying host functions.

Synchronous infection — When all cells in the culture are infected simultaneously. Cells are infected at a high MOI (>5) and at low temperatures (4°C). Then the temperature is raised to 37°C to permit entry and uncoating of all cell-bound viruses at the same time.

Syncytium (multinucleated giant cells) — Cells possessing multiple nuclei are formed due to fusion among a number of cells. Usually, viruses that can undergo fusion at a neutral pH (paramyxoviruses, retroviruses) produce syncytium.

Temperature-sensitive (ts) mutant — A mutant virus that will replicate at a permissive (low) temperature but not at the nonpermissive or restrictive (high) temperature. This phenotype is usually caused by missense mutations of one or more nucleotides, causing alteration of amino acid(s) of a protein that cannot assume the functional configuration at the nonpermissive (restrictive) temperature.

TGN — Trans-Golgi network.

Transmembrane proteins — These are membrane proteins that are anchored to the membrane by spanning the lipid bilayer of the membrane via transmembrane domains. These proteins can be classified as type I (e.g., influenza virus HA), type II (e.g., influenza virus NA), type III (e.g., influenza virus M2), or complex (e.g., coronaviral E1) depending on the orientation of the NH_2 and COOH termini (type I, II, or III), cleavage of signal peptide (type I), and multiple transmembrane spanning domains (complex).

Virion — The entire virus particle. It usually refers to infectious or complete virus particle as opposed to noninfectious or defective virus particles.

Viroids — These are small, circular, single-stranded infectious RNA molecules without a protein coat or capsid and cause a number of plant diseases, including potato spindle tuber disease, cucumber pale fruit disease, citrus exocortis disease, cadang-cadang (coconuts) disease, and so on.

VLPs or virus-like particles — These possess virus-like morphology but are noninfectious because they do not contain any viral genetic material (DNA or RNA). VLPs are produced by expression of viral structural proteins, such as envelope or capsid proteins, in a variety of cell culture systems, including mammalian cell lines, insect cell lines, yeast, and plant cells. VLPs are produced from a wide variety of virus families, including Orthomyxoviridae (influenza virus), Parvoviridae (e.g., adeno-associated virus), Retroviridae (e.g., HIV), Flaviviridae (e.g., hepatitis C virus), and so on. VLPs can be used as vaccine (e.g., hepatitis B virus, human papilloma virus) and as a delivery system for genes and therapeutics.

WSN/33 (H1N1) — A neurotropic variant of WS/33 (H1N1), a human influenza virus isolated in 1933 (Francis and Moore, 1940).

REFERENCES

Akarsu, H., Burmeister, W. P., Petosa, C., Petit, I., Muller, C. W., Ruigrok, R. W., and Baudin, F. (2003). Crystal structure of the M1 protein-binding domain of the influenza A virus nuclear export protein (NEP/NS2). *Embo J.* 22(18), 4646–4655.

Ali, A., Avalos, R. T., Ponimaskin, E., and Nayak, D. P. (2000). Influenza virus assembly: effect of influenza virus glycoproteins on the membrane association of M1 protein. *J. Virol.* 74(18), 8709–8719.

Avalos, R. T., Yu, Z., and Nayak, D. P. (1997). Association of influenza virus NP and M1 proteins with cellular cytoskeletal elements in influenza virus-infected cells. *J. Virol.* 71, 2947–2958.

Bancroft, C. T. and Parslow, T. G. (2002). Evidence for segment-nonspecific packaging of the influenza a virus genome. *J. Virol.* 76(14), 7133–7139.

Barman, S., Adhikary, L., Chakraborti, A. K., Bernas, C., Kawaoka, Y., and Nayak, D. P. (2004). Role of transmembrane domain and cytoplasmic tail amino acid sequences of influenza A virus neuraminidase in raft-association and virus budding. *J. Virol.* 78(10), 5258–5269.

Barman, S., Adhikary, L., Kawaoka, Y., and Nayak, D. P. (2003). Influenza A virus hemagglutinin containing basolateral localization signal does not alter the apical budding of a recombinant influenza A virus in polarized MDCK cells. *Virology* 305(1), 138–152.

Barman, S. and Nayak, D. P. (2000). Analysis of the transmembrane domain of influenza virus neuraminidase, a type II transmembrane glycoprotein, for apical sorting and raft association. *J. Virol.* 74, 6538–6545.

Barman, S. and Nayak, D. P. (2007). Lipid raft disruption by cholesterol depletion enhances influenza A virus budding from MDCK cells. *J. Virol.* 81(22), 12169–12178.

Baudin, F., Petit, I., Weissenhorn, W., and Ruigrok, R. W. H. (2001). *In vitro* dissection of the membrane and RNP binding activities of influenza virus M1 protein. *Virology* 281, 102–108.

Baumeister, W. (2002). Electron tomography: towards visualizing the molecular organization of the cytoplasm. *Curr. Opin. Struct. Biol.* 12, 679–684.

Burleigh, L. M., Calder, L. J., Skehel, J. J., and Steinhauer, D. A. (2005). Influenza A viruses with mutations in the m1 helix six domain display a wide variety of morphological phenotypes. *J. Virol.* 79(2), 1262–1270.

Calder, L. J., Wasilewski, S., Berriman, J.A., and Rosenthal, P. B. (2010). *Structural organization of a filamentous influenza A virus. Proc. Natl. Acad. Sci. USA.* 107(23), 10685–10690.

Carrasco, M., Amorom, M. J., and Digard, P. (2004). Lipid raft-dependent targeting of the influenza A virus nucleoprotein to the apical plasma membrane. *Traffic* 5(12), 979–992.

Chen, B. J., Leser, G. P., Jackson, D., and Lamb, R. A. (2008). The influenza virus M2 protein cytoplasmic tail interacts with the M1 protein and influences virus assembly at the site of virus budding. *J. Virol.* 82(20), 10059–10070.

Chen, B. J., Leser, G. P., Morita, E., and Lamb, R. A. (2007). Influenza virus hemagglutinin and neuraminidase, but not the matrix protein, are required for assembly and budding of plasmid-derived virus-like particles. *J. Virol.* 81(13), 7111–7123.

Davis, B. D., Dulbecco, R., Eisen. H. N., and Ginsberg, H. S. (1990). *Microbiology*, 4th edition. J. B. Lippincott, Philadelphia, PA.

Demirov, D. G. and Freed, E. O. (2004). Retrovirus budding. *Virus Res.* 106(2), 87–102.

Duhaut, S. D. and McCauley, J. W. (1996). Defective RNAs inhibit the assembly of influenza virus genome segments in a segment-specific manner. *Virology* 216(2), 326–337.

Elton, D., Digard, P., Tiley, L., and Ortin, J. (2006). Structure and function of the influenza virus RNP. *Influenza Virology; Current Topics*. Caister Academic Press, Wymondham, pp. 1–36.

Enami, M., Sharma, G., Benham, C., and Palese, P. (1991). An influenza virus containing nine different RNA segments. *Virology* 185(1), 291–298.

Fields, B. N., and Knipe, D. M. (1990). "*Fields' Virology*," Vols. 1 and 2. 2nd ed. Raven, New York.

Flint, J., Enquist, L., Krug, R., Racaniello, V. R., and Skalka, A. M. (1999). *Principles of Virology: Molecular Biology, Pathogenesis, and Control.* American Society of Microbiology, Washington, DC.

Francis, T. and Moore, H. E. (1940). A study of the neurotropic tendency in strains of virus of epidemic influenza. *J. Exp. Med.* 72, 717–728.

Fujii, K., Fujii, Y., Noda, T., Muramoto, Y., Watanabe, T., Takada, A., Goto, H., Horimoto, T., and Kawaoka, Y. (2005). Importance of both the coding and the segment-specific noncoding regions of the influenza A virus NS segment for its efficient incorporation into virions. *J. Virol.* 79(6), 3766–3774.

Fujii, Y., Goto, H., Watanabe, T., Yoshida, T., and Kawaoka, Y. (2003). Selective incorporation of influenza virus RNA segments into virions. *Proc. Natl. Acad. Sci. U. S. A.* 100(4), 2002–2007.

Fujii, K., Hurley, J. H., and Freed, E. O. (2007). Beyond Tsg101: the role of Alix in 'ESCRTing' HIV-1. *Nat. Rev. Microbiol.* 5(12), 912–916.

Ganser-Pornillos, B. K., Yeager, M., and Sundquist, W. I. (2008). The structural biology of HIV assembly. *Curr. Opin. Struct. Biol.* 18(2), 203–217.

Garoff, H., Wilschut, J., Liljestrom, P., Wahlberg, J. M., Bron, R., Suomalainen, M., Smyth, J., Salminen, A., Barth, B. U., Zhao, H., et al. (1994). Assembly and entry mechanisms of Semliki Forest virus. *Arch. Virol.* Suppl 9, 329–338.

Gujuluva, C. N., Kundu, A., Murti, K. G., and Nayak, D. P. (1994). Abortive replication of influenza virus A/WSN/33 in HeLa229 cells: defective viral entry and budding processes. *Virology* 204(2), 491–505.

Harris, A., Cardone, G., Winkler, D. C., Heymann, J. B., Brecher, M., White, J. M., and Steven, A. C. (2006). Influenza virus pleiomorphy characterized by cryoelectron tomography. *Proc. Natl. Acad. Sci. U. S. A.* 103(50), 19123–19127.

Hobman, T. C. (1993). Targeting of viral glycoproteins to the Golgi complex. *Trends Microbiol.* 1 (4), 124–130.

Holopainen, J. M., Angelova, M. I., and Kinnunen, P. K. (2000). Vectorial budding of vesicles by asymmetrical enzymatic formation of ceramide in giant liposomes. *Biophys. J.* 78(2), 830–838.

Hui, E. K. and Nayak, D. P. (2001). Role of ATP in influenza virus budding. *Virology* 290(2), 329–341.

Iwatsuki-Horimoto, K., Horimoto, T., Noda, T., Kiso, M., Maeda, J., Watanabe, S., Muramoto, Y., Fujii, K., and Kawaoka, Y. (2006). The cytoplasmic tail of the influenza A virus M2 protein plays a role in viral assembly. *J. Virol.* 80(11), 5233–5240.

Jin, H., Leser, G. P., Lamb, R. A., and Zhang, J. (1997). Influenza virus hemagglutinin and neuraminidase cytoplasmic tails control particle shape. *EMBO J.* 16(6), 1236–1247.

Khor, R., McElroy, L. J., and Whittaker, G. R. (2003). The ubiquitin-vacuolar protein sorting system is selectively required during entry of influenza virus into host cells. *Traffic* 4(12), 857–868.

Kozak, M. (1986). Point mutations define a sequence flanking the AUG initiator codon that modulates translation by eukaryotic ribosomes. *Cell* 44(2), 283–292.

Liang, Y., Hong, Y., and Parslow, T. G. (2005). *cis*-Acting packaging signals in the influenza virus PB1, PB2, and PA genomic RNA segments. *J. Virol.* 79(16), 10348–10355.

Liang, Y., Huang, T., Ly, H., Parslow, T. G., and Liang, Y. (2008). Mutational analyses of packaging signals in influenza virus PA, PB1, and PB2 genomic RNA segments. *J. Virol.* 82(1), 229–236.

Lin, S., Naim, H. Y., Rodriguez, A. C., and Roth, M. G. (1998). Mutations in the middle of the transmembrane domain reverse the polarity of transport of the influenza virus hemagglutinin in MDCK epithelial cells. *J. Cell Biol.* 142(1), 51–57.

Maisner, A., Klenk, H., and Herrler, G. (1998). Polarized budding of measles virus is not determined by viral surface glycoproteins. *J. Virol.* 72 (6), 5276–5278.

Mak, J. and Kleiman, L. (1997). Primer tRNAs for reverse transcription. *J. Virol.* 71, 8087–8095.

Marsh, G. A., Hatami, R., and Palese, P. (2007). Specific residues of the influenza A virus hemagglutinin viral RNA are important for efficient packaging into budding virions. *J. Virol.* 81 (18), 9727–9736.

Marsh, G. A., Rabadan, R., Levine, A. J., and Palese, P. (2008). Highly conserved regions of influenza A virus polymerase gene segments are critical for efficient viral RNA packaging. *J. Virol.* 82(5), 2295–2304.

McCown, M. F. and Pekosz, A. (2005). The influenza A virus M2 cytoplasmic tail is required for infectious virus production and efficient genome packaging. *J. Virol.* 79(6), 3595–3605.

McCown, M. F. and Pekosz, A. (2006). Distinct domains of the influenza A virus M2 protein cytoplasmic tail mediate binding to the M1 protein and facilitate infectious virus production. *J. Virol.* 80(16), 8178–8189.

Mora, R., Rodriguez-Boulan, E., Palese, P., and Garcia-Sastre, A. (2002). Apical budding of a recombinant influenza A virus expressing a hemagglutinin protein with a basolateral localization signal. *J. Virol.* 76(7), 3544–3553.

Muramoto, Y., Takada, A., Fujii, K., Noda, T., Iwatsuki-Horimoto, K., Watanabe, S., Horimoto, T., Kida, H., and Kawaoka, Y. (2006). Hierarchy among viral RNA (vRNA) segments in their role in vRNA incorporation into influenza A virions. *J. Virol.* 80(5), 2318–2325.

Murti, K. G., Webster, R. G., and Jones, I. M. (1988). Localization of RNA polymerases on influenza viral ribonucleoproteins by immunogold labeling. *Virology* 164(2), 562–566.

Nakajima, K., Ueda, M., and Sugiura, A. (1979). Origin of small RNA in von Magnus particles of influenza virus. *J. Virol.* 29(3), 1142–1148.

Nayak, D. P. (1997). Influenza virus infections. In: Encyclopedia of Human Biology (Dulbecco R, ed). Vol. 5; 67–80, Academic press.

Nayak, D. P. (2000). Virus morphology, replication, and assembly. In: Hurst, C. (ed.), *Viral Ecology*. Academic Press, New York, pp. 63–124.

Nayak, D. P., Balogun, R. A., Yamada, H., Zhou, Z. H., and Barman, S. (2009). Influenza virus morphogenesis and budding. *Virus Res.* 143, 147–161.

Nayak, D. P., Chambers, T. M., and Akkina, R. K. (1985). Defective-interfering (DI) RNAs of influenza viruses: origin, structure, expression, and interference. *Curr. Top. Microbiol. Immunol.* 114, 103–151.

Nayak, D. P., Chambers, T. M., and Akkina, R. M. (1989). Structure of defective-interfering RNAs of influenza viruses and their role in interference. In: Krug, R. M. (ed.), *The Influenza Viruses*. Plenum Press, New York, pp. 269–317.

Nayak, D. P. and Hui, E.-K. W. (2004). The role of lipid microdomains in virus biology. In: Quinn, P. J. (ed.), *Subcellular Biochemistry*, Vol. 37. Kluwer Academic/Plenum Publishers, New York, pp. 443–491.

Nayak, D. P., Hui, E.-K. W., and Barman, S. (2004). Assembly and budding of influenza virus. *Virus Res.* 106, 147–165.

Noton, S. L., Medcalf, E., Fisher, D., Mullin, A. E., Elton, D., and Digard, P. (2007). Identification of the domains of the influenza A virus M1 matrix protein required for NP binding, oligomerization and incorporation into virions. *J. Gen. Virol.* 88 (8), 2280–2290.

Noda, T., Sagara, H., Yen, A., Takada, A., Kida, H., Cheng, R. H., and Kawaoka, Y. (2006). Architecture of ribonucleoprotein complexes in influenza A virus particles. *Nature* 439(7075), 490–492.

Odagiri, T. and Tobita, K. (1990). Mutation in NS2, a nonstructural protein of influenza A virus, extragenically causes aberrant replication and expression of the PA gene and leads to generation of defective interfering particles. *Proc. Natl. Acad. Sci. U. S. A.* 87(15), 5988–5992.

Ozawa, M., Fujii, K., Muramoto, Y., Yamada, S., Yamayoshi, S., Takada, A., Goto, H., Horimoto, T., and Kawaoka, Y. (2007). Contributions of two nuclear localization signals of influenza A virus nucleoprotein to viral replication. *J. Virol.* 81(1), 30–41.

Pattnaik, A. K. and Wertz, G. W. (1991). Cells that express all five proteins of vesicular stomatitis virus from cloned cDNAs support replication, assembly, and budding of defective interfering particles. *Proc. Natl. Acad. Sci. U. S. A.* 88(4), 1379–1383.

Peter, B. J., Kent, H. M., Mills, I. G., Vallis, Y., Butler, P. J., Evans, P. R., and McMahon, H. T. (2004). BAR domains as sensors of membrane curvature: the amphiphysin BAR structure. *Science* 303(5657), 495–499.

Pralle, A., Keller, P., Florin, E. L., Simons, K., and Horber, J. K. (2000). Sphingolipid–cholesterol rafts diffuse as small entities in the plasma membrane of mammalian cells. *J. Cell Biol.* 148, 997–1008.

Roberts, P. C. and Compans, R. W. (1998). Host cell dependence of viral morphology. *Proc. Natl. Acad. Sci. U. S. A.* 95(10), 5746–5751.

Rossen, J. W., de Beer, R., Godeke, G. J., Raamsman, M. J., Horzinek, M. C., Vennema, H., and Rottier, P. J. (1998). The viral spike protein is not involved in the polarized sorting of coronaviruses in epithelial cells. *J. Virol.* 72(1), 497–503.

Sanger, C., Muhlberger, E., Ryabchikova, E., Kolesnikova, L., Klenk, H. D., and Becker, S. (2001). Sorting of Marburg virus surface protein and virus release take place at opposite surfaces of infected polarized epithelial cells. *J. Virol.* 75(3), 1274–1283.

Schroeder, C., Heider, H., Moncke-Buchner, E., and Lin, T. I. (2005). The influenza virus ion channel and maturation cofactor M2 is a cholesterol-binding protein. *Eur. Biophys. J.* 34(1), 52–66.

Simpson-Holley, M., Ellis, D., Fisher, D., Elton, D., McCauley, J., and Digard, P. (2002). A functional link between the actin cytoskeleton and lipid rafts during budding of filamentous influenza virions. *Virology* 301(2), 212–225.

Smith, G. L. and Hay, A. J. (1982). Replication of the influenza virus genome. *Virology* 118(1), 96–108.

Tamm, L. K. (2003). Hypothesis: spring-loaded boomerang mechanism of influenza hemagglutinin-mediated membrane fusion. *Biochim. Biophys. Acta* 1614(1), 14–23.

Tashiro, M. and Seto, J. T. (1997). Determinants of organ tropism of Sendai virus. *Frontiers Biosci.* 2, 588–591.

Taubenberger, J. K. (1998). Influenza virus hemagglutinin cleavage into HA1, HA2: no laughing matter. *Proc. Natl. Acad. Sci. U. S. A.* 95, 9713–9715.

Elton, D., Digard, P., Tiley, L., and Ortin, J. (2006). Virus RNP. In: Kawaoka, Y. (ed.), *Influenza Virology; Current Topics*. Caister Academic Press, Wymondham, pp. 1–36.

Watanabe, K., Handa, H., Mizumoto, K., and Nagata, K. (1996). Mechanism for inhibition of influenza virus RNA polymerase activity by matrix protein. *J. Virol.* 70(1), 241–247.

Watanabe, T., Watanabe, S., Noda, T., Fujii, Y., and Kawaoka, Y. (2003). Exploitation of nucleic acid packaging signals to generate a novel influenza virus-based vector stably expressing two foreign genes. *J. Virol.* 77(19), 10575–10583.

Whittaker, G. R. and Digard, P. (2006). Entry and intracellular transport of influenza virus. In: Kawaoka, Y. (ed.), *Influenza Virology; Current Topics*. Caister Academic Press, Wymondham, pp. 37–64.

Winkler, H. and Taylor, K. A. (2006). Accurate marker-free alignment with simultaneous geometry determination and reconstruction of tilt series in electron tomography. *Ultramicroscopy* 106, 240–254.

Ye, Z., Liu, T., Offringa, D. P., McInnis, J., and Levandowski, R. A. (1999). Association of influenza virus matrix protein with ribonucleoproteins. *J. Virol.* 73(9), 7467–7473.

Zhirnov, O. P. (1992). Isolation of matrix protein M1 from influenza viruses by acid-dependent extraction with nonionic detergent. *Virology* 186(1), 324–330.

Zurzolo, C., Polistina, C., Saini, M., Gentile, R., Aloj, L., Migliaccio, G., Bonatti, S., and Nitsch, L. (1992). Opposite polarity of virus budding and of viral envelope glycoprotein distribution in epithelial cells derived from different tissues. *J. Cell Biol.* 117(3), 551–564.

CHAPTER 4

THE (CO)EVOLUTIONARY ECOLOGY OF VIRUSES

MICHAEL J. ALLEN
Plymouth Marine Laboratory, Plymouth, UK

CONTENTS

4.1 VIR-OLUTION: SETTING THE SCENE

There is much debate on the precise status of viruses: Can they be considered alive? Do they have a place on the tree of life? How long have they existed? Do they predate the first living cells? Should the different types of viruses really be considered under the same "virus" banner? Regardless of the answer to these questions, it is undeniable that, whatever their status, viruses have had and continue to have a profound influence on the composition and function of the planet's living biota (Villarreal and Witzany, 2010). By their very definition, viruses, as obligate intracellular parasites, manipulate and selfishly hijack their host organisms purely for their own survival. This in itself leads to an interesting paradox: any virus that is too successful will ultimately go extinct since it will have no host left to infect. This paradox has effectively led to the field of viral ecology whereby viruses and their hosts are in a continuous, yet hugely dynamic and intricate relationship. These complex relationships between hosts and their viruses are at least as old as life on Earth itself. Clearly, the roots run deep in viral family trees and their

Studies in Viral Ecology: Animal Host Systems: Volume 2, First Edition. Edited by Christon J. Hurst.

interaction with their host(s) will run just as deep. The diverse nature of viral genomic material betrays their multiple and ancient ancestral origins (i.e., single- versus double-stranded, RNA versus DNA genomes). This polyphyletic group consists of many distinct lineages with independent origins that are all grouped under the "virus" banner by virtue of their lifestyle. Thus, a comprehensive tome on the topic of virus–host coevolution would need to encompass such a wide range of systems of such varying nature that it would justify at least a book all to itself, not merely a chapter! For this reason, I will attempt to provide an overview of the issues and processes associated with virus–host coevolution, using specific examples wherever necessary to illustrate points, but retaining a more generalist approach to the topic. Working with viruses has taught us many things in the life sciences, chiefly expect the unexpected and that there are exceptions to every rule. With this in mind, I invite the reader to read on with an open mind, never take anything at face value, question all ideas and hypotheses herein, but most importantly retain your wonder and amazement at the sheer audacity and beauty of this truly wonderful group of selfish and uncompromising biological replicators!

4.2 THE OBSESSION WITH DEATH: MORTALITY FROM A VIRAL PERSPECTIVE

From our human-centric perspective, viruses are associated with illness, disease, and often death. Yet, from a viral perspective every infection ultimately ends in death: either of the host cell or of the virus itself. A key difference between multicellular (the so-called "complex") organisms and their unicellular counterparts is that a successful infection in a multicellular organism does not usually lead to the death of the entire organism. This applies to hosts at all levels of complexity from fungi to mammals and trees. A successful infection in a single-cell organism will always lead to an untimely death of that cell. A successful infection in a multicellular organism will lead to the death of some cells, but usually leave the remainder of the host intact. Indeed, despite some incredibly virulent viruses ultimately causing the total death of their multicellular hosts (and not just the subpopulation of cells they actually infect), rarely does the physical loss of the infected cells cause death: the mortality is usually a product of "particularly" nasty viral dispersal mechanisms such as hemorrhage and diarrhea that are induced to aid the transfer of the virus to new hosts. A multicellular host offers a unique environment to a virus: a homogeneous population of cells within a contained system. Although a classic viral infection (e.g., by the influenza viruses, members of the family Orthomyxoviridae) is usually regarded by the patient as a single infection, the symptoms observed are actually a product of thousands of cells being infected. If a virus can successfully infect one type of cell within an organism, there is usually no reason for viral progeny to subsequently infect every other identical cell type within the organism. Clearly, this would have disastrous consequences, and this is why multicellular organisms have evolved defensive strategies (such as immune systems) against such an occurrence. These systems can actively seek out and destroy both virus and infected cells to stop the infection from spreading out of control. When viruses attack the cells involved in these processes, such as in the case of HIV infection (human immunodeficiency virus 1 and 2, both of the family Retroviridae, genus *Lentivirus*), the results are catastrophic to the host concerned. Crucially though, it is not HIV infection *per se* that causes mortality in such cases, but the compromised immune system function (and development of AIDS) that leads to susceptibility to opportunistic infections (sometimes other viruses, but usually from the other domains of life such as bacteria, fungi, or protozoan) and tumor growth. Thus, death of a multicellular organism through viral infection should be regarded as an exception to the rule. It is an unfortunate by-product of the viral infection of a subpopulation of cells within an organism. With the loss of their particular and specific function, goes the loss of whole organism integrity, leading to an untimely death.

It is important to realize that viral infection can be regarded on a cell-by-cell basis, regardless of whether that infection occurs to a single-cell organism or to a single cell *within* an organism. Viral infection is a fact of life. Despite our obsession with biology that is visible to the naked eye, it is a microbial/cellular world in which we live. In our oceans, virus-induced mortality is estimated to account for about 40% of the loss of microbial cells on a daily basis (Suttle, 2005). Microbial populations can withstand this sort of loss due to their rapid growth rates, a luxury not available to most complex multicellular organisms composed of a majority of cells that undergo irreversible differentiation and slower regeneration rates. Viruses that infect multicellular organisms are subjected to additional selection pressures that single-cell host viruses simply do not have to contend with.

4.3 A MARRIAGE MADE IN HELL

In dealing with the topic of viral ecology and evolution, we must always remember the polyphyletic nature displayed by viruses. The sheer wealth of diversity displayed by these "biological entities" makes any comprehensive study of the subject an almost impossible task. Yet, all viruses share one overarching property that defines them: they are entirely dependent upon the intracellular infection of their hosts for survival. This concept, despite the weird, wonderful and incredibly diverse strategies in which they act (and which you will read about within the pages of this book), binds all viruses together. The interaction between any virus and its host will be ingrained in the history of both lineages and, crucially, will have left and continues to leave, its mark on both host and virus. Put simply, the history of a host will help shape the future of both itself and any virus that infects it. Equally, this applies to a virus as well. Yet, even if a virus could have a memory of its illustrious past, it would have no care for this history. A virus, if it does "live" in the philosophical sense, lives only in the moment. At the population level, hosts and viruses are entwined in the closest marriage imaginable. It is far from a happy marriage though constant arguing ensues since only one partner (the virus) wants the marriage. Like all marriages, there are only two options available as a get-out clause: death (of either or both virus and host populations) or divorce. Ironically, divorce in this sense is always instigated by the virus (which wanted the marriage in the first place!), never the host and, crucially, the viral divorcee requires an immediate remarriage to whichever suitor (host) has turned its eye. Thus, from an evolutionary perspective, jumps by viruses across apparent species barriers (i.e., promiscuous extramarital activity) are mere examples of viruses taking advantage of an opportunity that has presented itself to them. If this new partnership is successful, it can be considered a divorce and immediate remarriage as the new selective pressures of interacting with a new system become applied. Crucially, the previous host will most likely still remain married to the original virus. The viruses really do have all the fun at their hosts' expense.

4.4 THE NUMBERS GAME

It is important to deal with the issue of species jumps early in this chapter since it is often the issue that most people mistakenly consider as the most important when thinking about viral ecology. For the vast majority of infections, a virus will infect a host that is similar to the host that it infected last. Despite the ease with which we mistakenly assign a conscious thought to the process, it is merely biochemical interaction and compatibility between host and virus that will determine if an infection occurs or not. The last host a virus successfully infected is most likely the most compatible future host. An increase in the abundance of a viral population will lead to an increase in the occurrence of physical interactions with potential hosts. An increase in viral diversity will lead to an increase in potential biochemical compatibility following a successful physical interaction. It is purely about numbers in a relentless game of chance. Despite viruses being highly specific for their hosts, a virus has only to infect a single cell of another host successfully to begin the natural selection process for the new host. Such opportunistic infections are spontaneous and the selection pressures against viruses are so strong they often fail to become established in their new host. These are incredibly rare events in relation to "normal" viral infections, but surprisingly common due to the sheer number of infections that take place. A useful analogy can be taken from the aviation industry: aeroplane crashes are very rare because of stringent safety regulations; however, due to the high number of flights made on a daily basis, aeroplane crashes do occur frequently. Usually, cross species barrier infections are associated with increased virulence (as would be expected for a host exposed to a new virus) and typically generate a disproportionate amount of attention from virologists and the media. The expanding human population coupled with globalization (which itself increases the chance of viral infection from interactions between human viruses and humans) has led to intensive farming methods to meet the increasing food demand. Intensive farming is exactly this: large populations of quite often genetically homogeneous animals (and plants) in relatively small areas, increasing not only the potential for viral infection within the population but also the chances of transmission of "animal" viruses to humans. Transmission of plant viruses to humans is possible but more unlikely: a human is more similar to a pig or chicken than wheat. Thus, according to the numbers game that viruses play so well, it is inevitable that species jumps will occur. While they are usually unwelcome, they are also inevitable and a product of simple viral ecology.

4.5 FIGHT TO DEATH: GENES ARE THE WEAPONS

We have already begun to touch on many of the issues associated with viral ecology. However, we should not be fooled into thinking that any host actively welcomes viral infection; a virus cares little for its host and serves only its own selfish requirements. It is not a one-way battle, and no cell takes a viral infection lying down. Furthermore, despite the subject matter of this volume, the viruses and hosts themselves should not be considered the lowest common dominator in the study of viral ecology. It is the genes that reside within them that are the driving force behind organic life. Hosts and viruses are examples of groups of genes clubbing together (into genetic lineages) for mutual success. Over time, these genetic groupings become so established that genes work together to produce (as a by-product) the weird and wonderful forms of life we see today. The longer the genes coevolve in these groups, the stronger the dependency that develops within the groups. Over time, complex regulation systems and multilayered interaction networks evolve as systems become increasingly more complex. Out of chaos, comes order. Viruses (as simple but ordered systems) ruthlessly exploit their hosts' ordered systems for their own benefit.

The host will, of course, develop countermeasures to ensure that this fails to occur. It has nothing to lose (except cellular integrity that will be lost to the virus anyway) by throwing every biochemical trick it has up its sleeve at the virus. Thus, an infection can be regarded as a winner-takes-it-all conflict between a virus and a host. This battle takes place both internally and externally to the cellular environment and provides scope and opportunity for natural selection and evolution to occur at a multitude of places.

This volume discusses such evolutionary selection. As stated previously, selection occurs at many different levels, each with widely varying degrees of pressure. Ultimately, these evolutionary processes are reflected in the genetic structure of both viruses and their hosts. Selection related to external environmental factors results in the *evolution* of systems in the classical sense, a process that all biological entities are subjected to (a host cell is an external factor to a virus and *vice versa* when no physical, biological, or chemical interaction is taking place between them). However, selection occurring within the biochemical components of the cellular environment when both host and virus genomes and their products are interacting directly can be considered a *coevolution* (Woolhouse et al., 2002), a process that symbiotic and parasitic organisms are subjected to and viruses take to the extreme by their very nature of being obligate intracellular parasites. Over the billions of years that organic life has been evolving on Earth, this has led the development of increasingly complex interactions between viruses and hosts. Let us not get distracted from the raison d'être of a virus' existence, which is to replicate at the expense of its host. Examples of beneficial effects to the host will be discussed within this chapter, but these examples are few and far between and could be considered accidents and quirks of biology. Quite simply, while a virus needs a host to replicate, no host ever relies on the virus that infects it. Hosts can survive without viruses, but the opposite is never true. Dinosaur viruses (and their genes) unable to infect other living organisms at the time of the alleged meteor strike went extinct alongside their dinosaur hosts.

Although viruses instigate the premature death of cells, this is not to say that viruses are not necessary to *support* life: without the constant virus-induced cellular mortality in our oceans in particular, entire ecosystems would undoubtedly collapse. A staggering 10^{23} viral infections are predicted to occur every second in our oceans, causing the constant cycling of nutrients through all trophic levels (Suttle, 2007). Quite simply, relentless and uncompromising viral infection should be considered the *status quo* in any system. Despite their small size, through their sheer abundance and activity viruses are the essential and unappreciated giants of the nutrient cycling realm. However, in addition, through their role as merciless predators, viruses play a crucial role in the natural selection and evolution of their hosts profoundly altering their appearance at the genomic, proteomic, and metabolic levels.

4.5.1 The Arms Race: Winner Takes it All in the Battle, But Not the War

Viruses are in a constant and ongoing battle with their hosts. Evolution by natural selection continually acts to tip the balance in favor of either host or virus. The direction and extent of this change is somewhat transient though, and depends on a plethora of variables that can change almost at will. Thus, a virus or host selected at one moment in time is by no means guaranteed to thrive under the next cycle of selection pressure. Yet, no host will ever take viral infection lying down, resistance mechanisms exist and are implemented ruthlessly. Indeed, there will usually be a natural population with an increased resilience and resistance that will be selected, thus tipping the balance, albeit momentarily, in the host's favor. It is crucial to remember though that population breeds disease: any resilient population that then flourishes has a greater chance of being destroyed by a future infection because it contains less diversity (having recently come through a selective bottleneck) compared to its abundance (Domingo et al., 1996).

Virulence leading to mass host death is no problem for a virus provided the host population is suitably abundant and future-proof. Virulence causing rapid decline in host population levels in turn leads to decreased infection rates (there are fewer hosts to infect), thus allowing host population recovery. Crucially, this selection can occur only after the host population has been decimated. It is about survival into the future, yet this is selected only after a virus' short-term fate has been sealed.

The ecological dynamics within a host–virus system can have vastly different outcomes depending on whether the host is single or multicellular. The dynamics also differ dramatically depending on whether a lytic or lysogenic lifestyle is adopted (or something in between). Further complications arise when factors such as virus dispersal have to be considered and selected for, for example, in microbial systems (perhaps exemplified by the marine environment); this mechanism is fairly equal for all viruses, in terrestrial plant or animal systems the pressures are somewhat higher.

Thus, there are conflicting, yet complementary, aspects of virus ecology. This ongoing arms race of adaptation and counteradaptation has been described in the Red Queen hypothesis. Taken from Lewis Carroll's *Through the Looking-Glass, and What Alice Found There* (1871, by Charles Lutwidge Dodgson, Macmillan Publishers Ltd., London), "Now, here, you see, it takes all the running you can do to keep in the same place." Applied to host–parasite systems, it can be translated as "For an evolutionary system, continuing development is needed just in order to maintain its fitness relative to the systems it is co-evolving with" (van Valen, 1973).

4.5.2 The War of the (Viral) World: the Battlegrounds

Before discussing individual examples of interacting and coevolving systems, we must consider the entire landscape of host–virus systems and identify the main sites where the battles are fought. To this end, the virus life cycle can be broken into two main parts: time spent within a cell and time spent outside the cell; this roughly correlates with being metabolically active and metabolically inactive (gray areas exist of course: for example, often whole and intact viruses need to be internalized and targeted to specific intracellular locations prior to release of the genomic contents). Where selection occurs on the host or virus when the virus is not metabolically active and has yet to instigate any biochemical influence over its host cell, this leads to evolutionary change for both host and virus that can be considered almost independent of each other. However, once the virus is metabolically active or is at least interacting with its host (through binding to the surface), any selection that occurs (for host or virus) can be considered a truly coevolutionary process since the two systems are inherently, intrinsically, and undeniably biochemically linked as just one system. At this stage, there can be only one winner: host or virus.

4.5.3 Without a Cell: the Vulnerability of Being in Limbo

When not battling within a cell for survival, metabolically inactive viruses are left exposed to the environment. The environment in this context varies hugely depending on the nature of the virus and the type of host it infects. The infection of multicellular organisms can also create a suite of additional environmental conditions that must be overcome by viruses. For example, a virus such as influenza virus has to cope with both biotic and abiotic environments in between infection cycles. Successful infection and release of influenza viruses within the body places the virus directly in contact with the human internal environment (with its specific temperature, biochemical composition, and immune system). One good sneeze can then catapult the virus from this relatively homogeneous and safe environment, yet biologically harsh due to the actions of the immune system, into the outer world where factors such as temperature, pressure, and UV exposure vary dramatically and where they represent a nutritious food packet for all manner of life forms. Crucially, when in this "limbo" state, viruses are completely at the mercy of whatever is thrown at them. Viral losses are thus nearly catastrophic at this stage and account for the large burst sizes observed when viruses infect cells (when one cell produces at least an order of magnitude more viruses, usually two to three orders more). However, the high wastage of viruses does ensure that only the most fit and robust viruses survive and ensures that evolutionary rates within viral populations are incredibly fast, far faster than in their hosts.

If we consider viral entry as the starting point in the infection process, there exists a range of strategies employed by viruses for ensuring their genomic material is safely delivered to the appropriate location. Binding to receptor sites (proteins, carbohydrates, lipids, and glycolipids) followed by injection of genomic material, absorption, and merging with the membrane, phagocytic engulfment, and many other mechanisms are all employed to pierce the hosts' outer armor. Even in the simple task of obtaining entry to the cell, there is enormous scope for evolutionary battle: modification of the receptors to avoid viral binding can ensure host resistance; conversely, modification of the viral binding receptor can ensure increased virus binding. It is these types of adaptation and subsequent counteradaptation that give rise to the previously mentioned Red Queen dynamics. Although cellular binding and internalization are essential for a successful infection under natural conditions, it can be bypassed in the laboratory with genetically compatible viruses that are "physically" incompatible; that is, the barrier represented by the cell surface only creates a physical spatial separation between where a virus is inactive and where it can safely and efficiently replicate. The artificial introduction of viruses to the intracellular regions of cells (effectively bypassing the membrane barrier) that under natural conditions are completely off limits by virtue of being "receptorless" often results in successful viral replication; for example, human poliovirus (human enterovirus C, family Picornaviridae, genus *Enterovirus*) can replicate happily when introduced artificially to the inside of mouse cells which, not having the CD155 receptor found in primates, would otherwise be resistant to infection (McLaren et al., 1959; Holland et al., 1959).

4.5.4 Within a Cell: Out of the Pan, Into the Fire

Before viral takeover, the host cell can represent the most hostile environment that a virus will encounter in its life cycle. Viruses suddenly become a huge threat to the long-term survival of a cell once they have breached the outer surface and find themselves inside the cell. Outside the cell, threats to a virus are almost all random and nonspecific. A slight exception to this are multicellular organisms with innate and adaptive immune responses, even though the immune system should be considered random: low immunogenic viruses will eventually be targeted by previously unexposed immune systems, but initially survival rates will be fairly high. As virus numbers increase, it will increase the opportunity for the "right" immune cells to interact with virus particles, which in turn triggers the specific immunogenic cascade.

However, within the cell, the intracellular host response to foreign DNA is harsh and uncompromising. Defense mechanisms involving RNA interference, RNases, and endonucleases are used to combat the invading viral genome. RNA interference works by using small RNA molecules to inhibit gene transcription (a defense that can be applied to all types of virus infection regardless of the nature of their genomic material) and can cause the direct degradation of dsRNA viral genomes (Marques and Carthew, 2007). Restriction endonucleases cleave DNA at specific recognition sequence sites (often found in viral genomes, but not present or protected in host genomes) and provide general protection from DNA viruses. Other cellular defenses include the apoptosis (programmed cell death) pathway in eukaryotes that can be induced in order to prevent the infection from spreading, a case of sacrificing a cell to save the larger population. With obvious advantages to multicellular organisms, this process has also been suggested to occur within single-cell systems whereby the sacrifice of single cells acting individually may be undertaken to prevent viral infection spreading to the neighboring natural population.

Following infection of a cell, in order to survive and to infect another cell, a virus must successfully replicate its genome and create functional virions. This intracellular infectious time should be regarded as the primary battleground where the majority of directly coevolving host–virus systems can be observed. Following successful virion production, the virus must then be able to exit the cell. This is done in a variety of ways from budding to total cellular rupture. The nature of the host cell will then determine what the virus is exposed to. For free-living single-cell hosts, the released viruses will be exposed directly to aquatic, aerosol, or solid surfaces. For hosts involved in symbiotic or parasitic relationships, their viruses may be exposed to contained biotic environments (e.g., the gut). Depending on the stage of infection and the scale of the host response, viruses infecting multicellular organisms will either continue to infect cells within the same organism or transfer to and attempt to infect a new host.

4.6 THE SILENCE OF THE VIRUSES

Viruses often lie dormant inside their hosts, in what are known as latent, lysogenic, or endogenous lifestyles. Provided their incorporation into the host genome induces no direct negative effects (e.g., the disruption or deregulation of useful gene function), this is usually a safe strategy for the virus with negligible impact on host fitness. When viruses undertake this strategy, it is a reflection of their close relationship with their host: they can afford to sit out, let the host take the strain until such time in the future when either conditions are favorable for mass viral production or, alternatively, the host is approaching cell death and has outlived its usefulness to the virus as a low copy number safe haven. It could be argued that when in a prolonged latent phase, since viruses are essentially part of their host genome, they cease to be viral in nature. Indeed, any host death that is not associated with the virus in question would result in the premature end of the "silent" virus, an outcome which is not uncommon and displays the deep-rooted "trust" shown by viruses in their hosts for ensuring their long-term survival. Furthermore, an inactive virus does not produce progeny and thus can be considered to be at an evolutionary standstill. This does not necessarily apply to the host organism over the same time period: multicellular organisms in particular will continue to evolve provided they remain reproductively active. However, if during a period of inactivity the virus is replicated as part of the host genomic cellular division (because either the virus has infected a stem cell or a single-cell organism), it will continue to diverge at a similar rate to the host. Only when the virus becomes active will selection occur and evolutionary rates accelerate. Viruses that become integrated into their host genomes can be considered as the crudest form of coevolving systems, where, by definition, every piece of their genetic makeup is coevolving alongside their hosts. The advent of the genomic era has heralded unique insights into this phenomenon. The human genome, for example, is thought to comprise up to a staggering 8% of its material from viral origin (Lander et al., 2001).

4.7 GIVING UP THE VIRAL GHOST

The success viruses gain from incorporating into their hosts' genome is perhaps exemplified by the endogenous retroviruses. These viruses lie dormant almost indefinitely within genomes after infecting the germ cells of many vertebrate genomes. While other viruses strive to replicate at their hosts' expense, these viruses have effectively stopped fighting the war and have become permanently incorporated into their host's genome: if you can't beat them, join them. This strategy has led to the permanent integration of huge amounts of previously viral genetic material into genomes. Most is deemed to be inactive or the so-called junk DNA. However, this may not be necessarily the case, diverging genes under no strong selection pressure over time can quite often assume new functions providing an advantage for either the host or for other active viruses (see Section 4.10.1).

4.8 THE MAKINGS OF VIRUS–HOST COMPATIBILITY

The total dependency and reliance that viruses have on their hosts is reflected in their genomic composition and metabolic potential. In order for viral genes (and proteins) to function correctly inside their hosts, they must be suitably adapted to and compatible with their host genetic background (e.g., composition, size, regulation, codon usage, folding, and post-translational modification). This intricate host–virus genetic compatibility creates the opportunity for genes to move between lineages in the process known as horizontal gene transfer. This can occur in either direction and provides an interesting aspect to viral ecology and evolution. Since viruses act as vectors for moving and shuttling genes between different lineages, no gene can or should ever be considered as being restricted indefinitely to a particular lineage. If a virus picks up a gene from its host, the gene becomes, by definition of its current location, a viral gene (despite its evolutionary history). The same applies vice versa or when transfer occurs between any genetic lineages. Where a gene hangs its hat is its home. In this context, viruses represent a mammoth hat stand, containing the largest reservoir of genes on the planet. It also adds a layer of complexity to the study of viruses and their hosts from an evolutionary perspective! Much has been written on the evolution of viruses and much more on the evolution of the organisms they infect. Yet, the subject of coevolution of viruses and hosts when considered as two intermingled parts of a whole is often neglected and forms the basis for the rest of this chapter. Quite often, the distinction between evolution and coevolution can become somewhat blurred, as is the distinction between what can be deemed viral or host with regard to genomic material with a shared history. Yet, it is at the interface that this shared and/or closely integrated biochemical machinery occupies where the fundamental selection occurs that is paramount to host–virus coevolution. I will attempt to dissect these issues and provide insights into and examples of the ever-changing landscape that is the ecology and (co)evolution of viruses.

4.9 THROWING LIGHT ON VIRUS–HOST EVOLUTION

Without doubt, the genomic era has instigated a change in attitude toward viruses. No longer thought of as merely bags of virus genes performing purely viral functions, it has become increasingly apparent that in reality many viruses harbor within their genomes homologous genes to their hosts. Debate still rages as to the nature of these genes, such as whether they originated as viral genes or host genes, the likelihood of further and ongoing recombination and transfer between host and viruses, their function in the viral system in relation to the "normal" host function. However, despite these issues, many of which are beginning to be resolved and many of which will never be resolved to satisfaction, there can be no doubt that these genes offer a unique insight into the process of virus–host coevolution. We are now blessed with an abundance of examples that can be utilized to illustrate our point from which I will take a select few to illustrate the types of evolutionary interaction that can occur between viruses and their hosts.

To begin with, we shall take our first example from the cyanobacteria *Synechococcus* and

Prochlorococcus. Viruses of the families Podoviridae and Myoviridae infecting these photosynthetically active organisms have been found to contain an assortment of photosynthesis-related gene products such as the photosystem II core reaction center proteins D1 and D2, a high light-inducible protein, plastocyanin and ferredoxin (Lindell et al., 2004; Mann et al., 2005). During infection, as the homologous host transcripts are in decline, viral transcripts become expressed and help to maintain the functioning of the photosynthetic system that in turn allows optimal viral production (Lindell et al., 2005, 2007). It remains to be determined whether the virus acquisition of the host photosynthetic genes came as a direct response to the host actively shutting photosynthesis down in response to infection or whether it is a mechanism used merely to increase the efficiency of the infection process, that is, providing more bang for the virus' buck. In addition to the photosynthesis-related genes, these interesting viruses possess homologues for stress response genes found in their hosts. Intriguingly, during infection, while the vast majority of host genes become downregulated as infection progresses, a few dozen are actually upregulated. These genes belong to two broad groups: stress response and nucleotide metabolism. It is likely that the stress response genes encoded by the viruses are involved in some aspect of this transcriptional regulation. Again, it is unclear whether this is a last ditch attempt by the host to slow or stop the infection or if they are actually induced by the virus and used against the host. This example shows the boundary between host defense and viral offense where true coevolutionary processes take place. This metabolic battleground consisting of both host and virus systems utilizing, exploiting, or manipulating the same processes, often through shared genes is a recurring theme mirrored in other host–virus systems. The cyanophage system does offer further insights that may shed light on other systems: there is a strong connection between the upregulated host genes, their position on the host genome (in hypervariable islands thought to be mobilized by phage), and the presence of viral homologues (Lindell et al., 2007). Thus, the cyanophage–cyanobacteria system provides clear directions to the site where biochemical confrontation and an intricate metabolic battle take place. Of course, different hosts and different viruses will all battle it out in different manners. Over time, the direct interaction and manipulation of metabolic pathways and processes lead to very interesting, intricate, and subtle host–virus coevolution dynamics. Indeed, the concept that viruses use the host systems against them (and vice versa) is becoming increasingly clear as full genomic sequencing lifts the lid on the Pandora's box of molecular evolution.

4.10 SOMETIMES IT TAKES MORE THAN THE ODD GENE

The previous example of a few genes involved in photosynthesis being acquired by a virus to aid infection provides an excellent illustration on how a few genes can be used by a virus to manipulate the host system. Yet, some viruses have taken the need to manipulate metabolic functions during infection to the extreme. The coccolithoviruses (family Phycodnaviridae, genus *Coccolithovirus*) are one such group of viruses. Amazingly, these viruses have acquired an almost complete metabolic pathway for the synthesis of sphingolipids from their algal host, the coccolithophore *Emiliania huxleyi* (see Chapter 7 for further information on this remarkable virus family) (Wilson et al., 2005). The reasons for the acquisition of this pathway are unclear at present, but clearly an important component of the battle between this host and virus is played out in the sphingolipid arena (Han et al., 2006; Monier et al., 2009). At this stage, the reasons behind these unique horizontal gene transfer events (genomic positioning suggests separate events were necessary) or why one gene is "missing" is not clear. However, crucially the sphingolipid pathway provides useful guidance on how the virus and host genome products interact during infection (Pagarete et al., 2009). For example, sphingolipids are well known for their role as signaling molecules in apoptosis (programmed cell death). As mentioned previously, apoptosis is a well-known antiviral defense mechanism, albeit more commonly used in multicellular organisms. Caspases (a type of proteinase) are usually the vehicles used to induce the process. Accordingly, caspase induction has been found upon viral infection of this system (Bidle et al., 2007). However, the story does not stop there. Caspase induction may be actually necessary for successful infection. Furthermore, many of the viral gene products are predicted to have caspase cleavage sites that presumably require cleavage (by the host system) before they become active. This is an excellent example of complex metabolic pathways becoming a site for host–virus coevolution and involves both gene products with a shared origin and gene products and metabolites that regulate and/or manipulate associated pathways. The cellular environment is composed of an intricate network of biochemical pathways, and during the early stages of infection, viral activity will target and impact the function of particular pathways. Importantly, the ripple effect will be felt in the closest interacting pathways first. The coccolithophore–coccolithovirus system provides an excellent example of this process (Allen et al., 2006). However, it is important to note that manipulation of these particular pathways is not limited to the coccolithoviruses. This battleground is common to many host–virus systems and is reflected in the examples of apoptosis inhibitors found in a diverse range of viral genomes such as those belonging to the baculoviruses (family Baculoviridae), adenoviruses (family Adenoviridae), human cytomegalovirus (family Herpesviridae, genus *Cytomegalovirus*), herpesviruses (family Herpesviridae), African swine fever virus (family Asfarviridae), poxviruses (family Poxviridae), human papillomaviruses (family Papillomaviridae), and myxoma virus (family Poxviridae, genus *Leporipoxvirus*) (Alcami and Koszinowski, 2000; Hanada, 2005). Although every host and virus interaction will be highly specific and niche adapted, there are strong themes running through the infection and coevolution process. Particular pathways and networks are continually targeted by widely diverse viruses. Ultimately, the entire biochemical network will break down and cellular integrity will be lost; but in the early stages of infection, the outcome of infection often depends upon the control and manipulation of just a few precise metabolic pathways, crucially either utilizing the same molecular machinery or manipulating existing function.

4.10.1 Immunity, Protection, and Infection

The previous examples show how viruses can manipulate their hosts, often utilizing and turning the host molecular machinery and biochemical pathways against itself. However, we should not think of this as a one-way battle. Hosts also pick up viral machinery and use it in the battle against viruses. A striking example is that of endogenous retroviruses. Consisting of just three gene products, group-specific antigen, polymerase, and envelope protein (known as gag, pol, and env, respectively) (Villesen et al., 2004), the endogenous retroviruses (commonly referred to as ERVs, of the family Retroviridae) differ from their "exogenous" retrovirus counterparts because they integrate into the genomes of the germ cells of their hosts, thus becoming transmitted to future generations (Arnaud et al., 2007). Comprising a staggering 8% of the genome, there are an estimated 450,000 copies of ERVs within the human genome (Lander et al., 2001). This genomic colonization is mirrored in all vertebrate genomes. Presumably, there is a strong selection against their integration into essential genomic loci or to positions with deleterious effects: their integration into primary germ cells allows for an easy selection process. Deleterious insertions will be immediately selected and only germ cells with stable inserts will be able to grow and develop normally following fertilization. Thus, despite our obsession with viruses being bad for our heath, it is undeniable that human (and animal) evolution is closely mired with that of our viruses.

Although the vast majority of ERVs are inactive after accumulating genetic defects, their sheer abundance ensures that some will be transcriptionally active and will be capable of producing functional gene products (Stoye, 2009). Indeed, their prevalence in genomes is thought to provide an advantage to their hosts. In a case of poacher-turned gamekeeper, endogenous (and therefore stably integrated) retroviruses are thought to provide protection from exogenous retrovirus infection. For example, expression of the ERV envelope glycoproteins provides protection from infection by exogenous retroviruses by blocking the entry through receptor competition (Malik and Henikoff, 2005). ERV Gag expression has been shown to protect mice against some murine leukemia virus strains (Villarreal, 1997).

But the story does not end here with viruses solely offering protection to their hosts from other viruses. We mentioned previously that apoptosis and cellular signaling is a pathway specifically targeted by a variety of viruses for manipulation during infection. Unchecked, apoptosis has the capacity to severely inhibit successful infection. Organisms with immune systems also provide a target that must be neutralized as efficiently as possible by the viruses for successful infection. Therefore, many viruses, including the retroviruses, harbor genes whose products can inhibit host immune responses. In similar fashion to the strategies employed to manipulate apoptosis, various viruses manipulate the immune response by either using virally encoded homologues of the host immune genes or by using genes whose products can interact with the molecular functioning of the host immune system (Alcami and Koszinowski, 2000). Targets include the humoral (antibody) response (poxviruses, coronavirus (family Coronaviridae), cytomegalovirus, herpesviruses, and HIV); interferon response (adenoviruses, poxviruses, reoviruses (family Reoviridae), baculoviruses, HIV, polioviruses, influenza viruses, rotaviruses (family Reoviridae, genus *Rotavirus*), and Sendai virus (family Paramyxoviridae, genus *Respirovirus*)); cytokine and chemokine response (African swine fever virus, adenoviruses, poxviruses, and Epstein–Barr virus (human herpesvirus 4, family Herpesviridae, genus *Lymphocryptovirus*)); and major histocompatibility complex (cytomegalovirus, HIV, herpesviruses, and adenoviruses).

Thus, as you can see, the incorporation of large amounts of viral material into a genome could potentially have severely deleterious effects and create a ticking time bomb especially with regard to crucial functions such as immunity. However, if harnessed correctly it does provide the host genome with the opportunity to have localized areas with inhibited immunological performance. While under most circumstances there would be little or no call for this situation, the evolution of the mammalian placenta has created the opportunity for the evolved products of stably integrated retrovirus genomes (which could/should actually be deemed host material because of their long-term integration) to perform such a function. The foreign fetus is thought to be protected from the maternal immune system through the actions of an immunosuppressive domain located on the envelope protein of an ERV (Villarreal, 1997). Furthermore, some ERV envelope glycoproteins, such as those of HERV-W group (family Retroviridae), have fusogenic effects and play a crucial role during the formation of the placental syncytium (Blaise et al., 2003). Despite being labeled as "syncytin" genes of the host, they are clearly coopted retroviral genes. A variety of syncytin genes have been identified suggesting that capture and utilization of retroviral genes is a recurrent theme in mammalian placental evolution (Heidmann et al., 2009). However, given the content of this chapter, it is not surprising that this event has occurred on so many occasions: it is almost inevitable given what we know about the effects of viruses on cells and the pathways they target to ensure that their infection is successful.

4.10.2 The End of the Concept of the Host Gene?

Our knowledge of host–virus coevolution derived from some of the examples described in this chapter has been entirely dependent on genomic sequencing. In particular, our knowledge is heavily biased toward human and animal viruses because these are the most economically relevant to justify the research on them. I have tried to avoid overly referring to these viruses since life is far more diverse than our human-centric obsession would have us believe. Nevertheless, I hope you have now obtained a taste of the issues and themes involved in the study of host and virus coevolution in any system. This is a field very much in its infancy, but is growing rapidly as we realize that viruses have and continue to shape the evolution of all living organisms in ways we are only just beginning to grasp. As more types of viruses become sequenced from diverse hosts, it will be unavoidable for us to realize that many of the genes that we have previously considered as being bacterial or eukaryotic in origin will actually reveal themselves to be viral in nature. The study of viral ecology will have to deal with the realization that all too often what is thought of as host function is actually virus in origin, what is virus function is actually host in origin, and there exists a large gray area in between where the quirkiness of nature expresses itself with beautiful intricacy. Evolutionarily, viruses and their hosts should no longer be considered separate entities, the boundaries between them have been exposed for what they are: an approximate line drawn in the sand, accurate at any given moment but never set in stone.

REFERENCES

Alcami, A. and Koszinowski, U. H. (2000). Viral mechanisms of immune evasion. *Immunol. Today* 21(9), 447–455.

Allen, M. J., Schroeder, D. C., Holden. M. T., and Wilson, W. H. (2006). Evolutionary history of the Coccolithoviridae. *Mol. Biol. Evol.* 23(1), 86–92.

Arnaud, F., Caporale, M., Varela, M., Biek, R., Chessa, B., Alberti, A., Golder, M., Mura, M., Zhang, Y. P., Yu, L., Pereira, F., Demartini, J. C., Leymaster, K., Spencer, T. E., and Palmarini, M. (2007). A paradigm for virus–host coevolution: sequential counter-adaptations between endogenous and exogenous retroviruses. *PLoS Pathog.* 3(11), e170.

Bidle, K. D., Haramaty, L., Barcelos, E., Ramos, J., and Falkowski, P. (2007). Viral activation and recruitment of metacaspases in the unicellular coccolithophore, *Emiliania huxleyi*. *Proc. Natl. Acad. Sci. U. S. A.* 104(14), 6049–6054.

Blaise, S., de Parseval, N., Bénit, L., and Heidmann, T. (2003). Genomewide screening for fusogenic human endogenous retrovirus envelopes identifies syncytin 2, a gene conserved on primate evolution. *Proc. Natl. Acad. Sci. U. S. A.* 100(22), 13013–13018.

Carroll, L. (1871). *Through the Looking-Glass, and What Alice Found There.* Macmillan Publishers Ltd., London.

Domingo, E., Escarmís, C., Sevilla, N., Moya, A., Elena, S. F., Quer, J., Novella, I. S., and Holland, J. J. (1996). Basic concepts in RNA virus evolution. *FASEB J.* 10(8), 859–864.

Han, G., Gable, K., Yan, L., Allen, M. J., Wilson, W. H., Moitra, P., Harmon, J. M., and Dunn, T. M. (2006). Expression of a novel marine viral single-chain serine palmitoyltransferase and construction of yeast and mammalian single-chain chimera. *J. Biol. Chem.* 281(52), 39935–39942.

Hanada, K. (2005). Sphingolipids in infectious diseases. *Jpn. J. Infect. Dis.* 58(3), 131–148.

Heidmann, O., Vernochet, C., Dupressoir, A., and Heidmann, T. (2009). Identification of an endogenous retroviral envelope gene with fusogenic activity and placenta-specific expression in the rabbit: a new "syncytin" in a third order of mammals. *Retrovirology* 6, 107.

Holland, J. J., McLaren, L. C., and Syverton, J. T. (1959). Mammalian cell—virus relationship. III. Poliovirus production by non-primate cells exposed to poliovirus ribonucleic acid. *Proc. Soc. Exp. Biol. Med.* 100(4), 843–845.

Lander, E. S. et al. (2001). Initial sequencing and analysis of the human genome. *Nature* 409 (6822), 860–921.

Lindell, D., Jaffe, J. D., Coleman, M. L., Futschik, M. E., Axmann, I. M., Rector, T., Kettler, G., Sullivan, M. B., Steen, R., Hess, W. R., Church, G. M., and Chisholm, S. W. (2007). Genome-wide expression dynamics of a marine virus and host reveal features of co-evolution. *Nature* 449(7158), 83–86.

Lindell, D., Jaffe, J. D., Johnson, Z. I., Church, G. M., and Chisholm, S. W. (2005). Photosynthesis genes in marine viruses yield proteins during host infection. *Nature* 438(7064), 86–89.

Lindell, D., Sullivan, M. B., Johnson, Z. I., Tolonen, A. C., Rohwer, F., and Chisholm, S. W. (2004). Transfer of photosynthesis genes to and from *Prochlorococcus* viruses. *Proc. Natl. Acad. Sci. U. S. A.* 101(30), 11013–11018.

Malik, H. S. and Henikoff, S. (2005). Positive selection of Iris, a retroviral envelope-derived host gene in *Drosophila melanogaster*. *PLoS Genet.* 1(4), e44.

Mann, N. H., Clokie, M. R., Millard, A., Cook, A., Wilson, W. H., Wheatley, P. J., Letarov, A., and Krisch, H. M. (2005). The genome of S-PM2, a "photosynthetic" T4-type bacteriophage that infects marine *Synechococcus* strains. *J. Bacteriol.* 187(9), 3188–3200.

Marques, J. T. and Carthew, R. W. (2007). A call to arms: coevolution of animal viruses and host innate immune responses. *Trends Genet.* 23(7), 359–364.

McLaren, L. C., Holland, J. J., and Syverton, J. T. (1959). The mammalian cell–virus relationship. I. Attachment of poliovirus to cultivated cells of primate and non-primate origin. *J. Exp. Med.* 109(5), 475–485.

Monier, A., Pagarete, A., de Vargas, C., Allen, M. J., Read, B., Claverie, J. M., and Ogata, H. (2009). Horizontal gene transfer of an entire metabolic pathway between a eukaryotic alga and its DNA virus. *Genome Res.* 19(8), 1441–1449.

Pagarete, A., Allen, M. J., Wilson, W. H., Kimmance, S. A., and de Vargas C. (2009). Host–virus shift of the sphingolipid pathway along an *Emiliania huxleyi* bloom: survival of the fattest. *Environ. Microbiol.* 11(11), 2840–2848.

Stoye, J. P. (2009). Proviral protein provides placental function. *Proc. Natl. Acad. Sci. U. S. A.* 106(29), 11827–11828.

Suttle, C. A. (2005). Viruses in the sea. *Nature* 437(7057), 356–361.

Suttle, C. A. (2007). Marine viruses: major players in the global ecosystem. *Nat. Rev. Microbiol.* 5(10), 801–812.

van Valen, L. (1973). A new evolutionary law. *Evol. Theory* 1, 1–30.

Villarreal, L. P. (1997). On viruses, sex, and motherhood. *J. Virol.* 71(2), 859–865.

Villarreal, L. P. and Witzany, G. (2010). Viruses are essential agents within the roots and stem of the tree of life. *J. Theor. Biol.* 262(4), 698–710.

Villesen, P., Aagaard, L., Wiuf, C., and Pedersen, F. S. (2004). Identification of endogenous retroviral reading frames in the human genome. *Retrovirology* 1, 32.

Wilson, W. H., Schroeder, D. C., Allen, M. J., Holden, M. T., Parkhill, J., Barrell, B. G., Churcher, C., Hamlin, N., Mungall, K., Norbertczak, H., Quail, M. A., Price, C., Rabbinowitsch, E., Walker, D., Craigon, M., Roy, D., and Ghazal, P. (2005). Complete genome sequence and lytic phase transcription profile of a Coccolithovirus. *Science* 309(5737), 1090–1092.

Woolhouse, M. E., Webster, J. P., Domingo, E., Charlesworth, B., and Levin, B. R. (2002). Biological and biomedical implications of the co-evolution of pathogens and their hosts. *Nat. Genet.* 32(4), 569–577.

SECTION II

VIRUSES OF MACROSCOPIC ANIMALS

CHAPTER 5

CORAL VIRUSES

WILLIAM H. WILSON
Bigelow Laboratory for Ocean Sciences, West Boothbay Harbor, ME

CONTENTS

5.1 INTRODUCTION

With up to 10^8 viruses mL^{-1} in open seawater (Bergh et al., 1989), viruses are "lubricants" of the Earth system "engine room" and are catalysts for global biogeochemical cycling by transforming planktonic cells to dissolved material. Viruses essentially act as biological transformers that accelerate the lysis of bacteria and phytoplankton (Fuhrman, 1999; Suttle, 2005; Wilhelm and Suttle, 1999). A direct consequence of this virus-induced transformation is short-circuiting the flow of carbon and nutrients to higher trophic levels and shunting the flux to the pool of dissolved and particulate organic matter. The net result is an increase in community respiration (Suttle, 2005). While such "black box" shunting exercises are extremely useful for modeling global nutrient cycling processes, they hide the enormous morphological, biological, and genetic diversity of viruses in marine systems. This is exemplified by the recent explosion of virus sequences obtained from environmental shotgun sequencing and genome sequencing projects (Breitbart and Rohwer, 2005; DeLong et al., 2006; Wilson et al., 2005b; Yooseph et al., 2007; Thurber et al., 2008). It is important to make sense of such diversity particularly in the context of global environmental change; the next big challenge for oceanic biogeochemistry is to use this genetic diversity to help understand the complexities of ecosystem functioning (Zak et al., 2006).

The coral reef ecosystem has an added level of complexity, almost a fourth dimension. Each coral holobiont contains a diverse assemblage of archaea, bacteria, algae, fungi, and protists as well as the coral animal (Knowlton and Rohwer, 2003). To add to the genetic and biogeochemical complexity, every microbial organism is likely to have a range of viruses that infect it. Simply looking at assemblages by transmission electron microscopy (TEM) will confirm this (Figure 5.1) (Wilson et al., 2005a; Davy et al., 2006; Davy and Patten, 2007; Patten et al., 2008a; Wilson and Chapman, 2001).

While TEM allows a morphological snapshot of viruses in corals, methodological

Studies in Viral Ecology: Animal Host Systems: Volume 2, First Edition. Edited by Christon J. Hurst.

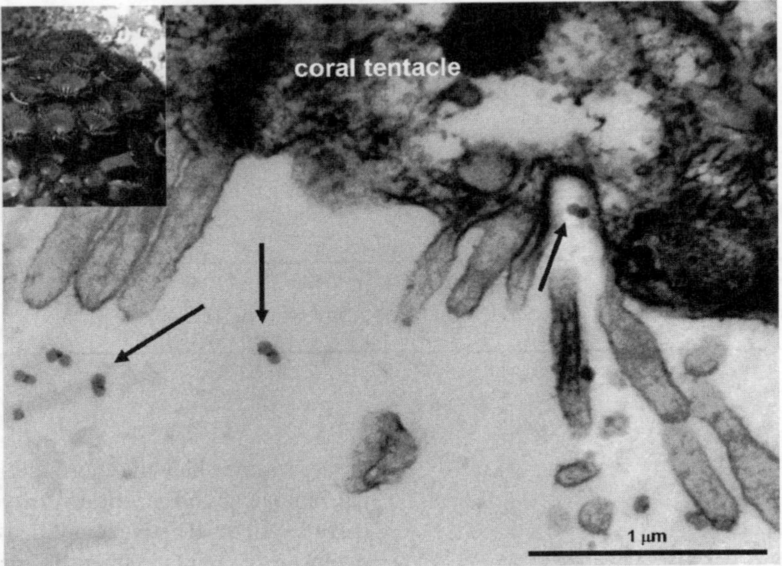

FIGURE 5.1 *Zoanthus* sp. (inset) virus-like particles (VLPs) adjacent to tentacles of a zoanthid after thermal shock. VLPs highlighted by arrows. Scale bar 1 μm. (*See the color version of this figure in Color Plates section.*)

advances such as flow cytometry (Figure 5.2) and genomics are crucial to get access to numerical (Patten et al., 2006) and diversity (Dinsdale et al., 2008; Marhaver et al., 2008; Thurber et al., 2008) data, respectively, for viruses. Arguably, one of the biggest failings in coral virus research has been lack of virus isolates to work with; indeed only a few moderately successful attempts have been recorded to date (Lohr et al., 2007; Wilson et al., 2001). This largely reflects the difficulty of culturing organisms from the coral holobiont. In our experience zooxanthellae (for example) are slow growing (not conducive to virus propagation) and in general there are very few dinoflagellate viruses known (Nagasaki, 2008;

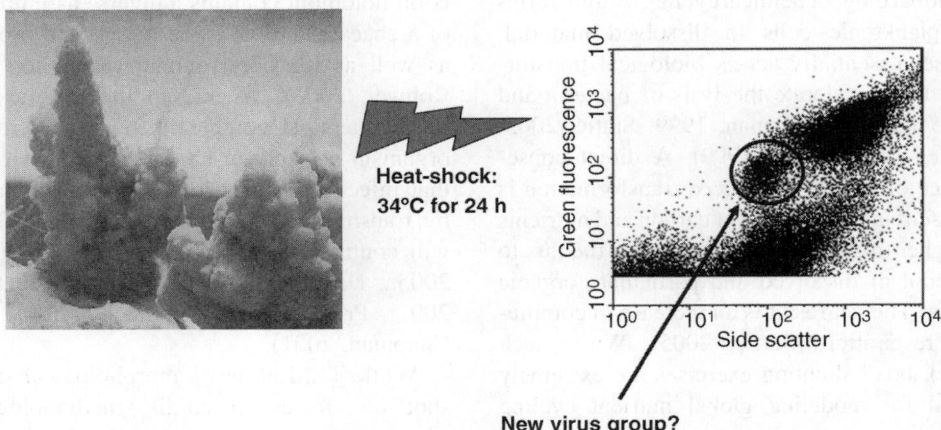

FIGURE 5.2 Flow cytometric analysis of seawater surrounding a nubbin of *Acropora formosa* following heat-shock. The virus group arrowed appeared after a 24 h heat-shock at 34 °C. (*See the color version of this figure in Color Plates section.*)

Nagasaki et al., 2006). In addition, there are no known cell lines of coral animals (cnidarians) to the author's knowledge, a prerequisite for propagation of true coral viruses.

5.2 WHY ARE VIRUSES NOT GIVEN PREVALENCE IN CORAL DISEASE DIAGNOSTICS?

It is only very recently that viruses have even been considered as potential pathogens in coral systems following the observations of virus-like particles (VLPs) in sea anemones (Wilson and Chapman, 2001; Wilson et al., 2001). An incidental report of VLPs from the sea anemone *Metridium senile* was recorded as far back as 1974 (Chapman, 1974). *M. senile* was collected and thin sectioned to provide material for a review of cnidarian histology and the VLPs were mentioned in passing on Figure 5.1 of this report (also see Figure 5.3).

The simple answer is that nobody ever looks for viruses routinely in diseased corals. It is clear that they are present, since they are invariably observed if appropriate procedures are used (Cervino et al., 2004; Patten et al., 2006). Recent metagenomics work comparing diseased and nondiseased corals have also revealed a prevalence of herpesvirus-like sequences in the metagenomes (Marhaver et al., 2008; Thurber et al., 2008). So there is increasing evidence that viruses are present and likely causative agents for some of the diseases observed. Arguably, the biggest issue is that virologists, microbiologists, or molecular biologists have not traditionally addressed the problem of coral disease despite the fact that these groups of scientists have the required toolboxes to determine the role of viruses in coral reef ecosystems. This trend has started changing with more studies focusing on microbial research questions related to corals (Rosenberg et al., 2007).

According to a virologist, it seems intuitive that many coral diseases, and the largely unexplained phenomenon of coral bleaching (we know it is caused by stress factors such as increased water temperature, and the associated elimination of symbiotic zooxanthellae, but little is known of the mechanisms behind the bleaching process), in part could be caused by viruses. The coral holobiont (host organisms plus its associated microorganisms) is a soup of microbes all likely infected by viruses with a wide range of propagation strategies. These are nondisease causing viruses (that statement is clearly debatable—disease is a human

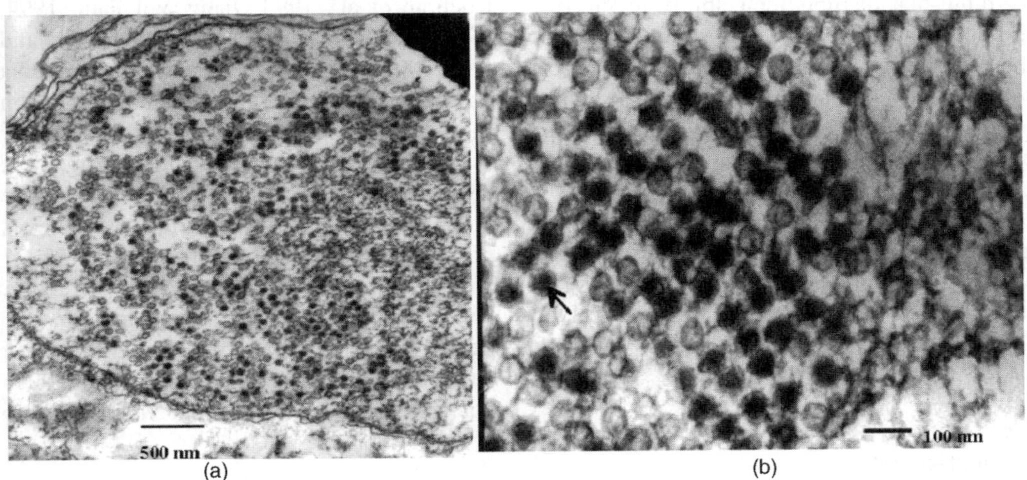

FIGURE 5.3 Electron micrograph of virus-like particles (VLPs) in a spiroblast nucleus of the plumose anemone *M. senile*. Scale bar approximately 500 nm (a). The VLPs are approximately 60 nm in diameter and the arrow indicates the electron dense core of approximately 40 nm of one VLP. Scale bar approximately 100 nm (b).

perception term, seen as something wrong), which are essentially part of a natural "healthy" microbial community that provide substrate for the coral animal. Viral action is a crucial part of a healthy coral ecosystem in the same way that viruses act as "lubricants" in pelagic systems (see above). The problem arises when the system starts to break down following the appearance of spots, bands, or "rashes" on the coral system in much the same way that microbial or viral pathogens do in humans. In coral systems, problems often occur when there is an environmental insult such as temperature increase, which stresses the system resulting in biological responses (bleaching is a clear example, McClanahan et al., 2007). *This may be a classic virus response*, particularly by a latent virus (often termed temperate viruses), although typically latent viral activity in response to systemic stress is more prevalently studied in prokaryote virus–host systems (Edgar and Lielausis, 1964). It seems reasonable to hypothesize that latent virus systems are prevalent in coral reef systems based on observations of disease and bleaching occurrence following environmental stress conditions. One obvious assumption would be that zooxanthellae harbor latent viruses. Zooxanthellae are the "life blood" of coral reefs and anything that disrupts their ability to survive through photosynthesis and symbiosis with the coral host will have an immediate effect on the whole coral reef system (as we see with bleaching). It is a strategically relevant hypothesis that temperature induction of latent viruses can control the bleaching and/or disease process, particularly given the clear evidence that temperature increases related to climate change are having devastating effects on coral reef ecosystems (Hoegh-Guldberg et al., 2007; Carpenter et al., 2008).

5.3 LATENT CORAL VIRUS HYPOTHESIS

With up to 10^8 viruses mL^{-1} in seawater (Bergh et al., 1989), and strong experimental evidence that these high concentrations are the result of lytic infection (Wilcox and Fuhrman, 1994), it is hardly surprising that the majority of research on marine viruses is conducted on lytic systems. Lytic infection is the process where a virus attaches to a host cell, injects its nucleic acid into the cell, and hijacks the host replication machinery to produce multiple copies of the virus. At this stage the cell bursts open (lysis), releasing viruses (and organic material) to start the cycle again.

However, there are two other generic types of virus propagation. The first is chronic infection where progeny viruses are released, typically via a budding mechanism, and the host cell, although weakened, survives the infection over several generations. The second mechanism is lysogeny (often termed latency[1] in eukaryotes), where virus nucleic acid is integrated into the host genome and replicates as part of the host-cell genetic complement. The replicated virus nucleic acid is referred to as a prophage or provirus. Crucially, lytic infection can then be induced via an environmental trigger, which is typically a host stress event (e.g., elevated temperature). Although there has been research conducted specifically on lysogeny in marine prokaryotes (Weinbauer and Suttle, 1996; Wilson and Mann, 1997; Cochran et al., 1998; Jiang and Paul, 1994, 1998; McDaniel et al., 2002; McDaniel and Paul, 2005), there is a surprising paucity of data on latency in marine eukaryotic phytoplankton in the recent literature.

So is latency even important? Arguably, latency represents the most important role that viruses play in regulating the dynamics of microbial (including phytoplankton or zoxanthellae) communities in seawater and or corals. This is contrary to what most people believe, that is, their role in mortality, but latency, as the term suggests, has hidden qualities and there are *strategic advantages to*

[1] Nonlytic infections (i.e., chronic, lysogeny, psuedolysogeny, persistent infection, and latency) are loosely defined here as "latency." These terms have different and specific definitions.

latent infections. Latency is a vehicle for lateral gene transfer, which fuels evolution and maintains high biodiversity of hosts. Proviruses can confer natural immunity to further infection by other lytic viruses (superinfection immunity) and they may even carry genes that are used by the host (phenotype conversion). This may either confer a selective advantage for the host (e.g., boost metabolism or confer toxicty) or the opposite is also thought to be true where a cost is associated with acquired immunity (e.g., reduced nutrient uptake efficiency). It has even been suggested that microalgae as symbionts serve as intermediary hosts for infection of other organisms (Dodds, 1979). Latency would provide the ideal refuge prior to infection of a secondary host (a secondary host could potentially be the coral or even grazing predators of the zooxanthellae).

A considerable amount of research was conducted on algal viruses in the early 1970s (largely not mentioned in recent literature), where viruses were observed in most eukaryotic algal phyla (Sherman and Brown, 1978; Dodds, 1979). Some of these early observations suggested latency (cf. lytic infection) as a mechanism of propagation, since the viruses were usually observed in otherwise healthy cell lines. The most convincing evidence of latency came from *Cylindrocapsa geminella* and *Uronema gigas* (both Chlorophytes), where virus production increased on heat-shock (Dodds and Cole, 1980; Hoffman and Stanker, 1976). It is amusing to note that in his (now 30-year old) review, Allen Dodds concluded that such observations "should encourage phycologists and virologists to join forces to make progress in the new field of eukaryotic algal virology" (Dodds, 1979). The review was written shortly before the discovery of *lytic* algal viruses (Mayer and Taylor, 1979; Meints et al., 1981; van Etten et al., 1982), which has been the focus of most of the subsequent research.

Arguably the most convincing evidence of latency in marine phytoplankton has come from the symbiotic zooxanthellae of cnidarians. Initial evidence came from research in our laboratory where viruses were induced following heat-shock of zooxanthellae isolated from the temperate sea anemone *Anemonia viridis* (Wilson et al., 2001). At the time we proposed that zooxanthellae harbor a latent virus infection that is induced by exposure to elevated temperatures. If such a mechanism also operates in the zooxanthellae harbored by reef corals, and these viruses kill the symbionts, then this could contribute to temperature-induced bleaching. This research was followed up using zooxanthellae from tropical reef corals with similar results (Davy et al., 2006; Wilson et al., 2005a). Further evidence using similar techniques was also reported by Cervino et al. (2004), who observed virus

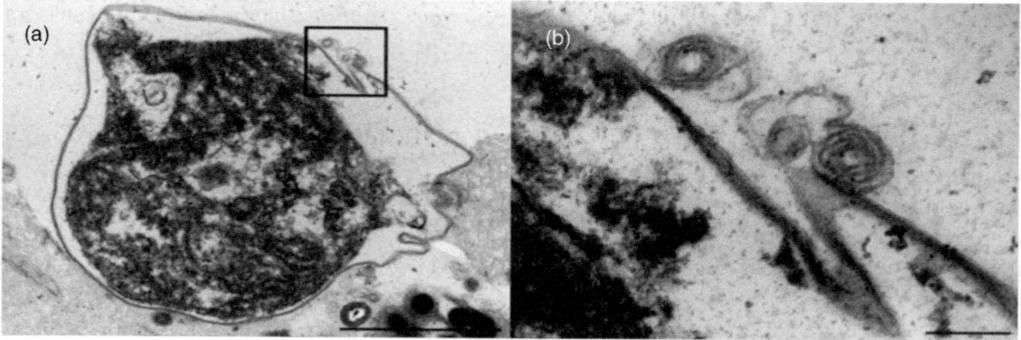

FIGURE 5.4 Electron micrographs showing the presence of filamentous VLPs curling off the surface of the cell within thin sections of zooxanthellae prepared 39 h after induction with UV light. The scale bars are 3 μm (a) and 200 nm (b). The black box in (a) highlights the area magnified in (b).

particles in zooxanthellae isolated from the Caribbean reef coral *Montastraea* sp. when exposed to either elevated temperature or bacterial pathogens. We went one-step further and isolated a latent filamentous virus induced from zooxanthellae cultures (Lohr et al., 2007) (Figure 5.4), however, we were unsuccessful in purifying nucleic acids from this virus (unpublished). Intriguingly, a recent widely reported study, revealed that sunscreens can induce viruses from symbiotic zooxanthellae in hard reef corals, resulting in rapid and complete bleaching, even at low concentrations (Danovaro et al., 2008). Viruses are known to be prevalent in coral reef systems as part of the coral holobiont (Dinsdale et al., 2008; Patten et al., 2006, 2008b; Davy and Patten, 2007; Seymour et al., 2005; Marhaver et al., 2008).

In a recent metagenomics paper by Marhaver et al. (2008), it was revealed that up to 8% of sequences from the virus metagenome of coral holobionts contained similarity to herpes-like viruses. Although there was no indication of what the hosts were (herpesviruses are typically latent viruses in animals), it remains a possibility that this classic latent virus (or a close relative) may also infect zooxanthellae (as well as the host cnidarians). Herpesviruses have evolved a highly efficient survival strategy, being able to establish life-long latency within the infected host, with occasional reactivation to allow virus replication and transmission to a new host (reviewed by Efstathiou and Preston, 2005). Latency occurs when the immediate early virus genes are not expressed: only the latency-associated transcripts (LATs) are transcribed. The LAT transcript is a piece of noncoding RNA, which is a precursor for four micro-RNAs in herpes simplex virus 1 (HSV-1, human herpesvirus 1, of the family Herpesviridae, genus *Simplexvirus*) infected cells. One of the functions of these micro-RNAs is thought to be prevention of apoptosis (cell death) through the process of RNA interference (Gupta et al., 2006). This strategy of keeping alive the host neuron cells harboring the latent virus is so successful that once infected with HSV-1 (usually in the first few years of life), people remain carriers for life: suffering recurrence of the characteristic facial sores that then shed virus particles, allowing virus transmission to new individuals. Common triggers for the reactivation of HSV-1 include stress and exposure to excessive sunlight: indeed UV irradiation is most often used in the laboratory to bring about the switch from latency to a lytic infection.

5.4 CORAL IMMUNITY AND ANTIVIRAL ACTIVITY

In a final note, it is worth briefly mentioning coral immunology. Although very little is known, there does appear to be a rudimentary immune system in cnidarians. Current evidence suggests highly conserved components of innate immunity and other cellular processes are more closely related to vertebrate homologues than more complex invertebrate model systems (Dunn, 2009). As clonal invertebrates, corals must rely on physiochemical barriers and cellular processes as a first line of defense against invading pathogens. Although, no studies have reported the role of pathogenic viruses to the coral immune system. Another important aspect of innate immunity includes proteins and compounds with direct antibacterial or antiviral activity. Corals have been shown to produce such bactericidal compounds that inhibit bacterial growth *in vitro* (Geffen et al., 2009) and metabolites from the soft coral *Sinularia capillosa* have demonstrated antiviral activity *in vitro* (Cheng et al., 2010).

5.5 SUMMARY

It is clear that viruses are central to the health of coral reefs (both good and bad health). Good health by acting as catalysts for nutrient cycling and, bad health through their ability to cause massive coral destruction. Many of the observed symptoms of coral disease and bleaching can easily be attributed to virus action. In addition, such symptoms are typically triggered by warmer waters (or other environmental insults) that are consistent with induction of latent viruses. Propagation of latent viruses in coral reefs is a sensible working hypothesis and, if it proves correct, will have serious ramifications with regard to current climate change scenarios. It seems likely that future research efforts will focus the effects of environmental factors associated with climate change (temperature increases and ocean acidification in particular) on diseased corals, the effectiveness of the immune response, and the contributions of the symbiotic relationships within the coral holobiont (Mydlarz et al., 2010).

ACKNOWLEDGMENTS

Current coral virus research in the WHW lab is supported through National Science Foundation grant OCE0851255 awarded to Dr. Susan Wharam and WHW. Thanks are given to various colleagues and students of the last 10 years who have contributed to the research (and figures) presented here, including Dr. David Chapman, Dr. Joanne Davy, Dr. Simon Davy, Dr. Jayme Lohr, Dr. Colin Munn, Dr. Susan Wharam, Sarah Burchett, Amy Dale, Piers Davies, Isobel Francis, and Cornelia Muncke.

REFERENCES

Bergh, O., Borsheim, K. Y., Bratbak, G., and Heldal, M. (1989). High abundance of viruses found in aquatic environments. *Nature* 340, 467–468.

Breitbart, M. and Rohwer, F. (2005). Here a virus, there a virus, everywhere the same virus? *Trends Microbiol.* 13, 278–284.

Carpenter, K. E., Abrar, M., Aeby, G., Aronson, R. B., Banks, S., Bruckner, A., Chiriboga, A., Cortes, J., Delbeek, J. C., DeVantier, L., Edgar, G. J., Edwards, A. J., Fenner, D., Guzman, H. M., Hoeksema, B. W., Hodgson, G., Johan, O., Licuanan, W. Y., Livingstone, S. R., Lovell, E. R., Moore, J. A., Obura, D. O., Ochavillo, D., Polidoro, B. A., Precht, W. F., Quibilan, M. C., Reboton, C., Richards, Z. T., Rogers, A. D., Sanciangco, J., Sheppard, A., Sheppard, C., Smith, J., Stuart, S., Turak, E., Veron, J. E. N., Wallace, C., Weil, E., and Wood, E. (2008). One-third of reef-building corals face elevated extinction risk from climate change and local impacts. *Science* 321, 560–563.

Cervino, J. M., Hayes, R., Goreau, T. J., and Smith, G. W. (2004). Zooxanthellae regulation in yellow blotch/band and other coral diseases contrasted with temperature related bleaching: *in situ* destruction vs. expulsion. *Symbiosis* 37, 63–85.

Chapman, D. M. (1974). *Cnidarian histology*. In: Lenhoff, H. and Muscatine, L. (eds.), *Coelenterate Biology*. Academic Press, New York/London, pp. 1–92.

Cheng, S.-Y., Huang, K.-J., Wang, S.-K., Wen, Z.-H., Chen, P.-W., and Duh, C.-Y. (2010). Antiviral and anti-inflammatory metabolites from the soft coral *Sinularia capillosa*. *J. Nat. Prod.* 73, 771–775.

Cochran, P. K., Kellogg, C. A., and Paul, J. H. (1998). Prophage induction of indigenous marine lysogenic bacteria by environmental pollutants. *Mar. Ecol. Prog. Ser.* 164, 125–133.

Danovaro, R., Bongiorni, L., Corinaldesi, C., Giovannelli, D., Damiani, E., Astolfi, P., Greci, L., and Pusceddu, A. (2008). Sunscreens cause coral bleaching by promoting viral infections. *Environ. Health Perspect.* 116, 441–447.

Davy, S. K., Burchett, S. G., Dale, A. L., Davies, P., Davy, J. E., Muncke, C., Hoegh-Guldberg, O., and Wilson, W. H. (2006). Viruses: agents of coral disease? *Dis. Aquat. Org.* 69, 101–110.

Davy, J. E. and Patten, N. L. (2007). Morphological diversity of virus-like particles within the surface microlayer of scleractinian corals. *Aquat. Microb. Ecol.* 47, 37–44.

DeLong, E. F., Preston, C. M., Mincer, T., Rich, V., Hallam, S. J., Frigaard, N.-U., Martinez, A., Sullivan, M. B., Edwards, R., Brito, B. R., Chisholm, S. W., and Karl, D. M. (2006). Community genomics among stratified microbial assemblages in the ocean's interior. *Science* 311, 496–503.

Dinsdale, E. A., Pantos, O., Smriga, S., Edwards, R. A., Angly, F., Wegley, L., Hatay, M., Hall, D., Brown, E., Haynes, M., Krause, L., Sala, E., Sandin, S. A., Thurber, R. V., Willis, B. L., Azam,

F., Knowlton, N., and Rohwer, F. (2008). Microbial ecology of four coral atolls in the Northern Line Islands. *PLoS ONE* 3, e1584.

Dodds, J. A. (1979). Viruses of marine algae. *Experientia* 35, 440–442.

Dodds, A. J. and Cole, A. (1980). Microscopy and biology of *Uronema gigas,* a filamentous eucaryotic green alga, and its associated tailed virus-like particle. *Virology* 100, 156–165.

Dunn, S. (2009). Immunorecognition and immunoreceptors in the Cnidaria. *Invert. Surv. J.* 6, 7–14.

Edgar, R. S. and Lielausis, I. (1964). Temperature sensitive mutants of phage T4D: their isolation and genetic characterisation. *Genetics* 49, 649–662.

Efstathiou, S. and Preston, C. M. (2005). Towards an understanding of the molecular basis of herpes simplex virus latency. *Virus Res.* 111, 108–119.

Fuhrman, J. A. (1999). Marine viruses and their biogeochemical and ecological effects. *Nature* 399, 541–548.

Geffen, Y., Ron, E. Z., and Rosenberg, E. (2009). Regulation of release of antibacterials from stressed scleractinian corals. *FEMS Microbiol. Lett.* 295, 103–109.

Gupta, A., Gartner, J. J., Sethupathy, P., Hatzigeorgiou, A. G., and Fraser, N. W. (2006). Anti-apoptotic function of a microRNA encoded by the HSV-1 latency-associated transcript. *Nature* 442, 82–85.

Hoegh-Guldberg, O., Mumby, P. J., Hooten, A. J., Steneck, R. S., Greenfield, P., Gomez, E., Harvell, C. D., Sale, P. F., Edwards, A. J., Caldeira, K., Knowlton, N., Eakin, C. M., Iglesias-Prieto, R., Muthiga, N., Bradbury, R. H., Dubi, A., and Hatziolos, M. E. (2007). Coral reefs under rapid climate change and ocean acidification. *Science* 318, 1737–1742.

Hoffman, L. R. and Stanker, L. H. (1976). Virus-like particles in green alga *Cylindrocapsa*. *Can. J. Bot.* 54, 2827–2841.

Jiang, S. C. and Paul, J. H. (1994). Seasonal and diel abundance of viruses and occurrence of lysogeny/bacteriocinogeny in the marine-environment. *Mar. Ecol. Prog. Ser.* 104, 163–172.

Jiang, S. C. and Paul, J. H. (1998). Significance of lysogeny in the marine environment: studies with isolates and a model of lysogenic phage production. *Microb. Ecol.* 35, 235–243.

Knowlton, N. and Rohwer, F. (2003). Multispecies microbial mutualisms on coral reefs: the host as a habitat. *Am. Nat.* 162, S51–S62.

Lohr, J., Munn, C. B., and Wilson, W. H. (2007). Characterization of a latent virus-like infection of symbiotic zooxanthellae. *Appl. Environ. Microbiol.* 73, 2976–2981.

Marhaver, K. L., Edwards, R. A., and Rohwer, F. (2008). Viral communities associated with healthy and bleaching corals. *Environ. Microbiol.* 10, 2277–2286.

Mayer, J. A. and Taylor, F. J. R. (1979). A virus which lyses the marine nanoflagellate *Micromonas pusilla*. *Nature* 281, 299–301.

McClanahan, T. R., Ateweberhan, M., Muhando, C. A., Maina, J., and Mohammed, M. S. (2007). Effects of climate and seawater temperature variation on coral bleaching and mortality. *Ecol. Monogr.* 77, 503–525.

McDaniel, L., Houchin, L. A., Williamson, S. J., and Paul, J. H. (2002). Plankton blooms: lysogeny in marine *Synechococcus*. *Nature* 415, 496.

McDaniel, L. and Paul, J. H. (2005). Effect of nutrient addition and environmental factors on prophage induction in natural populations of marine *Synechococcus* species. *Appl. Environ. Microbiol.* 71, 842–850.

Meints, R. H., Vanetten, J. L., Kuczmarski, D., Lee, K., and Ang, B. (1981). Viral-infection of the symbiotic chlorella-like alga present in *Hydra viridis*. *Virology* 113, 698–703.

Mydlarz, L. D., McGinty, E. S., and Harvell, C. D. (2010). What are the physiological and immunological responses of coral to climate warming and disease? *J. Exp. Biol.* 213, 934–945.

Nagasaki, K. (2008). Dinoflagellates, diatoms, and their viruses. *J. Microbiol.* 46, 235–243.

Nagasaki, K., Tomaru, Y., Shirai, Y., Takao, Y., and Mizumoto, H. (2006). Dinoflagellate-infecting viruses. *J. Mar. Biol. Assoc. U.K.* 86, 469–474.

Patten, N. L., Harrison, P. L., and Mitchell, J. G. (2008a). Prevalence of virus-like particles within a staghorn scleractinian coral (*Acropora muricata*) from the Great Barrier Reef. *Coral Reefs* 27, 569–580.

Patten, N. L., Mitchell, J. G., Middelboe, M., Eyre, B. D., Seuront, L., Harrison, P. L., and Glud, R. N. (2008b). Bacterial and viral dynamics during a mass coral spawning period on the Great Barrier Reef. *Aquat. Microb. Ecol.* 50, 209–220.

Patten, N. L., Seymour, J. R., and Mitchell, A. G. (2006). Flow cytometric analysis of virus-like particles and heterotrophic bacteria within coral-associated reef water. *J. Mar. Biol. Assoc. U.K.* 86, 563–566.

Rosenberg, E., Koren, O., Reshef, L., Efrony, R., and Zilber-Rosenberg, I. (2007). The role of microorganisms in coral health, disease and evolution. *Nat. Rev. Microbiol.* 5: 355–362.

Seymour, J. R., Patten, N., Bourne, D. G., and Mitchell, J. G. (2005). Spatial dynamics of virus-like particles and heterotrophic bacteria within a shallow coral reef system. *Mar. Ecol. Prog. Ser.* 288, 1–8.

Sherman, L. A. and BrownJr., R. M. (1978). Cyanophages and viruses of eukaryotic algae. In: Conrat, H. F. and Wagner, R. R. (eds.), *Comprehensive Virology: Newly Characterized Protist and Invertebrate Viruses*. Plenum Press, New York/London, pp. 145–234.

Suttle, C. A. (2005). Viruses in the sea. *Nature* 437, 356–361.

Thurber, R. L. V., Barott, K. L., Hall, D., Liu, H., Rodriguez-Mueller, B., Desnues, C., Edwards, R. A., Haynes, M., Angly, F. E., Wegley, L., and Rohwer, F. L. (2008). Metagenomic analysis indicates that stressors induce production of herpes-like viruses in the coral *Porites compressa*. *Proc. Natl. Acad. Sci. U.S.A.* 105, 18413–18418.

van Etten, J. L., Meints, R. H., Kuczmarski, D., Burbank, D. E., and Lee, K. (1982). Viruses of symbiotic chlorella-like algae isolated from *Paramecium bursaria* and *Hydra viridis*. *Proc. Natl. Acad. Sci. U.S.A.* 79, 3867–3871.

Weinbauer, M. G. and Suttle, C. A. (1996). Potential significance of lysogeny to bacteriophage production and bacterial mortality in coastal waters of the Gulf of Mexico. *Appl. Environ. Microbiol.* 62, 4374–4380.

Wilcox, R. M. and Fuhrman, J. A. (1994). Bacterialviruses in coastal seawater: lytic rather than lysogenic production. *Mar. Ecol. Prog. Ser.* 114, 35–45.

Wilhelm, S. W. and Suttle, C. A. (1999). Viruses and nutrient cycles in the sea: viruses play critical roles in the structure and function of aquatic food webs. *Bioscience* 49, 781–788.

Wilson, W. H. and Chapman, D. M. (2001). Observation of virus-like particles in thin sections of the plumose anemone, Metridium senile. *J. Mar. Biol. Assoc. U.K.* 81, 879–880.

Wilson, W. H., Dale, A. L., Davy, J. E., and Davy, S. K. (2005a). An enemy within? Observations of virus-like particles in reef corals. *Coral Reefs* 24, 145–148.

Wilson, W. H., Francis, I., Ryan, K., and Davy, S. K. (2001). Temperature induction of viruses in symbiotic dinoflagellates. *Aquat. Microb. Ecol.* 25, 99–102.

Wilson, W. H. and Mann, N. H. (1997). Lysogenic and lytic viral production in marine microbial communities. *Aquat. Microb. Ecol.* 13, 95–100.

Wilson, W. H., Schroeder, D. C., Allen, M. J., Holden, M. T. G., Parkhill, J., Barrell, B. G., Churcher, C., Hamlin, N., Mungall, K., Norbertczak, H., Quail, M. A., Price, C., Rabbinowitsch, E., Walker, D., Craigon, M., Roy, D., and Ghazal, P. (2005b). Complete genome sequence and lytic phase transcription profile of a coccolithovirus. *Science* 309, 1090–1092.

Yooseph, S., Sutton, G., Rusch, D. B., Halpern, A. L., Williamson, S. J., Remington, K., Eisen, J. A., Heidelberg, K. B., Manning, G., Li, W. Z., Jaroszewski, L., Cieplak, P., Miller, C. S., Li, H. Y., Mashiyama, S. T., Joachimiak, M. P., van Belle, C., Chandonia, J. M., Soergel, D. A., Zhai, Y. F., Natarajan, K., Lee, S., Raphael, B. J., Bafna, V., Friedman, R., Brenner, S. E., Godzik, A., Eisenberg, D., Dixon, J. E., Taylor, S. S., Strausberg, R. L., Frazier, M., and Venter, J. C. (2007). The Sorcerer II Global Ocean Sampling Expedition: expanding the universe of protein families. *PLoS. Biol.* 5, 432–466.

Zak, D. R., Blackwood, C. B., and Waldrop, M. P. (2006). A molecular dawn for biogeochemistry. *Trends Ecol. Evol.* 21, 288–295.

CHAPTER 6

VIRUSES INFECTING MARINE MOLLUSCS

TRISTAN RENAULT
Laboratoire de Génétique et Pathologie, Ifremer, La Tremblade, France

CONTENTS

6.1 INTRODUCTION

Although the association of shellfish transmitted infectious diseases with sewage contamination has been well documented since the late nineteenth and early twentieth centuries (Speirs et al., 1987; Jones and Bej, 1994; Lees, 2000; Potasman et al., 2002; Nishida et al., 2003) the discovery of viruses infecting marine molluscs is fairly recent. Marine bivalves through their filter feeder behavior, can accumulate well-known viruses that infect humans and mammals and act as transient reservoirs. Thus, there is considerable literature on the related implications on human health.

However, viruses have also been reported in association with massive mortality outbreaks of economically significant mollusc species worldwide. Irido-like viruses are believed to have led to the entire destruction of Portuguese oyster, *Crassostrea angulata*, stocks between 1967 and 1973 along the Atlantic coast in France (Comps, 1969, 1970, 1978; Comps and Duthoit, 1979; Comps and Bonami, 1977). Other viruses interpreted variously as members of the Iridoviridae, Herpesviridae, Papovaviridae, Togaviridae, Reoviridae, Birnaviridae, and Picornaviridae have been reported infecting different mollusc species (Farley, 1976, 1978, 1985; Elston, 1979, 1997; Meyer, 1979; Meyers and Hirai, 1980; Elston and Wilkinson, 1985; Rasmussen, 1986; Norton et al., 1993; McGladdery and Stephenson, 1994; Jones et al., 1996; Miyazaki et al., 1999; Novoa and Figueras, 2000; Bower, 2001; Carballal et al., 2003; Winstead and Courtney, 2003; Choi

Studies in Viral Ecology: Animal Host Systems: Volume 2, First Edition. Edited by Christon J. Hurst.

et al., 2004; Renault and Novoa, 2004). Finally, a putative retrovirus has been proposed as the etiological agent of a cell disorder of the circulatory system reported in *Mya arenaria*, termed disseminated neoplasia (Oprandy et al., 1981; Oprandy and Chang, 1983).

The interest of studying viruses infecting marine molluscs is twofold. Such studies are not only needed in order to increase basic knowledge on viruses in aquatic environments but, moreover, mollusc production is the second most important aquaculture activity in the world by quantity and by value (FAO, 2009). The world production in 2006 was estimated as 14.1 million tonnes, representing 27% of the total world aquaculture production valued at US$ 11.4 billion. By volume, oysters (Ostreidae) are the most important aquaculture taxonomic group second only to cyprinids at 4.6 million tonnes (FAO, 2006). The Pacific cupped oyster, *C. gigas*, itself had the greatest contribution among molluscs with a worldwide production volume of 4.4 million tonnes (FAO, 2006). Although mollusc culture is steadily growing in importance in the aquaculture sector, cultivated molluscs may suffer from severe mortality outbreaks. Among the possible causes is the occurrence of infectious diseases due to a variety of pathogens including viruses (Bower, 2001; Elston, 1997; Renault and Novoa, 2004).

Viral pathogens are often highly infectious and easily transmissible, and are commonly associated with mass mortalities among molluscs. In 2008 and 2009, massive mortality outbreaks were reported in several farming areas among *C. gigas* oysters in France, Ireland, and the Channel Islands. These were attributed to a combination of adverse environmental factors together with the presence of the *Ostreid herpesvirus 1* (OsHV-1) (Le Deuff and Renault, 1999; Davison et al., 2005) and the presence of bacteria belonging to the genus *Vibrio*. Mortalities have been considerable, particularly in seed stocks. The great oyster devastation of 2008 and 2009 will result in a shortage in supplies of the shellfish over the next 3 years.

Although mortality outbreaks have been reported among different mollusc species in association with the detection of viruses, little information is available on their exact affiliation and taxonomic position. This may be partly related to the lack of marine mollusc cell lines, a factor that restricts virus isolation capacities, and the predominant use of histology as the basic method for identification and examination of suspect mollusc samples. Although histology enables the identification of cellular changes associated with viral infections, other methods including transmission electron microscopy (TEM) and molecular approaches are needed in order to provide more conclusive identification of viruses. Since invertebrates lack antibody-producing cells, the direct detection of viral agents remains the only way for virus diagnosis. A variety of virus-like particles have been reported from various mollusc species based on transmission electron microscopy. Further investigations based on molecular approaches have been carried out for a small number of them including the herpes-like virus infecting Pacific oysters in France (Le Deuff and Renault, 1999; Davison et al., 2005).

This chapter tentatively summarizes the present knowledge on viruses infecting molluscs through two main examples (i) the herpes-like viruses focusing on the single member of the Malacoherpesviridae family, the *Ostreid herpesvirus 1* (OsHV-1) and (ii) the marine birnaviruses. Both virus groups have been studied more extensively than other mollusc viruses. They have been subjected to examination at molecular, epidemiological, and ecological levels. Moreover, herpesviruses and birnaviruses detected in molluscs demonstrated contrasts in terms of their infectivity. Although birnavirus have been detected in various marine molluscs, the infectivity of such viruses for molluscs should still be regarded as weak or hypothetical. On the contrary, herpesviruses and herpes-like viruses infecting molluscs clearly appear highly pathogenic.

6.2 HERPES-LIKE AND HERPESVIRUSES INFECTING MARINE MOLLUSCS

Herpes-like viruses have attracted particular research attention because of their economic and ecological impact on cultured and wild marine molluscs during the last 20 years. Farley et al. (1972) first reported viral particles morphologically similar to herpesviruses in an invertebrate, the Eastern oyster, *C. virginica*. A wide host range has since been reported for herpes-like viruses infecting various oyster species, clams, and scallops (Renault and Novoa, 2004) in different countries including the USA (Farley et al., 1972; Friedman et al., 2005; Meyers et al., 2009), New Zealand (Hine et al., 1992, 1998), France (Nicolas et al., 1992; Comps and Cochennec, 1993; Renault et al., 1994a, 1994b, 2000a, 2000b, 2001b; Arzul et al., 2001a, 2001b, 2001c), Australia (Hine and Thorne, 1997), French Polynesia (Comps et al., 1999), and Mexico (Vásquez-Yeomans et al., 2004). More recently, highly pathogenic herpes-like viruses have also been reported in different abalone species including *Haliotis diversicolor supertexta* (Chang et al., 2005), *H. laevigata*, and *H. rubra* (Hooper et al., 2007; Tan et al., 2008). All of these viruses have been reported associated with substantial mortality outbreaks.

6.2.1 *Ostreid herpesvirus 1*, A Virus Infecting Marine Bivalves

6.2.1.1 History and Classification The purification of herpes-like virus particles from French *C. gigas* larvae allowed the extraction of viral DNA (Le Deuff and Renault, 1999). The entire virus DNA was sequenced (GenBank accession number AY509253) and the virus classified under the name *Ostreid herpesvirus 1* (OsHV-1) as the single known species in the family Malacoherpesviridae (Davison et al., 2005, 2009; Mac Geoch et al., 2006).

The genome structure and sequence, and the capsid morphology (Davison et al., 2005) have been further studied in order to assess OsHV-1 phylogeny status in relation to vertebrate herpesviruses. The overall genome structure is similar to that of certain mammalian herpesviruses (e.g., herpes simplex virus and human cytomegalovirus). It consists of two invertible unique regions (U_L, 167.8 kbp and U_S, 3.4 kbp) each flanked by inverted repeats (TR_L and IR_L, 7.6 kbp; TR_S and IR_S, 9.8 kbp). A unique additional region (X, 1.5 kbp) is present between IR_L and IR_S. The coding potential of the genome sequence was analyzed allowing the identification of 132 unique protein-coding open reading frames (ORFs). Amino acid sequence comparisons demonstrated that OsHV-1 is not clearly related to vertebrate herpesviruses. However, a common origin between OsHV-1 and vertebrate herpesviruses may be suspected based on the identification of a gene coding the ATPase subunit of the terminase. Terminase is an enzyme complex involved with packaging DNA into preformed capsids, which is conserved among all herpesviruses (Davison, 2002). However, the fact that this herpesvirus contains a genetic sequence that is distantly related to a gene in bacteriophage T4 leaves open the possibility of convergent evolution.

In terms of morphological features, the OsHV-1 capsids (Davison et al., 2005) appear structurally similar to those of other herpesviruses that have been studied (Booy et al., 1994; Trus et al., 1999, 2001; Cheng et al., 2002). Tridimensional reconstructions of OsHV-1 capsids derived by cryoelectron microscopy revealed an icosahedral structure with a triangulation number of $T = 16$ (Davison et al., 2005), a structure observed only with herpesviruses.

The capsid structure and the presence of a putative terminase gene support assignment of OsHV-1 as a member of the herpesviruses despite its having very few other genetic relationships with the herpesviruses found in mammals, birds, amphibians, reptiles, and fish. The available data support the view that there may be three major lineages: herpesviruses of mammals and birds, herpesviruses of fish and amphibians, and herpesviruses of invertebrates

(Davison et al., 2005, 2009; McGeoch et al., 2006). They also indicate that OsHV-1 is the first identified member of that major group of herpesviruses and in this regard possibly is the single known representative of what may be a large number of invertebrate herpesviruses. However, in contrast with this concept, Gao and Qi (2007) showed through the composition vector method that OsHV-1 jumps out of the branch of Herpesviridae and groups with iridoviruses. This is interesting and unusual in that OsHV-1 capsids are formed in the nucleus of infected cells while particle synthesis takes place only in the cytoplasm for all iridoviruses.

6.2.1.2 Clinical Features and Pathology

Ostreid herpesvirus 1 infection causes mortality in the larvae and juveniles of several bivalve species including the Pacific cupped oyster *C. gigas*, the Portuguese oyster *C. angulata*, the European flat oyster *Ostrea edulis*, the carpet shell clam *Ruditapes decussatus*, and the Manila clam *R. philippinarum*, as well as the scallop *Pecten maximus* (Renault et al., 2000a, 2000b, 2001b; Arzul et al., 2001a, 2001b, 2001c). The virus can be found in adult bivalves most often in absence of mortality. Infected larvae show a reduction in feeding and swimming activities and mortality can reach 100% in few days. Affected spat show sudden and high mortalities mainly in summer time.

Histology reveals cell abnormalities on tissue sections from infected animals. The infection-associated lesions in spat are mainly observed in connective tissues in which fibroblastic-like cells exhibit enlarged nuclei with margined chromatin. Highly condensed nuclei were also reported in another cells interpreted as hemocytes (Figure 6.1). These cellular abnormalities are not associated with massive hemocyte infiltration.

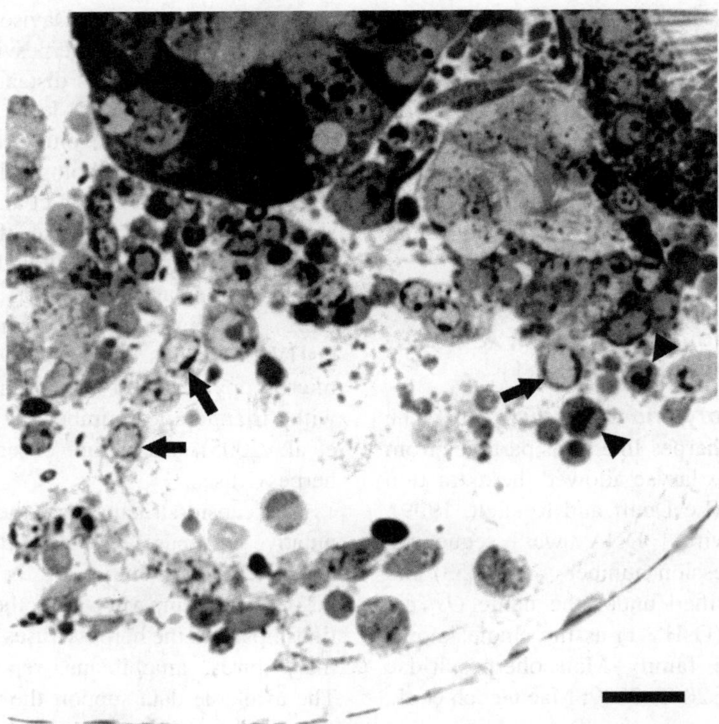

FIGURE 6.1 Toluidine blue-stained semithin section of a *C. gigas* larva infected with OsHV-1. Condensed hyperbasophilic nuclei (arrow heads) and enlarged nuclei presenting marginated chromatin are observed (arrows). Scale bar = 20 μm.

Ostreid herpesvirus 1 replication mainly takes place in fibroblastic-like cells throughout connective tissues especially in mantle, labial palps, gills, and digestive gland. Virogenesis begins in the nucleus of infected cells where capsids and nucleocapsids are observed (Figure 6.2). Viral particles then pass through the nuclear membrane into the cytoplasm and enveloped particles are released at the cell surface (Figure 6.3). Intranuclear and cytoplasmic capsids present a variety of morphological types including electron-lucent capsids, toroïdal core-containing capsids, and brick-shaped core-containing capsid (Figure 6.4). *Ostreid herpesvirus 1* resembles other herpesviruses in its morphological characteristics, cellular locations, and particle sizes.

6.2.1.3 Diagnosis Extraction and sequencing of OsHV-1 whole genome from purified particles infecting *C. gigas* larvae served as a platform for the development of specific molecular diagnosis tools. A nested-PCR using primers A3–A4 and A5–A6 combined with targeting (after the second amplification) 940 bp of a gene coding an unknown protein was the first attempt developed to detect the virus in *C. gigas* larvae and spat. A quick and convenient sample preparation first protocol using ground tissues allowed a sensitive detection of viral presence in infected oysters, with detection of up to 500 fg of viral DNA in 50 μL PCR reaction tubes. A primers do not amplify vertebrate herpesviruses (Renault et al., 2000b).

A simple PCR using primers C1–C6 (Renault and Arzul, 2001) was then developed targeting 896 bp of a part of the viral genome located in an inverted repeat (TR_L/IR_L) and coding unknown proteins. This second protocol allows detecting up to 10 fg of viral DNA using C1–C6 primers that do not amplify vertebrate herpesviruses. This second technique was

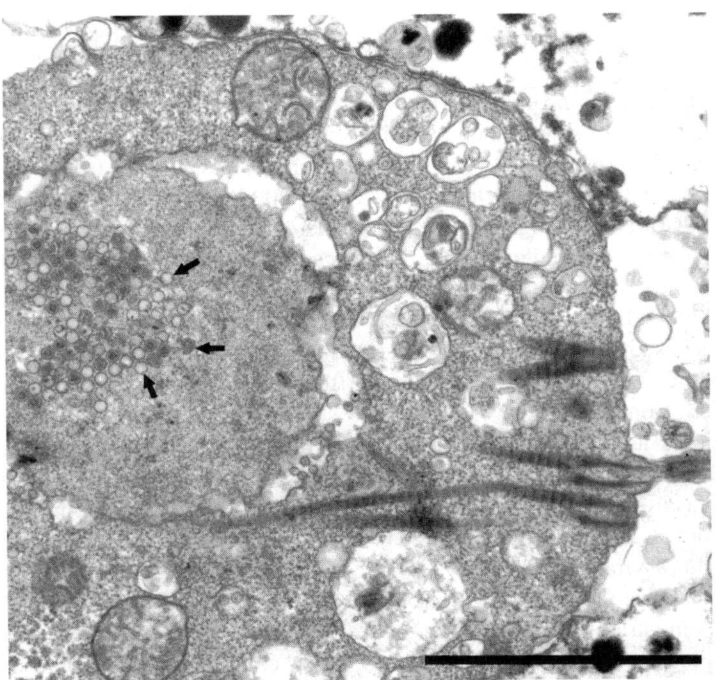

FIGURE 6.2 Transmission electron micrograph of OsHV-1 infected cells from Pacific oyster larvae. Spherical or polygonal virus particles are observed in the nucleus of an infected cell (arrows). Scale bar = 2 μm.

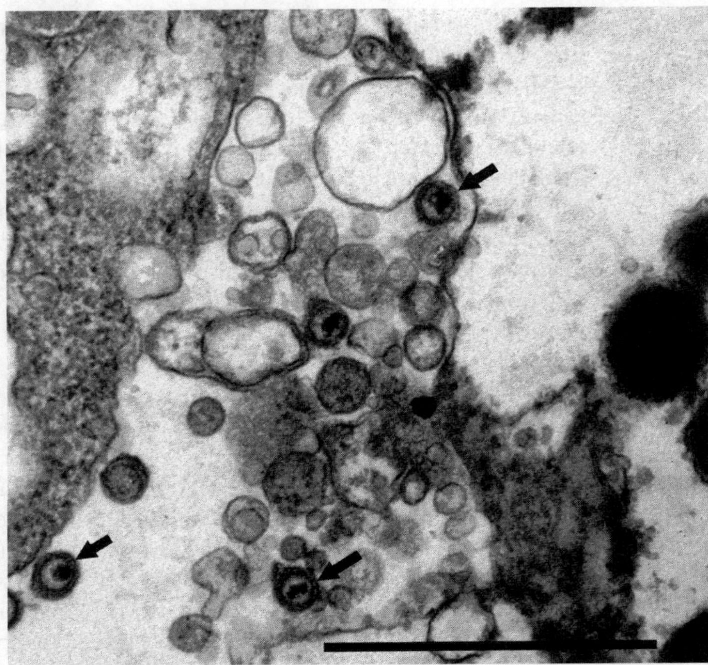

FIGURE 6.3 Transmission electron micrograph of OsHV-1 infected cells from Pacific oyster larvae. Extracellular enveloped particles (arrows). Virions are icosahedral in shape with a central electron-dense core, surrounded by an electron-lucent zone followed by another dense layer. Two unit membranes separated by a clear zone enclose the particle. Scale bar = 1 μm.

largely used for the detection of OsHV-1 and especially in the context of studying abnormal bivalve mortalities. A competitive PCR method was also developed using previously designed primer pairs, C2–C6, amplifying a 710 fragment of the viral genome located in an inverted repeat (TR_L/IR_L) and coding unknown proteins (Renault and Arzul, 2001; Renault et al., 2004). This technique is based on the use of OsHV-1-specific primers and an internal standard competitor that differs from the target DNA by a deletion of 76 bp. The assay allows detecting up to 1 fg of viral DNA in 0.5 mg of oyster tissues. This technique was used to check the presence of PCR inhibitors as well as performing a semiquantification of viral DNA.

Finally, PCR has been successfully used to detect viral DNA in various bivalve species at different stages of development and several PCR diagnostic protocols have been developed. Different methods of DNA extraction as well as various primer pairs have been designed and used to detect viral DNA using single-round or nested PCR (see review by Batista et al., 2007). A quantitative PCR protocol was also recently developed (Pépin et al., 2008) that can be used to quantify specifically OsHV-1 DNA.

In addition, the PCR methodology allows the study of OsHV-1 polymorphism through the detection of different OsHV-1 genotypes such as OsHV-1 var (Arzul et al., 2001a, 2001b) and OsHV-1 μvar (Segarra et al., 2010). Indeed, OsHV-1 exhibits several genomic differences from the reference genome (GenBank accession number AY509253), most notably deletions in different part of its genome (Arzul et al., 2001a, 2001b; Segarra et al., 2010). Van Regenmortel (2008) wrote "the genome of a virus cannot be defined by a unique sequence corresponding to a so-called wild type but consists of a distribution of mutant sequences,

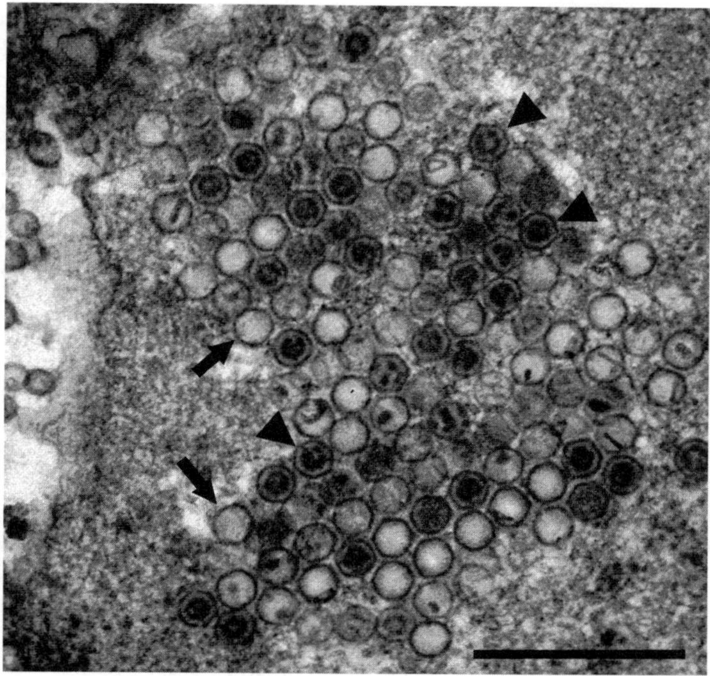

FIGURE 6.4 Transmission electron micrograph of OsHV-1 infected cells from Pacific oyster larvae. Intranuclear capsids presenting a variety of morphological types including electron-lucent capsids (arrows) and core-containing capsids (arrow heads). Scale bar = 500 nm.

each one differing from the sequence of the clone." Molecular approaches are thus needed to describe this diversity and molecular data are necessary in order to better understand virus ecology including host range, virulence, pathogenesis, host responses, vectors, and habitat.

An *in situ* hybridization protocol has also been developed using dig-labeled A5/A6 and C1/C6 PCR products as specific probes (Renault and Lipart, 1998; Lipart and Renault, 2002).

6.2.1.4 Epidemiological Investigations

PCR-based diagnostic methods have facilitated epidemiological investigations. To better understand the implication of OsHV-1 in *C. gigas* spat mortality outbreaks regularly reported both in the field and in the nurseries in France, samples were collected yearly through the French National Network for Surveillance of Mollusc Health since 1997. Analyses were carried out by PCR for OsHV-1 detection. Virus DNA was frequently detected in samples collected during mortality events. Data also demonstrated a particular seasonality and topography of spat oyster mortalities associated with OsHV-1 detection. In the field, mortality outbreaks appeared in summer, preferentially in sheltered environments. They mostly occurred in spots whereas they were rapid and massive in nursery. Data also suggested an influence of the seawater temperature on OsHV-1 detection (Garcia et al., unpublished data).

Transmission experiments demonstrated the pathogenicity of OsHV-1 at larval stages (Le Deuff et al., 1994, 1996) and more recently in Pacific oyster juveniles (Schikorski et al., 2011a). The incubation period can be as soon as 3 days, and mortality can rise rapidly to 40–100% (Le Deuff et al., 1994, 1996; Schikorski et al., 2011a). OsHV-1 is transmitted from infected oysters to healthy oysters and the virus DNA can be detected in healthy individuals as soon as 6 h after cohabitation with infected oysters (Schikorski et al., 2011a). Experimental trials also showed a similar genotype of OsHV-1 that can induce an infection at larval stages in different bivalve species and interspecies transmission occurred (Arzul et al., 2001a, 2001b, 2001c). This contrasts with vertebrate herpesviruses, which are generally confined to a single species in nature. Consequently, the true host of OsHV-1 is unknown and it is possible that the parental virus still resides in its natural host.

Herpesviruses have evolved with their hosts for long periods, capturing and modifying host genes in order to evade host immune defenses. As such, vertebrate herpesviruses tend to cause disease in young immunologically naïve individuals and then persist in cryptic forms (latency) for probably the lifetime of their hosts. Like vertebrates herpesviruses, OsHV-1 may persist in an inapparent form (true latency and/or replicating at low level) in adult oysters. Such healthy carriers may act as virus reservoirs and promote virus transmission to their progeny (Le Deuff et al., 1996; Arzul et al., 2002; Barbosa-Solomieu et al., 2004, 2005). Although bivalve larvae and juveniles appear more sensitive to OsVH-1 infections compared to adults (Arzul et al., 2001b, 2002), OsHV-1 DNA and proteins have been detected in asymptomatic *C. gigas* adults (Arzul et al., 2002; Barbosa-Solomieu et al., 2004, 2005). Similar observations have been noticed for vertebrate herpesviruses in aquatic environments. For example, fish herpesviruses including the *Oncorhynchus masou* virus (OMV) and the channel catfish virus (CCV), essentially induce severe diseases and high mortality rates among fingerlings and both of these viruses have also been detected in normal appearing adults (Plumb, 1973; Kimura et al., 1981; Wise et al., 1988).

6.2.1.5 OsHV-1 Detection in the Marine Environment Broadly available molecular tools have been used to detect OsHV-1 in various mollusc species at different stages of development and has especially been of importance when massive mortality outbreaks occurred. However, data on OsHV-1 detection in the open marine environment (e.g., in seawater) are very scarce. Such data may be useful to better understand the transmission of this disease. Improved knowledge about transmission of OsHV-1 may lead to practical epidemiological recommendations that would limit the impact of viral infection on the oyster industry. Vigneron et al. (2004) detected OsHV-1 DNA at 100 ng L^{-1} (4.4×10^8 OsHV-1 DNA copies assuming a genome size around 207 kbp) in seawater in bioassays using classical PCR. They also demonstrated that water parameters including temperature might influence virus DNA detection (Vigneron et al., 2004).

Moribund and dead infected oysters may act as a major source of the virus. Real-time PCR (RT-PCR) assay appears to be a valuable tool for investigating the abundance of OsHV-1 DNA in water samples and could therefore be used to assess virus presence in the environment. Recently, Sauvage et al. (2009) examined seawater samples collected during a mortality outbreak among Pacific oysters reared in an artificial pond. Virus DNA was detected using real-time PCR (Pépin et al., 2008) in the seawater surrounding moribund and dead oysters (around 10^3 viral DNA copies L^{-1}) (Sauvage et al., 2009). Another study was performed in 2009 in order to investigate the kinetics of OsHV-1 DNA detection during the course of a cohabitation assay between infected and healthy Pacific oysters (Schikorski et al., 2011a). Virus DNA quantification was carried out by real-time PCR in different oyster tissues and in seawater. Results indicated that OsHV-1 DNA amounts increased rapidly to approximately 10^8 DNA copies L^{-1} of seawater 6 h after moribund experimentally infected oysters have been added to aquaria (Schikorski et al., 2011b). The detection of viral DNA in seawater samples suggest that seawater may act as a source of OsHV-1 for horizontal transmission. Heavily infected oysters might be the site of intense viral replication until their death, and virus release from moribund and dead oysters could allow horizontal transmission to living ones.

Herpes-like particles have also been detected by chance in a unicellular organism interpreted as a marine fungoid protist (thraustochytrid-like organism) on the basis of ultrastructure features during an experimental trial (Arzul et al., 2001a; Renault et al., 2003). These virus particles demonstrated similar morphological characteristics to OsHV-1. Capsids and nucleocapsids were detected in the nucleus of infected cells (Figure 6.5). Moreover, enveloped virions accumulated between the cytoplasmic membrane and the outermost structure interpreted as the wall of the unicellular organism (Figure 6.6). This unicellular organism was detected in rearing tanks of Pacific oyster larvae, free in the water column or in larvae inside cell vacuoles suggesting a phagocytosis process (Figure 6.7). Although PCR analysis targeting OsHV-1 gave positive results, no capsids and nucleocapsids were detected in the nucleus of oyster larval cells. A herpes-like virus has been reported in an isolate of *Thraustochytrium* sp., a monocentric phycomycetous fungus, by Kazama and Schornstein (1972, 1973). The fungus was isolated from the York River Estuary (Virginia, USA). Viral particles resembling herpesviruses have also been described by Perkins in another fungus species, *Schizochytrium aggregatum* (Kazama, 1972). These findings suggest that some marine heterotrophic protists may be infected by herpesviruses related to OsHV-1 and a question arises about the possible involvement of such marine protists acting as vectors and/or as hosts for OsHV-1.

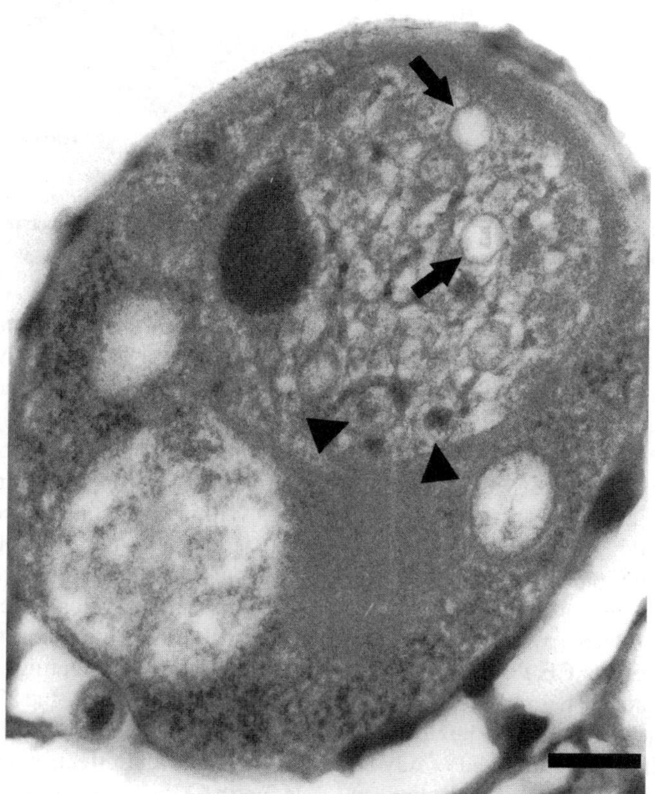

FIGURE 6.5 Transmission electron micrograph of infected cells from a fungus-like organisms. Capsids (arrows) and nucleocapsids (arrow heads) were detected in the nucleus of an infected cell. Scale bar = 100 nm.

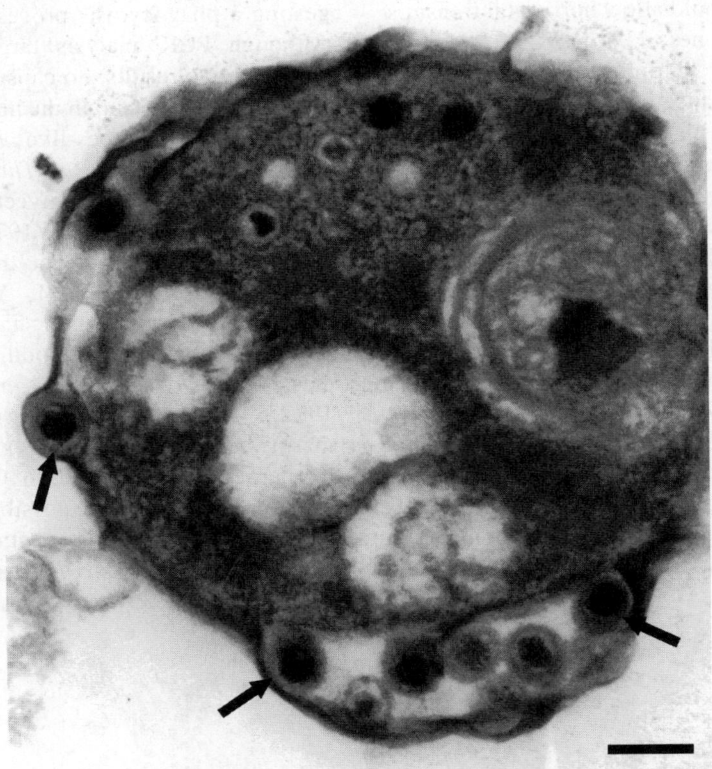

FIGURE 6.6 Transmission electron micrograph of infected cells from a fungus-like organisms. Enveloped virus particles (arrows) accumulated between the cytoplasmic membrane and the most outer structure interpreted as the wall of the fungus-like organism. Scale bar = 200 nm.

6.2.1.6 Immune Responses The suppressive subtraction hybridization (SSH) technology has been applied to *C. gigas* in order to identify OsHV-1-induced genes (Renault et al., unpublished data). This approach led to the first identification of several sequences presenting homologies with known genes involved in antiviral immunity in an oyster species (laccase, macrophage-expressed protein, molluscan defense protein, IK cytokine, IFI44, etc.). The related oyster genes were totally sequenced by RACE PCR (Renault et al., unpublished data). The antiviral immune responses remain a vast domain to be explored in bivalves. Identified encoding genes may be used as markers for improving resistance to infections and their products as therapeutic agents.

6.2.1.7 Prevention and Control As molluscs lack a "true" adaptive immune system and an immune memory, vaccination cannot be used to protect them against pathogens. Moreover, the use of drugs is highly limited for animals that are most often reared in the open sea. Within this context, controlling animal transfers appears to be one of the most suitable way to combat infectious diseases in molluscs and international rules have thus been developed. Minimum measures for the control of certain diseases affecting bivalve molluscs were established by the Council Directive 2006/88/EU.

Although the infection caused by OsHV-1 is neither listed at present as a notifiable disease by the EU legislation (2006/88/EC, Annex IV) nor by the Office International des

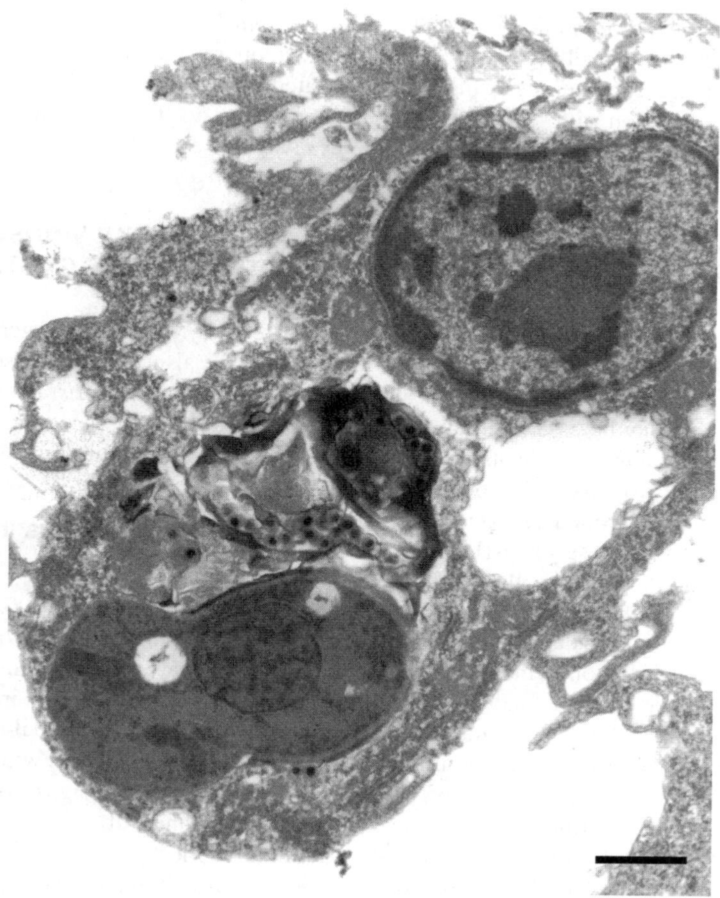

FIGURE 6.7 Transmission electron micrograph of infected cells from a fungus-like organisms. Infected fungus-like organisms detected in a Pacific oyster larva inside a cell vacuole. Scale bar = 600 nm.

Epizooties (OIE) (Aquatic Animal Health Code, 2009), a new EU regulation (Regulation 175/2010) was implemented in March 2010. This regulation lays down measures to be taken in the case of increased mortality in the species *C. gigas* in connection with the detection of OsHV-1 μvar. Increased mortalities of Pacific oysters were reported in several member states in 2008 and 2009. They were attributed to a combination of adverse environmental factors, the presence of bacteria of the genus *Vibrio* and the presence of OsHV-1 including a newly described genotype named OsHV-1 μvar. OsHV-1 μvar is defined on the basis of partial sequence data as exhibiting a systematic deletion of 12 base pairs in ORF4 of the genome in comparison with the "reference" genotype (GenBank accession number AY509253). In the case of abnormal mortality outbreaks among *C. gigas* oysters, a targeted surveillance should be carried out to detect or rule out the presence of OsHV-1 μvar. When the presence of that virus has been detected, disease control measures should be implemented including the establishment of a containment area and the restriction to the movements out of the containment areas of *C. gigas* oysters. The measures should apply until the end of 2010 and may be extended or amended as appropriate after that period, taking into account the knowledge and experience gained.

Rapid and accurate differential diagnosis is the key to success in controlling an OsHV-1 outbreak. Molecular techniques including real-time PCR, in combination with sequence analysis of amplicons enable a much more precise analysis of OsHV-1 epidemiology in the field and may help to define the most adapted measures for disease containment.

Biosecurity may be successfully applied in confined and controlled facilities such as hatcheries and nurseries in order to protect the facility and the surrounding environment from the introduction of the virus. As OsHV-1 is an enveloped virus, it may be assumed that it is fragile. High temperature, chemicals, or sunlight (UV) may destroy its lipid-containing envelope. Thus, in controlled rearing conditions (mollusc hatchery/nursery), OsHV-1 outbreaks may therefore be controlled through quarantine and hygienic measures including virus inactivation through adapted treatments such as ultraviolet irradiation of the recirculating water and water filtration technologies. However, it is necessary to keep in mind that reduction of virus load depends on the initial titer and the virus reduction capacity of the techniques used for inactivation. If there was an initial concentration of 1 million viruses L^{-1} and the inactivation method used allowed inactivation of 100,000 viruses L^{-1}; there would still numerous infective particles in the treated product. Moreover, it has been demonstrated that individual herpesvirus species may have widely different stability to inactivation treatments (Plummer and Lewis, 1965) and that inorganic salts such as Na_2SO_4 present in seawater may stabilize herpesviruses (Wallis and Melnick, 1965).

Finally, mollusc-breeding programs targeting the production of resistant animals to a particular disease appear as one of the most promising approaches for aquaculture development. Disease selection programs have already been successfully developed for *Haplosporidium nelsoni*, *Perkinsus marinus*, and Juvenile Oyster disease in *C. virginica* (Guo et al., 2003) and for *Bonamia ostreae* in *O. edulis* (Naciri-Graven et al., 1998). Based on recent data, it has been demonstrated that Pacific oyster strains resistant or tolerant to OsHV-1 can be developed (Sauvage et al., 2009).

6.3 HERPES-LIKE VIRUSES INFECTING MARINE GASTROPODS

More recently, highly pathogenic herpes-like viruses were also reported in several abalone species including *Haliotis diversicolor supertexta* in Taiwan, *H. laevigta*, *H. rubra*, and their hybrids in Australia.

6.3.1 Herpes-Like Virus Infecting Abalone in Taiwan

Commencing in January 2003, a mortality outbreak affecting abalone *H. diversicolor supertexta* was reported in northeastern Taiwan. Abalone cultivated in intensive culture ponds on land and in intertidal-out ponds were affected. Death occurred within 3 days of the onset of clinical signs. Both adult and juvenile abalone suffered from the disease. Cumulative mortality rates ranged from 70% to 80%. During the course of the outbreak, the water temperature varied between 16 and 19°C. These temperatures were abnormally low for the season in Taiwan.

Histological examination revealed lesions of nervous tissues and gills in diseased animals including hemocyte infiltration and necrosis (Chang et al., 2005). Analysis based on transmission electron microscopy demonstrated the presence of virus particles resembling herpesviruses (Chang et al., 2005). Capsids and nucleocapsids were observed in the nuclei of infected cells. Transmission assays based on intramuscular injection and bath trials demonstrated the capacity of the virus to induce high mortality rates in a few days (Chang et al., 2005). Mortality outbreaks in association with the viral infection were also reported in Taiwan in 2004 and 2005.

6.3.2 Herpes-Like Virus Infecting Abalone in Australia

Commencing in December 2005, unexplained mortalities of farmed abalone were reported in three farms in southeastern Australia (Victoria). The first case of a new abalone disease was detected in a land-based farm located near Portland in early December 2005. Two other farms, one at Port Fairy and the other in Westernport Bay were secondly affected. It had been noticed that from October to December 2005, a considerable amount of abalone translocation occurred (Hardy-Smith, 2006). On May 2006, the disease was reported for the first time in a wild population from a coastal lagoon from which one of the farms pumps its water. The presence of the virus was then assessed on wild stock abalone populations in different locations along the coast of western Victoria and more than 40 sites were surveyed during May 2006. In June 2006, the viral infection was detected in 19 sites. Spreading of the disease through wild abalone in western Victoria waters occurred very rapidly and ultimately the infected coastal area approached 200 km, with significant impacts on abalone populations. Although the virus was first detected in land-based farms and was only subsequently reported in wild abalone, the exact origin of the virus remains unknown.

In abalone farms, young stocks were more seriously affected by the virus outbreak than were older animals. Massive mortality occurred and a daily mortality rate of 5000 dead animals per day was recorded in one of the affected farms. Cumulative mortality rates varied from 5% to 90%. The role of stress factors including high densities, spawning period, and water temperature (up to 17.5–18°C) was highly suspected as contributing factors to the high mortality rates. Clinical signs included enlarged mouth parts, protruding radula, and that the animals were easily removed from their substrate. Both green-lip and black-lip abalone (*H. laevigata* and *H. rubra*) and their hybrids were affected. Lesions of the nervous tissue were reported in the Australian abalone outbreak just as had been reported in affected abalone in Taiwan. A herpes-like virus was identified within nerve cells of the affected animals by transmission electron microscopy (Hooper et al., 2007; Tan et al., 2008). The disease was determined and thus termed abalone viral ganglioneuritis (AVG) but was not previously described in Australia.

The Australian Animal Health Laboratory in Geelong (Victoria) conducted experimental assays and demonstrated that the virus could be transmitted from sick individuals to healthy ones by intramuscular injection of fresh and frozen (–80°C) abalone tissues. One hundred percent mortalities were observed within 5–6 days after intramuscular injection in the abalone foot (Corbeil et al., personal communication). The virus appeared highly pathogenic and remained infective after being frozen at –80°C. Experimental trials were also carried out using seawater as the source of the virus. Results showed that the presence of infected abalone is not necessary for virus transmission to healthy individuals indicating that the virus can also be spread through the water column.

The virus infecting Australian abalone shows all ultrastructure characteristics of herpesviruses, particularly hexagonal shaped capsids forming in infected cell nuclei. However, as the virus reported infecting abalone in Taiwan, it is identified at present as being a "herpes-like virus." Can the virus infecting Australian abalone be assigned to the order Herpesvirales, not only when its genetic sequence and its whole genome organization but also its capsid features are determined.

6.3.2.1 Diagnosis and Control

The detection of herpes-like viruses infecting abalone in Taiwan and in Australia raises a lot of questions including the possibility of a taxonomic relationship between OsHV-1 and viruses detected in abalone. Moreover, mortality outbreaks of abalone have also been reported in relation to the detection of a spherical virus in the eastern area of Guangdong province (China) (Wang et al., 2004). Specific molecular tools designed for the diagnosis of *Ostreid herpesvirus 1* (OsHV-1) (Lipart and Renault, 2002; Renault et al., 2000b) have been used on abalone samples collected in Taiwan. Although OsHV-1-specific primers did not allow the detection of amplicons from infected abalone (Chang et al., 2005; Renault et al., unpublished data), positive signals (nervous ganglia and connective tissue of different organs) were observed on histological sections from infected abalone using an OsHV-1-specific probe (Figures 6.8 and 6.9).

In this context, the necessity of studying herpes-like viruses infecting marine molluscs is confirmed. More knowledge on herpesviruses and herpes-like viruses infecting molluscs is needed in order to develop suitable strategies to minimize the impact of these viruses on shellfish production.

Although the infection caused by herpes-like viruses in abalone is not listed at present as a notifiable disease by the EU legislation (2006/88/EC, Annex IV), this purported infection of abalone by a herpes-like virus has been included in the most recent versions of the Aquatic Animal Health Code (OIE, 2009a) and in the Manual of Diagnostic Tests for Aquatic Animals (OIE, 2009b). Related codes of practice, standard operating procedures (SOP) and educational leaflets have also been written regarding the abalone herpes-like virus by Australian policy makers in order to propose that abalone farmers and divers follow appropriate biosecurity control measures.

6.4 MARINE BIRNAVIRUSES IN MOLLUSCS

6.4.1 History and Classification

Viruses interpreted as belonging to the Birnaviridae have been isolated *in vitro* from different bivalve mollusc species. Experimental infection trials in different bivalves species and molecular characterization studies have been carried out using these virus isolates.

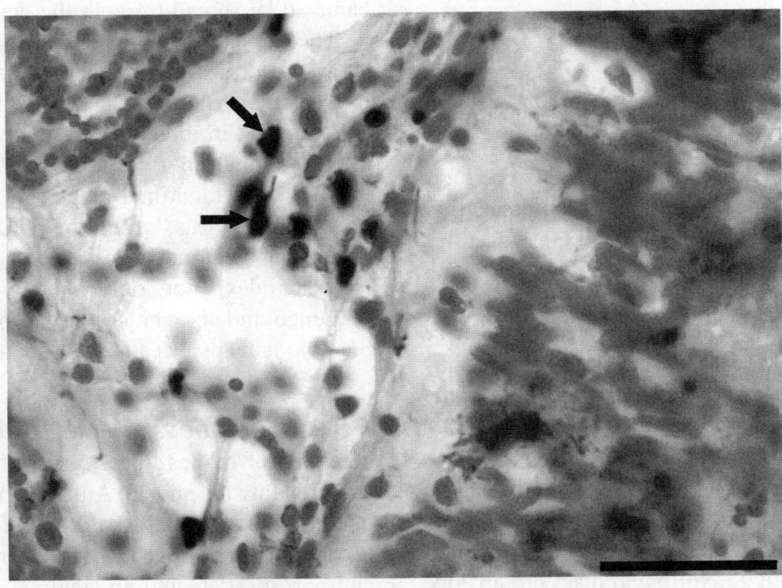

FIGURE 6.8 HIS stained cells (blue color, arrows) in the connective tissue of the digestive gland from an infected Pacific oyster collected in France using an OsHV-1-specific probe. Scale bar = 20 μm. (*See the color version of this figure in Color Plates section.*)

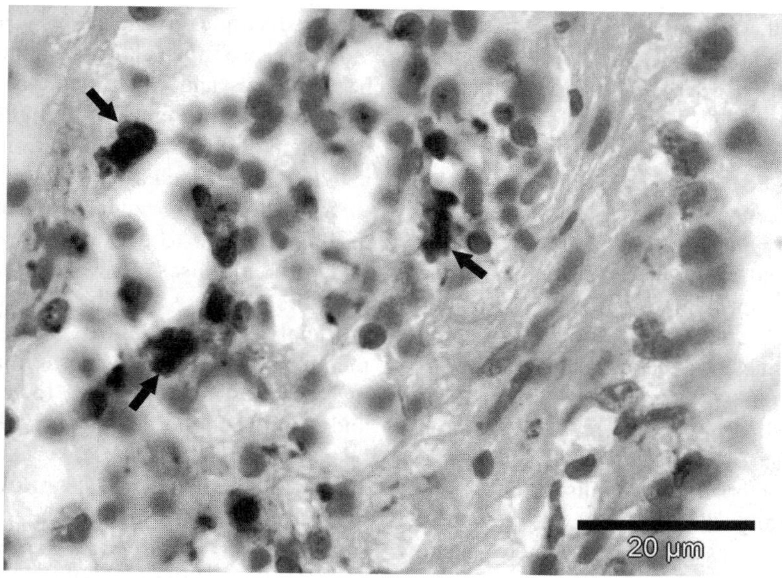

FIGURE 6.9 HIS stained cells (blue color, arrows) in the connective tissue of the digestive gland from an infected abalone collected in Taiwan using an OsHV-1-specific probe. Scale bar = 20 μm. (*See the color version of this figure in Color Plates section.*)

Birnaviruses were first isolated from bivalves in Europe (Hill, 1976) and Asia (Lo et al., 1988). Tellina virus 1 (TV-1) was isolated, on the BF-2 (bluegill fry) fish cell line, from *Tellina tenuis* in Great Britain (Hill, 1976) and assigned to the Birnaviridae family (Dobos et al., 1979). A similar virus was also isolated from *O. edulis* (Hill, 1976). Lo et al. (1988) reported the isolation of viruses assigned to the Birnaviridae from cultured hard clams, *Meretix lusoria*, in Taiwan, using fish cell lines. While examining a high mortality outbreak in Japanese pearl oysters, Suzuki et al. (1998a) isolated a presumably causative virus using the chinook salmon embryo cell line (CHSE-214) and that virus was tentatively named "marine birnavirus" (MABV) with the isolate being designated MABV strain JPO-96 (Suzuki et al., 1998a). The infectivity of the MABV strain JPO-96 against Japanese pearl oysters appeared to be weak when examined by Suzuki et al. (1998c). A birnavirus has also been isolated from Agemaki (Jack Knife Clam) *Sinovacura consticta* in Japan (Suzuki et al., 1998b). More recently, viruses interpreted as aquabirnaviruses were reported from Geoduck clams *Panope abrupta* and litteneck clams *Protothaca staminea* collected in Alaska (Meyers et al., 2009), although neither of the latter two viruses were associated with abnormal mortality nor lesions detected by histology in the collected native clams (Meyers et al., 2009).

Marine birnavirus have been defined as a group belonging to the genus *Aquabirnavirus* and forming a genogroup independent of the infectious pancreatic necrosis virus (IPNV) infecting salmonids. MABV are relevant fish pathogens. These viruses comprise the yellowtail ascites virus (YAV) first isolated from *Serriola quinqueradiata* in Japan (Sorimachi and Hara, 1985) and other similar viruses isolated from marine fish (Bonami et al., 1983; Schultz et al., 1984; Hedrick et al., 1986; Novoa et al., 1993; Novoa and Figueras, 1996). MABV have also been isolated from a variety of marine shellfish (Inaba et al., 2009). Based on serological and genomic properties, the strains isolated from shellfish and fish seem similar (Suzuki et al., 1997b, 1998c).

Specifically, the fish and shellfish isolates demonstrated high homologies in the VP2/NS junction region of the virus genome (Suzuki et al., 1998a, 1998b; Zhang and Suzuki, 2003, 2004; Inaba et al., 2009). This virus genome region is extensively used for genogrouping because of its variability. Zhang and Suzuki (2004) reported a total of seven genogroups in the Aquabirnaviruses with all of the MABV grouping in the same genogroup. Although MABV and IPNV resemble each other physically, genogrouping based on the nucleotide sequence of the VP2/NS junction region (A segment) separates them (Hosono et al., 1996). More recently, Nobiron et al. (2008) have demonstrated that genome and protein characterization establish TV-1 to be phylogenetically distant from all previously known birnaviruses and to define a new genetic cluster among the birnaviruses.

6.4.2 Experimental Trials and Virus Infectivity

Several studies have been carried out in order to address the question of whether birnaviruses isolated from bivalves using fish cell lines are subsequently able to induce infectious diseases in bivalves. The infectivity of virus particles isolated from *T. tenuis* and *O. edulis* in Great Britain was explored through experimental trials (Hill and Alderman, 1977) in an effort to help answer this question and by determining that the data obtained suggested that birnaviruses isolated in this way may be able to infect the European flat oyster inducing pathological effects. Virus-exposed flat oysters showed histological changes including necrosis of connective tissues and hemocyte infiltration into the digestive gland (Hill and Alderman, 1977). Other studies have been then carried out including that by Chou et al. (1994) that induced mortality of hard clams, *M. lusoria*, using birnavirus CV-TS-1, in experimental trials based on waterborne and injection exposures. The effect of heavy metal cations on the susceptibility of the hard clam to clam birnavirus infection has also been analyzed (Chou et al., 1998). The resulting indication was that clams contaminated with heavy metals and experimentally exposed to the birnavirus presented with higher mortality rates than did individuals either only contaminated or only infected (Chou et al., 1998).

In addition, birnaviruses isolated from molluscs appear to possess a weak pathogenicity for molluscs. However, physiological stressors such as spawning or insertion of an artificial pearl nuclei, exposure to heavy metals and changes in water temperature can increase host susceptibility to these viruses and result in higher levels of mortality as has been reported in clam, *M. lusoria* (Chou et al., 1994, 1998), Agemaki (jack knife clam), *Sinovacura constricta* (Suzuki et al., 1998c), and Japanese pearl oyster, *Pincada fucata* (Suzuki et al., 1997a, 1998a). MABV can be considered as opportunistic pathogens, which persistently infect marine organisms and can become pathogenic under stressful conditions (Suzuki et al., 1998b; Suzuki and Nojima, 1999; Kitamura et al., 2000).

6.4.3 Epidemiology and Ecology

The host range of MABV appears to be broad (Suzuki and Nojima, 1999). Furthermore, the fact of MABV isolation from a variety of apparently healthy marine shellfish (Hill, 1982; Suzuki and Nojima, 1999) and from environmental samples (Rivas et al., 1993) suggests that shellfish species may act as carriers and reservoirs for these viruses. Within this context, several studies have been carried out in order to search for and to define the kinetics of MABV detection in marine environments including the molluscs themselves and their surrounding seawater.

Kitamura et al. (2000) reported variations of the virus detection among Japanese pearl oysters depending upon the season. An increased level of MABV detection occurred in winter based on PCR analysis and was observed with virus isolation on CHSE-214 cells and the detection of virus particles by

TEM (Kitamura et al., 2000). During the rest of the year, MABV detection was reported only by PCR and indirect fluorescent antibody technique (IFAT). These authors concluded that oysters were persistently infected with MABV and that the virus remained in hemoctyes during the summer period (Kitamura et al., 2000). The results reported by these authors may also suggest that MABV was degraded while inside the hemocytes because, although viral proteins (IFAT) were detected in hemocytes during summer, neither were virus particles observable during this time period by means of TEM nor did successful virus isolation occur (Kitamura et al., 2000).

Additional studies on marine birnavirus occurrence in different ecological niches have been performed. The seasonal occurrence of MABV was monitored in seawater samples collected in 1997 and 1998 in Uwa Sea (Japan) (Kitamura and Suzuki, 2000). The authors showed that the MABV genome was detectable by PCR all along the course of the survey with the lowest levels of virus RNA reported in summer and increased amounts from fall to winter. Although they also detected virus proteins in some samples by using specific enzyme-linked immunosorbent assays (ELISA), infective particles were not isolated during the survey (Kitamura and Suzuki, 2000). These results may indicate that the virus can be released from infected animals including molluscs to seawater in winter, and that the virus can then be degraded in the marine environment.

Kitamura et al. (2002) carried out analysis of Japanese pearl oyster and seawater samples collected at two different depths (2 and 15 m) at one site (Uwa Sea, Ehime Prefecture, Japan). The results reported by Kitamura et al. (2002) are in accordance with previous results (Kitamura and Suzuki, 2000; Kitamura et al., 2000), which found that MABV was less detectable during the summer period in Japanese pearl oysters. Moreover, although virus RNA was present in seawater samples collected at 15 m depth all along the year, it was not detected in summer at 2 m depth (Kitamura et al., 2002). The authors suggested this to indicate that the stability of the virus is related to the depth, and that the virus is more stable in deeper waters. Although Kitamura et al. (2002) suggested that sunlight UV radiations may destroy the virus, this hypothesis was not confirmed though a specific study of that subject (Kitamura et al., 2004). Recently, the distribution of MABV was investigated in various marine organisms in different locations in Japan (Okinawa and Ishigaki Islands) (Inaba et al., 2009). Virus RNA was detected by RT-PCR in 20 gastropod species among the 25 collected species and in 9 bivalve species among the 16 collected species. These results confirm that the host range of MABV is extremely broad among marine molluscs. However, such results can be quite variable, with a case in point being that Inaba et al. (2009) reported successful virus isolation for only 2 mollusc species among a total of the 143 PCR positive samples that represented 41 different mollusc species.

A study has also been carried out by Kitamura et al. (2003) in order to determine whether phytoplankton and zooplankton may act as vectors for MABVs. Although MABV RNA was detected by PCR in some zooplankton samples collected from the Uwa Sea (Japan), the virus could not be isolated from PCR positive samples. As such, the infectivity of MABV detected in zooplankton should be regarded as hypothetical and the report of Kitamura et al. (2003) as a preliminary study.

These studies demonstrate that MABV is detectable in various shellfish species and in the marine environment (seawater) in several locations in Japan. They also suggest that shellfish may act as carriers and/or reservoirs of MABV and may have substantial effects on MABV distribution. Furthermore, these studies address the questions of whether MABV detection based on PCR corresponds to the presence of infective virus particles and whether marine molluscs may really act as sources of infective viruses.

6.5 CONCLUSION

Most of the currently recognized viruses of molluscs infect species that are farmed. Indeed, rearing conditions may enhance the likelihood of their detection. Keeping that in mind, we can assume that many more mollusc viruses await discovery.

Although preventing and controlling diseases has become a priority for the sustainability of mollusc aquaculture, viral pathogens remain difficult to diagnose, prevent, and control in open systems. Consequently, farmers are left with very limited disease management capacities. Advancement in the field of mollusc virology will require an increase of knowledge of virus/host interactions including antiviral immune responses, the refinement and greater use of molecular specific tools, and the development of continuous cell lines derived from marine molluscs.

Moreover, the ecology of viruses infecting marine molluscs is often not well known. The mechanisms that allow molluscan viruses to resist environmental conditions when outside their host need particular attention. We also need to define experimental criteria and methods to accurately measure pathogen survival, which would in turn enable us to search for infective viral pathogens in the field and to establish ecological characterizations that can help us to define less risky farming sites.

ACKNOWLEDGMENTS

Prof. P.H. Chang (National University of Taiwan, Taipei), Drs. M. Crane and S. Corbeil (CSIRO, Geelong, Victoria, Australia), Dr. M. Lancaster (PIRVic Attwood, Victoria, Australia), Dr. P. Harry-Smith (Panaquatic Health Solutions, Victoria, Australia), and H. Peeters (Western Abalone Divers Association Inc., Victoria, Australia) are thanked for their valuable support. Many thanks to C. J. Hurst for helping to improve this chapter through his very useful suggestions.

REFERENCES

Arzul, I., Nicolas, J. L., Davison, A. J., and Renault, T. (2001a). French scallops: a new host for ostreid herpesvirus-1. *Virology* 290, 342–349.

Arzul, I., Renault, T., and Lipart, C. (2001b). Experimental herpes-like viral infections in marine bivalves: demonstration of interspecies transmission. *Dis. Aquat. Organ.* 46, 1–6.

Arzul, I., Renault, T., Lipart, C., and Davison, A. J. (2001c). Evidence for inter species transmission of oyster herpesvirus in marines bivalves. *J. Gen. Virol.* 82, 865–870.

Arzul, I., Renault, T., Thébault, A., and Gérard, G. (2002). Detection of oyster herpesvirus DNA and proteins in asymptomatic *Crassostrea gigas* adults. *Vir. Res.* 84, 151–160.

Barbosa-Solomieu, V., Dégremont, L., Vazquez-Juarez, R., Ascencio-Valle, F., Boudry, P., and Renault, T. (2005). Ostreid herpesvirus 1 detection among three successive generations of Pacific oysters (*Crassostrea gigas*). *Vir. Res.* 107, 47–56.

Barbosa-Solomieu, V., Miossec, L., Vazquez-Juarez, R., Ascencio-Valle, F., and Renault, T. (2004). Diagnosis of Ostreid herpesvirus 1 in fixed paraffin-embedded archival samples using PCR and *in situ* hybridisation. *J. Virol. Methods* 119 (2), 65–72.

Batista, F. M., Arzul, I., Pepin, J. F., Ruano, F., Friedman, C. S., Boudry, P., and Renault, T. (2007). Detection of ostreid herpesvirus 1 DNA by PCR in bivalve molluscs: a critical review. *J. Virol. Methods* 39(1), 1–11.

Bonami, J. R., Cousserans, F., Weppe, M., and Hill, B. J. (1983). Mortalities in hatchery-reared sea bass fry associated with a birnavirus. *Bull. Eur. Assoc. Fish Pathol.* 3, 41.

Booy, F. P., Trus, B. L., Newcomb, W. W., Brown, J. C., Conway, J. F., and Steven, A. C. (1994). Finding a needle in a haystack: detection of a small protein (the 12-kDa VP26) in a large complex (the 200 MDa capsid of herpes simplex virus). *Proc. Natl. Acad. Sci. U.S.A.* 91, 5652–5656.

Bower, S. M. (2001). Synopsis of infectious diseases and parasites of commercially exploited shellfish: assorted viruses detected in oysters and of unknown significance. URL: http://www-sci.pac.dfo-mpo.gc.ca/shelldis/assortvirusoy_e.htm.

Carballal, M. J., Villalba, A., Iglesias, D., and Hine, P. M. (2003). Virus like particles associated with large foci of heavy hemocytic infiltration in cockles *Cerastoderma edule* from Galicia (NW Spain). *J. Invertebr. Pathol.* 84, 234–237.

Chang, P. H., Kuo, S. T., Lai, S. H., Yang, H. S., Ting, Y. Y., Hsu, C. L., and Chen, H. C. (2005). Herpes-like virus infection causing mortality of cultured abalone *Haliotis diversicolor supertexta* in Taiwan. *Dis. Aquat. Organ.* 65, 23–27.

Cheng, N., Trus, B. L., Belnap, D. M., Newcomb, W. W., Brown, J. C., and Steven, A. C. (2002). Handedness of the herpes simplex virus capsid and procapsid. *J. Virol.* 76, 7855–7859.

Choi, D. L., Lee, N. S., Choi, H. J., Park, M. A., McGladdery, S. E., and Park, M. S. (2004). Viral gametocytic hypertrophy caused by a papova-like virus infection in the Pacific oyster *Crassostrea gigas* in Korea. *Dis. Aquat. Organ.* 59, 205–209.

Chou, H. Y., Chang, S. J., Lee, H. Y., and Chiou, Y. C. (1998). Preliminary evidence for the effect of heavy metal cations on the susceptibility of hard clam (*Meretrix lusoria*) to clam birnavirus infection. *Fish Pathol.* 33, 213–219.

Chou, H. Y., Li, H. J., and Lo, C. F. (1994). Pathogenicity of a birnavirus to hard clam (*Meretrix lusoria*) and effect of temperature stress on its virulence. *Fish Pathol.* 29(3), 171–175.

Comps, M. (1969). Observations relatives à l'affection branchiale des huîtres portugaises (*Crassostrea angulata* Lmk). *Rev. Trav. Inst. Pêches Mari.* 33, 151–160.

Comps, M. (1970). La maladie des branchies chez les huîtres du genre *Crassostrea.* Caractéristiques et évolution des altérations, processus de cicatrisation. *Rev. Trav. Inst. Pêches Mari.* 34, 24–43.

Comps, M. (1978). Evolution des recherches et études récentes en pathologie des huîtres. *Oceanol. Acta* 1, 255–262.

Comps, M. and Bonami, J. R. (1977). Infection virale associée à des mortalités chez l'huître *Crassostrea angulata* Th. *C. R. Acad. Sci. D* 285, 1139–1140.

Comps, M. and Cochennec, N. (1993). A herpes-like virus from the European oyster *Ostrea edulis* L. *J. Invertebr. Pathol.* 62, 201–203.

Comps, M. and Duthoit, J. L. (1979). Infections virales chez les huîtres *Crassostrea angulata* (Lmk) et *C. gigas* (Th.). *Haliotis* 8(1977), 301–308.

Comps, M., Herbau, C., and Fougerousse, A. (1999). Virus-like particles in pearl oyster *Pinctada margaritifera. Bull. Eur. Assoc. Fish Pathol.* 19(2), 85–88.

Davison, A. J. (2002). Evolution of the herpesviruses. *Vet. Microbiol.* 86, 69–88.

Davison, A. J., Eberle, R., Ehlers, B., Hayard, G. S., McGeoch, D. J., Minson, A. M., Pellett, P. E., Roizman, B., Studdert, M. J., and Thiry, E. (2009). The order *Herpesvirales. Arch. Virol.* 154, 171–177.

Davison, A. J., Trus, B. L., Cheng, N., Steven, A. C., Watson, M. S., Cunningham, C., Le Deuff, R. M., and Renault, T. (2005). A novel class of herpesvirus with bivalve hosts. *J. Gen. Virol.* 86, 41–53.

Dobos, P., Hill, B. J., Hallett, R., Kells, D. T. C., Bescht, H., and Teninges, D. (1979). Biophysical and biochemical characterization of five animal viruses with bisegmented double-stranded RNA genomes. *J. Virol.* 32, 593–605.

Elston, R. A. (1979). Virus-like particles associated with lesions in larval Pacific oysters (*C. gigas*). *J. Invertebr. Pathol.* 33, 71–74.

Elston, R. A. (1997). Special topic review: bivalves mollusc viruses. *World J. Microbiol. Biotechnol.* 13, 393–403.

Elston, R. and Wilkinson, M. T. (1985). Pathology, management and diagnosis of oyster velar virus disease (OVVD). *Aquaculture* 48, 189–210.

FAO (2006). State of world aquaculture. FAO Fisheries Technical Paper 500, 129 pp.

FAO (2009). The state of world fisheries and aquaculture 2008. FAO Fisheries and Aquaculture Department, 176 pp.

Farley, C. A. (1976). Ultrastructural observations on epizootic neoplasia and lytic virus infection in bivalve mollusks. *Prog. Exp. Tumor Res.* 20, 283–294.

Farley, C. A. (1978). Viruses and virus-like lesions in marine molluscs. *Mar. Fish Rev.* 40, 18–20.

Farley, C. A. (1985). Viral gametocytic hypertrophy in oysters. In: Sindermann, C. J. (ed.), *Identification Leaflets for Diseases and Parasites of Fish and Shellfish*, Vol. 25. ICES, Copenhagen, pp. 3–5.

Farley, C. A., Banfield, W. G., Kasnic, J. R. G., and Foster, W. S. (1972). Oyster herpes-type virus. *Science* 178, 759–760.

Friedman, C. S., Estes, R. M., Stokes, N. A., Burge, C. A., Hargove, J. S., Barber, B. J., Elston, R. A., Burreson, E. M., and Reece, K. S. (2005). Herpes virus in juvenile Pacific oysters *Crassostrea gigas* from Tomales Bay, California, coincides with summer mortality episodes. *Dis. Aquat. Organ.* 63(1), 33–41.

Gao, L. and Qi, J. (2007). Whole genome molecular phylogeny of large dsDNA viruses using comparison vector method. *BMC Evol. Biol.* 7, 41–47.

Guo, X., Ford, S., De Brosse, G., and Smolowitz, R. (2003). Breeding and evaluation of eastern oyster strains selected for MSX, Dermo and JOD resistance. *J. Shellfish Res.* 22, 333–334.

Hardy-Smith, P. (2006). Report on the events surrounding the disease outbreak affecting farmed and wild abalone in Victoria, 29 August 2006. Panaquatic Health Solutions.

Hedrick, P. R., Eaton, W. D., Fryer, L. J., Groberg, W. G., and Bonnyaratapalin, S. (1986). Characteristics of a birnavirus isolated from cultured sand goby *Onyeleotris marmoratus*. *Dis. Aquat. Organ.* 1, 219–225.

Hill, B. J. (1976). Mollusc viruses: their occurrence, culture and relationships. In: Proceedings of the First International Colloquium on Invertebrate Pathology, pp. 25–29.

Hill, B. J. (1982). Infectious pancreatic necrosis virus and its virulence. In: Roberts, R. J. (ed.), *Microbial Diseases of Fish*. Academic Press, London, pp. 91–114.

Hill, B. J. and Alderman, D. J. (1977). Observations on the experimental infection of *Ostrea edulis* with two molluscan viruses. *Haliotis* 8, 297–299.

Hine, P. M. and Thorne, T. (1997). Replication of herpes-like viruses in haemocytes of adult flat oysters *Ostrea angasi* (Sowerby, 1871): an ultrastructural study. *Dis. Aquat. Organ.* 29(3), 197–204.

Hine, P. M., Wesney, B., and Besant, P. (1998). Replication of herpes-like viruses in larvae of the flat oyster *Tiostrea chilensis* at ambient temperatures. *Dis. Aquat. Organ.* 32(3), 161–171.

Hine, P. M., Wesney, B., and Hay, B. E. (1992). Herpesvirus associated with mortalities among hatchery-reared larval Pacific oysters, *C. gigas*. *Dis. Aquat. Organ.* 12(2), 135–142.

Hooper, C., Hardy-Smith, P., and Handlinger, J. (2007). Ganglioneuritis causing high mortalities in farmed Australian abalone (*Haliotis laevigata* and *Haliotis rubra*). *Aust. Vet. J.* 85(5), 188–193.

Hosono, N., Suzuki, S., and Kusuda, R. (1996). Genogrouping of birnaviruses isolated from marine fish: a comparison of VP2/NS junction regions on genome segment. *J. Fish Dis.* 19, 295–302.

Inaba, M., Suzuki, S., Kitamura, S. I., Kumazawa, N., and Kodama, H. (2009). Distribution of Marine Birnavirus (MABV) in marine organism from Okinawa, Japan, and a unique sequence variation of the VP2/NS region. *J. Microbiol.* 47, 76–84.

Jones, D. D. and Bej, A. K. (1994). Detection of foodborne microbial pathogens using polymerase chain reaction methods. In: Griffin, H. G. and Griffin, A. M. (eds.), *PCR Technology: Current Innovations*. CRC Press, Boca Raton, pp. 341–365.

Jones, J. B., Scotti, P. D., Dearing, S. C., and Wesney, B. (1996). Virus like particles associated with marine mussel mortalities in New Zealand. *Dis. Aquat. Organ.* 25, 143–149.

Kazama, F. (1972). Ultrastructure and phototaxis of the zoospores of *Phlyctochytrium* sp., an estuarine chytrid. *J. Gen. Microbiol.* 71, 555–566.

Kazama, F. and Schornstein, K. L. (1972). Herpes-type virus particles associated with a fungus. *Science* 177, 696–697.

Kazama, F. and Schornstein, K. L. (1973). Ultrastructure of a fungus herpes-type virus. *Virology* 52, 478–487.

Kitamura, S. I., Jung, S. J., and Suzuki, S. (2000). Seasonal change of infective state of marine birnavirus in Japanese pearl oyster *Pincada fucata*. *Arch. Virol.* 145, 2003–2014.

Kitamura, S., Kamata, S., Nakano, S., and Suzuki, S. (2003). Detection of marine birnavirus genome in zooplankton collected from the Uwa Sea, *Japan. Dis. Aquat. Organ.* 54(1), 69–72.

Kitamura, S., Kamata, S., Nakano, S., and Suzuki, S. (2004). Solar UV radiation does not inactivate marine birnavirus in coastal seawater. *Dis. Aquat. Organ.* 58(2–3), 251–254.

Kitamura, S. I. and Suzuki, S. (2000). Occurrence of Marine Birnavirus through the year in coastal seawater in the Uwa Sea. *Mar. Biotechnol.* 2 (2), 188–194.

Kitamura, S. I., Tomaru, Y., Kawabata, Z., and Suzuli, S. (2002). Detection of Marine birnavirus in the Japanese pearl oyster *Pinctada fucata* and seawater from different depths. *Dis. Aquat. Organ.* 50(3), 211–217.

Kimura, T., Yoshimizu, M., Tanaka, M., and Sannohe, H. (1981). Studies on a new virus (OMV) from *Oncorhynchus masou*: I. Characteristics and pathogenicity. *Fish Pathol.* 15(3/4) 143–147.

Le Deuff, R. M., Nicolas, J. L., Renault, T., and Cochennec, N. (1994). Experimental transmission of herpes-like virus to axenic larvae of Pacific oyster, Crassostrea gigas. *Bull. Eur. Assoc. Fish Pathol.* 14(2), 69–72.

Le Deuff, R. M. and Renault, T. (1999). Purification and partial genome characterization of a herpes-like virus infecting the Japanese oyster, Crassostrea gigas. *J. Gen. Virol.* 80, 1317–1322.

Le Deuff, R. M., Renault, T., and Gérard, A. (1996). Effects of temperature on herpes-like virus detection among hatchery-reared larval Pacific oyster *Crassostrea gigas*. *Dis. Aquat. Organ.* 24, 149–157.

Lees, D. (2000). Viruses and bivalve shellfish. *Int. J. Food Microbiol.* 59, 81–116.

Lipart, C. and Renault, T. (2002). Herpes-like virus detection *in Crassostrea gigas* spat using DIG-labelled probes. *J. Virol. Methods* 101, 1–10.

Lo, C. F., Hong, Y. W., Huang, S. Y., and Wang, C. H. (1988). The characteristics of the virus isolated from the gill of clam, *Meretrix lusoria. Fish Pathol.* 23, 147–154.

McGeoch, D. J., Rixon, F. J., and Davison, A. J. (2006). Topics in herpesvirus genomics and evolution. *Vir. Res.* 117, 90–104.

McGladdery, S. E. and Stephenson, M. F. (1994). A viral infection of the gonads of eastern oyster (*Crassostrea virginica*) from Atlantic Canada. *Bull. Aquac. Assoc. Can.* 94, 84–86.

Meyers, T. R. (1979). A reo-like virus isolated from juvenile American oysters (*Crassostrea virginica*). *J. Gen. Virol.* 46, 203–212.

Meyers, T. R., Burton, T., Evans, W., and Starkey, N. (2009). Detection of viruses and virus-like particles in four species of wild and farmed bivalve molluscs in Alaska, USA, from 1987 to 2009. *Dis. Aquat. Organ.* 88, 1–12.

Meyers, T. R. and Hirai, K. (1980). Morphology of a reo-like virus isolated from juvenile American oysters (*Crassostrea virginica*). *J. Gen. Virol.* 46(1), 249–253.

Miyazaki, T., Goto, K., Kobayashi, T., Kageyama, T., and Miyata, M. (1999). Mass mortalitiers associated with a virus disease in Japanese pearl oysters *Pinctada fucata martensii. Dis. Aquat. Organ.* 37, 1–12.

Naciri-Graven, Y., Martin, A. G., Baud, J. P., Renault, T., and Gérard, A. (1998). Selecting the flat oyster *Ostrea edulis* (L.) for survival when infected with the parasite Bonamia ostreae. *J. Exp. Mar. Biol. Ecol.* 224, 91–107.

Nicolas, J. L., Comps, M., and Cochennec, N. (1992). Herpes-like virus infecting Pacific oyster larvae, *C. gigas. Bull. Eur. Assoc. Fish Pathol.* 12(1), 11–13.

Nishida, T., Kimura, H., Saitoh, M., Shinohara, M., Kato, M., Fukuda, S., Munemura, T., Mikami, T., Kawamoto, A., Akiyama, M., Kato, Y., Nishi, K., Kozawa, K., and Nishio, O. (2003). Detection, quantitation, and phylogenetic analysis of noroviruses in Japanese oysters. *Appl. Environ. Microbiol.* 69(10), 5782–5786.

Nobiron, I., Galloux, M., Henry, C., Torhy, C., Boudinot, P., Lejal, N., Da Costa, B., and Delmas, B. (2008). Genome and polypeptides characterization of Tellina virus 1 reveals a fifth genetic cluster in the Birnaviridae family. *Virology* 371, 350–361.

Norton, J. H., Shepherd, M. A., and Prior, H. C. (1993). Papova-like virus infection of the golden-lipped pearl oyster, *Pinctada maxima,* from the Torres Strait, Australia. *J. Invertebr. Pathol.* 62, 198–200.

Novoa, B. and Figueras, A. (1996). Heterogeneity of marine birnaviruses isolated from turbot (*Scophtalmus maximus*). *Fish Pathol.* 31, 145–150.

Novoa, B. and Figueras, A. (2000). Virus-like particles associated with mortalities of the carpet-shell clam *Ruditapes decussatus. Dis. Aquat. Organ.* 39, 147–149.

Novoa, B., Figueras, A., Puentes, C. F., Ledo, A., and Toranzo, A. E. (1993). Characterization of a birnavirus isolated from diseased turbot cultured in Spain. *Dis. Aquat. Organ.* 15, 39–44.

OIE. (2009a). *International Aquatic Animal Health Code*, 9th edition. OIE, Paris.

OIE. (2009b). *Manual of Diagnostic Tests for Aquatic Animals*, 7th edition. OIE, Paris.

Oprandy, J. J. and Chang, P. W. (1983). 5-Brommodeoxyuridine induction of hematopoietic neoplasia and retrovirus activation in the soft-shell clam, *Mya arenaria. J. Invertebr. Pathol.* 38, 45–51.

Oprandy, J. J., Chang, P. M., Pronovost, A. D., Cooper, K. R., Brown, R. S., and Yates, V. J. (1981). Isolation of viral agent causing hemopoietic neoplasia in the soft shell clam, *Mya arenaria*. *J. Invertebr. Pathol.* 38, 45–51.

Pépin, J. F., Riou, A., and Renault, T. (2008). Rapid and sensitive detection of ostreid herpesvirus 1 in oyster samples by realtime PCR. *J. Virol. Methods* 149(2), 269–276.

Plumb, J. A. (1973). Effects of temperature on mortality of fingerling Channel Catfish (*Ictalurus punctatus*) experimentally infected with Channel Catfish Virus. *J. Fish. Res. Board Can.* 30(4), 568–570.

Plummer, G. and Lewis, B. (1965). Thermoinactivation of Herps Simplex Virus and Cytomegaloviurs. *J Bacteriol.* 89, 671–674.

Potasman, I., Paz, A., and Odeh, M. (2002). Infectious outbreaks associated with bivalve shellfish consumption: a worldwide prespective. *Clin. Infect. Dis.* 35(8), 921–928.

Rasmussen, L. P. D. (1986). Virus-associated granulocytomas in the marine mussel, *Mytilus edulis,* from three sites in Denmark. *J. Invertebr. Pathol.* 48, 117–123.

Renault, T. and Arzul, I. (2001). Herpes-like virus infections in hatchery-reared bivalve larvae in Europe: specific viral DNA detection by PCR. *J. Fish Dis.* 24, 161–167.

Renault, T., Arzul, I., and Lipart, C. (2004). Development and use of an internal standard for oyster herpesvirus 1 detection by PCR. *J. Virol. Methods* 121(1), 17–23.

Renault, T., Cochennec, N., Le Deuff, R. M., and Chollet, B. (1994a). Herpes-like virus infecting Japanese oyster (*C. gigas*) spat. *Bull. Eur. Assoc. Fish Pathol.* 14(2), 64–66.

Renault, T., Le Deuff, R. M., Cochennec, N., and Maffart, P. (1994b). Herpesviruses associated with mortalities among Pacific oyster, *C. gigas*, in France—comparative study. *Rev. Med. Vet.* 145(10), 735–742.

Renault, T., Le Deuff, R. M., Chollet, B., Cochennec, N., and Gérard, A. (2000a). Concomitant herpes-like virus infections in hatchery-reared larvae and nursery-cultured spat *Crassostrea gigas* and *Ostrea edulis*. *Dis. Aquat. Organ.* 42(3), 173–183.

Renault, T., Le Deuff, R. M., Lipart, C., and Delsert, C. (2000b). Development of a PCR procedure for the detection of a herpes-like virus infecting oysters in France. *J. Virol. Methods* 88, 41–50.

Renault, T. and Lipart, C. (1998). Diagnosis of herpes-like virus infections in oysters using molecular techniques. *EAS Special Publication* 26, 235–236.

Renault, T., Lipart, C., and Arzul, I. (2001a). A herpes-like virus infects a nonostreid bivalve species: virus replication in *Ruditapes philippinarum* larvae. *Dis. Aquat. Organ.* 45, 1–7.

Renault, T., Lipart, C., and Arzul, I. (2001b). A herpes-like virus infecting *Crassostrea gigas* and *Ruditapes philippinarum* larvae in France. *J. Fish Dis.* 24, 369–376.

Renault, T. and Novoa, B. (2004). Viruses infecting bivalve molluscs. *Aquat. Living Resour.* 17, 397–409.

Renault, T., Solliec, G., and Arzul, I. (2003). Détection d'un virus de type herpès chez un champignon présent dans les élevages larvaires d'huître creuse, Crassostrea gigas. *Virologie* 7, (numéro spécial) S37.

Rivas, C., Cepeda, C., Dopazo, C. P., Novoa, B., Noya, M., and Barja, J. L. (1993). Marine environment as reservoir for birnaviruses from poikilothermic animals. *Aquaculture* 115, 183–194.

Sauvage, C., Pépin, J. F., Lapègue, S., Boudry, P., and Renault, T. (2009). Ostreid herpes virus 1 infection in families of the Pacific oyster, *Crassostrea gigas,* during a summer mortality outbreak: differences in viral DNA detection and quantification using real-time PCR. *Vir. Res.* 142, 181–187.

Schikorski, D., Renault, T., Saulnie, D., Faury, N., Moreau, P. and Pepin, J. F. (2011a). Experimental infection of Pacific oyster Crassostrea gigas spat by ostreid herpesvirus 1: demonstration of oyster spat susceptibility. *Vet. Res.* 42(1), 27.

Schikorski, D., Faury, N., Pepin, J. F., Saulnier, D., Tourbiez, D. and Renault, T. (2011b). Experimental ostreid herpesvirus 1 infection of the Pacific oyster Crassostrea gigas: kinetics of virus DNA detection by q-PCR in seawater and in oyster samples. *Vir. Res.* 155(1), 28–34.

Schultz, M., May, E. B., Kraeuter, J. N., and Hetrick, F. M. (1984). Isolation of infectious pancreatic necrosis virus from an epizootic occurring in cultured striped bass, *Morone saxatilis* Walbaum. *J. Fish Dis.* 7, 505–507.

Segarra, A., Pepin, J. F., Arzul, I., Morga, B., Faury, N. and Renault T. (2010). Detection and description of a particular Ostreid herpesviurs A genotype associated with massive mortality outbreaks of pacific oysters Crassostrea gigas, in France in 2008. *Vir. Res.* 153(1), 92–99.

Sorimachi, M. and Hara, T. (1985). Characteristics and pathogenicity of a virus isolated from yellowtail fingerlings having ascites. *Gyobyo Kenkyu. Fish Pathol.* 19, 231–238.

Speirs, J. I., Pontefract, R. D., and Harwig, J. (1987). Methods for recovering poliovirus and rotavirus from oysters. *Appl. Environ. Microbiol.* 53, 2666–2670.

Suzuki, S., Hosono, N., and Kusuda, R. (1997a). Detection of aquatic birnavirus gene from marine fish using a combination of reverse transcription and nested PCR. *J. Mar. Biotechnol.* 5, 205–209.

Suzuki, S., Hosono, N., and Kusuda, R. (1997b). Production kinetics of antigenicity and serological analysis of viral polypeptides of yellowtail ascites virus. *Fish Pathol.* 30, 209–214.

Suzuki, S., Kamakura, M., and Kusuda, R. (1998a). Isolation of birnavirus from Japanese pearl oyster *Pinctada fucata. Fish Sci.* 64, 342–343.

Suzuki, S., Nakata, T., Kamakura, M., Yoshimoto, M., Furakawa, Y., Yamashita, Y., and Kusuda, R. (1998b). Isolation of birnavirus from Agemaki (Jack Knife Clam) *Sinovacura consticta* and survey of the virus using PCR technique. *Fish Sci.* 63, 563–566.

Suzuki, S. and Nojima, M. (1999). Distribution of a marine birnavirus in wild shellfish species from Japan. *Fish Pathol.* 34, 121–125.

Suzuki, S., Utsunomiy, I., and Kusuda, R. (1998c). Experimental infection of marine birnavirus strain JPO-96 to Japanese pearl oyster *Pinctada fucata. Bull. Mar. Sci. Fish. Kochi Univ.* 18, 39–41.

Tan, J., Lancaster, M., Hyatt, A., van Driel, D., Wong, F., and Warner, S. (2008). Purification of a herpes-like virus from abalone (*Haliotis* spp.) with ganglioneuritis and detection by transmission electron microscopy. *J. Virol. Methods* 149(2), 338–341.

Trus, B. L., Gibson, W., Cheng, N., and Steven, A. C. (1999). Capsid structure of simian cytomegalovirus from cryoelectron microscopy: evidence for tegument attachment sites. *J. Virol.* 73, 2181–2192.

Trus, B. L., Heymann, J. B., Nealon, K., Cheng, N., Newcomb, W. W., Brown, J. C., Kedes, D. H., and Steven, A. C. (2001). Capsid structure of Kaposi's sarcoma-associated herpesvirus, a gammaherpesvirus, compared to those of an alphaherpesvirus, herpes simplex virus type 1, and a betaherpesvirus, cytomegalovirus. *J. Virol.* 75, 2879–2890.

Van Regenmortal, M. H. V. (2008). Virus species. In: Mahy, B. W. J. and Van Regenmortel, M. (eds.), *Encyclopedia of Virology*, 3rd edition. Academic Press, pp. 401–406.

Vásquez-Yeomans, R., Cáceres-Martńez, J., and Huerta, A. F. (2004). Herpes-like virus associated with eroded gills of the Pacific oyster *Crassostrea gigas* in Mexico. *J. Shellfish Res.* 23(2), 417–419.

Vigneron, V., Solliec, G., Montanié, H., and Renault, T. (2004). Detection of Ostreid herpes virus 1 (OsHV-1) DNA in seawater by PCR: influence of water parameters in bioassays. *Dis. Aquat. Organ.* 62, 35–44.

Wallis, C. and Melnick, J. (1965). Thermostabilization and thermosensitization of herpesvirus. *J. Bacteriol.* 90, 1632–1637.

Wang, J., Guo, Z., Feng, J., Liu, G., Xu, L., Chen, B., and Pan, J. (2004). Virus infection in cultured abalone, *Haliotis diversicolor* Reeve in Guandong Province, *China. J. Shellfish Res.* 23, 1163–1168.

Winstead, J. T. and Courtney, L. A. (2003). Ovacystis-like condition in the eastern oyster *Crassostrea virginica* from the north-eastern Gulf of Mexico. *Dis. Aquat. Organ.* 53, 89–90.

Wise, J. A., Harrell, S. F., Busch, L. B., and Boyle, J. A. (1988). Vertical transmission of channel catfish virus. *Am. J. Vet. Res.* 49(9), 1506–1509.

Zhang, C. X. and Suzuki, S. (2003). Comparison the RNA polymerase genes of marine birnavirus strains and other birnaviruses. *Arch. Virol.* 148, 745–758.

Zhang, C. X. and Suzuki, S. (2004). Aquabirnaviruses isolated from marine organisms form a distinct genogroup form other aquabirnaviruses. *J. Fish Dis.* 27, 633–643.

CHAPTER 7

THE VIRAL ECOLOGY OF AQUATIC CRUSTACEANS

LEIGH OWENS
Discipline of Microbiology and Immunology, School of Veterinary and Biomedical Sciences, James Cook University, Townsville, Australia

CONTENTS

7.1 INTRODUCTION AND APPROACH

The ecology of viruses predominately involves the interaction of the virus at the animal and cellular level. When the virions are in the extracellular environment, they are quiescent waiting to infect a living cell. The virions bind to the cell receptors, undergo decapsidation in the phagolysosome, pass the nucleic acid to the cytoplasm or nucleus and begin replication. Evidently, the ecology of viruses starts with entry of the virus into the crustacean, evading the immune system, invasion of the cell, replication, and release of new virions. The crustacean immune system is critical to the functional ecology of their viruses, so we must start with an understanding of this system to have any hope of understanding the ecology of the viruses in crustaceans.

The experimental model for interactions between the crustacean immune system and pathogens has been, by and large, the excellent work by L. Cerenius, K. Soderhall, and coworkers. This model was initially developed on the interplay between the crayfish plague fungi *Aphanomyces astaci* and freshwater crayfish, particularly *Astacus astacus*. Much of the supplementary knowledge on decapod's immunity has come from the true crabs. Unfortunately, the Dendrobranchiata (including penaeids) last shared a common ancestor with the rest of the decapod crustacea during the Silurian epoch, approximately 437 million years ago (Figure 7.1) (Porter et al., 2005). Since that time the immune systems of these lineages have been evolving independently.

Furthermore, the viruses of freshwater crayfish have not been extensively researched because the relatively low economic value of crayfish has not allowed researchers to secure adequate funding. At present, the state of

Studies in Viral Ecology: Animal Host Systems: Volume 2, First Edition. Edited by Christon J. Hurst.

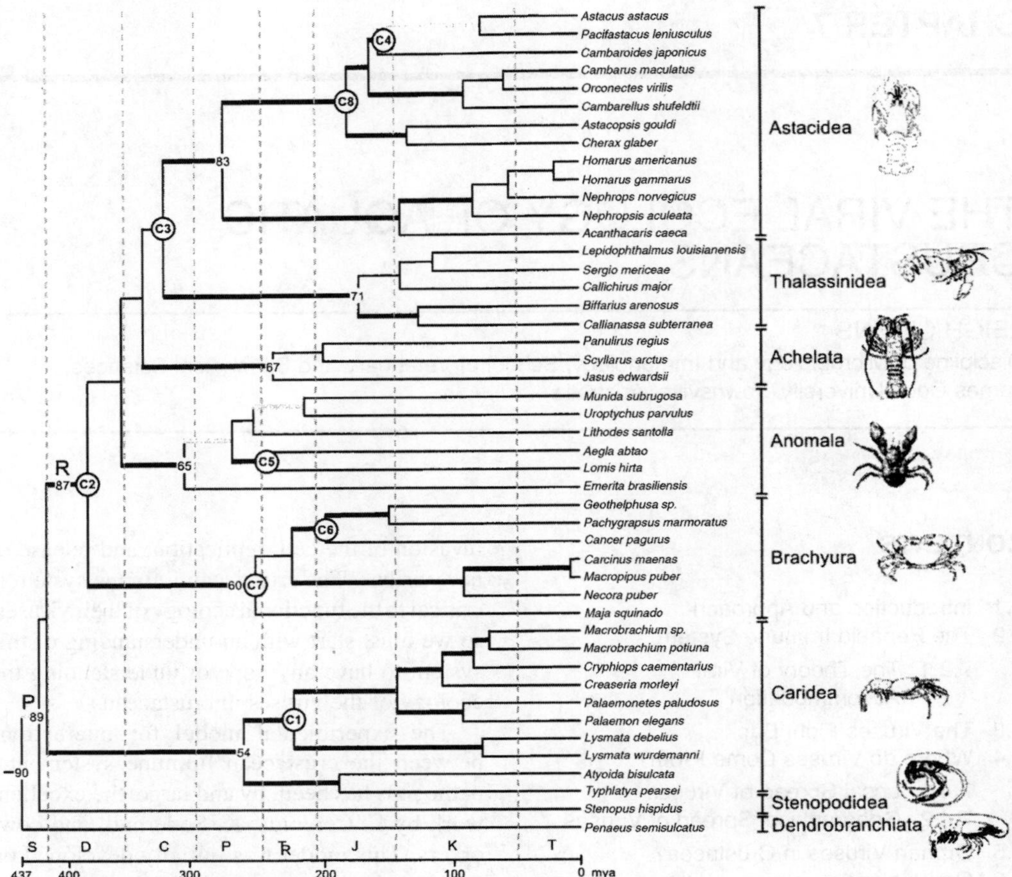

FIGURE 7.1 Decapod divergence time chronogram estimated using topology of ML (maximum likelihood) tree. On branches with both ML bootstrap values of >70% and BMCMC (Bayesian Marker Chain Monte Carlo sampling) $p = P > 0.95$, support is indicated by a thick black line; branches strongly supported by only one tree reconstruction method are indicated by thick gray lines. Fossil calibration nodes are indicated by C1–C8. Node numbers from divergence time estimations are included for reference on nodes of important decapod lineages. The decapod infraorders are delineated, and the nodes corresponding to the suborder Pleocyemata (P) and the informal Reptantia (R) are indicated on the phylogeny. The major geologic periods are also mapped onto the phylogeny, using the following standard symbols: S, Silurian; D, Devonian; C, Carboniferous; P, Permian; Tr, Triassic; J, Jurassic; K, Cretaceous; T, Tertiary. (Adapted from reference Porter et al., 2005.)

knowledge is little more than a catalogue of viruses in hosts and the methods of detection. Therefore, the approach taken in this chapter will be to use the information derived from the penaeids and their interaction with their viruses since the economic power of the aquaculture of penaeids has allowed a much more thorough investigation of their interaction with viruses. Due to the long geological separation of the penaeids from other crustacea, only when information is completely lacking from the penaeids will other comparative information be used.

The taxonomy of the penaeids is very controversial since the premature acceptance of the classification of Perez Farfante and Kensley (1997). Dall (2007) reviewed the controversy based on the morphological and available molecular evidence and largely followed the results of Lavery et al. (2004). He suggested that it is premature for the promotion of so many subgenera to full genus status. Based mostly on the evidence of Lavery et al. (2004), the facts to date suggests there should only be two genera within the old genus *Penaeus*: *Melicertus* for the old subgenus *Merlicertus* plus *Penaeus japonicus* and *Penaeus* for all other members of the genus. While early and the evidence is still accruing, it is the most up-to-date information available and it will be used in this chapter.

This chapter will not be an updated list of all the viruses found in their crustacean hosts or the methods for detecting the viruses. This approach to the topic has been under taken many times and dealt with in an excellent fashion by international experts such as Lightner (1996) and Flegel (2006) so the reader is referred to their publications for this information.

An alternative approach taken here is to concentrate on the *ecology* of the viruses of crustacea at the subcellular, cellular, and environmental levels.

7.2 THE PENAEID IMMUNE SYSTEM

It is clear that the immune system of decapods has three salient features. One, there is no production of antibodies by B-like cells. Two, the system is characterized by a cascade of cleavage of multiple inactive proteins into the active state by serine proteases. Three, the major triggering compounds are carbohydrates such as peptidoglycan (Gram-positive bacteria), beta 1,3-glucan (fungi), and lipopolysaccharide (LPS) (Gram-negative bacteria). The first of these features needs no further explanation.

The second cascade has been compiled by many authors. The preprophenoloxidase activating (PPA) protein has to be cleaved into an active form by prophenoloxidase activator (a serine protease itself) that cleaves prophenoloxidase into the active phenoloxidase. Phenoloxidase couples with superoxide dismutase that is probably membrane bound on the hemocytes. This turns oxygen-free radicals into hypochlorous acid that oxidizes the microbial invader. Recently, hemocyanin, the oxygen-carrying protein that makes up 95% of the hemolymph protein, has been shown to be cleaved under stress (Cimino et al., 2002) and microbial attack to produce an antimicrobial protein and a phenoloxidase (Lee et al., 2004) which have been shown to be active against a wide range of target microbes.

The triggering of the serine protease cascade by carbohydrates has been elegantly worked out over many years. However, the initial source of the serine protease has not been elucidated. The author hypothesizes that the initial source will be a mannose binding lectin (MBL) pathway. Lectins involved in the immune response in crustacea have been known for many years but their link to other components has not been demonstrated. However, the ability of MBL to be made active by mannose binding in peptidoglycan has been recently demonstrated. MBL in vertebrates has two moieties of two different serine proteases that become active on binding (Figure 7.2). It is hypothesized that the receptors binding to the pattern on mannose residues forces a conformational change to the other end of the lectin where the serine proteases are situated and they become active. This is an elegant way of activating the many cleaving enzymes of the immune system right at the surface of the microbe that needs to be destroyed.

The implications of understanding the functioning of the immune system in penaeids are astonishing. First of all, because the immune system is carbohydrate based rather than protein based as in the vertebrate lineages, the crustacean immune system did not evolve to deal with viruses that have limited, if any, carbohydrate moieties on their surface and where carbohydrates do occur they are not in the pattern necessary to change the conformational shape of the MBL-linked serine protease to activate them. Therefore, crustaceans have had to develop a second, totally independent immune system to deal with viruses separately.

Second, there is an implication that modern viruses of eukaryotes evolved after the

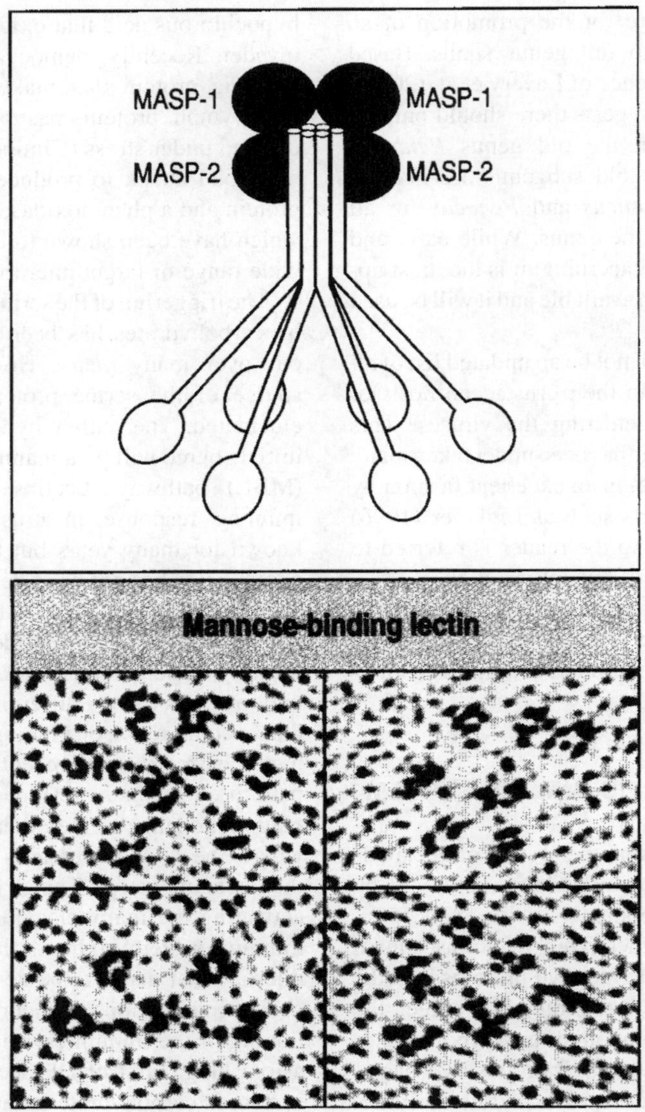

FIGURE 7.2 A schematic and electron micrograph of a mannose binding lectin from mammals that resembles the complement C1 complex. Mannose binding lectin forms clusters of two to six carbohydrate binding heads around a collagen-like stalk. This structure is discernable under the electron microscope (lower panels) (photograph courtesy of K. B. M. Reid). Associated with this complex are two serine proteases, MBL-associated serine protease 1 (MASP-1) and 2 (MASP-2). (Adapted from reference Janeway et al., 2005.)

crustaceans. Otherwise, it is likely that the antiviral immunity would have clear roots to the antipathogen system that was already there, that is, some use of the prophenoloxidase cascade to destroy viruses.

7.2.1 The Theory of Viral Accommodation

The theory of accommodation of viruses in crustacea has been championed by a series of publications by Flegel and his coworkers

(Flegel and Pasharawipas, 1998; Flegel, 2007; Flegel, 2009). In short, the viral accommodation theory suggests that after some generations, crustaceans tolerate viruses by locking the virus away in infected cells. The host stops the cells from lysing, preventing the release of virions or destroying large amount of host tissues. The theory grew out of a number of field observations. First, there was the lack of a vertebrate inflammation-like response around tissues which clearly showed viral inclusion bodies (Flegel and Pasharawipas, 1998). Second, there was the observation that after 2 or 3 years of epizootic mortalities in penaeids when a new virus was introduced to naïve populations, mortalities decreased to lower background levels even though the animals were demonstrably persistently infected with the virus (Flegel and Pasharawipas, 1998). Penaeids survived and grew to reproductive age, but they were more susceptible to environmental perturbations that could trigger mortality events. The control of apoptosis of viral infected cells was suggested as the mechanism that restricted mortality. However with excessive environmental fluctuations, the penaeids lost their ability to control apoptosis leading to systemic widespread cell death and subsequent animal mortality. The role of apoptosis in viral induced mortality is controversial and therefore not universally accepted (see Flegel, 2007). Nevertheless data from Midcrop Mortality Syndrome in *Penaeus monodon* from Australia supports the theory (Anggraeni and Owens, 2000).

One of the implications of the accommodation theory is that the tolerance to the virus must be passed on in a heritable manner so that the next one or two generations can also become tolerant to the virus. In recent years, it has been established that the interfering RNA (iRNA) pathway exists in all eukaryotes that have been investigated including crustacea and terrestrial crustacea (e.g., insects (see Regier et al., 2010)). Within the iRNA pathway, there are two separate components that operate a nonspecific dsRNA knockdown where the presence of any dsRNA enhances the RNA-induced silencing complex (RISC) and a specific dsRNA knockdown that is much more efficient at downregulating a viral gene (Robalino et al., 2004, 2005, 2007; La Fauce and Owens, 2009). Nonspecific dsRNA silencing maybe why prior infections with some viruses leads to some cross-protection against other viruses. For example, infection with infectious hypodermal and hematopoietic necrosis virus (IHHNV, *Penaeus stylirostris* densovirus, genus *Brevidensovirus*, family Parvoviridae, Figure 7.3) has been found to give protection to subsequent infection with white spot syndrome virus (WSSV, genus *Whispovirus*, family Nimaviridae, Figure 7.4) (Melena et al., 2006).

Flegel (2009) has used the presence of iRNA in crustacea to propose a mode of action for viral accommodation. It is proposed that nucleases, reverse transcriptase, and integrases that are common in the crustacean genome chop up the viral genome, convert RNA into DNA, and integrate the DNA into the crustacean genome as small fragments of viral ghost DNA. These subsequently act as template for iRNA pathway to knockdown mRNA of viral genes thus reducing the viral load below a critical threshold level that would cause disease. Through natural selection, those surviving animals having nondeleterious and beneficial inserts corresponding to viral ghost DNA contribute rapidly to the gene pool for the next generation leading to widespread tolerance.

There is some evidence that might suggest a mechanism of how the viral ghost DNA gets into the next generation's germ line. When crustacean hemocytes are not combating pathogens, they have a secondary function of shuttling lipoproteins from the hepatopancreas to the ovary (Dr. Ester Lubzens, National Institute of Oceanography, Israel, personal communication). The high-density lipoprotein 1 of *Penaeus semisulcatus* is a homologue of beta 1,3-glucan binding protein found in hemocytes, which probably binds to an ovarian lipoprotein receptor. If the hemocytes have phagocytosed virions and processed them to a short DNA structure in phagolysosomes, this would be a

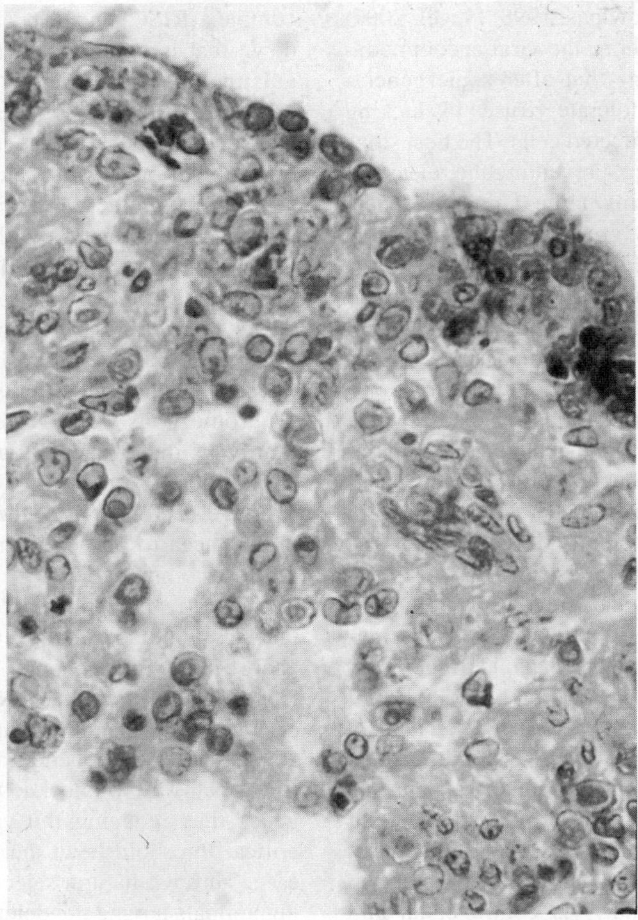

FIGURE 7.3 Infectious hypodermal and hematopoietic necrosis virus (IHHNV, *P. stylirostris* densovirus) infecting the lymphoid organ of a hybrid *P. monodon* crossed with *P. esculentus*. Note almost every cell has an eosinophilic Cowdrey A intranuclear inclusion body. (*See the color version of this figure in Color Plates section.*)

perfect way to shuttle the viral ghost DNA into the germ cells of the next generation as it shuttles the lipoprotein into the eggs. It also opens the door to inappropriately processed virus, that is, live virus also being shuttled into gametes.

Once a hemocyte has exocytosed its active components, then phagocytosed virions either have their nucleic acid processed or not as the case may be and then passed viral ghost template to the germ cells, there remains the problem of what to do with the spent hemocytes. If viral processing is not complete, then destruction of the hemocytes runs the risk of liberating unprocessed infectious virus or infectious nucleic acid such as with TSV (Taura syndrome virus, family Dicistroviridae, genus unassigned). So the penaeids sequester the spent hemocytes into the lymphoid organ to produce spheroids that lock the virus away out of circulation (Anggraeni and Owens, 2000, Figure 7.5). These spheroids can contain live infectious virus for some considerable time (Hasson et al., 1999), but spheroids eventually get encapsulated and disposed off via an unknown mechanism perhaps harmonized to the lunar cycle (Rusaini and Owens, 2010a). This is why so many viruses have been found

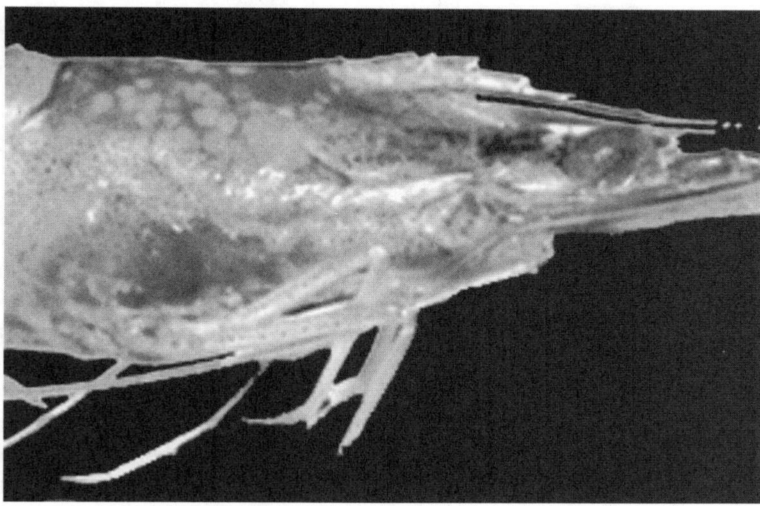

FIGURE 7.4 Clinical signs of white spot syndrome virus in *P. merguiensis*. Note the white lesions on the rear of the cephalothorax. Photograph by K. Claydon and L. Owens. (*See the color version of this figure in Color Plates section.*)

located in the lymphoid organs (Rusaini and Owens, 2010b). As the lymphoid organ is not found outside the penaeids, other tissues or individual spent hemocytes must be doing an equivalent role in other crustacea.

The theory on the accommodation of viruses leading to a heritable tolerance has now been tested both experimentally and by natural epizootics. While not all components are understood, it is proving to be a very robust and

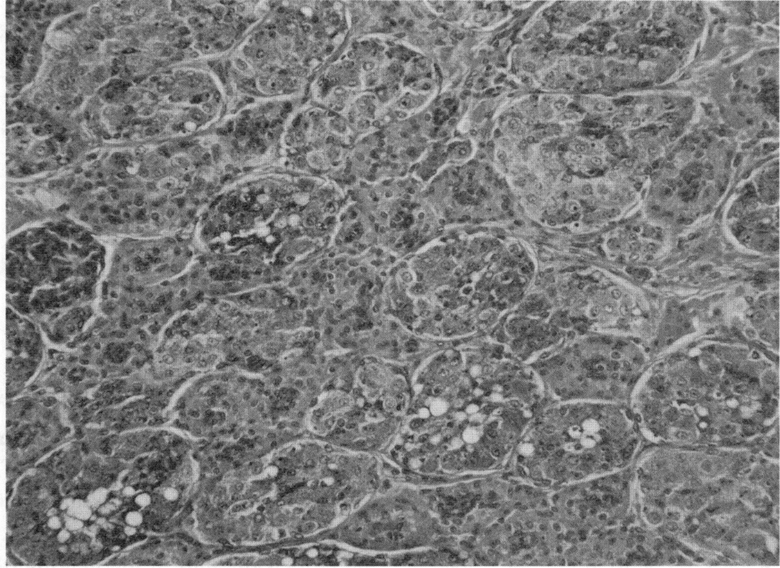

FIGURE 7.5 Lymphoid spheroids in the lymphoid organ of *P. merguiensis*. The round, more basophilic sections are the spheroids made up of spent hemocytes and the more eosinophilic tissue with the central hemolymph vessels are the normal stromal matrix areas of the lymphoid organ. (*See the color version of this figure in Color Plates section.*)

useful tool for understanding the dynamics of viral infections and tolerance of infections in crustacea.

7.3 THE VIRUSES FIGHT BACK

Viruses have evolved to side step some of the mechanisms that the crustaceans use to defeat them. At this stage, it is unclear if the viruses evolved these mechanisms in crustacea or having these mechanisms has allowed the viruses to infect crustacea more efficiently. Crustacean baculoviruses have the inhibition of apoptosis (IAP) genes, P35/38. The dicistrovirus, Taura syndrome virus has a biochemical ability similar to the baculoviral inhibitor of apoptosis protein repeat (BIR) gene (Mari et al., 2002). Both of these genes have the ability to stop the early apoptotic destruction of infected cells allowing full replication of the virus. These genes are transcribed in the immediate early phase of viral genome transcription ensuring they are functioning very early in the infection process. This is critical in the case of baculoviruses that must produce a massive polyhedral protein. The baculovirus cannot have the infected cell lysed early or the polyhedral protein will not be produced to protect the virions in the harsh intertidal zone at low tide from the ravages of UV light, desiccation, and high temperature.

Macrobrachium rosenbergii nodavirus (an unclassified member of the family Nodaviridae) produces the B2 protein from viral genome segment 1 that binds at multiple sites to the dsRNA intermediates of the virus. This then prevents the *Dicer* enzyme of the crustacean iRNA pathway from being able to cut up intermediates and therefore silence the virus.

The studying and understanding of the genes of crustacean viruses is in its infancy and severely hampered by the lack of crustacean cell lines that would allow manipulation of the viruses. As viral sequencing and classification of genes proceeds, it is anticipated that more information on how the viruses evade the host responses as well as the mechanism that some already discovered systems (i.e., IAP) genes function will be unveiled.

7.4 WHERE DO VIRUSES COME FROM?

Where does the crustacean index case that starts a viral epizootic get its infective load from? I believe most new viruses come from the practice of feeding rich maturation diets to broodstock. Broodstock need high levels of protein and particularly lipids to produce healthy eggs and larvae. Aquaculture has developed a number of maturation diets that include in particular, marine invertebrates. One practice that became particularly common before the worldwide WSSV outbreak was to feed frozen crabs broken up with a hammer, directly to the broodstock. This may explain why most of the other possible species of the genus *Whispovirus* (the genus of viruses that WSSV belongs to) are all found in crabs. There has been a tendency of late to restrict the feeding of crustacea to crustacean broodstock and a concomitant move toward the use of other invertebrates such as polychaetes, bivalves, and cephalopods as maturation feeds. Recently, the rate of emergence of new catastrophic viruses in crustaceans has seemed to have slowed and hopefully the two are related as a cause and beneficial effect of changing dietary feeding practices.

7.4.1 Local Spread of Viruses

Once the broodstock are infected, it is only a matter of time before the larvae become infected. Female broodstock defecate just prior to spawning. The eggs are then broadcast spawned into this milieu of enteric viruses, bacteria, and undigested food. The Tahitian method of separating contaminated egg shells and weaker larvae from healthy larvae has been instrumental in the decline in the importance as pathogens of baculoviruses, unassigned rod-shaped viruses, and perhaps the enteric densoviruses.

With the systematic viruses it is more difficult to prove the mode of infection. However, the huge successes using PCR-testing of broodstock or postlarvae (the postplanktonic stage of penaeids) and only stocking with viral free stock or lightly infected postlarvae has shown the power of the broodstock link. Perhaps systemic viruses are shed on spawning as the eggs are expelled, but one would expect the Tahitian method (see above) and surface sterilizing of the eggs would have reduced the impact of these systemic viruses. However, this does not seem to be the case. Therefore, it seems likely that the virions are shuttled to the ovaries with the lipids (see above) and are under the vitelline membranes when spawned.

Once a viral epizootic is underway it is easy for virions to transmit to the next host as densities of hosts in aquaculture are among the highest in any animal production system. With WSSV in experimental situations, waterborne virion loads have been shown to be sufficient to transmit disease to animals sharing only the same water. Unfortunately, many of the studies investigating waterborne infections were flawed in that the method of separating infected crustacea from animals being exposed to the water would not have stopped small pieces of pleopods, pereiopods, uropods, gills, antennae, and shredded tissue produced during cannibalization from being washed into the trial tanks where they could be consumed. While the concept of viruses being waterborne does seem logical for enteric viruses such as the baculoviruses and densoviruses, it does seem counterintuitive for those viruses that are systemic such as WSSV, yellow head virus (genus *Okavirus*, family Roniviridae), gill-associated virus (genus *Okavirus*, family Roniviridae, Figure 7.6), infectious myonecrosis virus (IMNV, presumably an unclassified member of the viral family Totiviridae) and IHHNV. So then we must ask, "How do the systemic virions escape the carcasses?" It is probable that during cannibalization the tissue is shredded enough to release virions.

There is no doubt that cannibalization is the main method of viral spread once an epizootic is underway. The most predatory species (e.g., *P. monodon*, *M. japonicus*) suffer more acute viral diseases, higher mortality, and suffer more viral diseases than other species. This has partially led to decreased tonnage of these species being recorded by farmers as they move away from these more problematic species.

Within the literature, there are a large number of publications that have identified animals living in the aquatic environments as alternate

FIGURE 7.6 *P. monodon* infected with gill-associated virus. Note the yellow gills similar to symptoms of prawns infected with the conspecific yellow head virus. (*See the color version of this figure in Color Plates section.*)

hosts or carriers for crustacean viruses. Unfortunately, there has not been a robust set of criteria applied to the generation of these lists. Most studies have used PCR-based detection that cannot tell if the virion is infectious or whether the PCR is amplifying ghost DNA. No confirmatory sequencing or infection studies were described in the majority of these studies. We have seen above how nucleases, *Dicer*, RISC, reverse transcriptase, and integrases process RNA into a DNA signature in the genome. Furthermore, other research in terrestrial crustacean cell lines, the mosquito cell line *Aedes albopictus* C6/36, has demonstrated the rapid accumulation of persistent interfering particles (PIP) from 3% in the original viral inoculum to 30% over four generations (Roekring et al., 2006). These persistent interfering particles were shown to be highly unlikely to be infectious due to frame shifts in the viral genome that encode for critical proteins. PIP may be empirical evidence of another method that crustaceans use to deal with viruses, or it might be an unknown precursor in the iRNA pathway or most likely, errors in matching the crustacean's cellular replication machinery with the viral transcripts. It is likely that the more unsuitable a host cell is for a virus, the more probable transcription errors will occur. This fact is used by vaccine manufacturers to produce mutant, weakened viruses for vaccine candidates. All of these modifications to viral genomes could give PCR signals of appropriate sizes but the virions are not, in fact, infectious. Therefore, many of the lists of alternate hosts for viruses are very suspect without confirmatory studies of some kind.

7.4.2 Geographical Spread of Viruses

The way that viruses spread from country to country, continent to continent, and from ocean to ocean is hotly debated as no country wants someone else's viral problem. Furthermore, some jurisdictions have viewed the threat to their own industries to be sufficiently severe as to apply restrictive conditions under the International Zoo-sanitary Code, which increases the level of global tensions. Nevertheless, it is impossible to argue that any process that allows crustaceans to arrive alive or in an unprocessed state fit for human consumption will not carry viable virions if they were present in the animals when harvested. The evidence is overwhelming that moving contaminated live broodstock or postlarvae has been responsible for transcontinental movement of crustacean viruses. The negligent movement of "clean" but untested broodstock or postlarvae is believed to have been responsible for the introduction of IHHNV and TSV into Hawaii, IHHNV into Mexico, IMNV into Indonesia, and Gill-associated virus into SE Asia. Lightner (1990) published a figure showing the then-known movement of live penaeids demonstrating the effectiveness of the "jumbo jet vector" in moving live crustaceans around the globe.

In every case tested experimentally, the viruses detected in frozen commodity shrimp were viable and caused disease and mortality in indicator crustaceans (e.g., Nunan et al., 1998; McColl et al., 2004). Processing of frozen commodity shrimp has been implicated in the transfer of WSSV to the American continents. Furthermore, birds have been implicated in the spread of viruses from rubbish dumps containing commodity shrimp wastes. This hypothesis has received support from studies that show that viable, nonenveloped viruses can pass through the gut of seagulls (Garza et al., 1997; Vanpatten et al., 2004) and chickens (Vanpatten et al., 2004).

7.5 ORPHAN VIRUSES IN CRUSTACEA?

Orphan viruses are those viruses found in a host but are not considered to cause any disease. While orphan viruses are believed to exist in crustacea, as evidence accrues it appears less and less likely that this is the case. The first piece of evidence was from hepatopancreatic parvovirus (HPV; taxonomic name, *P. monodon* densovirus). Apart from the original paper on HPV's discovery in wild *Penaeus merguiensis* and *Penaeus indicus* from Singapore (Chong and Loh, 1984), and in general review articles (e.g., Lightner, 1996) that attributed up to 100% mortalities during outbreaks, this virus was largely ignored because the industry believed it did not impact on production. However, Flegel et al. (1999) demonstrated statistically significant stunting caused by HPV. Recently Owens et al. (unpublished) have demonstrated a statistically significant 28% loss of production in *P. merguiensis* production due to a sister virus *P. merguiensis* densovirus.

Australian freshwater crayfish have in their hepatopancreas three so-called orphan viruses, Cherax intranuclear bacilliform virus (possibly an unclassified member of the viral family Baculoviridae), Cherax giardiavirus-like virus (presumably an unclassified member of the genus *Giardiavirus*, family Totiviridae), and Cherax reovirus (presumably an unclassified member of the family Reoviridae) that have been discounted by industry as unimportant. However, when hatchery technology was developed that allowed eggs to be surface sterilized thus producing eggs-specific pathogen free for these viruses, average size at harvest went from 35 to 70 g and the production cycle was shortened by 1 month.

If you consider that any virus must be at the very least removing cells from their normal function, upregulating immune functioning cells and diverting energy for nucleic acid processing (see above), then clearly there must always be a metabolic cost to any viral infection. There are no true orphan viruses in crustacea. However, whether it is economical to remove a virus from a growing system is another question.

7.6 CONCLUSIONS

To comprehend the ecology of viruses in aquatic crustacea, it is necessary to appreciate the immune system of crustacea because you cannot understand one without the other. Furthermore, most reader's backgrounds in immunology are from a mammalian viewpoint that does not necessarily set the scene for immediate understanding to the interrelationship of crustaceans and their viruses. As the main crustacean immune response is carbohydrate based, then crustaceans have taken on a different strategy for dealing with proteinaceous viruses. The steps of their strategy include isolating the viruses in cells, preventing the infected cells from being destroyed (control of apoptosis) thus not releasing progeny virions, passing memory iRNA molecules to germ cells, locking up infected hemocytes in lymphoid spheroid cells where the associated viruses can do no harm and then, breeding before either the internal containment system collapses via uncontrolled apoptosis induced by environmental fluctuations or predation kills the crustacean. Fundamentally, this has meant that most virus-exposed survivors are chronic carriers of the virus for life.

With the understanding of the antiviral immune system in crustacea, it is no longer correct to state that crustaceans do not have acquired immunity or immune memory as both of these operate within the iRNA and viral accommodation systems.

Broodstock practices and the global transport of shrimp as live and frozen commodities have been instrumental in the spread of viruses globally. Only the use of animals that truly are free of pathogens, coupled with attention to biosecurity, can bring back the heady days when crustacean aquaculture was rapidly approaching an industry generating US$ 20 billion per year. Intertwined with this is the necessity of a change in industry's attitude toward investment in antidisease research to take it from a "fire fighting exercise" to a progressive, structured program. This is more imperative than ever as research has demonstrated a very low heritability for resistance to viruses such as WSSV (e.g., Gitterle et al., 2005). Many researchers and farmers have promulgated that genetic selection was the answer to all the problems with viruses.

There is a need for a critical review of all the publications that list new crustacean hosts for viruses that have relied on only PCR as evidence of infection. The minimum methodology for any new publications in this area must be PCR signal plus a confirmation test that could include experimental exposure in a susceptible host or mRNA signal confirmed by sequencing or probing to show transcription of viral genes. In time as knowledge of ghost viral signatures increases, a multiple locus sequence analysis might be possible.

The future potentially includes the ability to transfect crustaceans with genes that can upregulate desirable abilities and transfect with nucleic acid constructs including iRNA that can downregulate viral proteins, that is, further blocking of the apoptosis cascade or the incorporation of betaine genes into the genome might be useful in situations of environmental stress. However, society's fear of genetically modified organisms, that is, "Frankenshrimp" will have to be overcome first or there will be a limited market for an expensive-to-produce commodity.

REFERENCES

Anggraeni, M. and Owens, L. (2000). Evidence for the haemocytic origin of lymphoidal spheroids in *Penaeus monodon*. *Dis. Aquat. Organ.* 40, 85–92.

Cimino, E. J., Owens, L., Bromage, E. S., and Anderson, T. A. (2002). A newly developed ELISA showing the effect of environmental stress on levels of hsp86 in *Cherax quadricarinatus* and *Penaeus monodon*. *Comp. Biochem. Physiol.: Part A Mol. Integr. Physiol.* 132, 591–598.

Chong, Y. and Loh, H. (1984). Hepatopancreas clamydial and parvoviral infections of farmed marine prawns in Singapore. *Singapore Vet. J.* 9, 51–56.

Dall, W. (2007). Recent molecular research on *Penaeus sensu lato*. *J. Crust. Biol.* 27, 380–382.

Flegel, T. W. (2006). Detection of major penaeids shrimp viruses in Asia, a historical perspective with emphasis on Thailand. *Aquaculture* 258, 1–33.

Flegel, T. W. (2007). Update on viral accommodation, a model for host–viral interaction in shrimp and other arthropods. *Dev. Comp. Immunol.* 31, 217–231.

Flegel, T. W. (2009). Hypothesis for heritable, anti-viral immunity in crustaceans and insects. *Biol. Direct* 4, 32.

Flegel, T. W. and Pasharawipas, T. (1998). Active viral accommodation: a concept for crustacean response to viral pathogens. In: Flegel, T. W. (ed.), *Advances in Shrimp Biotechnology*. National Center for Genetic Engineering and Biotechnology, Bangkok, pp. 245–250.

Flegel, T. W., Thamavit, V., Parawipas, T., and Alday-Sanz, V. (1999). Statistical correlation between severity of hepatopancreatic parvovirus infection and stunting of farmed black tiger prawn (*Penaeus monodon*). *Aquaculture* 174, 197–206.

Garza, J. R., Hasson, K. W., Poulos, B. T., Redman, R. M., White, B. L., and Lightner, D. V. (1997). Demonstration of infectious Taura syndrome virus in the feces of seagulls collected during an epizootic in Texas. *J. Aquat. Anim. Health* 9, 156–159.

Gitterle, T., Salte, R., Gjerde, B., Cock, J., Johansen, H., Salazar, M., Lozano, C., and Rye, M. T. (2005). Genetic (co)variation in resistance to White Spot Syndrome Virus (WSSV) and harvest weight in *Penaeus (Litopenaeus) vannamei*. *Aquaculture* 246, 139–149.

Hasson, K. W., Lightner, D. V., Mohney, L. L., Redman, R. M., and White, B. M. (1999). Role of lymphoid organ spheroids in chronic Taura syndrome virus (TSV) infections in *Penaeus vannamei*. *Dis. Aquat. Organ.* 38, 93–105.

JanewayJr., C. A., Travers, P., Walport, M., and Shlomchik, M. J. (2005). *Immunobiology*. Garland Science, New York.

La Fauce, K. and Owens, L. (2009). RNA interference reduces *Pmerg*DNV expression & replication in an *in vivo* cricket model. *J. Invertebr. Pathol.* 100, 111–115.

Lavery, S., Chan, T. Y., Tam, Y. K., and Chu, K. H. (2004). Phylogenetic relationships and evolutionary history of the shrimp genus *Penaeus* s.l. derived from mitochondrial DNA. *Mol. Phylogenet. Evol.* 31, 39–49.

Lee, S. Y., Lee, B. L., and Soderhall, K. (2004). Processing of crayfish hemocyanin subunits into phenoloxidase. *Biochem. Biophys. Res. Commun.* 322, 490–496.

Lightner, D. V. (1990). Viruses section: introductory remarks. In: Perkins, F. O. and Cheng, T. C. (eds.), *Pathology in Marine Science*. Academic Press, San Diego, pp. 3–6.

Lightner, D. V. (1996). *A Handbook of Pathology and Diagnostic Procedures for Diseases of Penaeid Shrimp*. World Aquaculture Society, Los Angeles.

Mari, J., Poulos, B. T., Lightner, D. V., and Bonami, J. R. (2002). Shrimp Taura syndrome virus: genomic characterization and similarity with members of the Cricket paralysis-like viruses. *J. Gen. Virol.* 83, 915–926.

McColl, K. A., Slater, J., Jeyasekaran, G., Hyatt, A. D., and Crane, M. S. (2004). Detection of White Spot Syndrome Virus and Yellowhead virus in prawns imported into Australia. *Aust. Vet. J.* 82, 69–74.

Melena, J., Bayot, B., Betancourt, I., Amano, Y., Panchana, F., Alday, V., Calderon, J., Stern, S., Roch, P., and Bonami, J. R. (2006). Pre-exposure to infectious hypodermal and haematopoietic necrosis virus or to inactivated white spot syndrome virus (WSSV) confers protection against WSSV in *Penaeus vannamei* (Boone) postlarvae. *J. Fish Dis.* 29, 589–600.

Nunan, L. M., Poulos, B. T., and Lightner, D. V. (1998). The detection of white spot syndrome virus (WSSV) and yellow head virus (YHV) in imported commodity shrimp. *Aquaculture* 160, 19–30.

Perez Farfante, I. and Kensley, B. (1997). *Penaeiod and Sergestoid Shrimps and Prawns of the World.* Editions du Museum national d'Histoire naturelle, Paris.

Porter, M. L., Perez-Losada, M., and Crandall, K. A. (2005). Model based multi-locus estimation of decapod phylogeny and divergence times. *Mol. Phylogenet. Evol.* 37, 355–369.

Regier, J. C., Shultz, J. W., Zwick, A., Hussey, A., Ball, B., Wetzer, R., Martin, J. W., and Cunningham, C. W. (2010). Arthropod relationships revealed by phylogenomic analysis of nuclear protein-coding sequences. *Nature* 1079–1084.

Robalino, J., Bartlett, T. C., Chapman, R. W., Grossa, P. S., Browdy, C. L., and Warra, G. W. (2007). Double-stranded RNA and antiviral immunity in marine shrimp: inducible host mechanisms and evidence for the evolution of viral counter-responses. *Dev. Comp. Immunol.* 31, 539–547.

Robalino, J., Bartlett, T. C., Shepard, E., Prior, S., Jaramillo, G., Scura, E., Chapman, R. W., Gross, P. S., Browdy, C. L., and Warr, G. W. (2005). Double-stranded RNA induces sequence-specific intiviral immunity silencing in addition to nonspecific immunity in a marine shrimp: convergence of RNA interference and innate immunity in the invertebrate antiviral response? *J. Virol.* 79, 13561–13571.

Robalino, J., Browdy, C. L., Prior, S., Metz, A., Parnell, P., Gross, P. S., and Warr, G. (2004). Induction of antiviral immunity by double-stranded RNA in a marine invertebrate. *J. Virol.* 78, 10442–10448.

Roekring, S., Flegel, T. W., Malasit, P., and Kittayapong, P. (2006). Challenging successive mosquito generations with a densonucleosis virus yields progressive survival improvements but persistent, innocuous infections. *Dev. Comp. Immunol.* 30, 878–892.

Rusaini and Owens, L. (2010a). Effect of moulting and lunar rhythms on the lymphoid organ spheroid cells of the black tiger prawn *Penaeus monodon. J. Exp. Mar. Biol. Ecol.* 389, 6–12.

Rusaini and Owens, L. (2010b). Insight into the lymphoid organ (LO) of penaeid prawns: a review. *Fish Shellfish Immunol.* 29, 367–377.

Vanpatten, K., Nunan, L. M., and Lightner, D. V. (2004). Seabirds as potential vectors of penaeid shrimp viruses and the development of a surrogate laboratory model utilizing domestic chickens. *Aquaculture* 241, 31–46.

CHAPTER 8

VIRUSES OF FISH

AUDUN HELGE NERLAND[1,2], AINA-CATHRINE ØVERGÅRD[1,2], and SONAL PATEL[2]
[1]The Gade Institute, University of Bergen, Bergen, Norway
[2]Institute of Marine Research, Bergen, Norway

CONTENTS

8.1 INTRODUCTION

Traditionally, fish have been considered to be very robust with regards to diseases as fishermen rarely observed any fish with signs of sickness. The obvious reason is that diseased fish, due to reduced health and ability of escaping, were eaten by predators, or just died and disappeared to the bottom of the lake or ocean. Consequently, there was little focus on fish diseases until the commencement of fish farming. Especially, when aquaculture became a large scaled industry and fish health thus became economically important. This economical aspect also reflects the fact that most of the fish viruses so far

Studies in Viral Ecology: Animal Host Systems: Volume 2, First Edition. Edited by Christon J. Hurst.

identified have been isolated from fish obtained from aquaculture, which are limited to a relatively few number of fish species.

Nowadays, the presence of viral agents in aquaculture is usually diagnosed by immunological assays or techniques involving molecular biology such as polymerase chain reaction (PCR); methods that are specific, sensitive, and work effectively as soon as they are developed and established. However, immunological assays need specific antibodies and PCR-based techniques require sequence information for designing specific primers. Consequently, these methods demand that the viral agent in question, at least to some extent, has been characterized beforehand.

The identification of novel virus is still based on electron microscopy (EM) or cell cultures. EM is very laborious and not very sensitive, as a relatively high number of viral particles have to be present in the sample in order to be detected at the high magnification in use. Furthermore, EM is more suited to detect viral particles with distinctive morphology such as bacteriophages. Viruses infecting eukaryotic cells usually have forms that are more difficult to distinguish from other particles that normally are present in these cells, such as ribosomes. Likewise, cell cultures require that permissive cell lines are available. As viruses usually are species and tissue specific, and relatively few cell lines from fish have been established so far, this method has its inherit limitation. Identification of viral agents is therefore normally far more cumbersome and expensive than identification of other infectious agents, such as bacteria. This may thus explain why compared to the number of existing fish species, relatively few different fish viruses have so far been identified.

In this overview of the ecology of viruses infecting fish, we will start with describing aspects of fish as host organism followed by more general aspects of viruses replicating in fish. Then, we will look at how fish viruses are transmitted, and how this may be influenced by various environmental factors. We will also give a brief introduction on the impact of viruses in the aquaculture industry, and the present status in development of vaccines against viral diseases. Finally, we will focus more specifically on some selected fish viruses belonging to the various groups of the Baltimore classification.

8.2 FISH AS VIRAL HOSTS

8.2.1 General Aspects

Taxonomically, fish belong to the vertebrates. The most primitive fish species, like hagfish and lampreys are classified as jawless vertebrates (agnatha). The majority of fish species, however, belong to jawed vertebrates (gnathostomata), which again are divided into cartilaginous fish (sharks, rays, skates) and bony fish. Classified under the bony fish we find the teleosts, which comprise the majority of the about 30,000 different fish species that so far have been described (FishBase, 2010). Almost all the fish species that are currently farmed belong to the teleosts, and this group has thus been best investigated biologically.

Fish is a rather heterogeneous group and the different species may differ enormously in many aspects. The largest now living fish species (the whale shark, *Rhincodon typus*) may reach a length of more than 10 m and a weight that can exceed 20,000 kg (Stevens, 2007), while the smallest known species, *Paedocypris progenetica*, is less than 1 cm in length, weighing just a few milligrams (Kottelat et al., 2006). The diversity of fish is also reflected in the size of their genomes, which range about 20-fold from the pufferfish (*Tetraodon fluviatilis*) to the green sturgeon (*Acipenser mediarostris*) (Smith and Gregory, 2009). The habitats of different fish may also differ considerably. Some live in freshwater, others live in seawater with varying salinity, while some change their habitat during the life cycle like anadromic fish that hatch in freshwater but later migrate to the sea. The vast majority of fish species are ectothermic, meaning that the body temperature is equal to the temperature of the surroundings. For some species, like the arctic cod

(*Boreogadus saida*) the temperature can be as low as –2°C. For others, like the desert pupfish (*Cyprinodon macularius*), living in hot springs, the temperatures might be higher than 45°C (FishBase, 2010). There are a few exceptions in being ectothermic, as tuna, swordfish, and some shark species can have body temperatures significantly higher than the temperature of their surrounding water (Graham and Dickson, 2001). Perhaps most typical for fish is the uptake of oxygen from the water by means of gills and having a streamlined body adapted for swimming. However, also in these aspects there are exceptions. The lungfish can under special circumstances take up oxygen from the air (Joss, 2006), and the body of the sea dragon (*Phycodurus eques*) looks more like seaweed (Connolly et al., 2002).

8.2.2 Antiviral Defense Mechanisms of Fish

There may be several routes of infections for viruses to the fish body such as skin, eyes, gills, or through the oral–gastrointestinal tract. In any case, the virus particle has to get across physical barriers as both exterior and interior body surfaces are covered by tight layers of epithelial cells. In addition, normally, there is also a thick mucous layer covering the epithelial cells, where the infectious agents may be trapped and subsequently excluded as the mucus is continuously excreted from the body (Cameron and Endean, 1973).

If the virus manages to enter the body of the fish, it will be met by a sophisticated immune system. As will be described below, the immune system can be divided into the innate immune system and the adaptive immune system. The borderline in evolution for the appearance of adaptive immunity is found at the jawless vertebrates (agnatha), which recently have been found to have some kind of adaptive immunity based on so-called variable lymphocyte receptors (VLR). This system is based on gene conversion (copy choice) for generating receptor diversity (Pancer et al., 2004), whereas the adaptive immune system we find from jawed vertebrates (gnathostomata) to mammalians is based on somatic recombination due to the presence of the recombination activating genes (RAG). This means that among the fish species we find the earliest hosts in the evolution where viruses face both innate and adaptive immunity, and this may have had a great impact on the evolution of the viruses.

Until a few decades ago there was little knowledge about the immune system of fish. The knowledge of the immune system of vertebrates was based on studies of mammalians (mice and man). However, information about the immune system of fish has increased rapidly during the past few years, and the knowledge gained has revealed that vertebrates to a large extent seem to share similar overall organization of their immune system. Below, we will give a brief description of the vertebrate immune system with focus on antiviral mechanisms, followed by a description of features that are special for fish. In this context, we will separate innate immunity and adaptive immunity. However, sometimes it is convenient to divide the immunity in another way, depending on if the effector molecules are present in the blood serum (humoral immunity) or not (cellular immunity).

8.2.3 The Vertebrate Immune System

8.2.3.1 Innate Immunity The innate immune system can be regarded as the first line of defense. It becomes activated when conserved structures of microbes, so-called pathogen-associated molecular patterns (PAMPs) are recognized by pattern recognition receptors (PRRs) present on (exterior or interior) membranes of the host cells (Ishii et al., 2008). For viral infections, the PAMPs may be recognized due to structures specific for the pathogen, such as triphosphate at the 5′-end of some viral RNA (Pichlmair et al., 2006), or due to abnormal location of molecules, such as the presence of DNA in the cytoplasm (Takaoka et al., 2007). As a consequence of sensing the invading viral pathogen by means of the PRR, distinct signaling pathways are activated, leading to activation and transcription of various genes. Some of the gene products will have a direct antiviral effect, while others work more indirectly by stimulation and activation of other cells in the immune system. Direct antiviral effects are executed by the interferons (IFNs), which can be expressed by any cell type as a response to viral infection, working both in an autocrine and a paracrine fashion. All types of cells also have the receptor for IFNs, and binding induces antiviral mechanisms such as degradation of viral genomes and inhibition of the protein synthesis machinery of the cell. These mechanisms will turn other cells in the body into an antiviral state, preventing spreading of the infection (Fensterl and Sen, 2006; Stetson and Medzhitov, 2006). More than 200 genes have been found to be activated as a consequence of interferon stimulation in mammalians (de Veer et al., 2001), among others the Mx-genes and genes encoding antimicrobial peptides (AMPs). The latter are small peptides that were first recognized due to their antibacterial activity, but recently some have been reported to execute antiviral mechanisms (Chia et al., 2010; Chiou et al., 2002; Wang et al., 2010a, 2010b). Direct antiviral activity may also be executed by the complement system. The membrane-attack complex, which is the final product of the complement cascade, may inactivate enveloped viruses by making lesions on their surrounding membranes (Carroll, 2008). Expressed as a consequence of interferon action there will also be cytokines and chemokines that call attention and stimulates cells such as neutrophils, natural killer (NK) cells and additional monocytes/macrophages, indirectly giving rise to other innate antiviral responses (Christensen and Thomsen, 2009). Neutrophils and macrophages may inactivate viruses by phagocytosis, followed by enzyme degradation or inactivation by means of free oxygen/nitrogen radicals (Lowenstein and Padalko, 2004). NK-cells are able to recognize stress markers (or the lack of markers like low expression of MHC molecules), which indicate viral infection, and then kill the cell by cytotoxic mechanisms (Eagle and Trowsdale, 2007). The innate immune system does not generate immunological memory, but gives the same intensity and duration of response, no matter how often a specific invader is encountered. However, as the innate immune system recognizes conserved patterns of the pathogens it can be considered to be part of a collective memory, accumulated through the evolution.

8.2.3.2 Adaptive Immunity Cells and components that are part of the innate immunity also play crucial roles of the adaptive immunity. Dendritic cells (DCs), which reside in almost all tissues, become activated upon viral infection as described above, and then migrate to secondary lymphoid organs (SLO), like the spleen (Angeli and Randolph, 2006; Johnson and Jackson, 2008). In SLO there is a collection of lymphocytes; naïve B-cells and T-cells expressing antigen receptors with a variety of binding properties, generated previously by somatic recombination. Here, the DCs act as professional antigen presenting cells (APCs) by presenting antigens deduced from the viral particles for binding to the T-cell receptor (TCR) of naïve T-cells if these cells have appropriate specificity (de Jong et al., 2005). Antigens need to be presented in association with the surface proteins MHC class II in order to activate helper T-cells (T_H-cells). For naïve B-cells to be stimulated, membrane-bound antibodies must bind to the viral antigens simultaneously as they get stimuli by cell-to-cell contact with T_H-cells as well as cytokines produced by T_H-cells (Christensen and Thomsen, 2009). Stimulated B-cells can either develop further to antibody-producing plasma cells or memory cells. The antibodies will start to circulate throughout the body and can bind to virus anywhere. Antibodies can bind to the virus particles and thereby prevent the virus from getting access to the receptors on the cells. If antibodies bind to enveloped virus particles, this can activate the complement system through the so-called classical pathway. Antibody–virus complexes are also more efficiently taken up and destroyed by phagocytic cells (Burton, 2002). However, antibodies do not pass the cell membrane and enter the interior of cells, and will consequently not be able to interfere with pathogens that replicate or hide inside cells. Detection and lysis of infected cells is therefore crucial for eradication of intracellular pathogens such as viruses. As mentioned above, NK-cells are able to detect and kill infected cells in an unspecific manner, while the cytotoxic T-cells (T_C-cells) can perform specific detection and lysis of infected cells. The activation of naïve T_C-cells also normally takes place in SLO (Christensen and Thomsen, 2009). In this case, antigens deduced from the virus have to be presented by the APC to the T_C-cells in association with MHC class I. At the same time stimulation by T_H-cells is required. After activation, both T_H-cells and T_C-cells can become either effector cells or memory cells. The effector cells will start to circulate in the body and will, due to the action of chemokines, reach the site of infection where they will execute their functions. The memory cells may stay in the body for a long time and get in action if the same or a similar pathogen infects the host at a later stage. However, the activation of the memory cells is different from the activation of naïve lymphocytes. For example, almost any type of cells in the body can express MHC class I and present viral antigens to memory T_C-cells, while only professional APCs are capable of presenting viral peptides to naïve T_C-cells. The cytotoxic effect of T_C-cells and NK-cells is achieved by inducing the target cells to enter apoptosis, a programmed cell death that occurs in a fashion that limits spreading of the virus (Lee et al., 2007; Fan and Zhang, 2008). An essential part of activation of naïve lymphocytes is a clonal expansion, which means that there exist many memory cells with the identical binding properties. Taken together, this means that a faster and more comprehensive response is a feature of immunological memory, which may be crucial to stop the infection before the virus will get time to propagate and spread in the body.

8.2.3.3 Special Features of the Fish Immune System There are differences between the immune system of different fish species. As mentioned, the jawless fish have a more primitive adaptive immune system. And the cartilaginous fishes have a somewhat different system for generating diversity of antibodies and T-cell receptors (Malecek et al., 2008). However, recent studies, including large-scaled sequencing projects, have revealed that genes for most of the components known from the mammalian immune system are also present in teleosts (Patel et al., 2009a; Park et al., 2009; Sarropoulou et al., 2009; Adzhubei et al., 2007), and it seems therefore that teleosts have an immune system where the principles of organization and functions are similar to what we have outlined above for higher vertebrates. As teleosts comprise the majority of fish species we will from now on focus particularly on their immune system.

Toll-like receptors (TLRs), which are members of PRRs, have been identified in teleosts (Rebl et al., 2010). Although interferon genes of teleosts show a poor sequence similarity with its mammalian counterparts, important structural features are identified within the genes, indicating a preserved function, and Mx-proteins have been shown to be upregulated in response to dsRNA and viral infections (Robertsen, 2006; Wu et al., 2010). Antiviral effect of the complement system has been shown in rainbow trout infected with membrane-coated viruses (Boshra et al., 2006). Furthermore, macrophage and granulocyte-like cells have been identified in the vicinity of virus infected fish tissue, suggesting that they could take part in the clearance of virus in teleosts (Johansen et al., 2004). NK-like cells have been found to have receptors likely to be functional orthologues of mammalian NK receptors, and virus infected cells were found to be killed nonspecifically by the action of these cells, as seen for the NK-cells in mammals (Fischer et al., 2006). A dendritic-like cell type has been described in Atlantic salmon and rainbow trout having an endocytic receptor, Langerin, that is specific for human Langerhans cells. These dendritic-like cells were observed in spleen and head kidney of healthy salmon, and in diseased gills (Lovy et al., 2009).

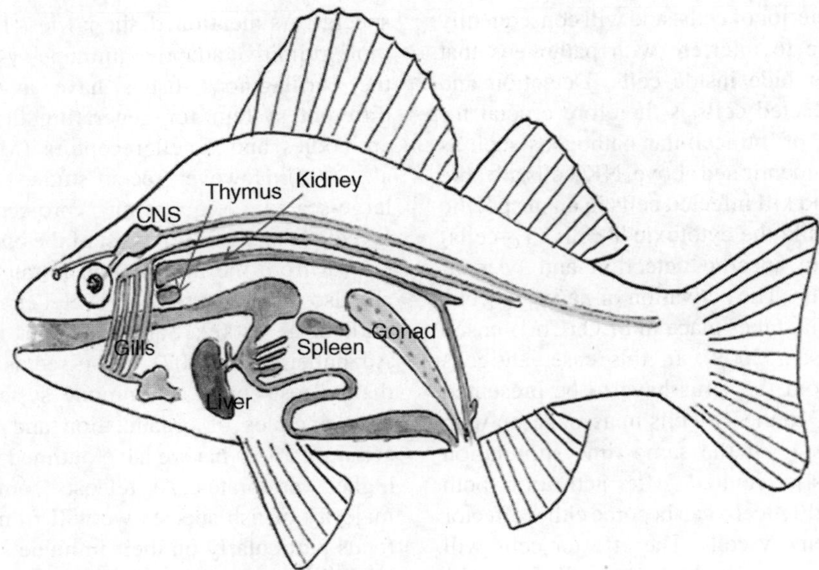

FIGURE 8.1 Schematic anatomic drawing of a teleost.

Both the T-cell receptor genes and the RAG genes encoding the enzymes responsible for TCR rearrangement are identified in teleost. The antigenic diversity of fish T-cells could thereby be equivalent to what is seen in higher vertebrates (Fischer et al., 2006).

However, comparing teleosts and mammalians we will find that there are some differences, especially, regarding organization of lymphoid tissues. In mammals, the bone marrow and the thymus are the primary lymphoid organs where immature lymphocytes are generated, become mature, and commit to a particular antigenic specificity. Lymph nodes, spleen, and various mucosa-associated lymphoid tissues (MALT), such as gut-associated lymphoid tissues (GALT), are secondary lymphoid organs that provide sites for mature lymphocytes to interact with antigen. Teleosts lack bone marrow, and the presence of lymph nodes, Peyer's patches, and germinal centers have not been demonstrated so far. Major immune tissues of teleost fish include the kidney (especially the anterior kidney) and the thymus. In addition, teleost fish possess a spleen and scattered immune areas within mucosal tissues (e.g., in the skin, gills, gut, and gonads) (Figure 8.1). There are also some differences at the molecular level. Teleosts have three isoforms of immunoglobulin, where two resemble IgD and IgM of higher vertebrates, and a third (IgT) that seems to be restricted to teleosts (Hansen et al., 2005). The other classes of immunoglobulins such as IgG, IgA, and IgE, found in higher vertebrates, have not been convincingly demonstrated in teleosts. Although, elevated secondary antibody responses have been demonstrated in several teleost species (Anderson et al., 1979; Lamers et al., 1985; Sinha and Chakravarty 1997; Palm et al., 1998), the level of enhancement is weaker in teleosts as compared to mammals (Lamers et al., 1985). In spite of these differences, it seems like the overall organization of the immune systems, and the functions of the mechanisms for protection against virus infection, are quite similar for the vertebrates from teleosts to mammalians.

8.2.3.4 RNA Interference Some organisms, such as plants and insects, defend themselves against viral infections by RNA interference (RNAi). This mechanism is based on similarity between sequences inherited within the host genome and sequences in the viral genome. By viral infection these host sequences are transcribed and the transcripts bind to the viral RNA, generating dsRNA, which is cut up by an enzyme called Dicer into fragments called small interfering RNA (siRNA). The siRNA then binds to the so-called RNA-induced silencing complex (RISC) where the small fragments help RISC to recognize viral RNA molecules, which then are degraded by the complex (Plasterk, 2002). As will be described later, a contra-defense system to RNAi has been detected in the betanodavirus, indicating that RNAi is acting to defend against viral infection in fish.

8.2.4 The Ontogeny of the Fish Immune System

Fish are mass spawners, where females release high numbers of eggs to be fertilized. The different fish species hatch at varying developmental stages. After hatching, the fish larva goes through a yolk-sac stage, before a live feed stage. Normally, and especially for marine fish, this takes place in an environment that is rich in microorganisms, such as algae, bacteria, and viruses. Many viruses are found to be causing mortality at these early stages of life. One of the reasons is that it takes some time before the immune system is developed and function properly. The development of the immune system may vary considerably between different fish species, but as an example, the ontogeny of the immune system of Atlantic halibut (*Hippoglossus hippoglusssus*), which is outlined in Figure 8.2, can illustrate the appearance of lymphoid organs and the timings of presence of immune markers. The complement factor C3 can be detected in halibut larvae sampled from 5 days posthatching (dph) and onwards (Lange et al., 2004a), but its presence or expression at an earlier stage is not known. Kidney-related tissues and anlage of thymus appear during the yolk stage, transcription of genes encoding parts of immunoglobulins is detected right before metamorphosis, but the immunoglobulin protein can be detected at the end of the metamorphosis at 94 dph (Patel et al., 2009b). At the same time, T-cell markers such as CD4 and CD8, start to appear (Øvergård et al. 2011). Thus, although lymphoid organs are developed, the presence of the immune markers such as IgM (present on B-cells), and CD4 and CD8 (present on subsets of T-cells) is often detected at much later state. In addition, as seen in Figure 8.2, the functionality of these markers, that is, their detection at the protein level is a delayed process that shows how vulnerable larvae are for a long period of time after hatching (Patel et al., 2009b). It is believed that, during the early life stages larvae/fry are protected by maternally transferred components such as lectins and antibodies, and the larvae are dependent on the innate defense mechanisms many of which are present already during early egg stages (Lange et al., 2004b, 2006; Swain and Nayak, 2009).

8.3 THE VIRUS

8.3.1 Viruses Infecting Fish

The number of different viruses infecting fish that have been identified is low compared to the number and variation of existing fish species. As already mentioned, most of the virus species have been isolated from teleosts, and particularly from teleost species used in aquaculture. From jawless fish there are, to our knowledge, no reports of virus identification so far. And from cartilaginous fish just a couple of virus species has been reported (Kent and Meyers, 2000). It is also interesting to note, that although to date no natural zebrafish virus has been characterized, experimental infection at the embryo/egg stages, with viruses obtained from other fish species, gave 100% mortality (Sanders et al., 2003; LaPatra et al., 2000). The current picture of virus affecting fish is most probably far from complete due to the lack of comprehensive surveys.

Table 8.1 summarizes the current status of virus affecting fish. Virus species belonging to five out of the seven groups of the Baltimore classification have been identified. So far, there

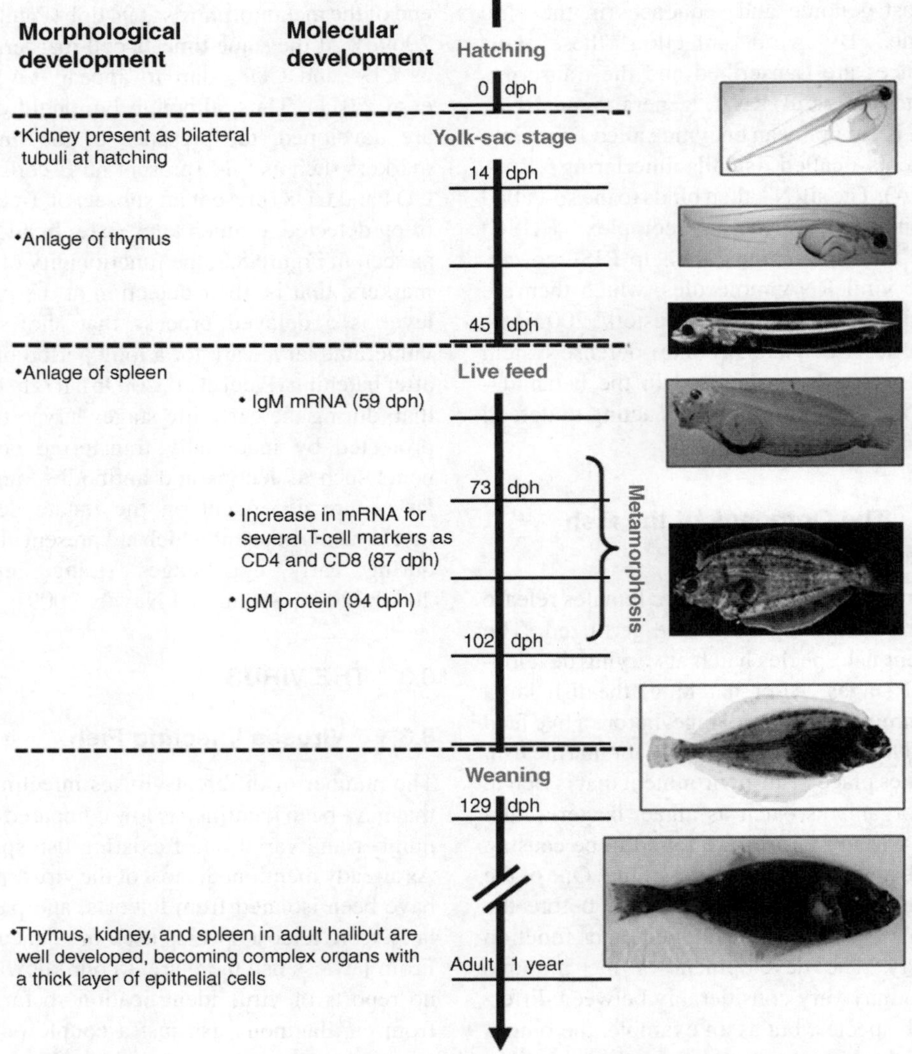

FIGURE 8.2 The ontogeny of the immune system of Atlantic halibut (see text). (*See the color version of this figure in Color Plates section.*)

have been no reports of fish viruses belonging to the Baltimore groups II and VII, which comprise ssDNA virus and dsDNA retrovirus, respectively. There are viruses representing 10 out of the 32 virus families that have been registered to infect vertebrates. Moreover, there are representatives of small simple viruses, like betanodavirus, to large and complex viruses, like iridovirus. Both viruses that are externally enveloped by membranes obtained from the host cells, and viruses without such membranes have been identified. Few fish viruses are restricted to a single host species, while many have a relatively broad host range.

Members within a family consist of viruses with somewhat similar organization, morphology, and life cycle. Sometimes, it may be uncertain if these features are due to convergent or divergent evolution. For example,

TABLE 8.1 Examples of Viruses from the Various Groups of the Baltimore Classifications (BC) That Have Been Identified in Fish

	Genome (BC)	Family	Example of Virus	Membrane Enveloped	Capsid Structure	Virion Size (nm)	Genome Size (kb)
DNA virus	dsDNA (I)	Alloherpesviridae	IcHV	+	Spherical	150–200	134
		Iridoviridae	LCDV	+	Polyhedral	120–350	100
	ssDNA (II)			Not detected in fish			
RNA virus	dsRNA (III)	Birnaviridae	IPNV	−	Icosahedral	60	6
		Reoviridae	GCRV	−	Icosahedral	75	20
	ssRNA(+) (IV)	Coronaviridae	WBV	+	Spherical	120–140	28
		Nodaviridae	Betanoda virus	−	Icosahedral	25–35	4.5
		Togaviridae	SAV	+	Icosahedral	45–75	12
	ssRNA(−) (V)	Rhabdoviridae	IHNV	+	Bullet shaped	75 × 180	11
		Orthomyxoviridae	ISAV	+	Spherical	90–130	13.5
Retrovirus	ssRNA(RT) (VI)	Retroviridae	WDSV	+	Spherical	80–100	13
	dsDNA(RT)			Not detected in fish			

Abbreviations: IcHV, Ictalurid herpesvirus; LCDV, lymphocystis disease virus; IPNV, infectious pancreatic necrosis virus; GCRV, grass carp reovirus; WBR, white bream virus; SAV, salmonid alphavirus; IHNV, infectious hematopoietic necrosis virus; ISAV, infectious salmon anemia virus; WDSV, Walleye dermal sarcoma virus.

considering alpha- and betanodaviruses, infecting insects and fish, respectively, we find that both have a genome consisting of two positive stranded RNA segments, with approximately the same length and both have a capsid with similar morphology. Moreover, the genomes of both viruses encode proteins of similar size and functions. However, sequence similarities between these two groups are low, both at nucleotide and amino acid levels. One can therefore speculate if this is simply an efficient organization of such small and simple viruses, and that they have appeared independent of each other.

8.3.2 The Life Cycle of Fish Viruses

Like other viruses, a fish virus has a life cycle including many different steps and stages. It must reach the host, find and enter the permissive cells, reach the intracellular site of replication, replicate and leave the cell, and finally leave the host organism and be transferred to another host. The transfer may be either to any suitable host organism in the environment (horizontal transmission) or from parents to offspring by means of eggs or sperms (vertical transmission). The transmission might occur via a vector organism, a nonliving vehicle, or from a prey to a predator. The latter can be either between different host species or between hosts belonging to the same species (cannibalism). In the following section, we will have a closer look at the different steps and stages in life cycle of viruses.

8.3.2.1 Virus Stability in the Environment Common for fish viruses is high stability in the environment, where viruses without membrane are more stable than membrane-coated viruses. The latter are usually dependent on the membrane for infectivity and cell membranes have a fragile structure. The birnavirus IPNV (infectious pancreatic necrosis virus), which does not have any membrane, was found to be stable in regards to infectivity for over 120 days both in freshwater and in seawater for temperatures up to 20°C (Mortensen et al., 1998), whereas the membrane-coated rhabdovirus VHSV (viral hemorrhagic septicaemia virus) is stable for up to 7 days at 10°C, but loses most of the infectivity within 5–10 days at 20°C (Hawley and Garver, 2008).

How critical the stability of the virus in the environment is, depends on the density of the host population. This may vary considerably between different fish host species. Herrings appear in vast schools, while there are deep sea species, such as some anglerfish, where the male lives as a parasite on the female in order to have a chance of mating during lifetime (Pietsch, 1975). As the water gives a considerable effect of dilution within a very short time due to a three-dimensional diffusion of the virus particles, most viral transmissions probably occur between individuals with rather close contact.

8.3.2.2 Vehicles Theoretically, if virus adheres to larger particles that may serve as vehicles, this may have great impact on the migration and distribution of the virus in water. If the larger particles have a certain density, for example, floating on the surface, the diffusion would be two-dimensional instead of three-dimensional. It may also be that a virus will be more stable in association with a vehicle or that vehicles have properties that guide the virus particles to the fish host as more specific targets.

In global aspect, water may itself act as a vehicle. Ballast water, used to stabilize ships, is transferred around the world for being discharged just before loading the cargoes, and may therefore act as a long-distance dispersal mechanism for microorganisms, including viruses affecting fish (Ruiz et al., 2000; Drake et al., 2007). More locally, virus may be transferred by the water flow in river systems. In fisheries and aquaculture, virus may also be moved from one place to another by contaminated nets, boats, or other equipments.

8.3.2.3 Vectors Although no fish virus are known to be obligate users of vectors for transmission, there are several reports that indicating that vectors might contribute to transmission. It has been shown that the lymphocystis disease virus 1 (LCDV-1), a member of the Iridoviridae family, can persist in an infective form in the brine shrimp *Artemia*, which is widely used as live feed for the larval stages of fish in aquaculture (Cano et al., 2009). Also, fish ectoparasites have been identified as vectors for viral transmission as the togavirus SAV (salmonid alphavirus) could be detected by PCR methods in sea lice (*Lepeophtheirus salmonis*) (Petterson et al., 2009) and the rhabdovirus IHNV (infectious hematopoietic necrosis virus) could be isolated from a leech (*Piscicola salmositica*) and a copepod (*Salmincola* sp.) (Mulcahy et al., 1990). The orthomyxovirus ISAV (infectious salmon anemia virus) has been reported to be accumulated by normal filtration in blue mussels (*Mytilus edulis*) and reported to persist for some time. Injection of hepatopancreas homogenate from the mussels caused ISAV positive salmon, indicating that the virus remains infectious for some time (Skår and Mortensen, 2007). Piscine virus has also been found in feces from seabirds such as herons and mallards that are fish predators, indicating a possibility of these as vectors (McAllister and Owens, 1992). However, in neither of the cases it has been shown that the viruses replicate in the potential vectors, so eventually the mentioned vectors can be so-called mechanical vectors only.

8.3.2.4 The Entrance of the Host Organism In fish, studies on the ports of entry have mainly been performed by elucidating where the virus can be found a short time after exposure. This is because fish lives in water and it is difficult to apply pathogens at specific sites or organs to find out if this causes infection. However, this approach has limitations as the virus not necessarily replicate in the cells or organ where it enters the host, but more or less will be in transition to the site for replication. So far, there are few viruses where the port of entry in fish is exactly known, but for some viruses, there are indications about likely sites.

In general, mucosal surfaces covering gills and the gastrointestinal tract (GIT) seem to be the most common port of entry for viral infections in fish. One of the reasons is that these surfaces, in order to carry out their functions to absorb oxygen or nutrients, contain just one single layer of epithelial cells. For the birnavirus IPNV, the GIT is considered to be the main entry site besides the gills (Wolf, 1988). For the orthomyxovirus ISAV, the gills most likely are the site of entry, although the GIT cannot be excluded (Mikalsen et al., 2001). Even for the betanodavirus, where the central nervous system (CNS) is the main organ of viral replication, the entry of virus to fish larvae probably is the intestinal epithelium, followed by an axonal transport to the brain through cranial nerves such as the vagus nerves (Grotmol et al., 1999). If viruses can penetrate undamaged skin is uncertain, but a study applying a recombinant IHNV expressing a luciferase reporter gene revealed that the fin bases are the major port of entry for this rhabdovirus (Harmache et al., 2006). Similar results have also been reported for the Koi herpesvirus (KHV) (Costes et al., 2009).

8.3.2.5 Receptors, Permissive Cells, and Tissue Tropism The first step in the replication cycle of a virus is entry of the host cell. Usually this is mediated by specific receptors, the expression of which will influence host range and tissue tropism. For some fish virus their receptors have been identified. The receptor for the rhabdovirus VHSV has been reported to be fibronectin (Bearzotti et al., 1999), which is an extracellular matrix glycoprotein. As fibronectin is involved in wound healing, this means that the virus will have high affinity for tissue damages such as skin ruptures. It has been shown that the orthomyxovirus ISAV binds to glycoproteins containing 4-*O*-acetylated sialic acids indicating that this represents a receptor determinant for the virus (Hellebø et al., 2004).

For some fish virus the subcellular sites of replication have been identified and been in accordance with what has been found for other members in the respective families. The nucleoprotein and the viral RNA of ISAV are localized in the nucleolus of infected cells like for other members of the Orthomyxovirus family (Goić et al., 2008). Betanodavirus has been found to replicate in association with the mitochondrions (Guo et al., 2004; Mézeth et al., 2007) as reported for alphanodavirus infecting insects (Van Wynsberghe and Ahlquist, 2009).

Viruses often exhibit tissue tropism, as there are specific cells or organs where virus propagation preferentially takes place. Which and where may depend on several factors such as accessibility of the virus to the cells, the presence of receptors on the cell surface, and the replication machinery in the cells. Failure to fulfill one of the requirements is enough to inhibit propagation of a given virus, as could be seen when the betanodavirus was injected into muscle tissue of turbots (*Scophthalmus maximus*). Multiplication of virus particles was shown to take place in the cells hit by the needle, but not in the adjacent cells, indicating that the necessary components for replication were present in the cells, but that the virus particles were either not released from the cell or taken up by neighboring cells (Sommerset et al., 2005).

The virus often undergoes an initial replication in the cells and tissues near its portal of entry into the host. For example, the epithelial cells in the mucosa will be a strategic place to replicate as the virus easily will be released from the host, exemplified by the rhabdovirus IHNV (Drolet et al., 1994; Harmache et al., 2006). Other strategic sites for replication will be organs such as kidney, liver, and pancreas, from where there are continuous secretions of progeny virus to the surroundings of the fish. An example is the birnavirus IPNV (Roberts and Pearson, 2005). Ability to replicate in endothelial cells, like the orthomyxovirus ISAV (Hovland et al., 1994) will then distribute the virus to most organs, including those providing secretions to the exterior of the fish. As discussed below, other organs may be preferred in order to avoid the immune system of the host.

8.3.2.6 Interaction with the Immune System There are differences in the extent to which any given virus will be affected by the various components of the immune system. It will depend on how accessible and susceptible the virus is for the respective effector subsystems, which will in turn depend on the morphology of the virus particle and the various steps of the viral life cycle. The interferon system has been shown to have antiviral effect against many fish viruses like the birnavirus IPNV, channel catfish herpesvirus (CCV) and betanodavirus (Jensen and Robertsen 2002; Robertsen et al., 2003; Long et al., 2004; Sommerset et al., 2003). Antiviral effect of the complement system has been shown in rainbow trout infected with the membrane-coated rhabdoviruses VHSV and IHNV. That antibodies can have antiviral effect is indicated by the fact that recombinant vaccines (see below) have been shown to give some protection against viral infections in fish. However, antibodies cannot cross the cell membrane. Virus particles that are inside a cell will accordingly not be reached by the antibodies unless the cell somehow is lysed. The protective effect of antibodies will therefore depend on how much the viral particles stay outside cells during the life cycle in the host, and especially during the time right after entering a new host when the number of virus particles are relatively low. There is a general agreement that antibodies contribute to the protection against most viral infections, but to get a better protection and clearance of viruses, a cellular immune response is necessary as well.

Interaction with the adaptive immune system will give a strong selective pressure to viruses. The B-cells recognize mostly epitopes on the surface of the virus particles. This means, as will be discussed below, that mutations in genes encoding for surface antigens may give the virus selective advantages. T-cells, on the other hand, recognize processed antigens, which can also be part of structural component interior of the virus particle. As discussed below, interaction with the adaptive immune system may under certain circumstances be advantageous for a virus.

In general virology, there are several examples how viruses can contra-interact with the immune system and thereby neutralize otherwise harmful effector systems. From viruses infecting mammalia there are examples of viral blocking of the effects of interferons, inhibiting antigen processing, downregulation of MHC expression, interference with the complement system, production of cytokine resembling molecules, and killing of lymphocytes (Iannello et al., 2006). Some contra-interactions have also been revealed for fish viruses. There are indications that proteins encoded by the orthomyxovirus ISAV counteract the interferon response (García-Rosado et al., 2008). Furthermore, a gene product from the birnavirus IPNV has been reported to exercise antiapoptosis activity (Hong et al., 2002). RNAi suppressor activity has been indicated for the B2 protein of betanodavirus, first discovered in plant cells (Iwamoto et al., 2005), and further extended to an animal cell line (Fenner et al., 2006, 2007; Ou et al., 2007). It was shown that B2 could be a sequence-non-specific dsRNA binding protein that could prevent the generation of siRNAs that would normally facilitate sequence-specific RNA silencing by coating the viral dsRNA. Also, the mutation of the B2 protein abolished the accumulation of both RNA1 and RNA2 in fish cells, indicating a preserved function of RNAi in fish (Fenner et al., 2007).

8.3.2.7 Persistence and Vertical Transmission To harm and/or destroy the host is not a goal for infectious agent like a virus, but sometimes pathogenic effects are necessary in order to optimize the life cycle where the ultimate goal is to multiply, be transferred to new hosts, and most of all—not to become extinct. Pathogenicity can be a result of the virus multiplication itself, where the membrane integrity of the infecting cells typically is lost, resulting in leakage of enzymes and cell damage. But it can also be a side effect of the immune response mounted against the virus, exemplified by cell lysis executed by cytotoxic T-cells or due to cell lysis accomplished by the complement system. Virulence, which is the relative ability to cause pathogenicity, will in addition to effectiveness of causing damage also depend on how successful the pathogen is throughout all the steps of the whole life cycle.

The virulence of a given virus species may vary considerably among the various fish species that a specific virus utilizes as hosts; from almost subclinical forms to massive mortality. It may also vary within a fish host species depending on age, living conditions, and genetic variations. In general, fish are more susceptible at the early stages of the life cycle. For example, we have experienced from our own studies that betanodavirus gives high mortality in larvae of Atlantic halibut, but when halibuts have gone through metamorphosis (around 2 g) they seem to become highly resistant against the virus. At that developmental stage, even injections with the virus generally do not give any symptoms of the disease, although fish test positive for the presence of virus. Challenge experiments with turbots using the same strain of betanodavirus also show age-dependent trends, although a somewhat different picture. Here, high mortality of fish up to 10 g, moderate mortality of around 20 g fish, but low mortality of fish larger than 40 g has been observed. On the other hand, there are also examples of fish species, such as spotted coralgroupers (*Plectropomus maculatus*), where adults are quite susceptible for betanodavirus (Nopadon et al., 2009). Similar trends of decreasing susceptibility with increasing age and size have been observed for the birnavirus IPNV (Roberts and Pearson, 2005), the rhabdovirus VHSV (Hershberger et al., 2007), and the orthomyxovirus ISAV (Rimstad and Mjaaland, 2002). For the togavirus SAV, however, outbreaks of disease have mostly been registered in adult individuals of Atlantic salmon (*Salmo salar*) in the seawater phase (Andersen et al., 2007).

The reason for these variations in susceptibility is not known. The explanation could be more efficient immunity in larger fish or that receptors for the virus are mainly expressed in young fish. From an evolutionary point of view, the explanation can be that there have not been any strong selective forces for fish species like halibut to acquire resistance for viral infections at the larvae stage. Adult halibuts live as single individuals and spawn in the oceans where egg and larvae will spread, soon giving very low densities. One can therefore say that the benefit of viral virulence will be inversely related with the density of the population of the host. Killing a host that lives as a deepwater hermit will be the dead end for the pathogen as well, while for host fish living in vast schools high mortality may be an advantage for the virus. This fact implicates that in aquaculture, where there is an artificial high density of fish hosts, there can be strong selective forces for viruses with high virulence.

A strategy for long-time survival of the virus is persistence, where the virus coexists with their hosts while causing little, if any pathology, or clinical symptoms. There are different kinds of persistence. In latent infections, the virus shuts down its replication and remains dormant for a period before becoming active again. In chronic infections, a low amount of viral particles may be detectable, while the clinical symptoms may be either mild or absent for long periods. Several mechanisms can cause viral persistence in vertebrates, from more active contra-immunological effects of the virus like downregulation of MHC class I in the host cells, to hiding somewhere in the hosts where the virus is less subjected to attacks by the immune system, like the central nervous system. Often, it is a fine balance where the virus tries to propagate but is kept in check in a restricted area by the immune system.

Persistence is more a rule than exception among virus species. For fish viruses, persistence has been reported for alloherpesvirus (van Nieuwstadt et al., 2001; Waltzek et al., 2009), birnavirus (Murray, 2006), betanodavirus (Johansen et al., 2004; Mézeth et al., 2009), togavirus (Graham et al., 2010), rhabdovirus (Kim et al., 1999), and orthomyxovirus (Devold et al., 2000). Persistence, as an integrated part of the host genome, is an obligatory stage in the life cycle of retroviruses. By examination for reverse transcriptase (RT) activity, followed by EM, the presence of retroviruses was found in all four fish cell lines examined. No cytopathic effects were observed on these cell lines, but such effects were seen when the same viruses were used to infect other cell lines (Frerichs et al., 1991). These findings indicate that persistent viruses, without any visible pathogenic effects, may be common in fish. However, such viruses may give synergetic effects if the fish is infected by other pathogens, including viruses.

A persistent state would end with extinction of the virus as a population if the virus did not produce any progeny, which could be transmitted further by one way or another. Such transmissions can occur in several ways. The virus can be reactivated due to some kind of immune suppression, for example, in connection with stressed situations such as maturation or spawning. There are few examples where virus reactivation has been shown to occur in fish, although such reactivations are well known for mammalian virus. For example, herpes simplex viruses infecting humans can persist for long-time periods in ganglia cells and then be reactivated by either UV-light or various other stress factors (Steiner et al., 2007).

Reactivation of virus in connection with maturation and spawning can be strategic as the viral particles can be associated with the gametes and thereby transferred to the offspring. If the virus is inside either the sperm or egg cells it is called true intraovum vertical transmission. If the virus is in association with gametes, like in the ovarian or seminal fluids, but not intracellular, then it is called extraovum vertical transmission. The result will be the same, transmission of the virus to the next generation. There are reports indicating vertical transmission for the birnavirus IPNV (Dorson et al., 1997), the orthomyxovirus ISAV (Totland et al., 1996), the rhabdovirus IHNV (Hattenberger-Baudouy et al., 1995), and the betanodavirus (Grotmol and Totland, 2000; Breuil et al., 2002).

Also, transmission of persistent virus can occur if the persistently infected fish is eaten by a predator that would be susceptible to the virus, either belonging to the same species (cannibalism) or to another species. Persistent viruses can be released if the host dies, disintegrating into the environment and the virus particles can then somehow be transferred to another host.

A persistently infected fish will therefore be a carrier that can spread the virus for a long time and over long distances. In addition, the specific immune response raised by the immune system of the fish may prevent other variants of the same virus from coinfecting that fish, giving a selective advantage for the persistently infective virus in question. However, this virus at the same time can be beneficial for the fish, as the immune response mounted against the persistent virus may protect the fish against other, subsequently encountered variants of the same virus that might be more virulent.

8.3.2.8 Genetic Stability and Adaptation Viruses have compact genomes, often with overlapping genes. The conservation of the viral genomes through thousands of generations may be astonishingly high. However, the conservation of the genome is not necessarily due to high fidelity of replication. RNA viruses do not have proofreading enzymes to correct mistakes in the replication process, resulting in a high frequency of mutations and genetic variability of the progeny. Likewise, some DNA viruses have inherited mechanisms to create permutations and genetic variations (Jakob and Darai, 2002). The explanation of high conservation must therefore be the strong selective forces that act upon the virus during the different steps in the life cycle. Of the vast amount of virus particles that are released from infected individuals, those with most similar genetics to the parents are most fitted to go through all the necessary steps of the life cycle if the living conditions are constant. This means, however, that if their life conditions are altered, viruses have a great capability for genetic changes. Especially one should be aware of this in aquaculture where the living conditions for the fish may be altered considerably as compared to natural conditions regarding parameters such as feed, stress, density, and temperature. The immune response of the host will also give a selective pressure for genetic changes, especially with respect to genes encoding the surface determinants of the virus. Such changes have taken place that can be seen by comparing different strains of the betanodavirus. The highest genetic variation is within the region of the capsid gene that has been suggested to encode the surface determinants (Nishizawa et al., 1999). This means that in a fish population that is persistently infected, there will be a selection for new variants of the virus that are not recognized by the immunity induced by their relative strains. The same affect will be achieved by vaccination of the fish.

In addition to genetic changes due to mutations, there may also be genetic interchanges of genome segments between different virus strains if they happen to infect the same individual, as is well known from influenza virus infecting mammalian hosts. Likewise, we might expect that similar interchanges might occur with ISAV, since it is an orthomyxovirus. Novel forms have been constructed in the laboratory by reassortment of the two RNA segments constituting the genome of betanodavirus. This was achieved by isolating RNA from two different strains of the virus, followed by cloning of the cDNA into a plasmid vector behind the T7-promoter, allowing *in vitro* synthesis of the RNA segments. By combining the RNA1 segment from one of the strains with the RNA2 segment from the other (and vice versa), followed by transfection into a cell line, functional novel virus particles were created (Iwamoto et al., 2001, 2004). Genetic analysis of the birnavirus IPNV from wild fish indicates that genetic interchanges have taken place in nature (Romero-Brey et al., 2009).

Genetic adaptation of viruses is also a plausible explanation for the reason why IPNV has been reported to have become commonly associated with larger fish in the seawater phase during the past 20 years, while it was originally a virus associated with disease of salmon fry and fingerlings in the freshwater phase (Smail et al., 1992; Roberts and Pearson, 2005). Similar observations have also been reported for the togavirus SAV (McLoughlin and Graham, 2007).

It is not unreasonable to assume that novel viruses will emerge due to mutations or interchange of viral and host genetic material. It can be that it is only mutations which are affecting one of the key steps in the viral life cycle that prevents a particular virus from being able to infect and replicate in a new given host. As an example, the Flock House virus (FHV), a member of the nodavirus family that infects insects, can multiply inside cells of yeast (*Saccharomyces cerevisiae*), but is not able to be transferred between yeast cells in a culture, either because the virus particles are not able to leave the cells or due to lack of receptors to enter new cells (Price et al., 2002). High density of hosts, and especially mixtures of host species living close together, may favor new genetic constitutions and emergence of viruses that may be able to infect novel host species.

8.4 THE IMPACT OF ENVIRONMENTAL FACTORS

Environmental factors such as temperature, salinity, oxygen tension, pollutions, photoperiods, and various kind of stress can influence the interaction between viruses and fish. Either the virulence of the virus or the susceptibility of the fish can be influenced. As discussed previously, different factors may influence the stability of virus particles in the environment and the presence of vehicles and vectors may be involved in the transmission of the viruses. It is likely that presence of either bacteria or ectoparasites may contribute indirectly to viral infections by causing lesions of skin or mucosal layers.

Temperature is a critical factor in all biology as it determines the kinetics of biological processes. In more extreme cases, it can influence the folding of proteins so much that enzyme activity is completely ceased or structural proteins are not able to assemble or otherwise function properly. Temperature will therefore be more of a challenge for viruses infecting ectothermic hosts like fish as compared to viruses infecting endothermic species. Even the simplest virus will, as a minimum, contain genes encoding a structural protein for building up the capsid and a polymerase responsible for replication of the viral genome. As polymerases have enzymatic activity they presumably will be more temperature influenced than the structural capsid proteins. One should therefore expect that viruses through evolution have adapted to different climatic zones or to the environmental temperatures of the habitat of their hosts. This seems to be the case for betanodavirus, where one can find some genotypes that are preferentially found in warm water fish while other are mostly found in cold water species (Nishizawa et al., 1995).

Temperature will heavily influence the immune system of the fish. In general, the adaptive immune system seems to be more temperature sensitive than the innate immune system (Magnadóttir, 2006). Part of the explanation can be that the activities of naïve T_H- and T_C-cells have been found to be downregulated at lower temperatures, which consequently will impair both humoral and cellular adaptive immunity. There are indications that memory T-cells are less temperature sensitive than naïve T-cells, as carp immunized and maintained at its optimal temperature for a period before transferred to a lower suboptimal temperature were able to mount an antibody response. Downregulation of the immune system also seems to be especially sensitive to rapid switches in temperature, as the immune response increases after an acclimation period (Bly and Clem, 1991).

The photoperiod also influences the immune response of fish, probably due to its effect on hormones such as testosterone, estradiol, and cortisol. Together with the effect of temperature this may explain the seasonal variations of fish immune parameters. The numbers of leukocytes in carp were found to be lower in winter and spring compared to summer and autumn

(Collazos et al., 1998), and the alternative pathway of the complement system to have a higher activity during the winter (Collazos et al., 1994). In goldfish (*Crassius auratur*) and rainbow trout (*Oncorhynchus mykiss*), the levels of IgM were found to decrease in winter season (Suzuki et al., 1996; Sánchez et al., 1993). Changes in hormone levels is probably the reason for the immune suppression that is observed in connection with sexual maturation. Reduction of lymphocytes have been reported in sexually mature brown trout (*Salmo trutta*), which is believed to be due to heightened level of cortisol. Furthermore, experimental data show that testosterone has an immune suppressing effect by reducing antibody production in chinook salmon (*Oncorhynchus tshawytscha*) (Slater et al., 1995). Likewise, increased cortisol level may also explain the immune suppression that is observed when fish encounter various stress factors.

The impact of the environmental factors is in accordance with the seasonal fluctuations of many infectious diseases affecting fish. Again, this can be explained either by the increased presence or virulence of the pathogen, or increased susceptibility of the fish.

8.5 IMPACT OF VIRUS FOR WILD FISH POPULATIONS

Surveillances have revealed that viruses are present in wild fish populations (Skall et al., 2005; Ruane et al., 2009; Gomez et al., 2009; Gagné et al., 2004; Hershberger et al., 2007). Most of the studies have been on randomly collected fish, where most of the infected individuals appeared healthy, indicating that the findings have been persistently infected fish. However, there have been observations of viral infections with great impact on wild fish populations. For example, rhabdovirus VHSV epizootics with high mortalities occurred in populations of Pacific herrings at the coast of British Colombia in 1993 and the coast of Alaska in 1994 (Traxler and Kieser, 1994; Meyers and Winton, 1995). Likewise, VHSV that recently introduced to the Great Lakes Basin in North America caused high mortality of many freshwater fish species (Gagné et al., 2007). Furthermore, great epizootics due to herpesvirus took place in the ocean between Australia and New Zealand during the 1990s, causing mortalities estimated to be as high as 75% in wild pilchard (*Sardinops sagax neopilchardus*) (Whittington et al., 2008).

The great impact of virus infection observed on pilchard and herrings can be explained by the fact that these are species living in vast schools. For fish populations with lower density, the mortalities most probably will not be so dramatic. However, it is not unreasonable to believe that viral infections will influence and have a regulatory effect on wild fish populations. The impact may not only be limited to species infected by the virus but also have further implications on, for example, predators higher up in the food chain of the ecosystem.

8.6 THE IMPACT OF VIRAL DISEASES FOR FISH FARMING

Infectious diseases have been one of the greatest obstacles of aquaculture, and control measurements of pathogens have been given high priority. Some of the bacterial diseases have successfully been brought under control. For viral diseases, the situation is far more problematic. One reason is that, unlike bacterial infections that can be treated by antibiotics, viral infections are not treated curatively. Appropriate antiviral medications are neither available, would be too expensive, nor allowed to be applied. The other reason is, as will be discussed later, that vaccines against viruses are difficult to develop. To date, the main control measures utilized have been stamping out the affected farms, and restricting transfer of either eggs, larvae, or fish between geographic areas. Broad host range is common among fish virus, and spread of viral disease to new aquaculture species have been reported for many viruses. An example is the rhabdovirus VHSV. Clinical signs caused by this virus were first registered in rainbow trout in Germany in 1938, and the disease was later also seen in other European countries. During the last decade (1990s and onwards), the disease had spread to the North Atlantic area, and also to the Asian pacific area. Not only has it spread to several parts of the world but the disease outbreaks have also been observed in several distinct species such as Atlantic salmon, rainbow trout (*O. mykiss*), striped bass (*Morone saxatilis*), stickleback (*Gasterosteus aculeatus*), Pacific herring (*Clupea pallasii*), Pacific hake (*Merluccius productus*), Walleye Pollock (*Theragra chalcogramma*), and Japanese flounder (*Paralichthys olivaceus*). Similarly, betanodavirus has been observed in several continents with a host range of around 30 species reported to date. So far, betanodavirus has affected mainly aquaculture of marine fish. However, recently a disease outbreak due to betanodavirus was reported on grouper reared in freshwater. It seems like different strains of betanodavirus are rather temperature dependent, where some are restricted to cold water environments, while others are restricted to warm water environments. Yet another virus, the orthomyxovirus ISAV was reported as a novel virus in 1984, causing problems in farming of Atlantic salmon in a restricted geographic area in Norway. However, around 1998, the virus was reported to affect fish farms in Scotland, the Faroe Islands, and United States; and in 2001 the virus was also isolated from Coho salmon (*Oncorhynchus kisutch*) in Chile.

Fortunately, most viruses affect the fish at early stage of life, which limits the economical burdens for the farmers. But there are exceptions. The togavirus SAV has in the past few years been causing disease outbreaks in Atlantic salmon during the second year at sea (McLoughlin and Graham, 2007). Mortalities at that stage endorse the farmers with high economical losses, as all the money invested in form of labor, feed, and other expenses is lost. Infectious diseases do not always cause high mortalities. However, the economical impact can still be considerable as the market quality of the fish can be greatly reduced. An example is the pathology caused by Walleye dermal sarcoma virus (WDSV), a retrovirus distantly related to other members of the family Retroviridae. The disease produced by WDSV is usually not fatal for the fish, but causes tumors, reducing the market value of the fish.

Most probably, diseases caused by viruses will be a great problem for fish farming for an indefinite period of time. The high density of fish may cause selection of more virulent virus strains, and close contact between different fish species may cause viruses to adapt to novel hosts. And as will be discussed below, the promises to develop efficient viral vaccines are not obvious.

8.7 VACCINES AND VACCINATION

Vaccination for protection against infectious diseases has become a prerequisite for large-scaled industrial fish farming. For example, every single smolt of salmon in Norway is given an injection to protect against bacterial diseases such as vibriosis and furunculosis before they are moved from the hatcheries to the sea pens. These vaccines are made by growing up the bacterial pathogen, killing it with formalin, and adding some kinds of adjuvants. Adjuvants are compounds that increase immunogenicity of antigens by i.g. inducing an inflammation or making a depot for slow release of the antigens. Several bacteria are normally combined in one vaccine in order to avoid more than one injection. However, most of the vaccines that are currently in use in aquaculture are against bacteria that prevail and grow extracellularly. For intracellular pathogens, including viral agents, the development of vaccines with an adequate level of protection has not been straightforward. As described under Section 8.2.3.2, the reason is that killed microorganisms mainly induce humoral immunity, while high protection against intracellular pathogens demands induction of both humoral and cellular immune

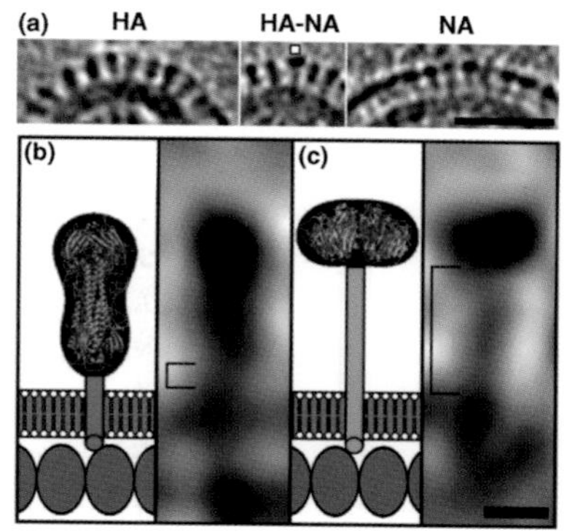

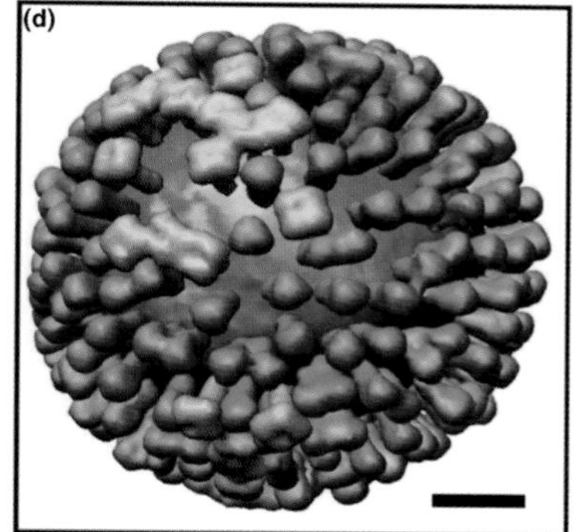

FIGURE 3.3 Model virus with HA and NA spikes by cryo-ET analysis. (*See text for full caption.*)

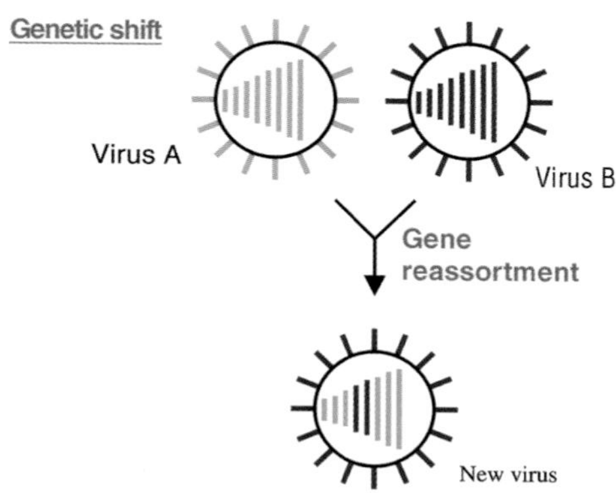

FIGURE 3.10 Reassortment of influenza virus RNA segments. (*See text for full caption.*)

Influenza virus

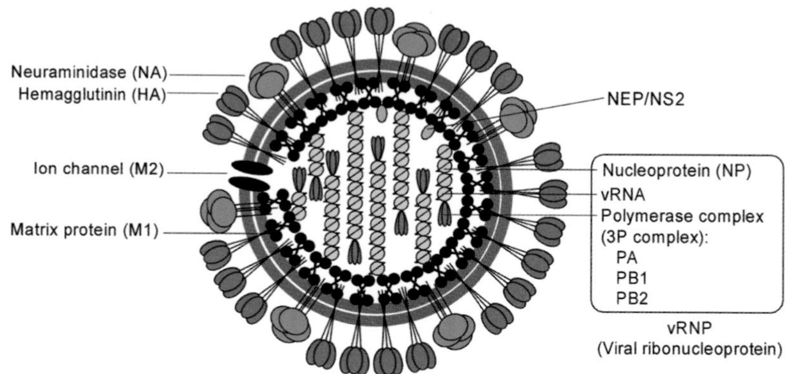

Family: Orthomyxoviridae

Genome: (–)ssRNA

Subtype: A (8 RNA segments) e.g., WSN
B (8 RNA segments)
C (7 RNA segments)

(a)

FIGURE 3.11 Schematic presentation of the infectious cycle of an influenza virus. (a) Schematic presentation of influenza virus structure.

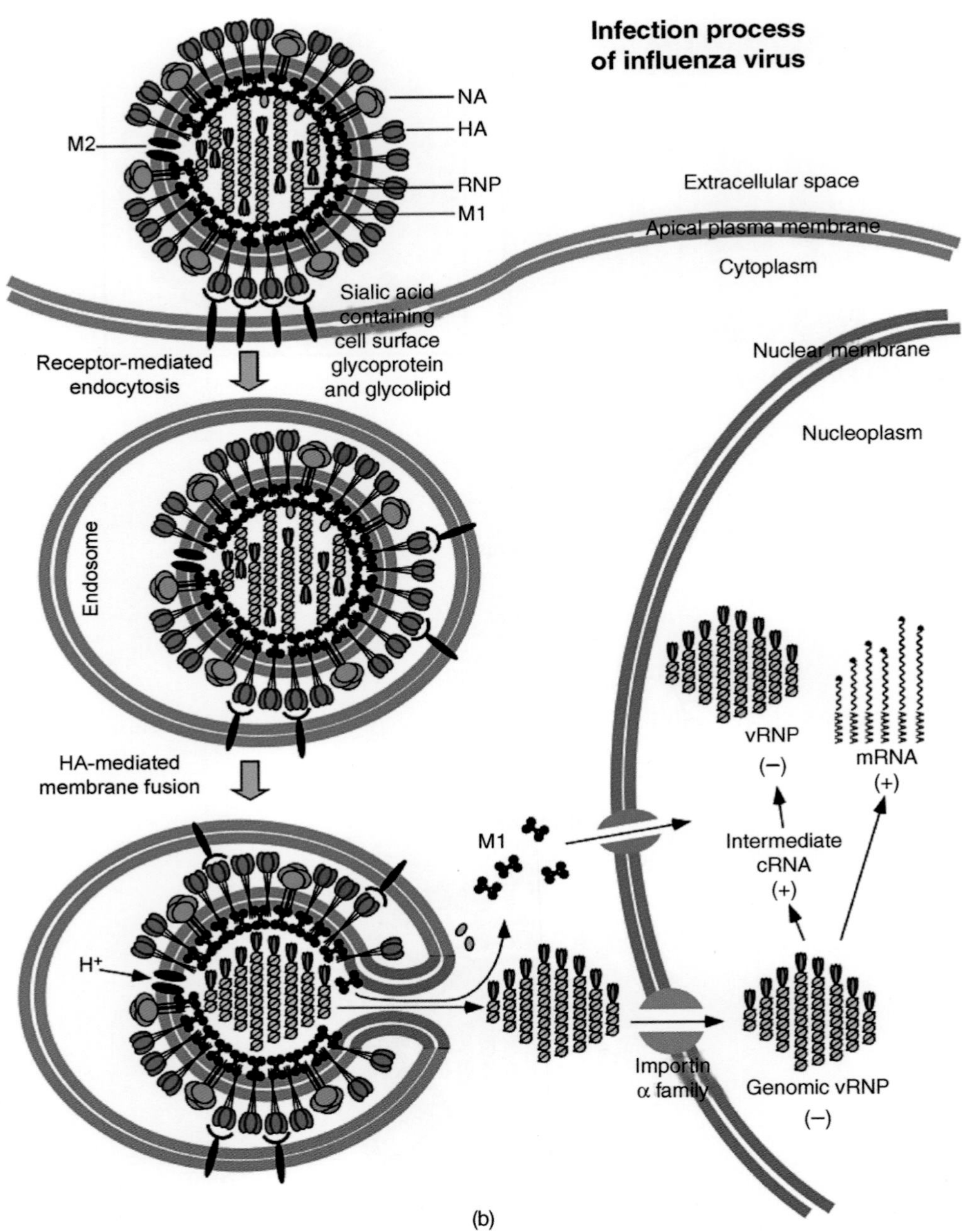

FIGURE 3.11 (*Continued*) (b) Schematic presentation of influenza virus infection showing attachment, entry, and uncoating of a virus particle. The steps in the replication cycle are attachment mediated through HA and sialic acid receptor, entry into the cell via endosome, HA-mediated fusion of virus membrane with endosomal membrane at low pH, release of vRNP, transport of vRNP into the nucleus, and transcription (mRNA synthesis) and replication (cRNA and vRNA synthesis) of vRNP in the nucleus.

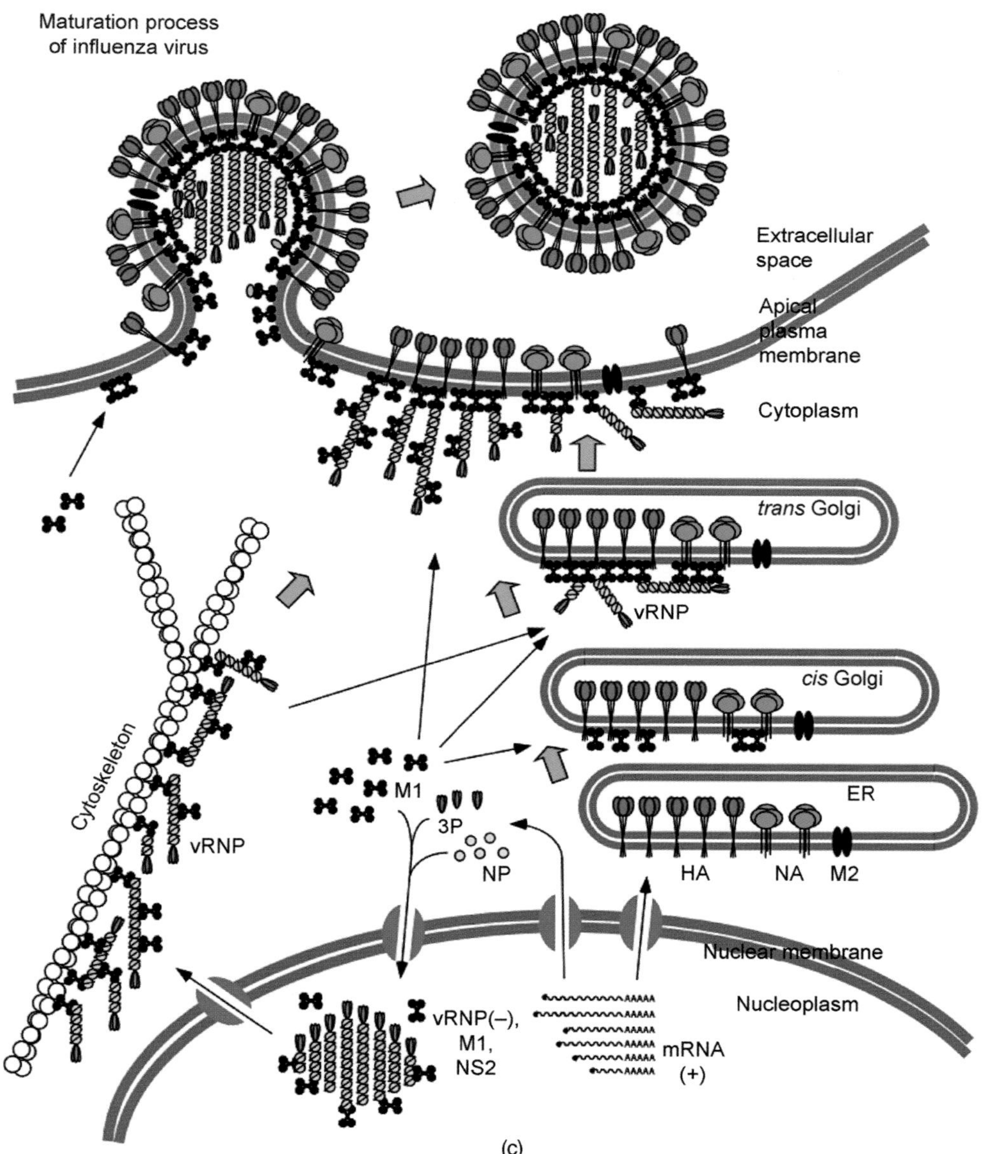

FIGURE 3.11 (*Continued*) (c) Schematic presentation of influenza virus infectious cycle showing export, assembly, and budding of a virus particle. The steps include export of vRNP from nucleus into cytoplasm, export of virus proteins, vRNP to the budding site, bud formation, and bud release by fusion and fission viral and cellular membranes.

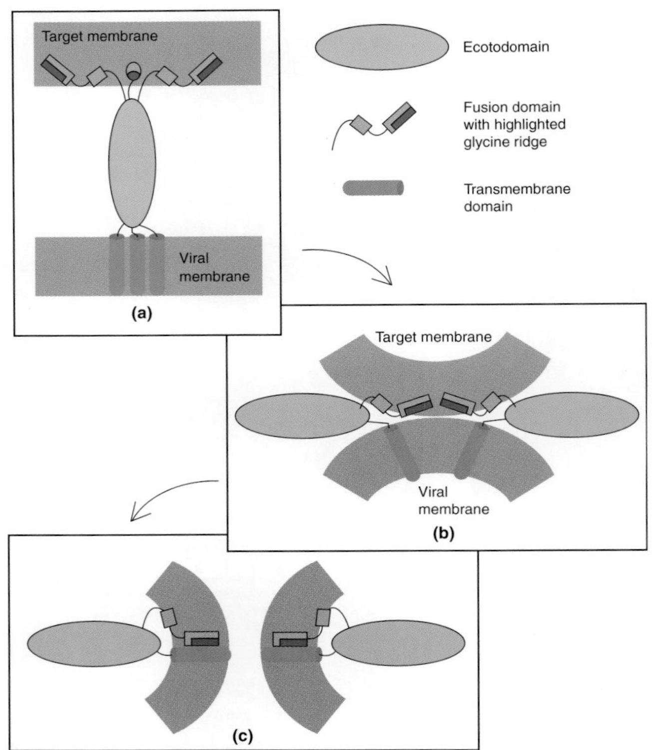

FIGURE 3.14 Boomerang model of influenza virus HA-mediated membrane fusion. (*See text for full caption.*)

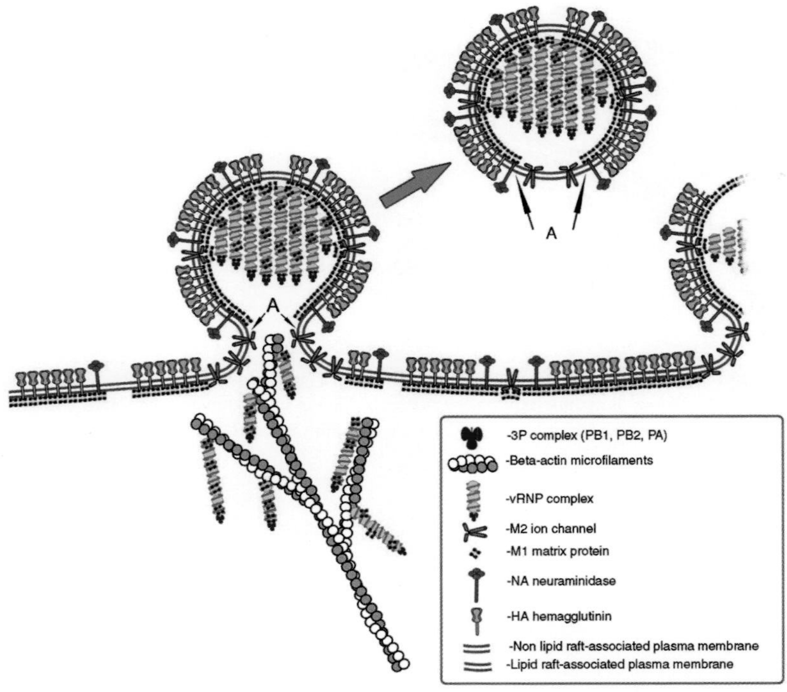

FIGURE 3.22 Schematic illustration of the pinching-off process of influenza virus bud. (*See text for full caption.*)

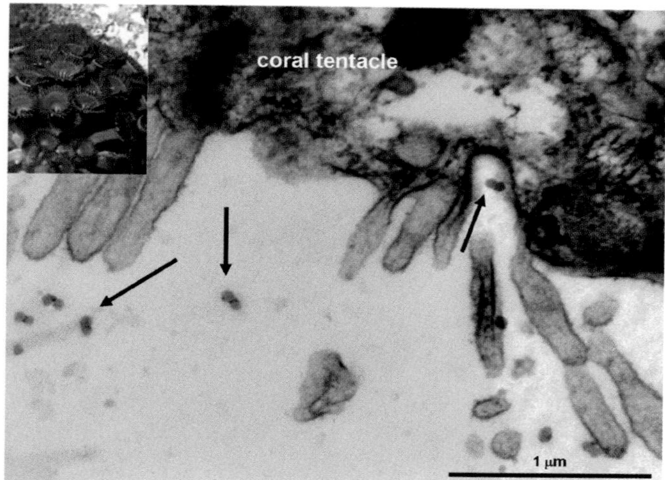

FIGURE 5.1 *Zoanthus* sp. (inset) virus-like particles (VLPs) adjacent to tentacles of a zoanthid after thermal shock. VLPs highlighted by arrows. Scale bar 1 μm.

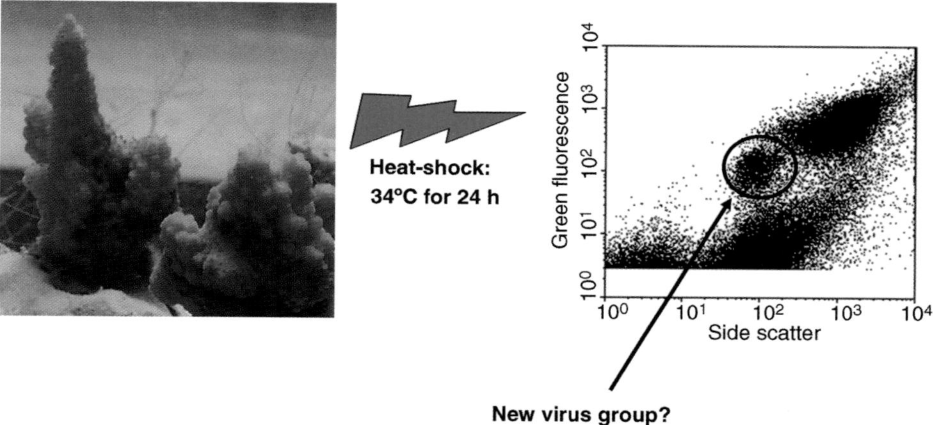

FIGURE 5.2 Flow cytometric analysis of seawater surrounding a nubbin of *Acropora formosa* following heat-shock. The virus group arrowed appeared after a 24 h heat-shock at 34 °C.

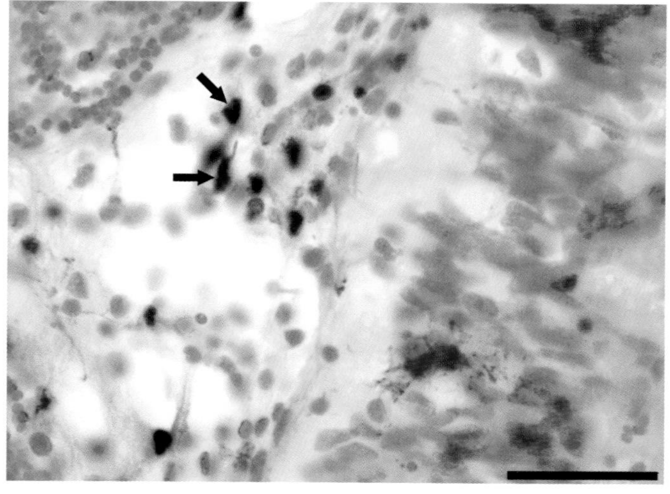

FIGURE 6.8 HIS stained cells (blue color, arrows) in the connective tissue of the digestive gland from an infected Pacific oyster collected in France using an OsHV-1-specific probe. Scale bar = 20 μm.

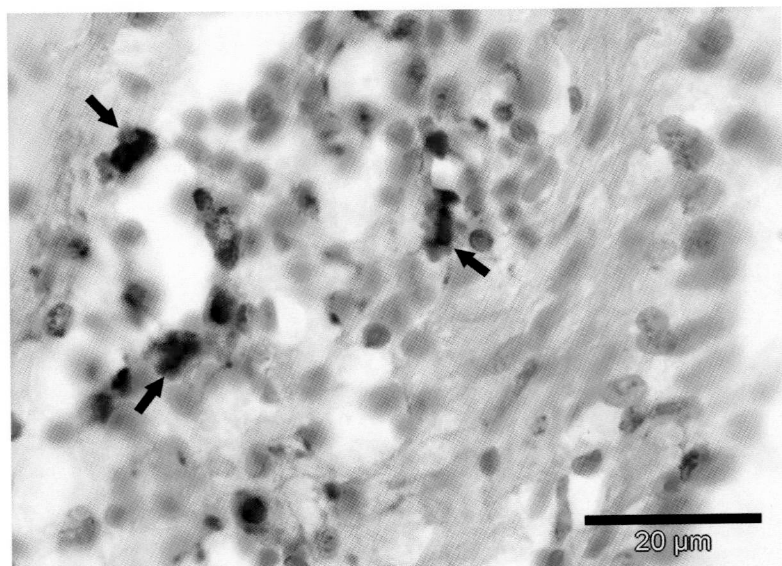

FIGURE 6.9 HIS stained cells (blue color, arrows) in the connective tissue of the digestive gland from an infected abalone collected in Taiwan using an OsHV-1-specific probe. Scale bar = 20 μm.

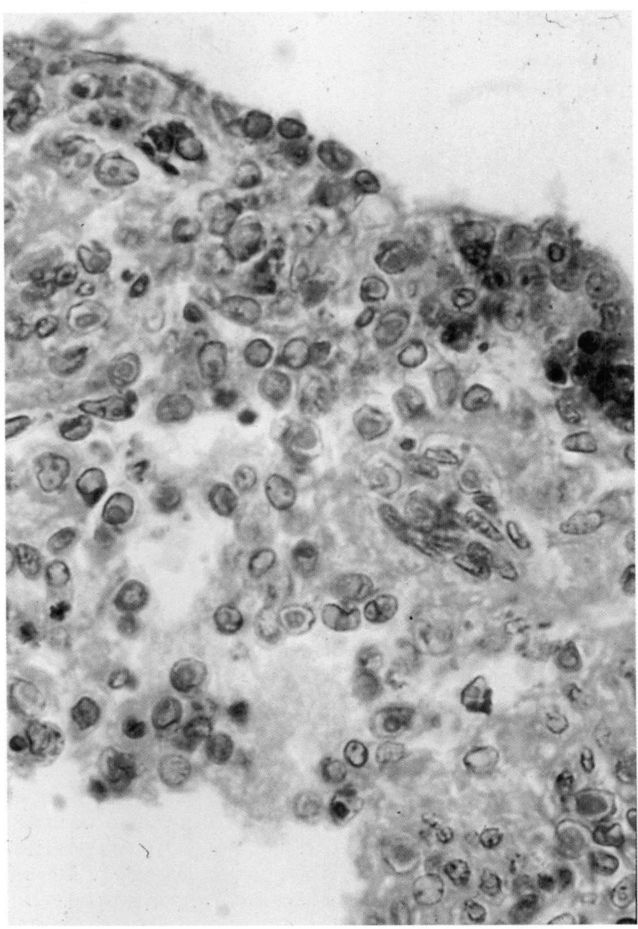

FIGURE 7.3 Infectious hypodermal and hematopoietic necrosis virus (IHHNV, *P. stylirostris* densovirus) infecting the lymphoid organ of a hybrid *P. monodon* crossed with *P. esculentus*. Note almost every cell has an eosinophilic Cowdrey A intranuclear inclusion body.

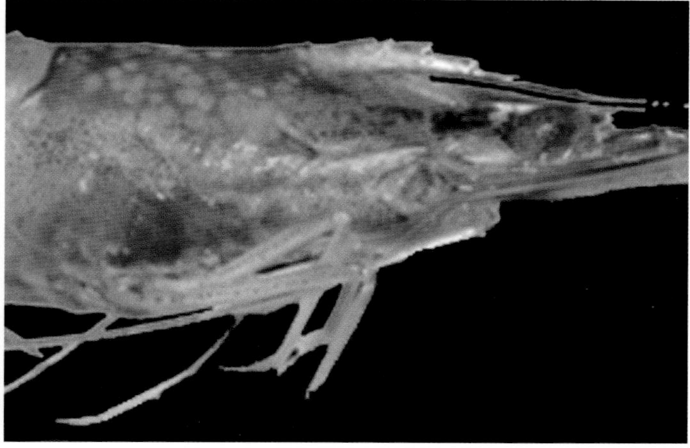

FIGURE 7.4 Clinical signs of white spot syndrome virus in *P. merguiensis*. Note the white lesions on the rear of the cephalothorax. Photograph by K. Claydon and L. Owens.

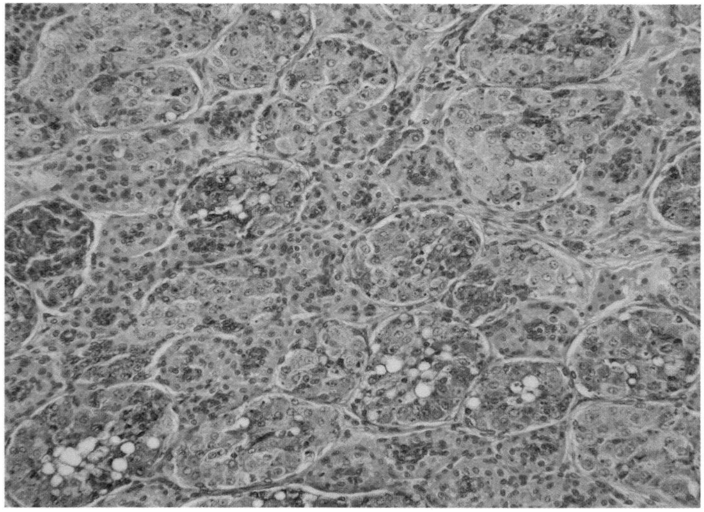

FIGURE 7.5 Lymphoid spheroids in the lymphoid organ of *P. merguiensis*. The round, more basophilic sections are the spheroids made up of spent hemocytes and the more eosinophilic tissue with the central hemolymph vessels are the normal stromal matrix areas of the lymphoid organ.

FIGURE 7.6 *P. monodon* infected with gill-associated virus. Note the yellow gills similar to symptoms of prawns infected with the conspecific yellow head virus.

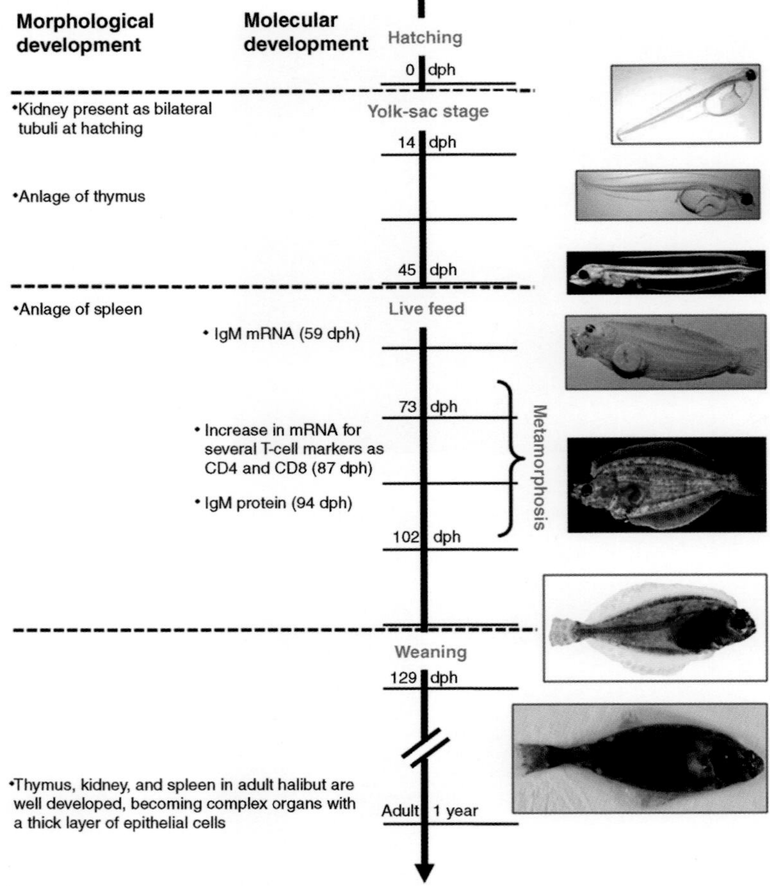

FIGURE 8.2 The ontogeny of the immune system of Atlantic halibut (see text).

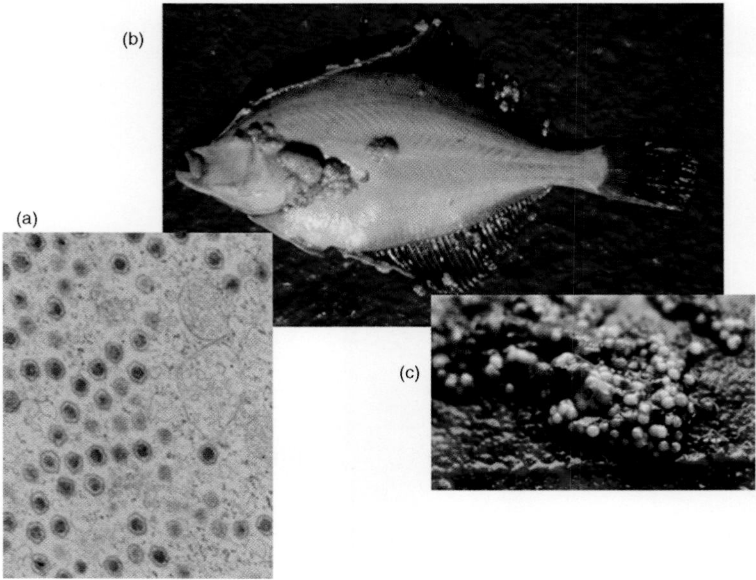

FIGURE 8.3 (a) Electron microscopic pictures of lymphocystis disease virus 1 (LCDV-1). (b) Picture of LCDV-1-affected flounder. (c) Close-up picture of the skin tumors. Photo: Dr. Tore Håstein, National Veterinary Institute, Oslo, Norway.

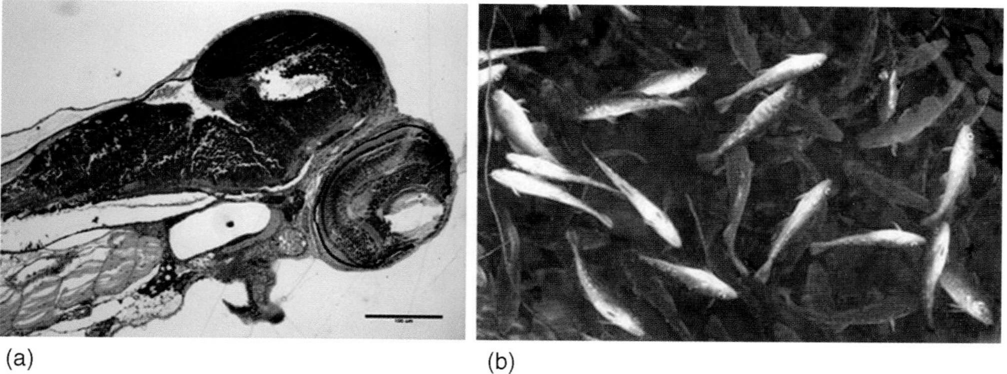

FIGURE 8.4 (a) Immunohistochemical (IHC) staining of paraffin wax section of Atlantic halibut larva infected by betanodavirus. Positive staining is seen as red while the blue is background that stains the tissue. Note the prominent immunolabeling in both eye and brain of the larva. (b) Farmed Atlantic cod (size 10–20 g) affected by betanodavirus. Typically the affected individuals are disorientated and swim in spiral with belly side up.

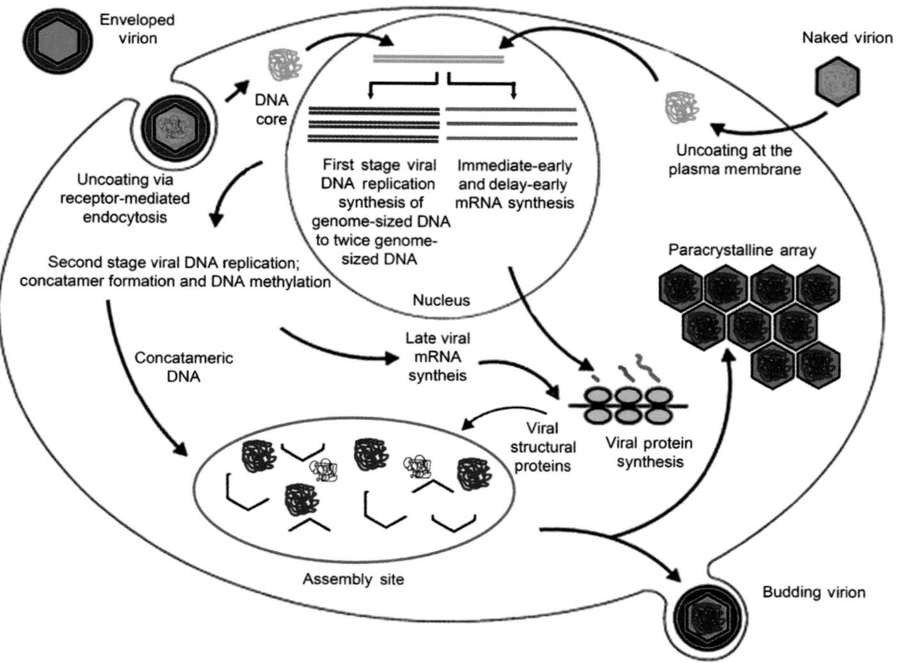

FIGURE 9.1 Life cycle of FV3. A schematic outline of events in FV3-infected cells is shown. See text for details. (Republished with permission from Springer, Chinchar, 2002.)

FIGURE 10.1 Snapshot of a comb within a husbanded honeybee colony. Arrow: honeybee with deformed wings. (Image courtesy of S. J. Martin.)

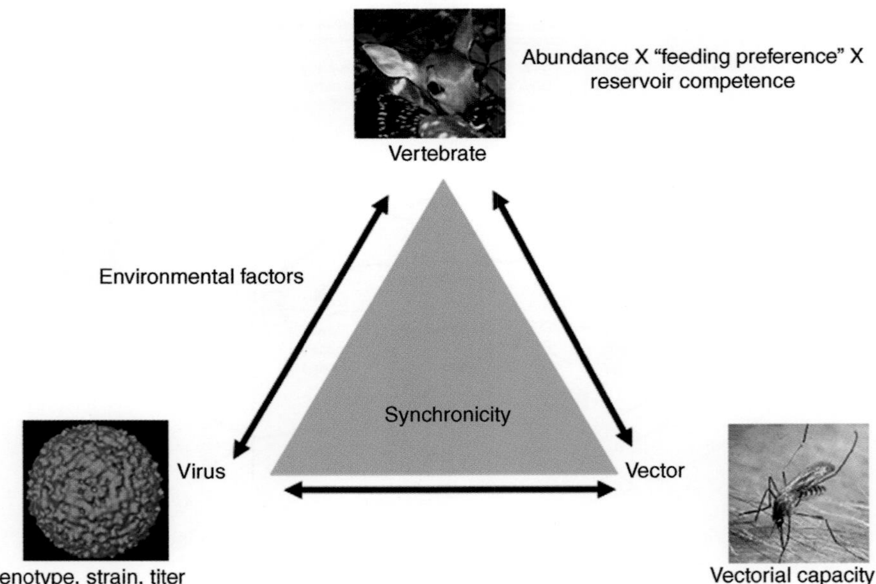

FIGURE 11.1 Factors affecting transmission and establishment of arboviruses.

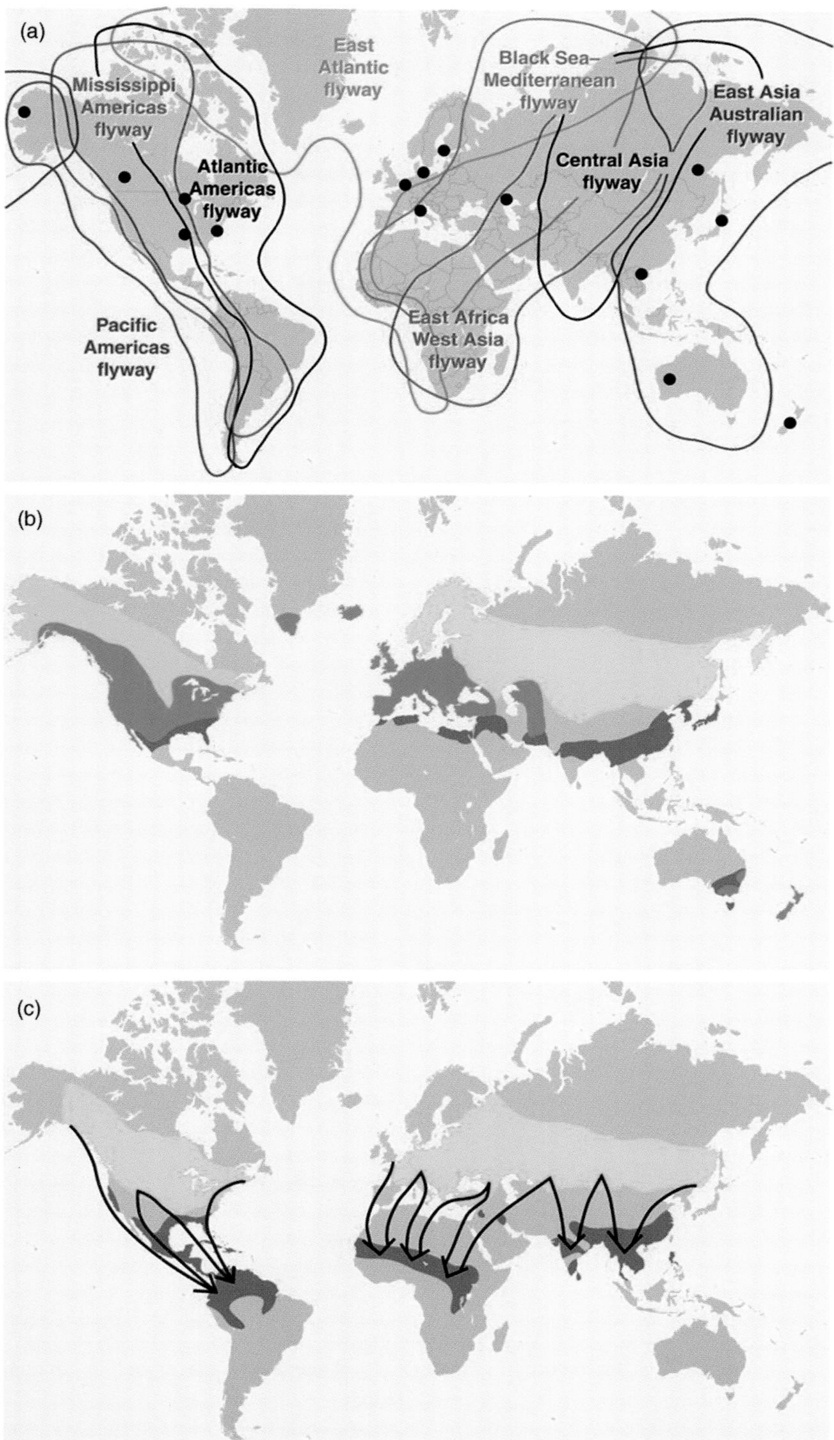

FIGURE 14.1 Migratory flyways of wild bird populations. (*See text for full caption.*)

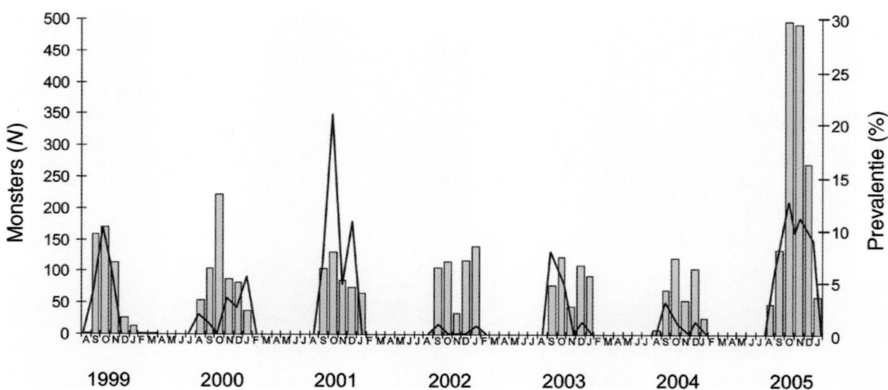

FIGURE 14.7 Annual influenza A virus prevalence in mallards during fall migration in The Netherlands from 1999 to 2005. Bars indicate the number of samples collected per month (left *y*-axis), and the red line indicates the number of samples positive for influenza A virus by RT-PCR (right *y*-axis). Clear seasonal patterns are observed, with peak prevalence occurring in September and October. Large variation in avian influenza prevalence between different years is observed. (Reproduced from Munster et al., 2007 with permission.)

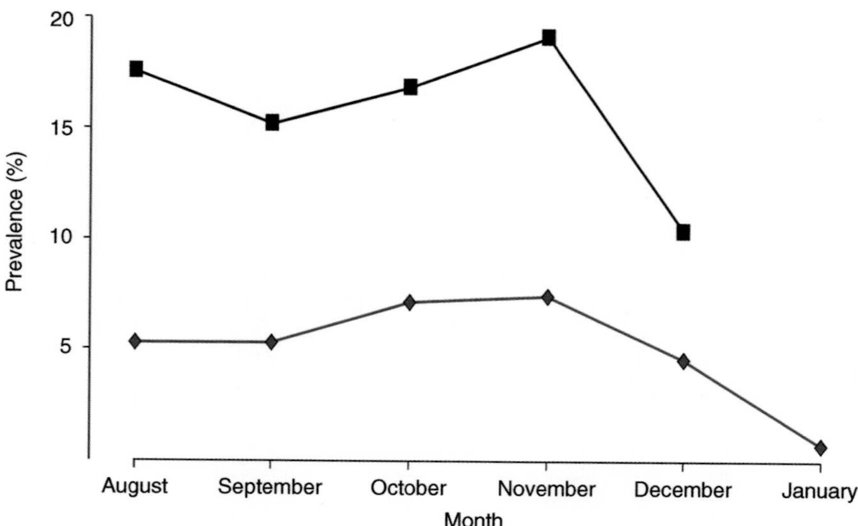

FIGURE 14.8 Trend lines for influenza A virus prevalence in Mallards caught in Sweden and The Netherlands during fall migration. The blue line and filled squares (■) represent the proportion (%) of influenza A virus positive mallards caught and sampled in Sweden between 2002 and 2005 at Ottenby bird Observatory and the red line and filled diamonds (◆) represent mallards caught at various locations in The Netherlands. (Reproduced from Munster et al., 2007 with permission.)

responses. The induction of humoral immunity might, however, give a fairly high protection against some viral diseases. This depends on the life cycle of the virus and to what extent the virus particles are accessible for antibodies at the various stages.

The ideal vaccine for the aquaculture should be cost effective for the farmer, easy to administer without any stress or side effects for the fish, and give protection during the entire production cycle. Furthermore, the vaccine should not be of any danger for either the consumers or the environment. However, it has turned out to be difficult to fulfill all these criteria at the same time, and it is therefore often necessary to compromise.

The vaccines against viral diseases that are currently in use are either inactivated whole virus or recombinant vaccines. The former are produced as described above for the bacterial vaccines, but with the difference that the virus particles are propagated on cell lines *in vitro*, which is rather expensive. Recombinant vaccines are made by cloning genes encoding protein antigens of the pathogen and express the recombinant protein in large quantities in a suitable system as bacteria, yeast, or insect cells. Adjuvants are then added to the recombinant proteins to formulate the vaccine. Such vaccines have much of the same properties as vaccines based on killed inactivated microorganisms, but for virus vaccines the production cost will be far less. Several recombinant subunit vaccines have been developed and tested to give some protection against viral diseases (Sommerset et al., 2005; Lecocq-Xhonneux et al., 1994; Christie, 1997), but so far only a vaccine against IPNV is commercially available.

In human medicine, many of the most successful vaccines consist of live attenuated microorganisms (e.g., vaccines against polio, measles, mumps). These vaccines are developed by manipulating the microorganisms in such a way that the pathogenic properties are lost while the infectivity and the antigenicity still remain. The microorganisms will propagate in the body and induce both humoral and cellular immunity. However, there are concerns about using such vaccine in the aquatic environment, as there is a certain probability that the microorganism way revert to a pathogenic form by either mutation or recombination with other viruses in the environment. In addition, as we are talking about live viruses and maybe genetic modifications, the vaccine producing company has to take into account the opinion of the consumers.

DNA vaccines are made by cloning genes encoding antigens into a plasmid vector behind a strong eukaryotic promoter. By injection of the plasmid into tissues there will be a local, transient expression of the antigen, which will be processed in a similar manner as during an infection, and thereby inducing both cellular and humoral immunity (Utke et al., 2008).

The first DNA vaccines that were tested on fish were against the rhabdoviruses VHSV and IHNV, and both vaccines were based on the viral glycoproteins that normally are located on the surface of the virus particles. The results were surprising. High protection was achieved only a few days after vaccination and it lasted at least for 1 year. Later studies have shown that the early protection was due to stimulation of the innate immune system, as the vaccines also give temporary short-term protection against unrelated viruses, whereas long-lasting adaptive immunity is rather effective after 3–4 weeks (Lorenzen et al., 2001; LaPatra et al., 2001; Kurath, 2005; Lorenzen and LaPatra, 2005; Sommerset et al., 2003).

DNA vaccines are usually given as intramuscular injections as muscle cells have turned out to be suited for expression of the antigen. A disadvantage is therefore that DNA vaccines are not compatible for administration together with the multivalent vaccines currently in use, which are given by intraperitonal injections.

Live vector vaccines are made by cloning the genes encoding antigens in such a way that it will be expressed by an apathogenic microorganism (preferentially on the surface), which can be applied to the fish either orally, by immersion, or injection. If this is a microorganism having an intracellular stage, both an humoral and a cellular immune response against the antigen will be induced. Also, it is possible to use live vectors to deliver DNA into tissues and cells where the genes express antigens as described for DNA vaccines above. The live vector may be bacteria as well as viruses. However, there are similar safety concerns about this kind of vaccines just as there are for the live attenuated virus vaccines.

The time of vaccination may be critical. It takes some time from hatching before the larvae are immunocompetent, which means that the immune system is developed to an adequate stage such that stimulation will induce a protective response. If the immune system is stimulated before this, tolerance instead of protection may be induced. Given this constraint, the vaccination has to take place as soon as possible after hatching to prevent infections and diseases. The time from hatching to competence may vary between the different fish species, and it depends on parameters such as nutrition and temperature (Pylkkö et al., 2002). Salmonids hatch in freshwater that contains rather few microorganisms and the larvae are rather developed when leaving the eggs. Consequently, diseases at the larval stage of salmonid fish have been relatively unproblematic. The farming of marine species is different as the larvae hatch at a far earlier stage and in seawater, which is more or less a soup of microorganisms due to the high contents of salts and organic material. Knowledge about the ontogeny of the immune system of the fish species in question is therefore crucial in order to make vaccination regimes that induce protection and not tolerance.

Another problem is that there is an interval following the time of vaccination until the adaptive immune system gives protection. This time period is temperature dependent and may in addition vary between different fish species. For cold water species like salmon and halibut it takes 4–6 weeks. As the fish usually are stressed by the vaccination process, this is a critical period for the fish where it is vulnerable for infections. An idea would be to add components to the vaccines that strongly stimulate the innate system including expression of interferons, which will give immediate protection against several viral diseases. Such a protection normally lasts for 5–8 weeks.

An indirect way to protect larvae at an early stage of life would be to vaccinate the broodfish to reduce vertical transmission. A recent publication indicates that vaccination of grouper before spawning will reduce the risk for transmission of virus to the offspring (Kai et al., 2010).

8.8 SELECTED VIRUS SPECIES FROM THE VARIOUS BALTIMORE GROUPS

8.8.1 Baltimore Group I: Double-Stranded DNA Viruses (Example: LCDV-1)

Virus Architecture: LCDV-1 belongs to the family Iridoviridae and the genus *Lymphocystisvirus*. The genome of LCDV-1 consists of a linear dsDNA molecule, 102,653 bp long and containing 195 open reading frames (ORFs), many of which are orthologues of cellular genes (Tidona and Darai, 1997, 2000). The genome is circularly permuted and terminally redundant (Schnitzler et al., 1987). The capsid is icosahedral, double-layered, around 200 nm in diameter (Wolf, 1988). The replication of the virus takes place both in the nucleus and in the cytoplasm of the cells (Sheng and Zhan, 2004). The viral particles are released from the host cells by budding and will then be surrounded by a membrane. However, progeny viral particles can also be observed in paracrystalline arrays in the cells, where they do not have any membranes (Sheng et al., 2007). Not much has been published about the stability of the LCDV-1 virus, but other members of the Iridoviridae family have been reported to be extremely stable in the environment (Langdon, 1989).

Host Range and Geographic Distribution: LCDV-1 has been reported from more than 100 different fish species in seawater and freshwater worldwide (Tidona and Darai, 1999).

Life Cycle: LCDV-1 seems mainly to be transmitted by direct contact with infected individuals. However, it has been speculated that the virus might be able to persist in vectors like *Artemia* (Cano et al., 2009). Persistent infected fish that probably act as carriers have also been described (Cano et al., 2006; Hossain et al., 2009).

Interaction with the Host Antiviral System: It has been demonstrated that LCDV-1 induces apoptosis of fish cell lines *in vitro*. However, the induction of apoptosis is limited *in vivo*. It has therefore been suggested that the virus somehow is able to prevent apoptosis *in vivo* (Hu et al., 2004). One of the ORFs in the LCDV-1 genome is encoding a molecule resembling the tumor necrosis factor receptor (TNFR), which has been shown to be involved in regulation of apoptosis by poxviruses (Essbauer et al., 2004).

Virulence and Pathogenicity: LCDV-1 infects dermal fibroblasts causing chronic benign tumors on skin and fins (Figure 8.3). The infected cells are hypertrophic and the individual cell is encapsulated by a hyaline extracellular matrix. Inside the cells there are usually massive accumulations of viral particles (Walker and Weissenberg, 1965). Sometimes the

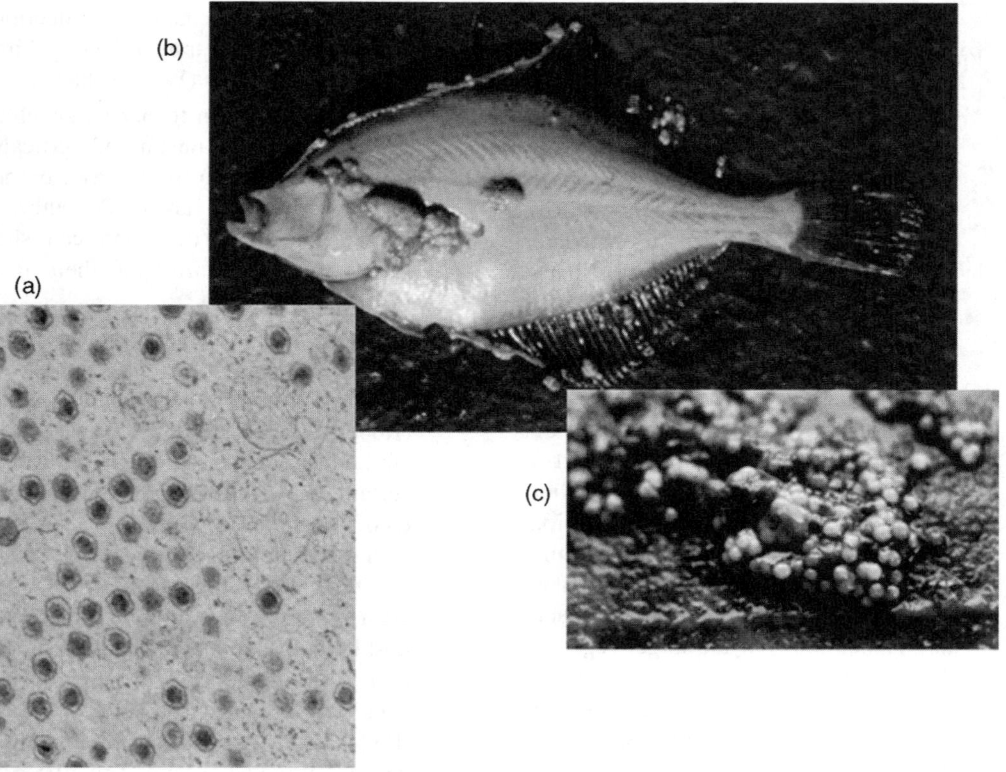

FIGURE 8.3 (a) Electron microscopic pictures of lymphocystis disease virus 1 (LCDV-1). (b) Picture of LCDV-1-affected flounder. (c) Close-up picture of the skin tumors. Photo: Dr. Tore Håstein, National Veterinary Institute, Oslo, Norway. (*See the color version of this figure in Color Plates section.*)

fish develop true epithelial tumors (Samalecos, 1986). Although the mortality is low, infected fish exhibit anemia, reduced growth, and are more susceptible to infection by other microorganism (Iwamoto et al., 2002). In addition, the disfigurement of the fish makes it unusable for sale.

Vaccines: No commercial vaccines against LCDV-1 are currently available. There are reports of experimental DNA vaccines where the antigens are expressed in fish tissues, but the protection levels of the vaccines has not been described in the reports (Zheng et al., 2006; Tian et al., 2008).

8.8.2 Baltimore Group III: Double-Stranded RNA Viruses (Example: IPNV)

Virus Architecture: IPNV belongs to the family Birnaviridae. The genome of IPNV consists of two segments of double-stranded RNA. The shorter segment (~2800 nt) encodes VP1, the RNA-dependent RNA polymerase (RdRp), which replicates the genome and later becomes incorporated in the virus particles by binding to the 5′-ends of the RNA strands. The larger segment (~3100 nt) encodes a protein that is cleaved cotranslationally by an endogenous protease into three components (VP2, VP3, and VP4), two of which will be part of the viral capsid (Duncan and Dobos, 1986; Duncan et al., 1987). In addition, a fifth short polypeptide (VP5) inhibiting apoptosis of the infected cell is transcribed from a different reading frame (Essbauer and Ahne, 2001) of the larger segment. The virions are nonenveloped icosahedrons, about 60 nm in diameter. The virus has been shown to be very stable in the environment (Mortensen et al., 1998).

Host Range and Geographic Distribution: IPNV variants apparently are widespread in the marine environment, having been isolated from many species of freshwater and marine fish. IPNV has also been isolated from crustaceans and molluscs, although replication of this virus has not been shown in these organisms. The geographical distribution is worldwide; being reported from Europe, Asia, New Zealand, South Africa, North America, and South America (Wolf, 1988).

Life Cycle: The intestine may be the primary organ for virus entry and replication (Biering and Bergh, 1996). The virus is shed via faeces, sexual fluids, and probably urine, and can therefore be transmitted horizontally through the water route by ingestion of infected material or by direct contact with infected fish. Likewise, infected transport water, contaminated nets, and other equipment may act as transmission channels. Infectious virus may also be transported by birds and other predators (Wolf, 1988).

IPNV has been shown to persist subclinically and later be transmitted vertically (via fertilized eggs) from carrier broodfish. Survivors of an IPNV outbreak become IPNV carriers and can shed the virus for the rest of their lives (Wolf, 1988).

Interaction with the Host Antiviral System: The interferon system has been shown to interact and protect against IPNV (Robertsen et al., 2003). Likewise, host antibodies protect against the virus to a certain degree as vaccines consisting of recombinant VP2 have been shown to confer some protection (Christie, 1997). The double-stranded genomic RNA has been reported to exhibit resistance for host RNases, and thereby avoid degradation in the cytoplasm of the host cells (Macdonald and Yamamoto, 1977). IPNV also confers contra-interactions with the immune system. The VP5 protein has been reported to delay induction of apoptosis during the early stages of infection, and thus slow down the antiviral effects of the host immune system (Hong et al., 2002).

Virulence and Pathogenicity: The external clinical signs of infection are abdominal distension, uncoordinated swimming and trailing, pale gills, dark pigmentation, and white fecal casts. Further examination of the fish often reveals that the spleen, kidneys, liver, and heart are abnormally pale and the peritoneal cavity may contain ascitic fluid (Biering et al., 1994). The disease caused by IPNV is principally associated with salmonid fish, in which acute infections occur in 4–16 weeks old fry and may cause mortalities up to 90%. Susceptibility decreases with increasing age. Factors such as stress appear to increase susceptibility. For example, Atlantic salmon smolts will often develop the disease shortly after transfer to seawater.

Vaccines: Vaccines based upon inactivated virus as well as recombinant antigens are currently available and are applied in the farming of salmonid fish. However, the degree of protection offered by the vaccines has been difficult to evaluate, as reproducible challenge tests have not been available until quite recently (Bowden et al., 2003).

8.8.3 Baltimore Group IV: Positive-Sense, Single-Stranded RNA Virus (Example: Betanodavirus)

Virus Architecture: Betanodavirus belongs to the family Nodaviridae, which comprises small, nonenveloped viruses (25–35 nm) with a very simple architecture. Viruses with this architecture were first isolated from insects (denoted alphanodavirus) (Scherer and Hurlbut, 1967), later from fish (denoted betanodavirus) (Mori et al., 1992), and more recently from Crustacea (Bonami et al., 1997). However, the homologies of the genomes at the nucleotide or amino acid level between alphanodaviruses, betanodaviruses, and those of nodaviruses isolated from Crustacea are remarkably low. The betanodavirus genome consists of two positive-sense and single-stranded RNA segments. RNA1 (3100 nt) encodes the RNA-dependent RNA polymerase responsible for the replication of the genome. RdRp is expected to have an RNA guanylyl and methyltransferase activity for capping of the viral RNA, as the viral RNAs are capped but not polyadenylated. RNA2 (1400 nt) encodes the capsid protein that builds up the icosahedral capsid into which the genome is packaged. The viral capsid is comprised of 180 capsid protein units, arranged in 60 triangular units (Tang et al., 2002), believed to be stabilized by disulfide bonds (Krondiris and Sideris, 2002) and calcium ion-mediated interactions (Wu et al., 2008b). A third RNA segment, RNA3, generated subgenomically from RNA1 encodes a polypeptide called B2. RNA3 is present only in the infected cells and is not packaged into the viral particles (Mori et al., 1992; Nishizawa et al., 1995; Grotmol et al., 2000; Sommerset and Nerland, 2004). Betanodavirus has been shown to be very stable under extreme environmental conditions (Frerichs et al., 2000).

Host Range and Geographic Distribution: Infections caused by betanodavirus have been described in around 30 different fish species from all over the world, including both farmed and wild fish species (Munday et al., 2002). Currently, infections by betanodavirus are considered to be among the most problematic diseases regarding the rearing of marine fish species. Betanodaviruses have been categorized into species and genotypes based on the formation of clusters in phylogenetic analyses of the variable region of the viral capsid gene (Nishizawa et al., 1995, 1997; Dalla Valle et al., 2001). The genotypes have been named according to the host species from which they were first isolated (like the genus Atlantic halibut nodavirus (AHNV) and the genus striped jack nervous necrosis virus (SJNNV)). This nomenclature may be a little misleading, as recent observations indicate that the viral strains are not host species-specific, but rather tend to be geographically distributed (Gagné et al., 2004), although it seems like the different viral genotypes can be divided according to whether they cause infection of cold water versus warm water fish species.

Virulence and Pathogenicity: Betanodavirus particularly affects the larval or juvenile stages of fish, in which mortality may be very high. For example, in Atlantic halibut disease outbreaks are mostly seen during first feeding and weaning (40–100 dph), although yolk-sac stages may also be affected (Grotmol et al., 1997, 1999). However, in recent years, significant mortalities caused by betanodavirus have occurred in older fish such as sevenband grouper (*Epinephelus septemfasciatus*) (Fukuda et al., 1996), sea bass (*Dicentrarchus labrax* (L.)) (Le Breton et al., 1997), and Atlantic cod (Patel et al., 2007). The main target organ for betanodavirus infections of fish is the central nervous system, including the brain, spinal cord, and retina (Figure 8.4a), where the virus causes extensive cellular vacuolation and neuronal degeneration (Mori et al., 1992). Cell death by apoptosis and secondary necrosis is believed to cause these pathological signs (Guo et al., 2003; Chen et al., 2006), with both the capsid protein and the B2 protein suggested to be involved (Chen et al., 2006, 2007; Wu et al., 2008a; Su et al., 2009). In general, the clinical signs relate to neurological distortion, abnormalities of movement such as uncoordinated swimming, tonic spasms of myotomal musculature, lethargy, change of pigmentation, and loss of appetite (Figure 8.4b).

Life Cycle: The entry of betanodavirus into cells has been suggested to be dependent on the endocytotic pathway (Liu et al., 2005; Adachi et al., 2007). The intracellular site of replication has been shown to be in the cytoplasma, where

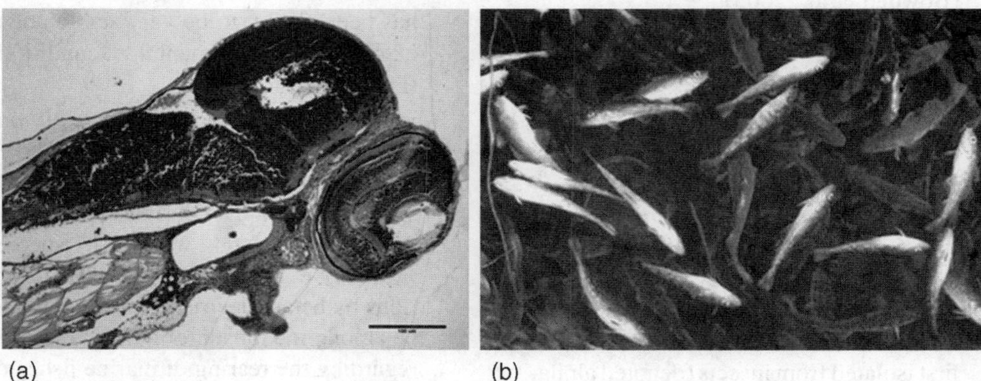

(a) (b)

FIGURE 8.4 (a) Immunohistochemical (IHC) staining of paraffin wax section of Atlantic halibut larva infected by betanodavirus. Positive staining is seen as red while the blue is background that stains the tissue. Note the prominent immunolabeling in both eye and brain of the larva. (b) Farmed Atlantic cod (size 10–20 g) affected by betanodavirus. Typically the affected individuals are disorientated and swim in spiral with belly side up. (*See the color version of this figure in Color Plates section.*)

RdRp is predicted to contain amino acid sequences signals for mitochondrial localization (Guo et al., 2004; Mézeth et al., 2007).

How betanodaviruses are transmitted is not fully understood. Recent reports have shown that the virus can persist for a long time in subclinically infected fish and still be infectious (Johansen et al., 2004). The virus has been detected in gonads and in association with ovarian fluids (Grotmol and Totland, 2000), and experimental vertical transmission has been demonstrated in sea bass (Breuil et al., 2002). It is therefore tempting to speculate that usual transmission mode is vertical, from subclinically infected parents to the egg/larvae, but obviously the transmission by this route alone is ineffective, as only a few of the offspring acquire the virus before birth. Very high concentrations of betanodavirus have been reported in rearing units, where infected halibut larvae were held, indicating that waterborne horizontal transmission may also take place (Nerland et al., 2007). Theoretically, such a combination of vertical and horizontal transmission would effectively spread the virus.

Interaction with the Host Antiviral System: An increase in cytokine genes and interferon stimulated genes has been seen in sea bass and sea bream after injection with betanodavirus, where Mx protein seems to have an important antiviral function in terms of limiting the replication of the virus (Poisa-Beiro et al., 2008; Scapigliati et al., 2010). The same has also been noted for turbot (Montes et al., 2010), grouper (Lin et al., 2006; Chen et al., 2008), and barramundi (Wu et al., 2010). This is also in accordance with the observation that induction of the interferon system with either poly I:C (Nishizawa et al., 2009) or a VHSV-based DNA vaccine (Sommerset et al., 2003) gave high protection against betanodavirus. An antibody response was induced through the activation of B-cells in the kidney and spleen of Atlantic halibut experimentally infected with betanodavirus, with migration of antibody-secreting cells to infected peripheral tissue (Grove et al., 2006). The fact that recombinant vaccines based on the capsid protein induce protection against these viruses (Husgard et al., 2001; Sommerset et al., 2005), indicate that antibodies alone inhibit the virus to some extent. The presence of leukocytes in relation to infection has been reported, but direct evidence of their function is still lacking. In Atlantic halibut larva infected with betanodavirus, a large numbers of leukocytes were observed in the eye chamber (Johansen et al., 2004) and a detectable increase in both B- and T-cells was found in experimentally infected sea bass (Scapigliati et al., 2010).

Antimicrobial peptides as epinecidin-1, hepcidin 1–5, and cyclic shrimp antilipopolysaccharide factor (cSALF) have been found to play a role in viral agglutination and induction of Mx expression (Chia et al., 2010; Wang et al., 2010a, 2010b). As already discussed, recent studies indicate that the B2 protein encoded by betanodavirus is involved in inhibition of RNAi.

Vaccines: No commercial vaccines against betanodavirus are currently available, although experimental vaccination with recombinant vaccines has shown high protection (Húsgard et al., 2001; Sommerset et al., 2003, 2005; Liu et al., 2006; Pakingking et al., 2010). The vaccination of larvae or juvenile fish is difficult for two reasons: the virus affects the larvae/juveniles at a stage in which the immune system is not well developed and the small size of the fish at that point in their physical development makes injections impractical. However, one possibility could be to immunize the broodfish before spawning, in order to prevent vertical transmission (Kai et al., 2010).

8.8.4 Group V: Negative-Sense, Single-Stranded RNA Viruses (Example: ISAV)

Virus Architecture: ISAV is an aquatic representative of the Orthomyxoviridae family. The genome contains eight negative-stranded ssRNA segments (Mjaaland et al., 1997), ranging in size from 1 to 2.4 kb, comprising totally 14.3 kb (Clouthier et al., 2002). Together the segments encode for at least 10 proteins (Kibenge et al., 2001). Morphologically, the virus is slightly pleiomorphic with a diameter of around 100 nm, and is enveloped by a membrane obtained when budding out of the cells. On the virus surface there is a hemagglutinin-esterase protein, being responsible for both receptor binding and receptor destroying enzyme activities. Another surface protein is responsible for fusion of viral and cellular membranes (Müller et al., 2010). The virus has been reported to be of moderate environmental stability, where the titer was reduced 1000 times when kept in seawater at 4°C for 4 months (Rimstad and Mjaaland, 2002).

Host Range and Geographic Distribution: The disease was described for the first time as affecting juvenile Atlantic salmon in Norway in 1984 (Thorud and Djupvik, 1988). During the next years it spread geographically. At the end of 1990s the disease was reported from most of the North Atlantic, and later in Chile. ISAV has a relatively narrow host range. It mainly infects Atlantic salmon. Even at sites where both Atlantic salmon and rainbow trout (*O. mykiss*) have been farmed, it has been reported that salmons have been infected while the trout remained unaffected. The majority of disease outbreaks have been during the seawater stage. ISAV has been detected in wild fish populations (Raynard et al., 2001; Plarre et al., 2005).

Virulence and Pathogenicity: Diseased fish are lethargic, suffering severe anemia and hemorrhagic necrosis of the liver and kidney. The mortality rate generally can be more than 90%, but a slowly developing form of the disease with low mortality may also occur that can progress into a long-termed infection and persistent carrier state (Rimstad and Mjaaland, 2002).

Life Cycle: The port of entry seems to be the gills, but infection through the GIT cannot be excluded (Mikalsen et al., 2001). The preferential site for replication is endothelial cells (Hovland et al., 1994). The receptor on the cell surface is sialic acid (Hellebø et al., 2004), as it is for the orthomyxoviruses infecting mammalians. Replication of the genome probably takes place in the nucleus of the cells (Falk et al., 1997). The virus particles are released from the cell by budding (Koren and Nylund, 1997).

Interaction with the Host Antiviral System: ISAV has been shown to induce the interferon system in cell lines, but the induction did not protect the cells from ISAV infection. It has therefore been speculated if ISAV confer some kind of activity that neutralize the effect of the interferon system (Jensen and Robertsen, 2002; Kileng et al., 2007). Passive immunization has indicated that antibodies confer some protection against ISAV infection (Falk and Dannevig, 1995), as do the protection achieved by vaccination with inactive virus (see below). Expression studies of immune-related genes following ISAV infection have shown that a T-cell response is involved. However, a change in expression of cell surface markers indicates that the virus interferes with protein expression and circumvents the host immune response (Leblanc et al., 2010; Hetland et al., 2010).

Vaccines: Vaccines based on inactivated virus, which give some protection, are commercially available.

8.8.5 Baltimore Group VI: Positive-Sense ssRNA Viruses with DNA Intermediate in Life Cycle due to Reverse Transcriptase (Example: WDSV)

Virus Architecture: WDSV belongs to the genus *Epsilonretrovirus* in the family Retroviridae. The genome consists of two, in principal identical, copies of positive-sense ssRNA molecules with 5′-cap and 3′-poly A-tails. Each copy has a length of 12.8 kb, encoding the proteins Gag (group-specific antigens), Pol (polymerase activities), and Env (envelope glycoproteins), which are common for retrovirus. In addition, there are three other genes, *orf-A*, *orf-B*, and *orf-C*, where the first two seem to have arisen by gene duplication and encode retroviral cyclin (rv-cyclin), and the last encodes a protein that has been shown to induce apoptosis. Like other retrovirus, there are also long terminal repeats (LTRs) at each ends of the ssRNA molecules (Quackenbush et al., 1997). The viral particles are spherical, 80–100 nm in diameter, and enveloped.

Host Range and Geographic Distribution: WDSV seems to have a very narrow host range. It has only been reported to infect walleye (*Sander vitreus*) and some closely related species (Bowser et al., 1999; Holzschu et al., 1998). The virus was first described in 1969 in association with epidermal hyperplasia of fish from Oneida Lake in New York (Holzschu et al., 2003). Although the virus is endemic in parts of North America, it has not been reported from elsewhere in the world.

Life Cycle: After entering the cells, the ssRNA genome is copied into linear dsDNA molecules by means of RT. The dsDNA molecules enter the nucleus where a random integration into the host cell's genome takes place. Later, transcription of the integrated genome generates both structural and nonstructural viral proteins, as well as the RNA molecules that will be packaged into mature viral particles. The progeny virus particles are released from the cells by budding.

Virulence and Pathogenicity: WDSV causes multifocal, benign skin tumors. There are no reports of WDSV causing death of feral walleyes, although the generation of invasive tumors has been observed after experimental infections in the laboratory (Earnest-Koons et al., 1996). There is a seasonal prevalence of the disease, as the tumors develop during the autumn and regress in the spring. Tumors are seldom observed in the summer. The regressive tumors in the spring are shed in the environmental water (Bowser et al., 1990). As they contain a high number of viral particles, they probably contribute to the transmission of the virus. The Orf-C protein, enable to induce apoptosis, may play a role in regression of the tumors, and thereby be important for the viral transmission (Bowser et al., 1996). Host cell rv-cyclin stimulates cell proliferation and it has been demonstrated that the rv-cyclin encoded by WDSV is able to influence cell proliferation much as does the natural host cell gene (LaPierre et al., 1998). It has therefore been speculated that WDSV has acquired the rv-cyclin gene from host cellular counterparts, and that this acquisition has given the virus selective advantages (Holzschu et al., 2003).

Interaction with the Host Antiviral System: The seasonal induction and regression of tumors is likely due to interaction mechanisms between host, pathogen, and environment. The role of the immune system is not known. However, there have been indications that fish that have been through this cycle are less likely to develop the disease in successive years (Getchell et al., 2000). The induction of apoptosis by the viral Orf-C protein has been mentioned above.

Vaccines: As WDSV mainly affects feral walleyes the development of vaccines has not been of commercial interest.

8.9 SUMMARY

The current picture of viruses infecting fish is probably far from complete, as relatively few fish species have been investigated thoroughly. Most of the virus species that have been detected to date were identified because they cause diseases of fish in aquaculture. Those viruses thus far identified in fish represent 5 of the 7 groups of the Baltimore classification, and 10 of the 32 virus families registered to infect vertebrates. During the most recent years, our knowledge about viral biology and the immune system of fish has increased rapidly, giving us increased insight into how piscine viruses are transmitted, reproduce, and eventually cause disease. Such information about host–pathogen interactions will help us both to understand the impact of viruses in the ecology of wild fish and to make efficient measures to control viral diseases in aquaculture.

REFERENCES

Adachi, K., Ichinose, T., Takizawa, N., Watanabe, K., Kitazato, K., and Kobayashi, N. (2007). Inhibition of betanodavirus infection by inhibitors of endosomal acidification. *Arch. Virol.* 152(12), 2217–2224.

Adzhubei, A. A., Vlasova, A. V., Hagen-Larsen, H., Ruden, T. A., Laerdahl, J. K., and Høyheim, B. (2007). Annotated expressed sequence tags (ESTs) from pre-smolt Atlantic salmon (*Salmo salar*) in a searchable data resource. *BMC Genom.* 8, 209.

Andersen, L., Bratland, A., Hodneland, K., and Nylund, A. (2007). Tissue tropism of salmonid alphaviruses (subtypes SAV1 and SAV3) in experimentally challenged Atlantic salmon (*Salmo salar* L.). *Arch. Virol.* 152(10), 1871–1883.

Anderson, D. P., Dixon, O. W., and Roberson, B. S. (1979). Kinetics of the primary immune response in rainbow trout after flush exposure to *Yersinia ruckeri* O-antigen. *Dev. Comp. Immunol.* 3(4), 739–744.

Angeli, V. and Randolph, G. J. (2006). Inflammation, lymphatic function, and dendritic cell migration. *Lymphat. Res. Biol.* 4(4), 217–228.

Bearzotti, M., Delmas, B., Lamoureux, A., Loustau, A. M., Chilmonczyk, S., and Bremont, M. (1999). Fish rhabdovirus cell entry is mediated by fibronectin. *J. Virol.* 73(9), 7703–7709.

Biering, E. and Bergh, O. (1996). Experimental infection of Atlantic halibut, *Hippoglossus hippoglossus* L., yolk-sac larvae with infectious pancreatic necrosis virus: detection of virus by immunohistochemistry and *in situ* hybridization. *J. Fish Dis.* 19, 405–413.

Biering, E., Nilsen, F., Rødseth, O., and Glette, J. (1994). Susceptibility of Atlantic halibut *Hippoglossus hippoglossus* to infectious pancreatic necrosis virus. *Dis. Aquat. Organ.* 20, 183–190.

Bly, J. E. and Clem, L. W. (1991). Temperature-mediated processes in teleost immunity: *in vitro* immunosuppression induced by *in vivo* low temperature in channel catfish. *Vet. Immunol. Immunopathol.* 28(3–4), 365–377.

Bonami, J. R., Hasson, K. W., Mari, J., Poulos, B. T., and Lightner, D. V. (1997). Taura syndrome of marine penaeid shrimp: characterization of the viral agent. *J. Gen. Virol.* 78(2), 313–319.

Boshra, H., Li, J., and Sunyer, J. O. (2006). Recent advances on the complement system of teleost fish. *Fish Shellfish Immunol.* 20(2), 239–262.

Bowden, T. J., Lockhart, K., Smail, D. A., and Ellis, A. E. (2003). Experimental challenge of post-smolts with IPNV: mortalities do not depend on population density. *J. Fish Dis.* 26(5), 309–312.

Bowser, P. R., Martineau, D., and Wooster, G. A. (1990). Effects of water temperature on experimental transmission of dermal sarcoma in fingerling walleyes (*Stizostedion vitreum*). *J. Aquat. Anim. Health* 2, 157–161.

Bowser, P. R., Wooster, G. A., and Getchell, R. G. (1999). Transmission of walleye dermal sarcoma and lymphocystis via water-borne exposure. *J. Aquat. Anim. Health* 11, 158–161.

Bowser, R. N., Wooster, G. A., Quackenbush, S. L., Casey, R. N., and Casey, J. W. (1996). Comparison of fall and spring tumors as inocula for experimental transmission of walleye dermal sarcoma. *J. Aquat. Anim. Health* 8, 78–81.

Breuil, G., Pépin, J. F. P., Boscher, S., and Thiéry, R. (2002). Experimental vertical transmission of nodavirus from broodfish to eggs and larvae of the sea bass, *Dicentrarchus labrax* (L.). *J. Fish Dis.* 25(12), 697–702.

Burton, D. R. (2002). Antibodies, viruses and vaccines. *Nat. Rev. Immunol.* 2(9), 706–713.

Cameron, A. M. and Endean, R. (1973). Epidermal secretions and the evolution of venom glands in fishes. *Toxicon* 11, 401–410.

Cano, I., Alonso, M. C., Garcia-Rosado, E., Saint-Jean, S. R., Castro, D., and Borrego, J. J. (2006). Detection of lymphocystis disease virus (LCDV) in asymptomatic cultured gilt-head seabream (*Sparus aurata,* L.) using an immunoblot technique. *Vet. Microbiol.* 113(1–2), 137–141.

Cano, I., Lopez-Jimena, B., Garcia-Rosado, E., Ortiz-Delgado, J. B., Alonso, M. C., Borrego, J. J., Sarasquete, C., and Castro, D. (2009). Detection and persistence of Lymphocystis disease virus (LCDV) in *Artemia* sp. *Aquaculture* 291, 230–236.

Carroll, M. C. (2008). Complement and humoral immunity. *Vaccine* 26 (Suppl. 8), I28–I33.

Chen, Y. M., Su, Y. L., Shie, P. S., Huang, S. L., Yang, H. L., and Chen, T. Y. (2008). Grouper Mx confers resistance to nodavirus and interacts with coat protein. *Dev. Comp. Immunol.* 32(7), 825–836.

Chen, S. P., Wu, J. L., Su, Y. C., and Hong, J. R. (2007). Anti-Bcl-2 family members, zfBcl-x(L) and zfMcl-1a, prevent cytochrome c release from cells undergoing betanodavirus-induced secondary necrotic cell death. *Apoptosis* 12(6), 1043–1060.

Chen, S. P., Yang, H. L., Her, G. M., Lin, H. Y., Jeng, M. F., Wu, J. L., and Hong, J. R. (2006). Betanodavirus induces phosphatidylserine exposure and loss of mitochondrial membrane potential in secondary necrotic cells, both of which are blocked by bongkrekic acid. *Virology* 347(2), 379–391.

Chia, T. J., Wu, Y. C., Chen, J. Y., and Chi, S. C. (2010). Antimicrobial peptides (AMP) with antiviral activity against fish nodavirus. *Fish Shellfish Immunol.* 28(3), 434–439.

Chiou, P. P., Lin, C. M., Perez, L., Chen, T. T. (2002). Effect of cecropin B and a synthetic analogue on propagation of fish viruses *in vitro*. *Mar. Biotechnol. (N.Y.).* 4(3), 294–302.

Christensen, J. E. and Thomsen, A. R. (2009). Coordinating innate and adaptive immunity to viral infection: mobility is the key. *APMIS* 117(5–6) 338–355.

Christie, K. E. (1997). Immunization with viral antigens: infectious pancreatic necrosis. *Dev. Biol. Stand.* 90, 191–199.

Clouthier, S. C., Rector. T., Brown, N. E., and Anderson, E. D. (2002). Genomic organization of infectious salmon anaemia virus. *J. Gen. Virol.* 83(Pt. 2), 421–428.

Collazos, M. E., Barriga, C., and Ortega, E. (1994). Optimum conditions for the activation of the alternative complement pathway of a cyprinid fish (*Tinca tinca* L.). Seasonal variations in the titres. *Fish Shellfish Immunol.* 4, 499–506.

Collazos, M. E., Ortega, E., Barriga, C., and Rodrìguez, A. B. (1998). Seasonal variation in haematological parameters in male and female *Tinca tinca*. *Mol. Cell Biochem.* 183(1–2), 165–168.

Connolly, R. M., Melville, A. J., and Keesing, J. K. (2002). Abundance, movement and individual identification of leafy seadragons, *Phycodurus eques* (Pisces: Syngnathidae). *Mar. Freshwater Res.* 53, 777–780.

Costes, B., Raj, V. S., Michel, B., Fournier, G., Thirion, M., Gillet, L., Mast, J., Lieffrig, F., Bremont, M., and Vanderplasschen, A. (2009). The major portal of entry of koi herpesvirus in *Cyprinus carpio* is the skin. *J. Virol.* 83(7), 2819–2830.

Dalla Valle, L., Negrisolo, E., Patarnello, P., Zanella, L., Maltese, C., Bovo, G., and

Colombo, L. (2001). Sequence comparison and phylogenetic analysis of fish nodaviruses based on the coat protein gene. *Arch. Virol.* 146, 1125–1137.

de Jong, E. C., Smits, H. H., and Kapsenberg, M. L. (2005). Dendritic cell-mediated T cell polarization. *Springer Semin. Immunopathol.* 26(3), 289–307.

de Veer, M. J., Holko, M., Frevel, M., Walker, E., Der, S., Paranjape, J. M., Silverman, R. H., and Williams, B. R. (2001). Functional classification of interferon-stimulated genes identified using microarrays. *J. Leukoc. Biol.* 69(6), 912–920.

Devold, M., Krossøy, B., Aspehaug, V., and Nylund, A. (2000). Use of RT-PCR for diagnosis of infectious salmon anaemia virus (ISAV) in carrier sea trout *Salmo trutta* after experimental infection. *Dis. Aquat. Organ.* 40(1), 9–18.

Dorson, M., Rault, P., Haffray, P., and Torhy, C. (1997). Water-hardening rainbow trout eggs in the presence of an iodophor fails to prevent the experimental egg transmission of infectious pancreatic necrosis virus. *Bull. Eur. Assoc. Fish Pathol.* 17(1), 13–16.

Drake, L. A., Doblin, M. A., and Dobbs, F. C. (2007). Potential microbial bioinvasions via ships' ballast water, sediment, and biofilm. *Mar. Pollut. Bull.* 55 (7–9), 333–341.

Drolet, B. S., Rohovec, J. S., and Leong, J. C. (1994). The route of entry and progression of infectious hematopoietic necrosis virus in *Oncorhynchus mykiss* (Walbaum)—a sequential immunohistochemical study. *J. Fish Dis.* 17, 337–347.

Duncan, R. and Dobos, P. (1986). The nucleotide sequence of infectious pancreatic necrosis virus (IPNV) dsRNA segment A reveals one large ORF encoding a precursor polyprotein. *Nucleic Acids Res.* 14, 5934.

Duncan, R., Nagy, E., Krell, P. J., and Dobos, P. (1987). Synthesis of the infectious pancreatic necrosis virus polyprotein, detection of a virus-encoded protease, and fine structure mapping of genome segment A coding regions. *J. Virol.* 61, 3655–3664.

Eagle, R. A. and Trowsdale, J. (2007). Promiscuity and the single receptor: NKG2D. *Nat. Rev. Immunol.* 7(9), 737–744.

Earnest-Koons, K., Wooster, G. A., and Bowser, P. A. (1996). Invasive walleye dermal sarcoma in laboratory-maintained walleyes (*Stizostedion vitreum*). *Dis. Aquat. Organ.* 24, 227–232.

Essbauer, S. and Ahne, W. (2001). Viruses of lower vertebrates. *J. Vet. Med. Ser. B* 48, 403–475.

Essbauer, S., Fischer, U., Bergmann, S., and Ahne, W. (2004). Investigations on the ORF 167L of lymphocystis disease virus (Iridoviridae). *Virus Genes* 28(1), 19–39.

Falk, K. and Dannevig, B. H. (1995). Demonstration of a protective immune response in infectious salmon anaemia (ISA)-infected Atlantic salmon *Salmo salar*. *Dis. Aquat. Organ.* 21, 1–5.

Falk, K., Namork, E., Rimstad, E., Mjaaland, S., and Dannevig, B. H. (1997). Characterization of infectious salmon anemia virus, an orthomyxo-like virus isolated from Atlantic salmon (*Salmo salar* L.). *J. Virol.* 71(12), 9016–9023.

Fan, Z. and Zhang, Q. (2005). Molecular mechanisms of lymphocyte-mediated cytotoxicity. *Cell. Mol. Immunol.* 2(4), 259–264.

Fenner, B. J., Goh, W., and Kwang, J. (2006). Sequestration and protection of double-stranded RNA by the betanodavirus B2 protein. *J. Virol.* 80(14), 6822–6833.

Fenner, B. J., Goh, W., and Kwang, J. (2007). Dissection of double-stranded RNA binding protein B2 from betanodavirus. *J. Virol.* 81(11), 5449–5459.

Fensterl, V. and Sen, G. C. (2009). Interferons and viral infections. *Biofactors* 35(1), 14–20.

Fischer, U., Utke, K., Somamoto, T., Köllner, B., Ototake, M., and Nakanishi, T. (2006). Cytotoxic activities of fish leucocytes. *Fish Shellfish Immunol.* 20(2), 209–226.

FishBase. (2010). http://www.fishbase.org/home.htm

Frerichs, G. N., Morgan, D., Hart, D., Skerrow, C., Roberts, R. J., and Onions, D. E. (1991). Spontaneously productive C-type retrovirus infection of fish cell lines. *J. Gen. Virol.* 72(Pt. 10), 2537–2539.

Frerichs, G. N., Tweedie, A., Starkey, W. G., and Richards, R. H. (2000). Temperature, pH and electrolyte sensitivity, and heat, UV and disinfectant inactivation of sea bass (Dicentrarchus labrax) neuropathy nodavirus. *Aquaculture* 185(1–2), 13–24.

Fukuda, Y., Nguyen, H. D., Furuhashi, M., and Nakai, T. (1996). Mass mortality of cultured sevenband grouper, *Epinephelus septemfasciatus,* associated with viral nervous necrosis. *Fish Pathol.* 31, 165–170.

Gagné, N., Johnson, S. C., Cook-Versloot, M., MacKinnon, A. M., and Olivier, G. (2004). Molecular detection and characterization of nodavirus in several marine fish species from the northeastern Atlantic. *Dis. Aquat. Organ.* 62(3), 181–189.

Gagné, N., Mackinnon, A. M., Boston, L., Souter, B., Cook-Versloot, M., Griffiths, S., and Olivier, G. (2007). Isolation of viral haemorrhagic septicaemia virus from mummichog, stickleback, striped bass and brown trout in eastern Canada. *J. Fish Dis.* 30(4), 213–223.

García-Rosado, E., Markussen, T., Kileng, O., Baekkevold, E. S., Robertsen, B., Mjaaland, S., and Rimstad, E. (2008). Molecular and functional characterization of two infectious salmon anaemia virus (ISAV) proteins with type I interferon antagonizing activity. *Virus Res.* 133(2), 228–238.

Getchell, R. G., Wooster, G. A., Rusdstm, L. G., Van De Valk, A. J., Brookings, T. E., and Bowser, P. R. (2000). Prevalence of walleye dermal sarcoma by age class in walleyes (*Stizostedion vitreum*) from Oneida Lake, New York. *J. Aquat. Anim. Health* 12, 220–223.

Goić, B., Bustamante, J., Miquel, A., Alvarez, M., Vera, M. I., Valenzuela, P. D., and Burzio, L. O. (2008). The nucleoprotein and the viral RNA of infectious salmon anemia virus (ISAV) are localized in the nucleolus of infected cells. *Virology* 379(1), 55–63.

Gomez, D. K., Matsuoka, S., Mori, K., Okinaka, Y., Park, S. C., and Nakai, T. (2009). Genetic analysis and pathogenicity of betanodavirus isolated from wild redspotted grouper *Epinephelus akaara* with clinical signs. *Arch. Virol.* 154(2), 343–346.

Graham, J. B. and Dickson, K. A. (2001). Anatomical and physiological specializations for endothermy. In: Block, B. A. and Stevens E. D. (eds.), *Tuna: Physiology, Ecology and Evolution, Fish Physiology*, Vol. 19 Academic Press, San Diego, pp. 121–165.

Graham, D. A., Fringuelli, E., Wilson, C., Rowley, H. M., Brown, A., Rodger, H., McLoughlin, M. F., McManus, C., Casey, E., McCarthy, L. J., and Ruane, N. M. (2010). Prospective longitudinal studies of salmonid alphavirus infections on two Atlantic salmon farms in Ireland; evidence for viral persistence. *Fish Dis.* 33(2), 123–135.

Grotmol, S., Bergh, Ø., and Totland, G. K. (1999). Transmission of viral encephalopathy and retinopathy (VER) to yolk-sac larvae of the Atlantic halibut *Hippoglossus hippoglossus*: occurrence of nodavirus in various organs and a possible route of infection. *Dis. Aquat. Organ.* 36, 95–106.

Grotmol, S., Nerland, A. H., Biering, E., Totland, G., and Nishizawa, T. (2000). Characterisation of the capsid protein gene from a nodavirus strain affecting the Atlantic halibut *Hippoglossus hippoglossus* and design of an optimal reverse-transcriptase polymerase chain reaction (RT-PCR) detection assay. *Dis. Aquat. Organ.* 39, 79–88.

Grotmol, S. and Totland, G. K. (2000). Surface disinfection of Atlantic halibut *Hippoglossus hippoglossus* eggs with ozonated sea-water inactivates nodavirus and increases survival of the larvae. *Dis. Aquat. Organ.* 39(2):89–96.

Grotmol, S., Totland, G. K., Thorud, K., and Hjeltnes, B. K. (1997). Vacuolating encephalopathy and retinopathy associated with a nodavirus-like agent: a probable cause of mass mortality of cultured larval and juvenile Atlantic halibut *Hippoglossus hippoglossus. Dis. Aquat. Organ.* 29, 85–97.

Grove, S., Johansen, R., Reitan, L. J., Press, C. M., and Dannevig, B. H. (2006). Quantitative investigation of antigen and immune response in nervous and lymphoid tissues of Atlantic halibut (*Hippoglossus hippoglossus*) challenged with nodavirus. *Fish Shellfish Immunol.* 21(5), 525–539.

Guo, Y. X., Chan, S. W., and Kwang, J. (2004). Membrane association of greasy grouper nervous necrosis virus protein A and characterization of its mitochondrial localization targeting signal. *J. Virol.* 78(12), 6498–6508.

Guo, Y. X., Wei, T., Dallmann, K., and Kwang, J. (2003). Induction of caspase-dependent apoptosis by betanodaviruses GGNNV and demonstration of protein alpha as an apoptosis inducer. *Virology* 308(1), 74–82.

Hansen, J. D., Landis, E. D., and Phillips, R. B. (2005). Discovery of a unique Ig heavy-chain isotype (IgT) in rainbow trout: implications for a distinctive B cell developmental pathway in teleost fish. *Proc. Natl. Acad. Sci. U.S.A.* 102(19), 6919–6924.

Harmache, A., LeBerre, M., Droineau, S., Giovannini, M., and Brémont, M. (2006). Bioluminescence imaging of live infected salmonids reveals that the fin bases are the major portal of entry for Novirhabdovirus. *J. Virol.* 80(7), 3655–3659.

Hattenberger-Baudouy, A. M., Danton, M., Merle, G., and de Kinkelin, P. (1995). Epidemiology of infectious hematopoietic necrosis (IHN) of salmonid fish in France—study of the course of natural infection by combined use of viral examination and seroneutralization test and eradication attempts. *Vet. Res.* 26, 256–275.

Hawley, L. M. and Garver, K. A. (2008). Stability of viral hemorrhagic septicemia virus (VHSV) in freshwater and seawater at various temperatures. *Dis. Aquat. Organ.* 82(3), 171–178.

Hellebø, A., Vilas, U., Falk, K., and Vlasak, R. (2004). Infectious salmon anemia virus specifically binds to and hydrolyzes 4-*O*-acetylated sialic acids. *J. Virol.* 78(6), 3055–3062.

Hershberger, P. K., Gregg, J., Pacheco, C., Winton, J., Richard, J., and Traxler, G. (2007). Larval Pacific herring, *Clupea pallasii* (Valenciennes), are highly susceptible to viral haemorrhagic septicaemia and survivors are partially protected after their metamorphosis to juveniles. *J. Fish Dis.* 30(8), 445–458.

Hetland, D. L., Jørgensen, S. M., Skjødt, K., Dale, O. B., Falk, K., Xu, C., Mikalsen, A. B., Grimholt, U., Gjøen, T., and Press, C. M. (2010). *In situ* localisation of major histocompatibility complex class I and class II and CD8 positive cells in infectious salmon anaemia virus (ISAV)-infected Atlantic salmon. *Fish Shellfish Immunol.* 28(1), 30–39.

Holzschu, D. L., LaPierre, L. A., and Lairmore, M. D. (2003). Comparative pathogenesis of epsilonretroviruses. *J. Virol.* 77(23), 12385–12391.

Holzschu, D. L., Wooster, G. A., and Bowser, P. R. (1998). Experimental transmission of dermal sarcoma to the sauger *Stizostedion canadense*. *Dis. Aquat. Organ.* 32, 9–14.

Hong, J. R., Gong, H. Y., and Wu, J. L. (2002). IPNV VP5, a novel anti-apoptosis gene of the Bcl-2 family, regulates Mcl-1 and viral protein expression. *Virology* 295(2), 217–229.

Hossain, M., Kim, S. R., Kitamura, S. I., Kim, D. W., Jung, S. J., Nishizawa, T., Yoshimizu, M., and Oh, M. J. (2009). Lymphocystis disease virus persists in the epidermal tissues of olive flounder, *Paralichthys olivaceus* (Temminch and Schlegel), at low temperatures. *J. Fish Dis.* 32(8), 699–703.

Hovland, T., Nylund, A., Watanabe, K., and Endresen, C. (1994). Observation of infectious salmon anaemia virus in Atlantic salmon, *Salmo salar* L. *J. Fish Dis.* 17(3), 291–296.

Hu, G. B., Cong, R. S., Fan, T. J., and Mei, X. G. (2004). Induction of apoptosis in a flounder gill cell line by lymphocystis disease virus infection. *J. Fish Dis.* 27(11), 657–662.

Húsgard, S., Grotmol, S., Hjeltnes, B. K., Rødseth, O. M., and Biering, E. (2001). Immune response to a recombinant capsid protein of striped jack nervous necrosis virus (SJNNV) in turbot *Scophthalmus maximus* and Atlantic halibut *Hippoglossus hippoglossus,* and evaluation of a vaccine against SJNNV. *Dis. Aquat. Organ.* 45, 33–44.

Iannello, A., Debbeche, O., Martin, E., Attalah, L. H., Samarani, S., and Ahmad, A. (2005). Viral strategies for evading antiviral cellular immune responses of the host. *J. Leukoc. Biol.* 79(1), 16–35.

Ishii, K. J., Koyama, S., Nakagawa, A., Coban, C., and Akira, S. (2008). Host innate immune receptors and beyond: making sense of microbial infections. *Cell Host Microbe.* 3(6), 352–363.

Iwamoto, R., Hasegawa, O., Lapatra, S., and Yoshimizu, M. (2002). Isolation and characterization of the Japanese flounder (*Paralichthys olivaceus*) lymphocystis disease virus. *J. Aquat. Anim. Health* 14, 114–123.

Iwamoto, T., Mise, K., Mori, K., Arimoto, M., Nakai, T., and Okuno, T. (2001). Establishment of an infectious RNA transcription system for Striped jack nervous necrosis virus, the type species of the betanodaviruses. *J. Gen. Virol.* 82(Pt. 11), 2653–2662.

Iwamoto, T., Mise, K., Takeda, A., Okinaka, Y., Mori, K., Arimoto, M., Okuno, T., and Nakai, T. (2005). Characterization of Striped jack nervous necrosis virus subgenomic RNA3 and biological activities of its encoded protein B2. *J. Gen. Virol.* 86(Pt. 10), 2807–2816.

Iwamoto, T., Okinaka, Y., Mise, K., Mori, K., Arimoto, M., Okuno, T., and Nakai, T. (2004). Identification of host-specificity determinants in betanodaviruses by using reassortants between striped jack nervous necrosis virus and sevenband grouper nervous necrosis virus. *J. Virol.* 78(3), 1256–1262.

Jakob, N. J. and Darai, G. (2002). Molecular anatomy of Chilo iridescent virus genome and the evolution of viral genes. *Virus Genes* 25(3), 299–316.

Jensen, I. and Robertsen, B. (2002). Effect of double-stranded RNA and interferon on the antiviral activity of Atlantic salmon cells against infectious salmon anemia virus and infectious pancreatic necrosis virus. *Fish Shellfish Immunol.* 13(3), 221–241.

Johansen, R., Grove, S., Svendsen, A. K., Modahl, I., and Dannevig, B. (2004). A sequential study of pathological findings in Atlantic halibut, *Hippoglossus hippoglossus* (L), throughout one year after an acute outbreak of viral encephalopathy and retinopathy. *J. Fish Dis.* 27(6), 327–341.

Johnson, L. A. and Jackson, D. G. (2008). Cell traffic and the lymphatic endothelium. *Ann. N. Y. Acad. Sci.* 1131, 119–133.

Joss, J. M. (2006). Lungfish evolution and development. *Gen. Comp. Endocrinol.* 148(3), 285–289.

Kai, Y. H., Su, H. M., Tai, K. T., and Chi, S. C. (2010). Vaccination of grouper broodfish (*Epinephelus tukula*) reduces the risk of vertical transmission by nervous necrosis virus. *Vaccine* 28(4), 996–1001.

Kent, M. L. and Myers, M. S. (2000). Hepatic lesions in a redstriped rockfish (*Sebastes proriger*) suggestive of a herpesvirus infection. *Dis. Aquat. Organ.* 41, 237–239.

Kibenge, F. S., Gárate, O. N., Johnson, G., Arriagada, R., Kibenge, M. J., and Wadowska, D. (2001). Isolation and identification of infectious salmon anaemia virus (ISAV) from Coho salmon in Chile. *Dis. Aquat. Organ.* 45(1), 9–18.

Kileng, Ø., Brundtland, M. I., and Robertsen, B. (2007). Infectious salmon anemia virus is a powerful inducer of key genes of the type I interferon system of Atlantic salmon, but is not inhibited by interferon. *Fish Shellfish Immunol.* 23(2), 378–389.

Kim, C. H., Dummer, D. M., Chiou, P. P., and Leong, J. A. (1999). Truncated particles produced in fish surviving infectious hematopoietic necrosis virus infection: mediators of persistence? *J. Virol.* 73(1), 843–849.

Koren, C. W. R. and Nylund, A. (1997). Morphology and morphogenesis of infectious salmon anaemia virus replicating in the endothelium of Atlantic salmon *Salmo salar*. *Dis. Aquat. Organ.* 29, 99–109.

Kottelat, M., Britz, R., Hui, T. H., and Witte, K. E. (2006). Paedocypris, a new genus of Southeast Asian cyprinid fish with a remarkable sexual dimorphism, comprises the world's smallest vertebrate. *Proc. Biol. Sci.* 273(1589), 895–899.

Krondiris, J. V. and Sideris, D. C. (2002). Intramolecular disulfide bonding is essential for betanodavirus coat protein conformation. *J. Gen. Virol.* 83(Pt. 9), 2211–2214.

Kurath, G. (2005). Overview of recent DNA vaccine development for fish. *Dev. Biol. (Basel)* 121, 201–213.

Lamers, C. H., De Haas, M. J., and Van Muiswinkel, W. B. (1985). Humoral response and memory formation in carp after injection of *Aeromonas hydrophila* bacterin. *Dev. Comp. Immunol.* 9(1), 65–75.

Langdon, J. S. (1989). Experimental transmission and pathogenicity of epizootic haematopoietic necrosis virus (EHNV) in redfin perch, *Perca fluviatilis* L., and 11 other teleosts. *J. Fish Dis.* 12, 295–310.

Lange, S., Bambir, S. H., Dodds, A. W., Bowden, T., Bricknell, I., Espelid, S., and Magnadóttir, B. (2006). Complement component C3 transcription in Atlantic halibut (*Hippoglossus hippoglossus* L.) larvae. *Fish Shellfish Immunol.* 20(3), 285–294.

Lange, S., Bambir, S., Dodds, A. W., and Magnadóttir, B. (2004a). An immunohistochemical study on complement component C3 in juvenile Atlantic halibut (*Hippoglossus hippoglossus* L.). *Dev. Comp. Immunol.* 28(6), 593–601.

Lange, S., Bambir, S., Dodds, A. W., and Magnadóttir, B. (2004b). The ontogeny of complement component C3 in Atlantic cod (*Gadus morhua* L.)—an immunohistochemical study. *Fish Shellfish Immunol.* 16(3), 359–367.

LaPatra, S. E., Barone, L., Jones, G. R., and Zon, L. I. (2000). Effects of infectious hematopoietic necrosis virus and infectious pancreatic necrosis virus infection on hematopoietic precursors of the zebrafish. *Blood Cells Mol. Dis.* 26(5), 445–452.

LaPatra, S. E., Corbeil, S., Jones, G. R., Shewmaker, W. D., Lorenzen, N., Anderson, E. D., and Kurath, G. (2001). Protection of rainbow trout against infectious hematopoietic necrosis virus four days after specific or semi-specific DNA vaccination. *Vaccine* 19(28–29), 4011–4019.

LaPierre, L. A., Casey, J. W., and Holzschu, D. L. (1998). Walleye retroviruses associated with skin tumors and hyperplasias encode cyclin D homologs. *J. Virol.* 72(11), 8765–8771.

Leblanc, F., Laflamme, M., and Gagné, N. (2010). Genetic markers of the immune response of Atlantic salmon (*Salmo salar*) to infectious salmon anemia virus (ISAV). *Fish Shellfish Immunol.* 29(2), 217–232.

Le Breton, A., Grisez, L., Sweetman, J., and Ollevier, F. (1997). Viral nervous necrosis (VNN) associated with mass mortalities in cage-reared sea bass, Dicentrarchus labrax (L). *J. Fish Dis.* 20, 145–151.

Lecocq-Xhonneux, F., Thiry, M., Dheur, I., Rossius, M., Vanderheijden, N., Martial, J., and de Kinkelin, P. (1994). A recombinant viral haemorrhagic septicaemia virus glycoprotein expressed in insect cells induces protective immunity in rainbow trout. *Gen. Virol.* 75(Pt. 7), 1579–1587.

Lee, S. H., Miyagi, T., and Biron, C. A. (2007). Keeping NK cells in highly regulated antiviral warfare. *Trends Immunol.* 28(6), 252–259.

Lin, C. H., Christopher John, J. A., Lin, C. H., and Chang, C. Y. (2006). Inhibition of nervous necrosis virus propagation by fish Mx proteins. *Biochem. Biophys. Res. Commun.* 351(2), 534–539.

Liu, W., Hsu, C. H., Chang, C. Y., Chen, H. H., and Lin, C. S. (2006). Immune response against grouper nervous necrosis virus by vaccination of virus-like particles. *Vaccine* 24(37–39), 6282–6287.

Liu, W., Hsu, C. H., Hong, Y. R., Wu, S. C., Wang, C. H., Wu, Y. M., Chao, C. B., and Lin, C. S. (2005). Early endocytosis pathways in SSN-1 cells infected by dragon grouper nervous necrosis virus. *J. Gen. Virol.* 86(Pt. 9), 2553–2561.

Long, S., Wilson, M., Bengten, E., Bryan, L., Clem, L. W., Miller, N. W., and Chinchar, V. G. (2004). Identification of a cDNA encoding channel catfish interferon. *Dev. Comp. Immunol.* 28(2), 97–111.

Lorenzen, N. and LaPatra, S. E. (2005). DNA vaccines for aquacultured fish. *Rev. Sci. Tech.* 24(1), 201–213.

Lorenzen, N., Lorenzen, E., and Einer-Jensen, K. (2001). Immunity to viral haemorrhagic septicaemia (VHS) following DNA vaccination of rainbow trout at an early life-stage. *Fish Shellfish Immunol.* 11(7), 585–591.

Lovy, J., Savidant, G. P., Speare, D. J., and Wright, G. M. (2009). Langerin/CD207 positive dendritic-like cells in the haemopoietic tissues of salmonids. *Fish Shellfish Immunol.* 27(2), 365–368.

Lowenstein, C. J. and Padalko, E. (2004). iNOS (NOS2) at a glance. *J. Cell Sci.* 117(Pt. 14), 2865–2867.

Macdonald, R. D. and Yamamoto, T. (1977). The structure of infectious pancreatic necrosis virus RNA. *J. Gen. Virol.* 34(2), 235–247.

Magnadóttir, B. (2006). Innate immunity of fish (overview). *Fish Shellfish Immunol.* 20(2), 137–151.

Malecek, K., Lee, V., Feng, W., Huang, J. L., Flajnik, M. F., Ohta, Y., and Hsu, E. (2008). Immunoglobulin heavy chain exclusion in the shark. *PLoS Biol.* 6(6), e157.

McAllister, P. E. and Owens, W. J. (1992). Recovery of infectious pancreatic necrosis virus from the faeces of wild piscivorous birds. *Aquaculture* 106, 227–232.

McLoughlin, M. F. and Graham, D. A. (2007). Alphavirus infections in salmonids—a review. *J. Fish Dis.* 30(9), 511–531.

Meyers, T. R. and Winton, J. R. (1995). Viral haemorrhagic septicaemia virus in North America. *Annu. Rev. Fish Dis.* 5, 3–24.

Mézeth, K. B., Nylund, S., Henriksen, H., Patel, S., Nerland, A. H., and Szilvay, A. M. (2007). RNA-dependent RNA polymerase from Atlantic halibut nodavirus contains two signals for localization to the mitochondria. *Virus Res.* 130(1–2), 43–52.

Mézeth, K. B., Patel, S., Henriksen, H., Szilvay, A. M., and Nerland, A. H. (2009). B2 protein from betanodavirus is expressed in recently infected but not in chronically infected fish. *Dis. Aquat. Organ.* 83(2), 97–103.

Mikalsen, A. B., Teig, A., Helleman, A. L., Mjaaland, S., and Rimstad, E. (2001). Detection of infectious salmon anaemia virus (ISAV) by RT-PCR after cohabitant exposure in Atlantic salmon *Salmo salar*. *Dis. Aquat. Organ.* 47(3), 175–181.

Mjaaland, S., Rimstad, E., Falk, K., and Dannevig, B. H. (1997). Genomic characterization of the virus causing infectious salmon anemia in Atlantic salmon (*Salmo salar* L.): an orthomyxo-like virus in a teleost. *J. Virol.* 71(10), 7681–7686.

Montes, A., Figueras, A., and Novoa, B. (2010). Nodavirus encephalopathy in turbot (*Scophthalmus maximus*): inflammation, nitric oxide production and effect of anti-inflammatory compounds. *Fish Shellfish Immunol.* 28(2), 281–288.

Mori, K. I., Nakai, T., Muroga, K., Arimot, M., Mushiake, K., and Furusawa, I. (1992). Properties of a new virus belonging to Nodaviridae found in larval striped jack (*Pseudocaranx dentex*) with nervous necrosis. *Virology* 187, 368–371.

Mortensen, S. H., Nilsen, R. K., and Hjeltnes, B. (1998). Stability of an infectious pancreatic necrosis virus (IPNV) isolate stored under different laboratory conditions. *Dis. Aquat. Organ.* 33(1), 67–71.

Mulcahy, D., Klaybor, D., and Batts, W. N. (1990). Isolation of infectious hematopoietic necrosis virus from a leech (*Piscicola salmositica*) and a copepod (*Salmincola* sp.), ectoparasites of sockeye salmon *Oncorhynchus nerka*. *Dis. Aquat. Organ.* 29–34.

Müller, A., Markussen, T., Drabløs, F., Gjøen, T., Jørgensen, T. O., Solem, S. T., and Mjaaland, S. (2010). Structural and functional analysis of the hemagglutinin-esterase of infectious salmon anaemia virus. *Virus Res.* 151(2), 131–141.

Munday, B., Kwang, J., and Moody, N. (2002). Betanodavirus infections of teleost fish: a review. *J. Fish Dis.* 25, 127–142.

Murray, A. G. (2006). Persistence of infectious pancreatic necrosis virus (IPNV) in Scottish salmon (*Salmo salar* L.) farms. *Prev. Vet. Med.* 76(1–2), 97–108.

Nerland, A. H., Skaar, C., Eriksen, T. B., and Bleie, H. (2007). Detection of nodavirus in seawater from rearing facilities for Atlantic halibut *Hippoglossus hippoglossus* larvae. *Dis. Aquat. Organ.* 73(3), 201–205.

Nishizawa, T., Furuhashi, M., Nagai, T., Nakai, T., and Muroga, K. (1997). Genomic classification of fish nodaviruses by phylogenetic analysis of the coat protein gene. *Appl. Environ. Microbiol.* 63, 1633–1636.

Nishizawa, T., Mori, K., Furuhashi, M., Nakai, T., Furusawa, I., and Muroga, K. (1995). Comparison of the coat protein genes of five fish nodaviruses, the causative agents of viral nervous necrosis in marine fish. *J. Gen. Virol.* 76(Pt. 7), 1563–1569.

Nishizawa, T., Takami, I., Kokawa, Y., and Yoshimizu, M. (2009). Fish immunization using a synthetic double-stranded RNA Poly(I:C), an interferon inducer, offers protection against RGNNV, a fish nodavirus. *Dis. Aquat. Organ.* 83(2), 115–122.

Nishizawa, T., Takano, R., and Muroga, K. (1999). Mapping a neutralizing epitope on the coat protein of striped jack nervous necrosis virus. *J. Gen. Virol.* 80(Pt. 11), 3023–3027.

Nopadon, P., Aranya, P., Tipaporn, T., Toshihiro, N., Takayuki, K., Masashi, M., and Makoto, E. (2009). Nodavirus associated with pathological changes in adult spotted coralgroupers (*Plectropomus maculatus*) in Thailand with viral nervous necrosis. *Res. Vet. Sci.* 87(1), 97–101.

Ou, M. C., Chen, Y. M., Jeng, M. F., Chu, C. J., Yang, H. L., and Chen, T. Y. (2007). Identification of critical residues in nervous necrosis virus B2 for dsRNA-binding and RNAi-inhibiting activity through by bioinformatic analysis and mutagenesis. *Biochem. Biophys. Res. Commun.* 361(3), 634–640.

Øvergård, A. C., Fiksdal, I. U., Nerland, A. H. and Patel, S. (2011). Expression of T-cell markers during Atlantic halibut (Hippoglossus hippoglossus L.) ontogenesis. *Dev. Comp. Immunol.* 35(2), 203–13

Pakingking Jr., R., Bautista, N. B., de Jesus-Ayson, E. G., and Reyes, O. (2010). Protective immunity against viral nervous necrosis (VNN) in brown-marbled grouper (*Epinephelus fuscogutattus*) following vaccination with inactivated betanodavirus. *Fish Shellfish Immunol.* 28(4), 525–533.

Palm Jr., R. C., Landolt, M. L., and Busch, R. A. (1998). Route of vaccine administration: effects on the specific humoral response in rainbow trout *Oncorhynchus mykiss*. *Dis. Aquat. Organ.* 33(3), 157–166.

Pancer, Z., Amemiya, C. T., Ehrhardt, G. R., Ceitlin, J., Gartland, G. L., and Cooper, M. D. (2004). Somatic diversification of variable lymphocyte receptors in the agnathan sea lamprey. *Nature* 430(6996), 174–180.

Park, K. C., Osborne, J. A., Montes, A., Dios, S., Nerland, A. H., Novoa, B., Figueras, A., Brown, L. L., and Johnson, S. C. (2009). Immunological responses of turbot (*Psetta maxima*) to nodavirus infection or polyriboinosinic polyribocytidylic acid (pIC) stimulation, using expressed sequence tags (ESTs) analysis and cDNA microarrays. *Fish Shellfish Immunol.* 26(1), 91–108.

Patel, S., Korsnes, K., Bergh, Ø., Vik-Mo, F., Pedersen, J., and Nerland, A. H. (2007). Nodavirus in farmed Atlantic cod (*Gadus morhua*) in Norway. *Dis. Aquat. Organ.* 77, 169–173.

Patel, S., Malde, K., Lanzén, A., Olsen, R. H., and Nerland, A. H. (2009a). Identification of immune related genes in Atlantic halibut (*Hippoglossus hippoglossus* L.) following *in vivo* antigenic and *in vitro* mitogenic stimulation. *Fish Shellfish Immunol.* 27(6), 729–738.

Patel, S., Sørhus, E., Uglenes, I. F., Espedal, P. G., Bergh, Ø., Rødseth, O.-M., Morton, H. C., and Nerland, A. H. (2009b). Ontogeny of lymphoid organs and development of immunocompetance for Ig bearing cells in Atlantic halibut (*Hippoglossus hippoglossus*). *Fish Shellfish Immunol.* 26(3), 385–395.

Petterson, E., Sandberg, M., and Santi, N. (2009). Salmonid alphavirus associated with *Lepeophtheirus salmonis* (Copepoda: Caligidae) from Atlantic salmon, *Salmo salar* L. *J. Fish Dis.* 32(5), 477–479.

Pichlmair, A., Schulz, O., Tan, C. P., Näslund, T. I., Liljeström, P., Weber, F., and Reis e Sousa, C. (2006). RIG-I-mediated antiviral responses to single-stranded RNA bearing 5′-phosphates. *Science* 314(5801), 997–1001.

Pietsch, T. W. (1975). Precocious sexual parasitism in the deep sea ceratioid anglerfish, *Cryptopsaras couesi* Gill. *Nature* 256, 38–40.

Plarre, H., Devold, M., Snow, M., and Nylund, A. (2005). Prevalence of infectious salmon anaemia virus (ISAV) in wild salmonids in western Norway. *Dis. Aquat. Organ.* 66(1), 71–79.

Plasterk, R. H. (2002). RNA silencing: the genome's immune system. *Science.* 296(5571), 1263–1265.

Poisa-Beiro, L., Dios, S., Montes, A., Aranguren, R., Figueras, A., and Novoa, B. (2008). Nodavirus increases the expression of Mx and inflammatory cytokines in fish brain. *Mol. Immunol.* 45(1), 218–225.

Price, B. D., Ahlquist, P., and Ball, L. A. (2002). DNA-directed expression of an animal virus RNA for replication-dependent colony formation in *Saccharomyces cerevisiae*. *J. Virol.* 76(4), 1610–1616.

Pylkkö, P., Lyytikäinen, T., Ritola, O., Pelkonen, S., and Valtonen, E. T. (2002). Temperature effect on the immune defense functions of Arctic charr *Salvelinus alpinus*. *Dis. Aquat. Organ.* 52(1), 47–55.

Quackenbush, S. L., Holzschu, D. L., Bowser, P. R., and Casey, J. W. (1997). Transcriptional analysis of walleye dermal sarcoma virus (WDSV). *Virology* 237(1), 107–112.

Raynard, R. S., Murray, A. G., and Gregory, A. (2001). Infectious salmon anaemia virus in wild fish from Scotland. *Dis. Aquat. Organ.* 46(2), 93–100.

Rebl, A., Goldammer, T., and Seyfert, H. M. (2010). Toll-like receptor signaling in bony fish. *Vet. Immunol. Immunopathol.* 134(3–4), 139–150.

Rimstad, E. and Mjaaland, S. (2002). Infectious salmon anaemia virus. *APMIS* 110(4), 273–282.

Roberts, R. J. and Pearson, M. D. (2005). Infectious pancreatic necrosis in Atlantic salmon, *Salmo salar* L. *J. Fish Dis.* 28(7), 383–390.

Robertsen, B. (2006). The interferon system of teleost fish. *Fish Shellfish Immunol.* 20(2), 172–191.

Robertsen, B., Bergan, V., Røkenes, T., Larsen, R., and Albuquerque, A. (2003). Atlantic salmon interferon genes: cloning, sequence analysis, expression, and biological activity. *J. Interferon Cytokine Res.* 23(10), 601–612.

Romero-Brey, I., Bandín, I., Cutrín, J. M., Vakharia, V. N., and Dopazo, C. P. (2009). Genetic analysis of aquabirnaviruses isolated from wild fish reveals occurrence of natural reassortment of infectious pancreatic necrosis virus. *J. Fish Dis.* 32(7), 585–595.

Ruane, N. M., McCarthy, L. J., Swords, D., and Henshilwood, K. (2009). Molecular differentiation of infectious pancreatic necrosis virus isolates from farmed and wild salmonids in Ireland. *J. Fish Dis.* 32(12), 979–987.

Ruiz, G. M., Rawlings, T. K., Dobbs, F. C., Drake, L. A., Mullady, T., Huq, A., and Colwell. R. R. (2000). Global spread of microorganisms by ships. *Nature* 408(6808), 49–50.

Samalecos, C. P. (1986). Analysis of the structure of fish lymphocystis disease virions from skin tumours of pleuronectes. *Arch. Virol.* 91(1–2), 1–10.

Sánchez, C., Babin, M., Tomillo, J., Ubeira, F. M., and Domínguez, J. (1993). Quantification of low levels of rainbow trout immunoglobulin by enzyme immunoassay using two monoclonal antibodies. *Vet. Immunol. Immunopathol.* 36(1), 65–74.

Sanders, G. E., Batts, W. N., and Winton, J. R. (2003). Susceptibility of zebrafish (*Danio rerio*) to a model pathogen, spring viremia of carp virus. *Comp. Med.* 53(5), 514–521.

Sarropoulou, E., Sepulcre, P., Poisa-Beiro, L., Mulero, V., Meseguer, J., Figueras, A., Novoa, B., Terzoglou, V., Reinhardt, R., Magoulas, A., and Kotoulas, G. (2009). Profiling of infection specific mRNA transcripts of the European seabass *Dicentrarchus labrax*. *BMC Genom.* 10, 157.

Scapigliati, G., Buonocore, F., Randelli, E., Casani, D., Meloni, S., Zarletti, G., Tiberi, M., Pietretti, D., Boschi, I., Manchado, M., Martin-Antonio, B., Jimenez-Cantizano, R., Bovo, G., Borghesan, F., Lorenzen, N., Einer-Jensen, K., Adams, S., Thompson, K., Alonso, C., Bejar, J., Cano, I., Borrego, J. J., and Alvarez, M. C. (2010). Cellular and molecular immune responses of the sea bass (*Dicentrarchus labrax*) experimentally infected with betanodavirus. *Fish Shellfish Immunol.* 28(2), 303–311.

Scherer, W. F. and Hurlbut, H. S. (1967). Nodamura virus from Japan: a new and unusual arbovirus resistant to diethyl ether and chloroform. *Am. J. Epidemiol.* 86, 271–285.

Schnitzler, P., Delius, H., Scholz, J., Touray, M., Orth, E., and Darai, G. (1987). Identification and nucleotide sequence analysis of the repetitive DNA element in the genome of fish lymphocystis disease virus. *Virology* 161(2), 570–578.

Sheng, X. Z. and Zhan, W. B. (2004). Occurrence, development and histochemical characteristics of lymphocystis in cultured Japanese flounder (*Paralichthys olivaceus*). *High Technol. Lett.* 10(2), 92–96.

Sheng, X. Z., Zhan, W. B., Xu, S., and Cheng, S. (2007). Histopathological observation of lymphocystis disease and lymphocystis disease virus (LCDV) detection in cultured diseased *Sebastes schlegeli*. *J. Ocean Univ. Chin. (Oceanic Coastal Sea Res.)* 6(4), 378–382.

Sinha, A. and Chakravarty, A. K. (1997). Immune responses in an air-breathing teleost *Clarius batrachus*. *Fish Shellfish Immunol.* 7, 105–114.

Skall, H. F., Olesen, N. J., and Mellergaard, S. (2005). Prevalence of viral haemorrhagic septicaemia virus in Danish marine fishes and its occurrence in new host species. *Dis. Aquat. Organ.* 66(2), 145–151.

Skår, C. K. and Mortensen, S. (2007). Fate of infectious salmon anaemia virus (ISAV) in experimentally challenged blue mussels *Mytilus edulis*. *Dis. Aquat. Organ.* 74(1), 1–6.

Slater, C. H., Fitzpatrick, M. S., and Schreck, C. B. (1995). Characterization of an androgen receptor in salmonid lymphocytes: possible link to androgen-induced immunosuppression. *Gen. Comp. Endocrinol.* 100(2), 218–225.

Smail, D. A., Bruno, D. W., Dear, G., McFarlane, L. A., and Ross, K. (1992). Infectious pancreatic necrosis (IPN) virus Sp serotype in farmed Atlantic salmon, *Salmo salar* L., post-smolts associated with mortality and clinical disease. *J. Fish Dis.* 15(1), 77–83.

Smith, E. M. and Gregory, T. R. (2009). Patterns of genome size diversity in the ray-finned fishes. *Hydrobiologia* 625, 1–25.

Sommerset, I., Lorenzen, E., Lorenzen, N., Bleie, H., and Nerland, A. H. (2003). A DNA vaccine directed against a rainbow trout rhabdovirus induces early protection against a nodavirus challenge in turbot. *Vaccine* 21(32), 4661–4667.

Sommerset, I. and Nerland, A. H. (2004). Complete sequence of RNA1 and subgenomic RNA3 of Atlantic halibut nodavirus (AHNV). *Dis. Aquat. Organ.* 58, 117–125.

Sommerset, I., Skern, R., Biering, E., Bleie, H., Fiksdal, I. U., Grove, S., and Nerland, A. H. (2005). Protection against Atlantic halibut nodavirus in turbot is induced by recombinant capsid protein vaccination but not following DNA vaccination. *Fish Shellfish Immunol.* 18(1), 13–29.

Steiner, I., Kennedy, P. G., and Pachner, A. R. (2007). The neurotropic herpes viruses: herpes simplex and varicella-zoster. *Lancet Neurol.* 6(11), 1015–1028.

Stetson, D. B. and Medzhitov, R. (2006). Antiviral defense: interferons and beyond. *J. Exp. Med.* 203(8), 1837–1841. Epub 2006, Jul 31. Erratum in: *J. Exp. Med.* 203(9), 2215.

Stevens, J. D. (2007) Whale shark (*Rhincodon typus*) biology and ecology: a review of the primary literature. *Fish. Res.* 84, 4–9.

Su, Y. C., Wu, J. L., and Hong, J. R. (2009). Betanodavirus non-structural protein B2: a novel necrotic death factor that induces mitochondria-mediated cell death in fish cells. *Virology* 385(1), 143–154.

Suzuki, Y., Orito, M., Iigo, M., Kenzuka, H., Kobayashi, M., and Aida, K. (1996). Seasonal changes in blood IgM levels in goldfish, with special reference to water temperature to water temperature and gonadal maturation. *Fish. Sci.* 62, 754–759.

Swain, P. and Nayak, S. K. (2009). Role of maternally derived immunity in fish. *Fish Shellfish Immunol.* 27(2), 89–99.

Takaoka, A., Wang, Z., Choi, M. K., Yanai, H., Negishi, H., Ban, T., Lu, Y., Miyagishi, M., Kodama, T., Honda, K., Ohba, Y., and Taniguchi, T. (2007). DAI (DLM-1/ZBP1) is a cytosolic DNA sensor and an activator of innate immune response. *Nature* 448(7152), 501–505.

Tang, L., Lin, C. S., Krishna, N. K., Yeager, M., Schneemann, A., and Johnson, J. E. (2002). Virus-like particles of a fish nodavirus display a capsid subunit domain organization different from that of insect nodaviruses. *J. Virol.* 76(12), 6370–6375.

Thorud, K. E. and Djupvik, H. O. (1988). Infectious anaemia in Atlantic salmon (*Salmo salar* L.). *Bull. Eur. Assoc. Fish Pathol.* 8, 109–111.

Tian, J., Sun, X., Chen, X., Yu, J., Qu, L., and Wang, L. (2008). The formulation and immunisation of oral poly(DL-lactide-*co*-glycolide) microcapsules containing a plasmid vaccine against lymphocystis disease virus in Japanese flounder (*Paralichthys olivaceus*). *Int. Immunopharmacol.* 8(6), 900–908.

Tidona, C. A. and Darai, G. (1997). The complete DNA sequence of lymphocystis disease virus. *Virology* 230(2), 207–216.

Tidona, C. A. and Darai, G. (1999). Lymphocystis disease virus (Iridoviridae) In: Granoff, A. and Webster, R. G. (eds.), *Encyclopedia of Virology*. Academic Press, New York, pp. 908–911.

Tidona, C. A. and Darai, G. (2000). Iridovirus homologues of cellular genes—implications for the molecular evolution of large DNA viruses. *Virus Genes* 21(1–2), 77–81.

Totland, G. K., Hjeltnes, B. K., and Flood, P. R. (1996). Transmission of infectious salmon anaemia (ISA) through natural secretions and excretions from infected smolts of Atlantic salmon *Salmo salar* during their presymptomatic phase. *Dis. Aquat. Organ.* 26, 25–31.

Traxler, G. S. and Kieser, D. (1994). Isolation of the North American strain of viral haemorrhagic septicaemia virus (VHSV) from herring (*Clupea harengus pallasi*) in British Columbia. *Fish Health Sect. Am. Fish. Soc. Newslett.* 22, 8.

Utke, K., Kock, H., Schuetze, H., Bergmann, S. M., Lorenzen, N., Einer-Jensen, K., Köllner, B., Dalmo, R. A., Vesely, T., Ototake, M., and Fischer, U. (2008). Cell-mediated immune responses in rainbow trout after DNA immunization against the viral hemorrhagic septicemia virus. *Dev. Comp. Immunol.* 32(3), 239–252.

Van Nieuwstadt, A. P., Dijkstra, S. G., and Haenen, O. L. (2001). Persistence of herpesvirus of eel *Herpesvirus anguillae* in farmed European eel *Anguilla anguilla*. *Dis. Aquat. Organ.* 45(2), 103–107.

Van Wynsberghe, P. M. and Ahlquist, P. (2009). 5′ *cis* elements direct nodavirus RNA1 recruitment to mitochondrial sites of replication complex formation. *J. Virol.* 83(7), 2976–2988.

Walker, R. and Weissenberg, R. (1965). Conformity of light and electron microscopic studies on virus particle distribution in lymphocystis tumor cells of fish. *Ann. N. Y. Acad. Sci.* 126(1), 375–385.

Waltzek, T. B., Kurobe, T., Goodwin, A. E., and Hedrick, R. P. (2009). Development of a polymerase chain reaction assay to detect cyprinid herpesvirus 2 in goldfish. *J. Aquat. Anim. Health* 21(1), 60–67.

Wang, Y. D., Kung, C. W., and Chen, J. Y. (2010a). Antiviral activity by fish antimicrobial peptides of epinecidin-1 and hepcidin 1–5 against nervous necrosis virus in medaka. *Peptides* 31(6), 1026–1033.

Wang, Y. D., Kung, C. W., Chi, S. C., and Chen, J. Y. (2010b). Inactivation of nervous necrosis virus infecting grouper (*Epinephelus coioides*) by epinecidin-1 and hepcidin 1–5 antimicrobial peptides, and downregulation of Mx2 and Mx3 gene expressions. *Fish Shellfish Immunol.* 28(1), 113–120.

Whittington, R. J., Crockford, M., Jordon, D., and Jones, B. (2008). Herpesvirus that caused epizootic mortality in 1995 and 1998 in pilchard, *Sardinops sagax neopilchardus* (Steindachner), in Australia is now endemic. *J. Fish Dis.* 31, 97–105.

Wolf, K. (1988). Infectious pancreatic necrosis virus. In: Wolf, K. (ed.), *Fish Viruses and Fish Viral Diseases*. Cornell University Press, Ithaca, NY, pp. 115–157.

Wu, H. C., Chiu, C. S., Wu, J. L., Gong, H. Y., Chen, M. C., Lu, M. W., and Hong, J. R. (2008a). Zebrafish anti-apoptotic protein zfBcl-xL can block betanodavirus protein alpha-induced mitochondria-mediated secondary necrosis cell death. *Fish Shellfish Immunol.* 24(4), 436–449.

Wu, Y. M., Hsu, C. H., Wang, C. H., Liu, W., Chang, W. H., and Lin, C. S. (2008b). Role of the DxxDxD motif in the assembly and stability of betanodavirus particles. *Arch Virol.* 153(9), 1633–1642. Epub 2008, Jul 15.

Wu, Y. C., Lu, Y. F., and Chi, S. C. (2010). Antiviral mechanism of barramundi Mx against betanodavirus involves the inhibition of viral RNA synthesis through the interference of RdRp. *Fish Shellfish Immunol.* 28(3), 467–475.

Zheng, F. R., Sun, X. Q., Liu, H. Z., and Zhang, J. X. (2006). Study on the distribution and expression of a DNA vaccine against lymphocystis disease virus in Japanese flounder (*Paralichthys olivaceus*). *Aquaculture* 261, 1128–1134.

CHAPTER 9

ECOLOGY OF VIRUSES INFECTING ECTOTHERMIC VERTEBRATES—THE IMPACT OF RANAVIRUS INFECTIONS ON AMPHIBIANS

V. GREGORY CHINCHAR
Department of Microbiology, University of Mississippi Medical Center, Jackson, MS

JACQUES ROBERT
Department of Microbiology and Immunology, University of Rochester Medical Center, Rochester, NY

ANDREW T. STORFER
School of Biological Sciences, Washington State University, Pullman, WA

CONTENTS

9.1 INTRODUCTION

Although isolates from at least 15 viral families infect one or more classes of ectothermic vertebrates, only representatives of three families (Adenoviridae, Alloherpesviridae (order Herpesvirales), and Iridoviridae), naturally infect amphibians. With few exceptions most viruses within the order Herpesvirales and the family Adenoviridae infect mammals and birds (Fauquet et al., 2005). In contrast, all viruses within the family Alloherpesviridae infect either fish (e.g., Ictalurid herpesvirus 1) or frogs (ranid herpesvirus 1), whereas viruses within three genera of the family Iridoviridae (*Ranavirus*, *Megalocytivirus*, and *Lymphocystivirus*) infect

Studies in Viral Ecology: Animal Host Systems: Volume 2, First Edition. Edited by Christon J. Hurst.

ectothermic vertebrates. Within those three genera, megalocytiviruses and lymphocystiviruses infect only fish, whereas individual ranaviruses target amphibians, reptiles, or fish, and a few viral species appear able to infect members of different taxonomic classes (Moody and Owens, 1994; Mao et al., 1999; Chinchar et al., 2009; Jancovich et al., in press). For example, although Bohle iridovirus (BIV) naturally infects anurans (Cullen et al., 1995), it has also been shown to infect fish following intraperitoneal injection (Moody and Owens, 1994). Likewise, whereas *Ambystoma tigrinum virus* (ATV, family Iridoviridae, genus *Ranavirus*), and *Frog virus 3* (FV3, family Iridoviridae, genus *Ranavirus*) infections are limited, respectively, to salamanders and frogs in nature, both will infect heterologous species, albeit at low levels, under experimental conditions (Clark et al., 1968; Schock et al., 2008).

In addition to these well-documented natural infections, amphibians have been suggested to serve as hosts for arboviruses of human concern, for example, Eastern equine encephalitis virus, West Nile virus, St. Louis encephalitis virus, and so on (Whitney et al., 1968; Klenk and Komar, 2003; Cupp et al., 2004; Densmore and Green, 2007; Burkett-Cadena et al., 2008). Although field and serological studies, as well as experimental infections, suggest at least limited amounts of virus replication take place in some frog species (Kostyukov et al., 1986), it is not clear whether infections of amphibians with Flaviviruses or Togaviruses lead to clinical disease, or whether infected amphibians act as virus reservoirs or amplifying hosts and contribute to disease in humans and other mammals. Partly this uncertainty may reflect the fact that some mosquito species that feed on amphibians do not readily feed on mammals or birds, and vice versa (Cupp et al., 2004), and the observation that viral titers in experimentally infected frogs are very low and transient (Klenk and Komar, 2003) and unlikely to result in mosquito infection following feeding. Because of this uncertainty, only viruses within the three aforementioned families that naturally infect amphibians and contribute to disease in those animals are listed in Table 9.1.

TABLE 9.1 Viruses Infecting Amphibians[a]

Virus Family	Viral Genus	Viral Species/Isolates	Host
Alloherpesviridae[b]		*Ranid herpesvirus 1* (Lucke tumor herpesvirus)	Leopard frogs (*Rana pipiens*)
		Ranid herpesvirus 2 (frog virus 4)	Leopard frogs
Adenoviridae	*Siadenovirus*	*Frog adenovirus*	Leopard frogs
Iridoviridae	*Ranavirus*	*Frog virus 3* (FV3)	Leopard frogs, other frog species, several species of turtles
		Ambystoma tigrinum virus (ATV)	Tiger salamanders (*Ambystoma tigrinum*), axolotls (*A. mexicanum*)
		Bohle iridovirus (BIV)	Ornate burrowing frog (*Limnodynastes ornatus*)
		Tiger frog virus (TFV)[c]	*R. tigrina rugulosa*
		Rana grylio virus (RGV)[c]	*R. grylio*
		Rana catesbeiana virus-Z (RCV-Z)[d]	*R. catesbeiana*

[a]Fauquet et al. (2005).

[b]The taxonomy of the herpesviruses has been reorganized and the order Herpesvirales established. Ranid herpesviruses have been classified as unassigned species within the family (Alloherpesviridae) that includes *Ictalurid herpesvirus 1* as the type species of the genus *Ictalurivirus* (Davison et al., 2009).

[c]TFV and RGV (Zhang et al., 2001; Weng et al., 2002; Chinchar et al., 2009) likely represent strains/isolates of FV3 rather than separate viral species.

[d]Based on sequence and restriction fragment length polymorphism analyses, RCV-Z (Majji et al., 2006) may represent a novel virus species, rather than a strain/isolate of FV3.

Since ranaviruses represent the primary viral genus infecting amphibians, they will be the focus of this review. While ranaviruses are not the most important pathogen infecting amphibians (that distinction goes to the chytrid fungus *Batrachochytrium dendrobatidis*, Bd), ranaviruses have triggered frog die-offs in natural and aquacultural settings and localized die-offs of salamanders both in the wild and in a laboratory colony (Chinchar, 2002; Green et al., 2002; Greer et al., 2005; Wake and Vredenburg, 2008; Chinchar et al., 2009; Gray et al., 2009; Robert, 2010). Moreover, because study of ranavirus replication and host–virus immune interactions have markedly advanced our understanding of amphibian immunity and iridovirid (a generic term referring to all members of the family) biology, these areas will also be considered in this review.

9.2 RANAVIRUS TAXONOMY

The genus *Ranavirus* constitutes one of five genera within the family Iridoviridae. Of these five genera, three (*Lymphocystivirus, Megalocytivirus*, and *Ranavirus*) infect cold-blooded vertebrates such as bony fish (class Osteichthyes), amphibians, and reptiles, whereas two (*Iridovirus* and *Chloriridovirus*) infect invertebrates, primarily insects and crustaceans (Chinchar et al., 2009; Jancovich et al., in press). Although most evidence supports the division of the family Iridoviridae into two subfamilies, one infecting vertebrates and the other invertebrates, recent work suggests that at least one member of the family may infect both insects and reptiles (Weinmann et al., 2007). If correct, this observation might explain the persistence of ranavirus infections in vertebrate populations via invertebrate reservoirs.

As shown in Table 9.2, the genus *Ranavirus* is composed of six recognized species (FV3, ATV, BIV, European catfish virus (ECV), epizootic hematopoietic necrosis virus (EHNV), and Santee-Cooper ranavirus (which includes the fish pathogen, largemouth bass virus, LMBV)) and several tentative species (Singapore grouper iridovirus (SGIV) and Rana

TABLE 9.2 Taxonomy of the Family Iridoviridae

Family: Iridoviridae
- Tentative subfamily: Chordavirinae
  - Genus: *Ranavirus*
    - Species: *Frog virus 3* (FV3), *Ambystoma tigrinum virus* (ATV), *Bohle iridovirus* (BIV), *epizootic hematopoietic necrosis virus* (EHNV), *European catfish virus* (ECV), *Santee-Cooper ranavirus* (SCRV)
    - Tentative species/isolates: Singapore grouper iridovirus (SGIV), Grouper iridovirus (GIV), soft-shelled turtle iridovirus (STIV), Rana catesbeiana virus-Z (RCV-Z); tiger frog virus (TFV), Rana grylio virus (RGV)
  - Genus: *Lymphocystivirus*
    - Species: *Lymphocystis disease virus-1* (LCDV-1), *lymphocystis disease virus-China* (LCDV-C)
    - Tentative species: LCDV-2
  - Genus: *Megalocytivirus*
    - Species: *infectious spleen and kidney necrosis virus* (ISKNV)
    - Tentative species: red sea bream iridovirus (RSIV), rock bream iridovirus (RBIV), orange-spotted grouper iridovirus (OSGIV)
- Tentative subfamily: Invertavirinae
  - Genus: *Iridovirus*
    - Species: *invertebrate iridescent virus 6* (IIV-6), *invertebrate iridescent virus 1* (IIV-1)
    - Tentative species: IIV-2, -9, -16, -22, -23, -24, -29, -30, -31
  - Genus: *Chloriridovirus*
    - Species: *invertebrate iridescent virus 3* (IIV-3)

This table is based on the current 8th Report and the pending 9th Report of the International Committee on Taxonomy of Viruses (Chinchar et al., 2005; Jancovich et al., in press). Taxonomic names currently approved by the ICTV are italicized, whereas tentative superfamily, species, and strain designations are shown in standard face type.

catesbeiana virus-Z (RCV-Z)). Four species (FV3, ATV, BIV, and RCV-Z) infect frogs or salamanders, whereas the other four infect various species of fish. Although ranaviruses likely infect reptiles, these isolates have not been fully characterized and as a result it is not clear if they represent separate viral species or strains of currently isolated viruses. In support of the latter view, sequence analysis of several viral isolates from various turtle species were found to be remarkably similar to FV3 (Mao et al., 1997; Huang et al., 2009). At present, the most attention has been focused on two species within the family: FV3 and ATV. With reference to FV3, considerable effort has been directed at understanding replicative events in FV3-infected cells and antiviral immunity in FV3-infected *Xenopus laevis* (see below). In contrast, studies with ATV have dealt primarily with ecological concerns such as the impact of infection on native salamander populations and the mechanism of virus spread (Brunner et al., 2004; Storfer et al., 2007; Schloegel et al., 2009; Picco and Collins, 2008). For these reasons, the remainder of this review will concentrate on FV3 and ATV with discussion of other "frog" and "salamander" ranaviruses as appropriate.

9.3 RANAVIRUS MORPHOLOGY AND REPLICATION

As a family, iridovirids are readily distinguished from other animal viruses by virion size (~150–300 nm depending upon the genus), genomic structure (dsDNA that is circularly permuted and terminally redundant), capsid symmetry (icosahedral), and the presence of morphologically distinct assembly sites and paracrystalline arrays within the cytoplasm of infected cells (Chinchar et al., 2009; Jancovich et al., in press). Ranavirus virions are ~150 nm in diameter; their genomes range in size from ~100–145 kb (Table 9.3) and contain between ~100 and 140 open reading frames (ORFs, Eaton et al., 2007). Twenty-six genes are common to all members of the family and most have known or putative functions (Eaton et al., 2007). The remainders are poorly conserved between genera and, for the most part, have not been assigned definitive functions.

Most of what is known about ranavirus replication has resulted from studies using FV3, the type species of the genus. Because ranavirus replication has recently been reviewed (Williams, 1996; Williams et al., 2005; Chinchar et al., 2009) only a brief summary will be presented here. Ranavirus

TABLE 9.3 Iridovirid Genomic Features

Genus	Virus[a]	Host	Genome Size (bp)	% GC Content	No. of ORFs
Ranavirus	ATV	Salamanders	106,332	54	92
	EHNV	Fish	127,011	54	100
	FV3	Frog	105,903	55	97
	TFV	Frog	105,057	55	103
	STIV	Turtles	105,890	55	105
	SGIV	Fish	140,131	48	139
	GIV	Fish	139,793	49	139
Megalocytivirus	ISKNV	Fish	111,362	55	117
	OSGIV	Fish	112,636	54	116
	RBIV	Fish	112,080	53	116
Lymphocystivirus	LCDV-1	Fish	102,653	29	108
	LCDV-C	Fish	186,247	27	178
Iridovirus	IIV-6	Insects	212,482	29	211
Chloriridovirus	IIV-3	Insects	190,132	48	126

[a]The abbreviations referred to here are defined in Table 9.2. GenBank accession numbers for all 14 viral species/strains are found within Jancovich et al. (2010).

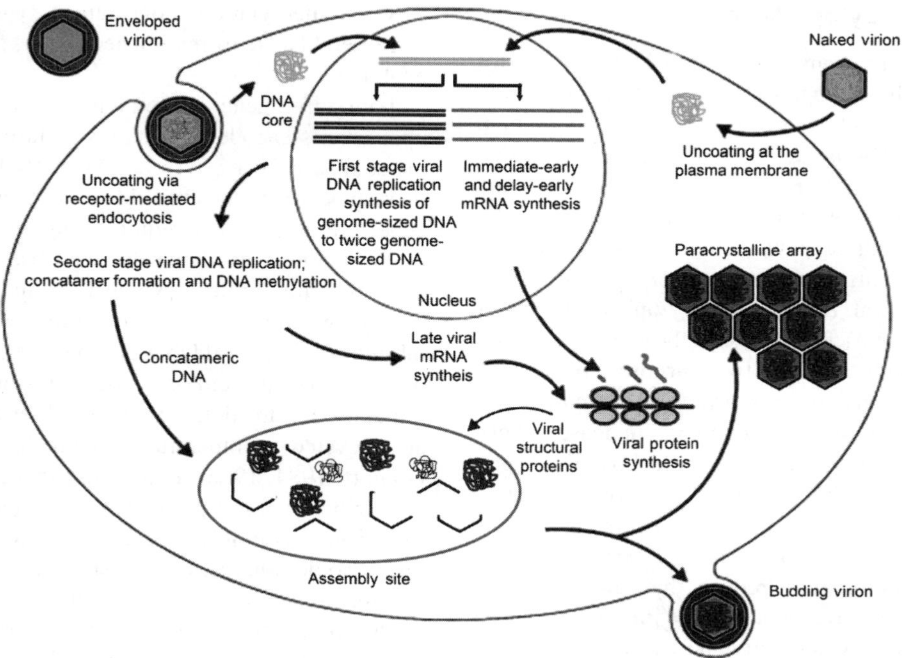

FIGURE 9.1 Life cycle of FV3. A schematic outline of events in FV3-infected cells is shown. See text for details. (Republished with permission from Springer, Chinchar, 2002.) (*See the color version of this figure in Color Plates section.*)

replication begins with the binding of virions to receptor(s), currently unknown, on the surface of susceptible cells (Figure 9.1). Interestingly, the receptor(s) present on cultured cells may be different from that used *in vivo* since the broad host range observed in culture is not seen when challenge assays are conducted with whole animals. Virion uncoating takes place at the plasma membrane (for nonenveloped virions) or within endocytic vesicles (for enveloped virions) and is followed by the transport of viral DNA into the nucleus where early viral replicative events occur.

9.3.1 Nuclear Events

Within the nucleus viral genomic DNA is transcribed by host RNA polymerase II to yield ~33 immediate-early (IE) messages, one of which is required for the subsequent transcription of delayed-early (DE) messages (Majji et al., 2009). Interestingly, naked viral DNA is not infectious and at least one virion-associated protein is required for the transcription of IE genes by host RNA polymerase II (Willis and Granoff, 1985). As with other large DNA viruses, IE messages likely encode regulators of the virus replication cycle as well as genes that impair the host's antiviral response, whereas the 22 DE messages encode catalytic proteins such as the viral DNA polymerase, subunits of the viral transcriptase, a D5-NTPase, and so on (Tan et al., 2004; Eaton et al., 2007; Majji et al., 2009). Collectively, IE and DE transcripts are considered early (E) viral messages and their synthesis, at least initially, likely takes place in the nucleus. Following the synthesis of the viral DNA polymerase, genome-sized copies of the viral genome are generated within the nucleus and subsequently transported to the cytoplasm (Goorha, 1982).

9.3.2 Cytoplasmic Events

Following transport of viral DNA molecules into the cytoplasm, two additional DNA-related synthetic events take place. In the first, viral DNA is methylated by a virus-encoded DNA methyltransferase (DMTase) (Willis and Granoff, 1980; Willis et al., 1984). However, in an alternative view, methylation takes place in the nucleus prior to transport of newly synthesized viral DNA to the cytoplasm (Schetter et al., 1993). Regardless of the site, methylation is thought to serve two purposes. In the first place, it may protect viral DNA from degradation by a virus-encoded endonuclease, whose function may be to degrade host DNA and provide additional nucleotides for subsequent high levels of viral DNA synthesis (Willis et al., 1984; Agarkova et al., 2006). In addition, methylation may reduce the possibility that viral DNA triggers an innate immune response mediated by interaction between unmethylated viral DNA and Toll-like receptor 9 (TLR9) or cytoplasmic DNA sensors such as DAI, RNA polymerase III, and AIM2 (Krug et al., 2004; Hoelzer et al., 2008; Kawai and Akira, 2009, 2010; Rathinam et al., 2010). The second cytoplasmic DNA synthetic event is the formation of large concatamers comprised of multiple copies of genome-sized viral DNA (Goorha, 1982). Concatamers are thought to be required for DNA packaging. It is not clear whether methylation and/or concatamer formation take place within, or outside, viral assembly sites. The observation that viral assembly sites are perinuclear structures suggests that DNA transport, methylation, concatamer formation, and assembly site formation may be linked. However, the ability to form assembly sites in the absence of methylation and full DNA synthesis, that is, within cells infected by temperature-sensitive mutants that synthesize markedly reduced levels of viral DNA at nonpermissive temperatures or in the presence of azacytidine, an inhibitor of methylation, suggests that only E proteins may be needed for assembly site formation (Goorha et al., 1984; Chinchar et al., 1984; Sample, 2010). Clearly the relationship between methylation, concatamer synthesis, and assembly site formation needs to be further explored.

Late viral gene expression is dependent upon a virus-encoded RNA transcriptase (Sample et al., 2007) and in its absence only E viral messages are synthesized. Moreover, full late gene expression also requires ongoing viral DNA synthesis as shown by the observation that late viral protein synthesis is markedly reduced when DNA synthesis is blocked using inhibitors of the viral DNA polymerase such as phosphonoacetic acid or temperature-sensitive mutants that are defective in DNA synthesis at elevated temperature (Chinchar and Granoff, 1984, 1986). Late transcripts encode primarily structural proteins and are required for virion assembly. Genetic and electron microscopic analyses suggest that at least a dozen viral gene products are likely involved in virion morphogenesis either as structural components of the virion or as catalytic or scaffolding proteins (Chinchar and Granoff, 1986; Sample, 2010). Virion formation is thought to proceed in a manner similar to that of African swine fever virus (Andres et al., 1998; Tulman et al., 2009). Based on that model, the FV3 major capsid protein (MCP) likely binds to one surface of fragmented cellular membranes that have been recruited to assembly sites, while other viral proteins, for example, 53R, bind the opposite face (Whitley et al., 2010). The continued addition of the MCP on one face of the membrane and p53, and perhaps other viral proteins, on the opposite face result in the formation of crescents and partially formed icosahedrons that eventually enclose (or acquire) the viral genome and form mature virions. Virus particles display (from inside out) an electron-dense DNA–protein core, an inner protein shell, an internal cell-derived lipid membrane, and an outer protein layer composed primarily of the MCP. Viral DNA packaging is thought to occur via a headful mechanism and results in genomes that are circularly permuted and terminally redundant. After their formation, mature virions either accumulate within cytoplasmic paracrystalline

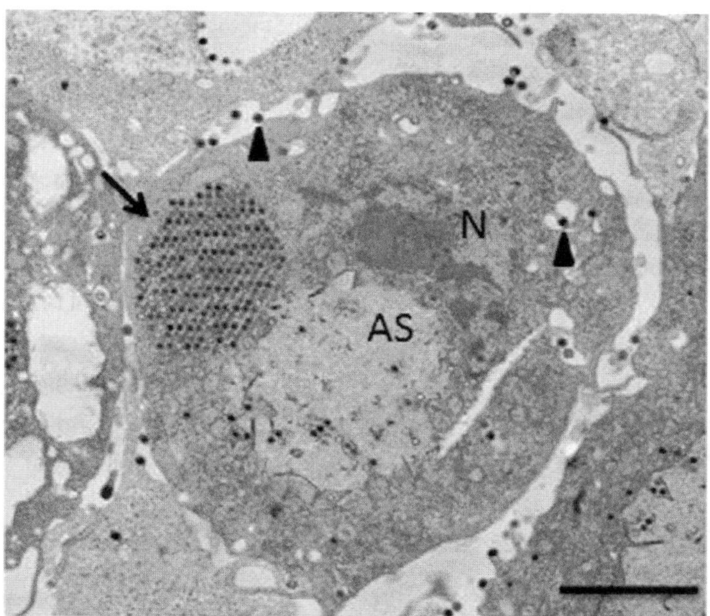

FIGURE 9.2 Transmission electron micrograph of FV3-infected fathead minnow (FHM) cells. Fathead minnow cells were infected with FV3 at an MOI of 20 PFU/cell and processed for electron microscopy 24 h later. Nucleus, N, showing chromatin condensation indicative of apoptosis; viral assembly site, AS, containing a small number of full and empty virus particles; a large paracrystalline array of virus particles, arrow; and individual virus particles budding from the plasma membrane or into an intracytoplasmic vesicle, arrow head. (Republished with permission from Elsevier, Whitley et al., 2010.)

arrays or migrate to the plasma membrane where they bud from infected cells and acquire a viral envelope (Figure 9.2). In cultured cells, the vast majority of FV3 virions remain cell associated and are released by cell lysis. In contrast to many other viruses, the presence of a viral envelope, although enhancing infectivity, is not specifically required for infectivity *in vitro*, as interaction can occur between nonenveloped (naked) virions and cellular receptors and lead to productive infections.

9.4 VIRAL GENOMES

9.4.1 Phylogenetics

To date, the genomes of 14 members of the family Iridoviridae (seven ranaviruses, three megalocytiviruses, two lymphocystiviruses, and one species each from the *Iridovirus* and *Chloriridovirus* genera) have been fully sequenced (see Jancovich et al. (2010) and Huang et al. (2009) for sequence information regarding these isolates). Within the genus *Ranavirus*, the complete genomes of FV3, ATV, EHNV, SGIV, and soft-shelled turtle iridovirus (STIV) have been determined as well as the genomes of tiger frog virus (TFV) and grouper iridovirus (GIV), likely strains of FV3 and SGIV, respectively. Although isolated from different classes (Reptilia and Amphibia), STIV and FV3 are strikingly similar. In fact, STIV, like FV3, is the only other known ranavirus that contains a truncated version of the viral homologue of eukaryotic translational initiation factor 2α (vIF-2α, Tan et al., 2004; Huang et al., 2009). These results suggest that FV3, STIV, and related isolates (e.g., TFV, turtle virus 3, turtle virus 5, etc.) may constitute a single viral species capable of infecting frogs, turtles, and other cold-blooded animals (Clark et al., 1968; Mao et al., 1997; Huang et al., 2009).

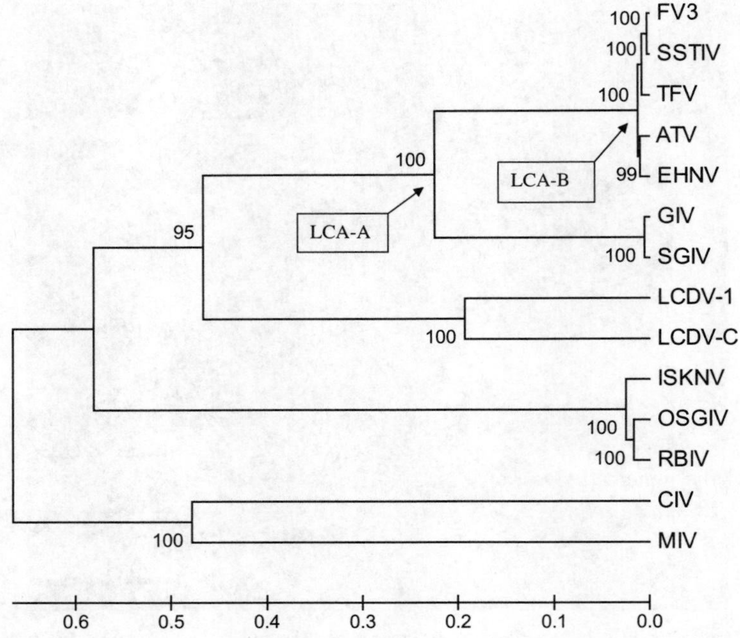

FIGURE 9.3 Phylogenetic relationships within the family Iridoviridae. The evolutionary history of iridovirid species/strains was inferred using the neighbor-joining method (Saitou and Nei, 1987) as accessed through MEGA4 (Tamura et al., 2007). The optimal tree with a branch length = 3.38276371 is shown. The percentage of replicate trees in which taxa clustered together after bootstrap analysis (1000 replicates) is shown next to the branches (Felsenstein, 1985). The phylogenetic tree was linearized assuming equal evolutionary rates in all lineages (Takezaki et al., 1995) and drawn to scale, with branch lengths in the same units as those of the evolutionary distances used to infer the phylogenetic tree. Evolutionary distances were computed using the Poisson correction method (Zuckerandl and Pauling, 1965) and the units indicate the number of amino acid substitutions per site. All positions containing gaps and missing data were eliminated from the data set (complete deletion option). There were a total of 9769 positions in the final data set representing a set of 26 concatenated genes common to all iridovirus species and strains sequenced to date. The last common ancestors (LCA) that gave rise to the GIV/SGIV and FV3/SSTIV/TFV/ATV/EHNV (LCA-A) and FV3/SSTIV/TFV and EHNV/ATV lineages (LCA-B) are indicated. Abbreviations are identical to those in Table 9.2 with the exception of soft-shelled turtle iridovirus (SSTIV).

Phylogenetic analysis (Figure 9.3) based on a concatenated set of the 26 iridovirus core genes supports the organization of the family into five genera, and indicates that the genus *Ranavirus* can be separated into three groups comprising ATV/EHNV, FV3/STIV/TFV, and GIV/SGIV (Jancovich et al., 2010). The ATV and FV3 groups, while clearly distinct, share greater than 90% amino acid identity, whereas SGIV/GIV are only distantly related. Jancovich et al. (2010) argue that phylogenetic and dot-plot analyses (which examine genomic organization) support the view that the last common ancestor (LCA) of extent ranaviruses was a virus (LCA-A) that infected fish (Figure 9.3). Furthermore, colinearity of the EHNV and ATV genomes suggests that ATV may have recently evolved from an EHNV-like fish virus (LCA-B, Figure 9.3), and that host shifts into frogs and salamanders subsequently took place giving rise to the FV3 and ATV lineages. Interestingly, the EHNV genome is larger than the genomes of amphibian-specific ranaviruses (127 kb versus 105–106 kb), but smaller than SGIV/GIV (140 kb). Despite the fact that the EHNV genome is 19% larger than ATV, there is only a 9% increase in the number (8) of ORFs suggesting that most of the extra sequences are noncoding. The marked sequence conservation between ATV and EHNV (i.e., >95%

sequence similarity) suggests that the jump from fish to salamanders may have required mutation of only a few critical amino acids within the major capsid protein or another key protein. This scenario is supported by the observation that feline parvovirus successfully made the jump from cats to dogs by a change in the capsid protein of as few as two amino acids (Hueffer et al., 2003; Shackelton et al., 2005). Once established in salamanders, it is possible that the subsequent loss of genes not needed for replication in an amphibian resulted in the more compact genome currently seen with ATV. The loss of "nonessential" genes as the EHNV-ancestor became adapted to salamanders may reflect the generally "weaker" salamander immune system compared to fish or frogs (Tournefier et al., 1998). If so, the lost genes might be those encoding immune evasion proteins that are no longer needed for virus replication in salamanders.

9.4.2 Gene Function

Twenty-six core genes are present within all 14 sequenced members of the family Iridoviridae and putative functions have been assigned to 21 of them (Eaton et al., 2007; Jancovich et al., 2010). An additional 40 genes are common to the amphibian-specific ranaviruses (i.e., ATV, FV3, TFV), but in contrast to the former, the functions of most are unknown. The 21 core genes with known or presumed function appear to play key roles in viral DNA and RNA metabolism and virion assembly. Among the remaining 40 genes, only 4 have been linked with any proteins outside the family Iridoviridae, that is, a viral caspase activation and recruitment domain (CARD) motif-containing protein (vCARD), a homologue of dUTPase, a neurofilament triplet H1-like protein, and a viral homologue of eIF-2α (vIF-2α). We speculate that, like poxviruses (Alcami and Koszinowski, 2000; Tortorella et al., 2000; Johnston and McFadden, 2003; Findlay and McFadden, 2006; Mohamed et al., 2009.), iridoviruses likely encode multiple proteins , e.g., vCARD, vIF-2 alpha, cytosine DNA methyltransferase, beta-hydroxysteroid dehydrogenase, and an RNase III-like protein (R3LP), that inhibit host cell immune responses, whereas other proteins, such as homologues of dihydrofolate reductase (DHFR), thymidine kinase, ribonucleotide reductase, and perhaps dUTPase, may play roles in nucleotide synthesis and host range (Chinchar et al., 2009). A brief description of the role of four putative viral immune evasion genes follows.

9.4.2.1 vCARD In vertebrate cells virus-templated or synthetic dsRNA (e.g., poly [I:C]) induces the synthesis of interferon (IFN) and over 200 IFN-stimulated genes (ISGs) such as major histocompatibility complex (MHC) class I/II proteins, ISG-15, protein kinase R (PKR), 2′5′-oligoadenylate synthetase (OAS), RNase L, and so on, following interaction with one or more cellular receptors such as Toll-like receptor 3 (TLR3), PKR, retinoic acid-induced gene 1 (RIG-I) or MDA-5 (Levy and Marie, 2004; Yoneyama et al., 2004; Sen and Sarkar, 2005; Hausmann et al., 2008; Kawai and Akira, 2009, 2010). PKR, OAS, and RNase L are activated by dsRNA and block protein synthesis in virus-infected cells, MHC I/II present viral antigens to CD8+ cytotoxic T cells (CTLs) and CD4+ T helper cells, and ISG-15 covalently modifies a large number of cellular proteins and alters their properties (Loeb and Haas, 1992; Kerscher et al., 2006; Majji et al., 2006; Sadler and Williams, 2008; Harty et al., 2009). Induction via RIG-I and MDA-5 is mediated by interaction between CARD motifs within these cytoplasmic dsRNA receptors and CARD motifs in downstream signaling molecules such as MAVS/Cardiff/IPS-1 (Biacchesi et al., 2009). It is possible that the ranavirus CARD motif-containing protein, designated vCARD, interacts with either RIG-I/MDA-5 or MAVS/IPS-1 and short circuits the signaling pathway and, as a consequence, blocks the induction of IFN. In addition, because RIG-I and MDA-5 also participate in the induction of apoptosis, vCARD may also inhibit programmed cell death (Hofmann et al., 1997; Bouchier-Hayes and Martin, ; Werts et al., 2006; Besch et al., 2009; Lei et al., 2009). Finally, because CARD motifs are also found in other immune-related signaling proteins such as CARD9, interruption of these pathways may also lead to immune suppression (Colonna, 2007; Hara et al., 2007).

9.4.2.2 RNase III-Like Protein (R3LP)

FV3, along with other ranaviruses, encodes a protein with marked similarities to bacterial RNase III, an enzyme that degrades dsRNA (Eaton et al., 2007). This observation is not unique to ranaviruses since several other viruses encode proteins that bind dsRNA, prevent its interaction with PKR or other cellular sensors, and block translational shutoff, IFN induction, and a proinflammatory response (Kreuze et al., 2005; Fenner et al., 2006; Hussain et al., 2010). One of the best-known examples is the Vaccinia virus (VV) protein E3L that binds dsRNA, blocks PKR activation, and prevents the turnoff of protein synthesis in virus-infected cells (Langland and Jacobs,). In addition, there is evidence, albeit conflicting, that E3L binds and sequesters small interfering RNA (siRNA) and thus blocks RNA interference in Vaccinia virus-infected cells (Li et al., 2004; Lantermann et al., 2007). Supporting this hypothesis, Hussain et al. (2010) showed that an ascovirus RNase III-like protein suppressed siRNA-mediated knockdown by degrading dsRNA. This latter study is especially relevant since ascoviruses and iridoviruses are phylogenetically linked (Federici et al., 2009). Lastly, Zenke and Kim (2008) showed that an RNase III-like protein encoded by Rock bream iridovirus (genus *Megalocytivirus*) specifically cleaves dsRNA. Collectively, these results indicate that proteins capable of binding dsRNA may play multiple roles in viral metabolism.

9.4.2.3 3β-Hydroxysteroid Dehydrogenase

β-HSD is required for the synthesis of progesterone, mineralocorticoids, and glucocorticoids (GCs) (Sun et al., 2006). Glucocorticoids are potent immunosuppressive and anti-inflammatory agents that display marked beneficial effects in organ transplantation and the control of autoimmune and inflammatory disease (Rhen and Cidlowski, 2005). Conversely, GCs increase the severity of VV infections *in vivo* (Moore and Smith, 1992; Reading et al., 2003). To determine the role of the VV homologue of β-HSD in viral replication and pathogenesis, Reading et al. (2003) examined the *in vitro* and *in vivo* impact of infections initiated by wild type (wt) VV and a β-HSD knockout (KO) mutant. They found that mice infected with the β-HSD KO mutant showed fewer signs of illness, lower virus titers, higher levels of CD4 and CD8 cells, higher levels of IFN-γ, enhanced VV-specific CTL activity, and decreased cortisone levels than those infected with wt virus. Despite these marked differences *in vivo*, wt and KO virus replicated equally well in both single- and multicycle infections *in vitro* indicating that β-HSD was not required for replication in cell culture. Collectively these results suggest that VV β-HSD expression leads to elevated glucocorticoid levels that suppress immunity and lead to more severe disease. If ranavirus β-HSD behaves the same way, similar results should be seen *in vitro* and *in vivo*.

9.4.2.4 vIF-2α

To date, most ranavirus genes have been assigned functions based on homology to known proteins from other species. While this approach is often correct, the marked sequence diversity between viruses infecting lower vertebrates and mammals makes such a strategy open to error. For example, the ranavirus vIF-2α protein shows limited sequence similarity to the K3L protein of Vaccinia virus and to zebrafish eIF-2α suggesting that, like K3L, it plays a role in binding protein kinase R and maintaining viral protein synthesis during infection (Langland and Jacobs, ; Majji et al., 2006). While this is an attractive hypothesis, K3L (88 amino acids) and vIF-2α (256–281 amino acids depending upon the isolate) differ markedly in size, and the ~200 amino acids that make up the C-terminus of vIF-2α do not match zebrafish eIF-2α or any other sequences within the database except those from other ranaviruses. Thus, the ranavirus vIF-2α protein may have two functions: a K3L-like activity encoded by its NH-terminal end, and a completely novel function located in the C-terminal two-thirds of the protein.

Recently, attempts to determine the function of various ranavirus proteins have been undertaken using antisense approaches. Sample et al. (2007) showed through the use of antisense morpholino oligonucleotides (asMOs) that the largest subunit of the FV3 homologue of RNA polymerase II is required for the synthesis of late viral messages and is likely a critical component of the viral transcriptase. Other studies utilizing asMOs and siRNA will likely extend these studies and confirm gene function by linking gene knockdown to changes in phenotype (Xie et al., 2005; Kim et al., 2010; Sample, 2010). Ultimately, the introduction of site-directed mutations, or the deletion of specific viral genes using homologous recombination will allow us to assess the phenotype of specific ranavirus mutations *in vivo*. This approach, which has proven so powerful with Vaccinia virus and African swine fever virus (Ward et al., 2003; Johnston and Ward, 2009; Epifano et al., 2006a, 2006b), has proven difficult to adapt to the ranavirus system, but preliminary evidence suggests that it is possible given the use of the appropriate constructs and selective agents (B. Ward, J. Robert, and J. Jancovich, personal communication).

9.5 THE ROLES OF INNATE AND ACQUIRED IMMUNITY IN DETERMINING THE OUTCOME OF RANAVIRUS INFECTION

Discussions of viral ecology are markedly different from discussions of plant and animal ecology. Whereas most considerations of ecology involve the relationship of an organism to external living (e.g., competitors, predators, or prey,) and/or inanimate (e.g., climate, water, stressors, toxins, food supply, etc.) entities, viral ecology involves an intimate relationship between a parasitic organism (the virus) and its host. In view of that, the susceptibility of an organism to infection (i.e., the permissiveness of the host) depends upon the presence of receptors and other cellular gene products required for a productive infection, and the immunocompetence of the host, which can be influenced by the above-mentioned extrinsic factors as well as by the developmental stage, that is, very young and old animals are more susceptible to ranavirus infection. While considerable information exists concerning the immune response of humans and other mammals to viral infections, relatively little is known about immunity to viral infections in lower vertebrates. Here we will summarize the nature of amphibian immunity in general and then describe events in a model organism, *X. laevis*, following infection with FV3.

9.5.1 Amphibian Immunity

Although showing marked divergence in the amino acid sequence of key immune molecules, the immune responses of model amphibians such as *X. laevis* are markedly similar to those of mammals (DuPasquier, 2001; Hughes and Yeager, 1997; Zou et al., 2007; Bird et al., 2006; Robert and Ohta, 2009; Guselnikov et al., 2010). Frogs (anurans) generate both humoral (antibody) and cell-mediated responses to virus infection, produce a panel of bioactive cytokines, and possess NK-like cells, antimicrobial peptides, T-cell receptors α, β, γ, δ, Toll-like receptors, MHC class I and II surface receptors, TNFα, interferon (IFN), and other features of a mammalian-like response (Robert and Ohta, 2009; Qi et al., 2010). The frog immune system is not, however, simply a cold-blooded version of the mammalian immune system. For example, amphibians lack efficient affinity maturation of an antibody response. As in mammals, immunogloblulin switching and hypermutation take place, but the increase in affinity is only about 10-fold, much lower than that seen in mammals. Selection among hypermutations apparently does not occur and may be due to the absence of germinal centers, and as a result much lower antibody titers are produced following secondary immunization (Charlemagne and Tournefier, 1998; DuPasquier, 2001; Kasahara et al., 2004). In addition, anuran larvae lack MHC class I and thus do not possess cytotoxic T lymphocytes (CTLs). Moreover, the profound restructuring of frogs during metamorphosis leaves them especially vulnerable to infection (Rollins-Smith, 1998; Carey et al., 1999). In contrast to frogs, in which the principal features of the immune system are reasonably well described, little is known of the immune system of caudate amphibians. What is clear, however, is that they do not respond as well to allogeneic transplants as their tailless brethren, and that even adults are targets for life-threatening ATV-induced disease. A brief description of immunity in the frog *X. laevis* and the Mexican axolotl follows.

9.5.2 FV3 Infection in *X. laevis*: Antiviral Responses in a Model Anuran

Elucidation of antiviral immunity in amphibians has been profitably examined in the FV3 *Xenopus* model established by Robert and his coworkers. Gantress et al. (2003) showed that immunocompetent *Xenopus* were productively infected following injection of 5×10^7 PFU of FV3. Within 8–9 days post infection (dpi) virus was detected within multiple organs (kidney, spleen, intestine, skeletal muscle, and lung) with the kidney appearing to be the major organ affected. In immunocompetent adults, infection was not life threatening and although infected adults showed signs of clinical illness (e.g., lethargy, loss of appetite, erythema) these symptoms disappeared within 2–3 weeks and by 30 dpi were resolved as indicated by the absence of viral DNA. Moreover, anti-FV3 IgY antibodies were detected with 1 week after subsequent reinfection, which occurred 1 month after initial exposure, and peaked at 3 weeks p.i. Further evidence of long lasting immunological memory was obtained by challenge with FV3 4 months after the initial infection. In this case, there was a marked acceleration of viral clearance, no clinical illness, and a strong anti-FV3 IgY response. In contrast to immunocompetent adults, *Xenopus* larvae were extremely sensitive to infection and as few as 100 PFU of FV3 were capable of killing 80–100% of infected animals. However, despite their marked susceptibility to FV3 infection, larvae are to some extent immunocompetent. For example, they generate an antibody response to DNP-KLH and other antigens. They also possess T cells and demonstrate a mixed lymphocyte response. In addition, both B-cell and T-cell memory last through metamorphosis. FV3-infected tadpoles upregulate AID (activation-induced cytidine deaminase), which is involved in class switching, and hypermutations are increased in infected larvae (Marr et al., 2007). Finally, tadpoles from different genetic backgrounds are differentially susceptible to FV3 (Gantress et al., 2003).

In subsequent studies, Robert et al. (2005) showed that FV3 targets the renal proximal tubular epithelium. However, in immunocompetent adults, FV3 infections are confined to discrete foci that rapidly decrease in size by 2 weeks p.i. Antiviral CTL responses were shown to play a major role in protection from severe disease since sublethal γ-irradiation or treatment with a monoclonal antibody (mAb) targeted to CD8 resulted in a dramatic increase in susceptibility and mortality due to FV3 infection. Moreover, γ-irradiated FV3-infected frogs (but not immunocompetent, FV3-infected frogs) released sufficient virus into the water to infect cohabiting uninfected γ-irradiated adults and larvae. Treatment with a mAb targeting IgM did not result in increased susceptibility to FV3 infection suggesting antiviral antibodies may not be as important as functional CTLs in protecting frogs from clinically apparent FV3 disease.

A more detailed examination of the humoral response to FV3 showed that immunization of *Xenopus* on days 0 and 14 did not result in an appreciable antibody response over the next 28 days (Maniero et al., 2006). However, challenge on day 42 resulted in a marked antibody response 7–14 days later. Corresponding to the marked rise in Ab levels, an examination of lymphocyte populations in the spleen showed that there was an overall increase in the number of splenocytes by day 3, and specific increases in CD5+, CD8+, MHC class II$^+$ as well as in IgM+ B cells. Lastly, Maniero et al. (2006) showed that at a 1:10 dilution, serum from immunized animals protected A6 monolayers from infection, reduced virus titers by 10-fold, and provided partial protection of susceptible tadpoles against *in vivo* challenge with FV3. Their work showed that protective neutralizing Abs were generated during secondary FV3 infection and suggest that immunological memory in anurans may be sufficient to provide significant defense against subsequent viral infection.

To explore the role of CD8+ CTL in FV3 infections, Morales and Robert (2007) performed *in vivo* proliferation assays using BrdU incorporation to monitor the expression of activation-induced CD8+ cells. Frogs were infected with FV3 and assayed for proliferating CD8+ cells 3, 6, 9, and 14 days after infection. Two days prior to assay, frogs were incubated in water containing BrdU and on the day of assay splenocytes were isolated and examined by two-color flow cytometry to detect expansion of CD8+ cells. A significant proliferation of CD8+ cells was detected by days 6 and 9, and migration of CD8+ cells into infected kidneys was observed from day 3 to day 6. Upon secondary infection, proliferation of splenic CD8+ cells took place earlier (day 3) than during a primary infection. Interestingly, the number of CD8+ cells that infiltrated infected kidneys following a secondary infection was significantly higher than in uninfected controls, but not as high as seen during a primary response.

Taken together, the results of these last two studies suggest that during a primary FV3 infection, CD8+ anti-FV3 CTLs play a major role in resolving the infection, whereas anti-FV3 Abs play little, if any, part. In contrast, secondary infection with FV3 is characterized by a robust anti-FV3 Ab response, but a much more restrained CTL response. This type of response is similar to that seen following infection with Vaccinia virus in which CTL activation is marked during a primary infection, whereas B-cell responses involving the generation of neutralizing Ab appear early after secondary infection (Buller and Pallumbo, 1991). Likewise, in another poxvirus, ectomelia virus, CD8+ T cells were shown to be unnecessary for resistance of mice to subsequent reinfection (Panchanathan et al., 2005). Thus, antiviral CTL and Ab responses may also play complementary roles in protecting *Xenopus* during primary and secondary infections with FV3.

In addition to the induction of anti-FV3 humoral and cell-mediated responses, Morales et al. (2010) showed that *Xenopus* also mounted an innate antiviral response characterized by an expansion of activated mononucleated macrophage-like cells, NK cells, and the rapid upregulation of proinflammatory genes such as arginase 1, interleukin-1β, and TNFα. Specifically they showed that during a primary infection there were two to three times more peripheral leukocytes (PLs) in frogs at 6 dpi than in control frogs and that the relative fraction of macrophages within the PL population increased to 40% of total by 6 dpi. Following a secondary FV3 infection, the macrophage population showed a markedly quicker expansion than after a primary infection. In addition to an increase in the number and activation state of macrophages, Morales et al. (2010) observed an increase in NK cells and B cells following a primary infection. Interestingly, while the B-cell population was also markedly upregulated after secondary FV3 infection, the NK population remained the same as in control cells. Examination of T-cells numbers showed that following both primary and secondary infections with FV3, the numbers of T cells declined by 3 dpi and returned to baseline, or near baseline levels, by 6 dpi. Collectively, these data suggest that a potent innate immune response involving macrophages and NK cells is triggered by primary infection. The authors also examined the permissiveness of PLs to FV3 infection and found that a small fraction of the PL population could support FV3 replication. Moreover, whereas viral RNA was not detected in FV3-infected PLs beyond day 6, viral DNA could be detected as late as day 20 suggesting that transcriptionally inactive FV3 could persist and perhaps, following various stresses, reactivate and give rise to a productive infection. If virus could remain inactive within PL for several months, it may explain how infections could be passed from generation to generation and may establish *Xenopus*, or other frog species, as reservoirs of virus infection. In support of this view, Robert et al. (2007) have detected FV3-like DNA in a large fraction of apparently healthy *Xenopus* from various US suppliers.

9.5.3 Immune Responses in Urodeles

Compared to better studied *Xenopus* and fish systems, the immune system of the Mexican axolotl is considered "immunodeficient," a result consistent with the marked susceptibility of adult axolotls and the related tiger salamander to ATV infection. Compared to mammals, axolotls possess fewer immunoglobulin classes, display poor mixed lymphocyte reactions, and do not respond to soluble antigens (reviewed in Sammut et al., 1999; Cotter et al., 2008). Studies conducted more than 30 years ago suggested that antibody titers in a variety of urodeles were considerably lower than in mammals and reached peaks levels much later (i.e., 7–12 weeks) after exposure (Houdayer and Fougereau, 1972; Tournefier, 1975; Tournefier and Charlemagne, 1975; Charlemagne and Tournefier, 1977). However, new Ig isotypes have recently been detected in urodeles and the observation made that different families of urodeles express different Ig isotypes (Schaerlinger and Frippiat, 2008). In view of this, Schaerlinger and Frippiat (2008) suggested that antibody responses in urodeles need to be reinvestigated because new techniques could reveal different results and suggest a more robust humoral response than previously thought possible. Likewise, similar concerns arise in regard to the T-cell response. For example, the axolotl T-cell receptor and MHC repertoires are as wide as that seen in vertebrates (Fellah et al., 2002; Richman et al., 2007; Bos and DeWoody, 2005) and both MHC class I and II are polymorphic (Sammut et al., 1999). Given this, it may be that, as with the aforementioned antibody response, part of the inability to detect a vigorous mixed lymphocyte reaction or to demonstrate *in vitro* proliferation was due to the suboptimal culture conditions. In light of this, both the T- and B-cell aspects of salamander immunity will likely need to be revisited.

Recently urodele immunity was monitored using a first-generation axolotl microarray. Cotter et al. (2008) examined the transcriptional response in spleens and lungs of axolotls 1–6 days after ATV infection and identified 158 up- and 105 downregulated genes. As early as 24 h p.i., they observed evidence of a marked innate and antiviral immune response, and by 144 h p.i., the authors detected changes in the expression of cell death, inflammatory, and cytotoxicity-related genes. However, they did not observe evidence of lymphocyte proliferation in the spleen, which is associated with clearance of FV3 infections in adult *Xenopus*. Although suggestive, this finding may reflect the sensitivity of the assay used to detect the proliferation of virus-specific T cells. Among the genes upregulated at one or more time points after infection were IFN regulatory factor 1, OAS, Mx1, RNase L, CXCR4, and poly-ADP ribose polymerase. OAS and RNase L are two of the three principal proteins (PKR is the third) that are induced by IFN and that play critical roles in blocking viral protein synthesis (Sadler and Williams, 2008). These results suggest that, at least in the initial stages of infection, axolotls mount an antiviral defense. Why the axolotl immune response eventually proves inadequate to resolve an ongoing infection remains to be seen, but will be explored using a more complete second-generation axolotl array.

9.6 RANAVIRUS INFECTIONS OF AMPHIBIANS

9.6.1 ATV Infection

From an ecological point of view, the most extensively studied *Ranavirus*–host system is the *Ambystoma tigrinum* virus-tiger salamander model (Jancovich et al., 1997; Collins et al., 2004). Ranavirus strains infecting tiger salamanders are monophyletic relative to other ranaviruses and are designated *Ambystoma tigrinum virus* after the original isolate (Jancovich et al., 2003). Strains of ATV have been isolated from tiger salamander epizootics throughout western North America, ranging from south-central Arizona, USA, to Saskatchewan and Manitoba, Canada (Jancovich et al., 1997, 2005; Bollinger et al., 1999; Docherty et al., 2003; Storfer et al., 2007).

ATV is highly infectious and horizontal transmission occurs easily from sick to healthy tiger salamanders via water exposure, fomites, direct contact, or ingestion of infected individuals (Jancovich et al., 1997). ATV virions remain infectious in moist substrates, at least for a few days, and are a likely source of infection (Brunner et al., 2007). Vertical transmission is unknown, but ATV has been isolated from salamander testes (Docherty et al., 2003). Exposed salamanders can become infectious in as little as 2 days (Brunner et al., 2007), and highly infected animals show symptoms within 10–14 days and often die within 2–3 weeks of exposure (Forson and Storfer, 2006a, 2006b). Individuals that survive this critical window can harbor sublethal infections for months (Brunner et al., 2004) and may serve as a source of infection for naïve animals. Diseased animals have high viral loads in skin, spleen, lungs, and liver (Jancovich et al., 1997).

Die-offs in tiger salamander populations appear to be density dependent, making localized extinctions unlikely (Brunner et al., 2004; Greer et al., 2008). Transmission appears to occur primarily in the larval stage where densities are highest (Collins et al., 2004), with peak infection levels likely lagging behind peak larval densities due to the viral incubation period (Brunner et al., 2004). In addition to density, temperature is also critical for disease dynamics because the replication rate of ATV is temperature-dependent and temperature also affects host susceptibility (Rojas et al., 2005; Gray et al., 2007). Thus, although ATV grows fastest *in vitro* at 26°C, infection levels were 2.5 times higher in salamanders held at 18°C, probably due to immunosuppression at lower temperatures (Rojas et al., 2005). In addition, laboratory experiments suggest that the outcome of ATV exposure and infection is dosage dependent, with higher doses more likely to result in infection and disease than lower doses (Brunner et al., 2005).

Epizootics appear to be maintained by infected postmetamorphic individuals that overwinter in the terrestrial landscape and return to breeding ponds in the spring (Brunner et al., 2004). Indeed, sublethally infected metamorphic individuals were experimentally shown to harbor virus for over 7 months, exceeding overwintering duration (Brunner et al., 2004). Other species may also act as biotic reservoirs for the pathogen, as ATV has a relatively broad host range, including other ambystomid salamanders, eastern spotted newts (*N. v. viridescens*), and some anuran species (Jancovich et al., 2001; Forson and Storfer, 2006a, 2006b; Schock et al., 2008). However, infection of heterologous hosts is not universally achieved as Jancovich et al. (2001) found that ATV was not infectious to leopard frogs, bullfrogs, sunfish, mosquito fish, rainbow trout, and predaceous insects.

Because tiger salamanders are commonly found in the absence of other amphibians, salamanders themselves are thought to be the primary reservoir for ATV, serving as a key source of virus transmission both within and between years (Brunner et al., 2004). This type of intimate pathogen–host connection is expected to lead to a tight coevolutionary relationship, which is supported by four lines of research.

First, there is a negative correlation between disease frequency and cannibal frequency among salamander populations throughout Arizona (Pfennig et al., 1991; Collins, unpublished). Although cannibals enjoy a performance and fitness advantage by preying on conspecifics (Collins et al., 1993), a major fitness cost of cannibalism is enhanced risk of acquiring ATV due to increased contact rates with congeners. As a result, selection should favor non-cannibalistic individuals in populations with high disease prevalence because reduced cannibalism might prevent disease spread (Pfennig et al., 1991; Bolker et al., 2008). Common garden experiments support this view and suggest that cannibalism is least common where disease is most prevalent and reflects a genetic basis that is likely the result of past selection (Parris et al., 2005). Animals were not plastic in the development of the cannibalistic phenotype among virus present versus virus absent treatments, and observed differences of cannibalistic frequency in the field were replicated in the lab (Parris et al., 2005).

Secondly, viral dynamics within host populations suggest a coevolutionary history. Although larvae, branchiate adults and metamorphosed adults are all susceptible to ATV, larvae are significantly more likely to recover than metamorphs (Brunner et al., 2004). The larval stage is most important for transmission because epizootics occur at peak larval densities.

Thirdly, analysis of phylogenetic concordance suggests salamander–virus coevolution (Storfer et al., 2007). Without three host switches, a term that refers to the alteration of host range and that has been attributed to the movement of infected salamanders as fishing bait (Jancovich et al., 2005), there is complete concordance, with high correlation of nodal depths, between phylogenetic trees for both salamanders and virus (Storfer et al., 2007). This finding provides strong support for a coevolutionary history of ATV and its hosts (see Huelsenbeck et al., 1997).

Lastly, ATV strains vary in the amino acid sequence of putative virulence genes among different host populations suggesting that ATV is evolving differently in different host populations due to localized variation in host susceptibility and resistance (Ridenhour and Storfer, 2008).

In general, ATV-induced tiger salamander epizootics appear to result in larval population fluctuations. Long-term studies of tiger salamander populations on the Kaibab Plateau and San Rafael Valley, Arizona, USA, indicate populations most commonly recover from die-offs within 1–2 years (Greer et al., 2009). However, populations in Saskatchewan, Canada have failed to recover (Schock and Bollinger, 2005). The effects of long-term and frequent ATV-induced epizootics on natural tiger salamander populations are thus unclear, but preliminary genetic evidence suggests disease reduces effective population size by reducing the number of breeding adults and thus genetic variability (Storfer, unpublished). ATV is thought to be one factor implicated in the endangerment of the Sonora tiger salamander that faces several other threats to population viability including cattle grazing, introduction of nonnative invasive species (including bullfrogs and predatory fish), and introgression due to introduction of bait tiger salamanders (*A. t. mavortium*; Collins et al., 1988; Storfer et al., 2004).

9.6.2 Infections Due to FV3 and FV3-Like Viruses

The other main body of ranavirus work has focused on FV3-like viruses in amphibians. The discovery of FV3 in *Rana pipiens* (leopard frog) in the 1960s (Granoff et al., 1965), predates the description of ATV by over 30 years and follows the discovery of insect iridoviruses by 11 years (Williams, 1996). Granoff and other investigators (Granoff et al., 1966; Clark et al., 1968) isolated several "frog viruses" from ostensibly healthy and tumor-bearing leopard frogs (*Rana pipiens*). While most of these viruses likely represented reisolation of the same viral species, one isolate, FV3, was singled out for further study due to its putative association with kidney tumors. Although FV3 was later found to play no role in tumor development, intense study of its life cycle lead to its selection as the type species of the genus *Ranavirus*. Moreover, because FV3 could be isolated from "healthy" animals, it and similar viruses were viewed as agents of low pathogenicity. However, Wolf et al. (1968) challenged that benign view and reported the isolation and characterization of tadpole edema virus, a virus that now would likely be classified as a ranavirus, and showed it had the potential to cause marked disease in bullfrog tadpoles. Since that time, additional outbreaks of iridoviral disease have been detected in a variety of frog species in Europe, North and South America, Asia, and Australia (e.g., see Gray et al., 2009; Mazzoni et al., 2009; Une et al., 2009). Some isolates, for example, the Australian BIV and the North American RCV-Z isolates, may be distinct from FV3 (Majji et al., 2006; Hyatt et al., 2000), whereas others, for example, TFV and *Rana grylio* virus (RGV), are likely strains of FV3 (Zhang et al., 2001; Weng et al., 2002). While some of these outbreaks occurred in commercial settings and might be attributable to disease brought on by over-crowding or other types of stresses, other outbreaks have occurred in natural or agricultural settings where the proximal trigger is much less clear.

FV3-like ranavirus strains have been described in both anurans and urodeles, and appear to be as highly transmissive as ATV (Pearman et al., 2004; Harp and Petranka, 2006; Cunningham et al., 1996, 2007, 2008; Pasmans et al., 2008). Infections involve primarily the kidney, spleen, and gastrointestinal tract, but ocular infections have been noted (Chinchar, 2002; Burton et al., 2008; Gray et al., 2009). In general, larvae and recently metamorphosed individuals are more susceptible than adults, with the apparent exception of common frogs (*Rana temporaria*) and common toads (*Bufo bufo*) in the United Kingdom (Cunningham et al., 2007). The egg stage appears least susceptible (see Gray et al., 2009), but eggs have tested positive when collected in the wild (Duffus et al., 2008). In addition to variation in susceptibility among life history stages, susceptibility also varies among anuran species, with mortality rates varying at least threefold (Hoverman et al., 2010).

Similar to ATV, FV3-induced die-offs are frequent (Cunningham et al., 2007; Gahl and Calhoun, 2008), and in some cases, annual (Duffus et al., 2008). However, unlike ATV, a relationship between anuran densities and ranavirus presence is not supported in either wood frogs (Harp and Petranka, 2006) or a series of Tennessee farm ponds (Gray et al., 2007). In fact, one study suggests that dynamics are frequency-independent due to the presence of ranavirus reservoirs (Duffus et al., 2008).

Among native populations, animals most at risk may be those at high elevation or those that are geographically isolated. A study of a series of ponds in Acadia National Park (Maine, USA) showed that high catchment position was most significantly correlated with the propensity for a pond to experience ranavirus-induced die-offs (Gahl and Calhoun, 2008). Populations of the agile frog (*Rana latastei*) that were most geographically isolated, and consequently had the least genetic diversity, were more susceptible than more genetically variable populations (Pearman and Garner, 2005). Previous work on FV3 supports this result, showing that inbred *X. laevis* are more susceptible than outbred lines (Gantress et al., 2003).

In contrast, recent work of Teacher et al. (2009a) showed no appreciable declines in allelic richness following ranavirus epizootics in the common frog (*R. temporaria*).

However, decreases in heterozygosity as well as relatedness were observed suggesting epizootics led to behavioral changes and consequent assortative mating (Teacher et al., 2009a). Furthermore, comparison of MHC class I variability in frogs previously infected with FV3 versus uninfected frogs showed higher frequencies of particular alleles in previously exposed populations (Teacher et al., 2009b). In comparison, neutral molecular loci did not show similar patterns among common frog populations, reinforcing the apparent selection favoring particular MHC alleles (Teacher et al., 2009b). In sum, although (potentially naïve) inbred populations may be at higher risk than outbred populations, selective sweeps that reduce overall genetic variability in populations with higher exposure may confer some level of disease resistance. Clearly further work on the relationship between host genetic variability and ranavirus susceptibility is needed.

It should be noted that in many cases of ranavirus disease, the precise identification of the etiological agent is often lacking. Viruses have been tentatively identified as ranaviruses by ultrastructural analysis of virus-infected cells, and as FV3-like agents by sequence analysis of the major capsid protein or another conserved gene (Pasmans et al., 2008). While this approach is useful in identifying ranavirus-infected cells, high morphological and sequence similarities can lead to misidentification or identification at the genus level only. For example, the sequences of FV3 and RCV-Z are markedly similar within the MCP gene, but FV3, in contrast to RCV-Z, lacks a full-length vIF-2α gene, is less pathogenic in tadpoles, and shows a markedly different restriction fragment profile (Majji et al., 2006).

Overall, ranaviruses, when present, appear to be a significant part of the ecology of amphibian populations. In some cases, viruses appear to have been introduced via anthropogenic means, while in others, they may be endemic. In either case, ranaviruses appear to cause population fluctuations and in some cases, declines, thereby imposing a strong selective force on their amphibian hosts. In at least one case, host populations have evolved in response to this threat, with increases in frequencies of apparently adaptive antigen recognition (i.e., MHC) alleles (Teacher et al., 2009b). In another system, it appears as though the virus has responded to variation in host resistance via evolution of putative immune evasion and virulence genes (Ridenhour and Storfer, 2008). Given the wide host and geographic range of these viruses, combined with their propensity to spread via anthropogenic and other means, further research is needed on ranavirus–host coevolution, as well as their effects on host population dynamics.

9.7 CONSERVATION ISSUES

Because amphibians are declining worldwide and pathogens have been hypothesized as a leading cause of these declines (Daszak et al., 2000; Collins and Storfer, 2003; Stuart et al., 2004), ranaviruses are clearly of concern to conservation biologists. Although the chytrid fungus (*B. dendrobatidis*) is likely more important overall for amphibian declines and extinctions (Stuart et al., 2004), ranaviruses have been associated with a number of amphibian population declines throughout North America and Europe (Cunningham et al., 1996; Schock and Bollinger, 2005; Teacher et al., 2009a, 2009b; Ariel et al., 2009). Ranaviruses may also have adverse indirect effects on population viability in addition to directly increasing mortality rates. Recent research suggests that ranavirus infections may alter fitness by causing developmental stress, as measured by a higher degree of fluctuating asymmetry among individuals in infected populations relative to those in uninfected populations (St-Amour et al., 2010).

To date, applied ranavirus research has focused on three major concerns regarding viral emergence in wild populations: introduction of ranaviruses via human transport of fishing bait or possibly via fish themselves; introduction of ranaviruses via infected nonnative food species such as bullfrogs and fish; and anthropogenic stressors, such as pesticides, that may facilitate disease emergence by compromising amphibian immunity.

The first major concern stems from recent evidence suggesting long-distance movement of ATV within infected salamanders that are used as fishing bait (Jancovich et al., 2005; Storfer et al., 2007). Salamanders, collected from bait shops, have repeatedly tested positive for ATV infection, and the introduction by anglers of infected bait salamanders, either accidentally or intentionally, into heretofore ATV-free waters has resulted in the spread of ATV (Collins et al., 1988; Jancovich et al., 2005; Picco et al., 2007; Picco and Collins, 2008). Phylogenetic analysis of ATV strains collected from throughout western North America suggests that whereas ATV is likely endemic in most tiger salamander populations, infected bait salamanders have the potential to introduce novel virus strains into naïve populations (Storfer et al., 2007).

A follow-up experiment showed that the bait ATV strain was significantly more virulent (i.e., caused higher mortality) than endemic strains (Storfer et al., 2007). These results suggest that high pathogen virulence can evolve in high-density host environments where transmission can easily take place because virulence is usually correlated with the within-host replication rate and subsequent transmission (Ewald, 1994). Thus, bait ATV strains are expected to be more virulent than endemic strains because tiger salamander densities are artificially high in bait shops and transmission is virtually guaranteed. These results raise concern that introduction of infected bait tiger salamanders may introduce novel, highly virulent viral strains into areas with naïve hosts or into areas where hosts have been previously exposed but are adapted to other ranavirus strains. The implications of this work are particularly pertinent for conservation of two federally endangered species where genetic introgression by bait subspecies has been documented—the Sonora tiger salamander (*A. t. stebbinsi*, Lowe) and the California tiger salamander (*A. californiense*), both of which are susceptible to ATV infection (Jancovich et al., 1997; Picco et al., 2007; Storfer et al., 2004). Mandatory testing for ATV in salamander bait colonies or a complete moratorium on the use of tiger salamanders as bait should also be considered, as has been passed in California (Fitzpatrick and Shaffer, 2007). Because researchers can also be a source of virus spread, field equipment should be disinfected between sampling sites using chemicals such as 10% bleach that inactivate ranavirus Brunner and Sesterhenn (2001).

Various fish species have also been found to be positive for ranaviruses, and evidence suggests that ranaviruses that infect fish can cause disease in amphibians, or vice versa (Mao et al., 1999). As such, introduction of fish may present a serious disease threat to native amphibians. For example, rainbow trout have been introduced throughout much of North America for fishing purposes, including via truck and plane drops to remote areas (Kats and Ferrer, 2003). As mentioned above, recent genomic evidence suggests a close relationship between EHNV (isolated from rainbow trout) and ATV (Jancovich et al., 2010). In support of this hypothesis, and in contrast to ATV and FV3 where large sequence inversions have taken place, the genomes of ATV and EHNV are colinear. However, the genome of ATV is smaller than EHNV, perhaps reflecting the loss of genes unnecessary for replication in salamanders. Previous phylogeographic evidence also supports the possible host switch, as EHNV is evolutionarily closer to ATV than other amphibian ranaviruses (Jancovich et al., 2005).

The second major concern involves movement and introduction of invasive, nonnative species such as bullfrogs (*R. catesbeiana*). Bullfrogs have tested positive for FV3 and FV3-like ranaviruses (Gray et al., 2007; Majji et al., 2006; Miller et al., 2007), are thought to be resistant carriers, and may act as vectors to spread ranaviruses to caudates. Indeed, ranavirus strains isolated from bullfrog colonies appear highly virulent (Gray et al., 2007, 2009), as expected for viral strains evolving under artificial selection in high-density aquaculture environments and consistent with studies of ATV in salamander bait colonies. Schloegel et al. (2009) reported that over 28 million amphibians were imported into the United States during 2000–2005, with an 8.5% prevalence rate of ranavirus infection. The bullfrog trade has a global market value of over 48 million U.S. dollars, with France, Belgium, and the USA as leading importers. The bullfrog trade also occurs in South America, Africa, and Asia, making their worldwide transport a potential threat for global introduction (Schloegel et al., 2010). Once introduced, bullfrogs are often invasive and can coexist, outcompete, or prey upon native amphibians, thereby magnifying this threat (Lannoo et al., 2005). As a result, ranaviruses are now listed as a notifiable pathogen by the World Organization for Animal Health (OIE) Aquatic Animal Health Code (Schloegel et al., 2010).

The third major concern regards anthropogenic stressors as cofactors in ranavirus emergence. Infectious diseases and environmental contaminants are two of the leading hypotheses for global amphibian declines, particularly "enigmatic" declines in which straightforward causes such as habitat loss are not implicated (Collins and Storfer, 2003; Stuart et al., 2004). Environmental contamination may influence disease emergence by compromising the innate or adaptive immune system of the host or by disrupting homeostasis (Blaustein et al., 2003; Carey et al., 1999). A number of recent studies have investigated the effects of ecologically relevant doses of common pesticides on ATV infections in tiger salamanders. In long-toed salamanders, ATV resulted in lower mortality at intermediate levels of the widely used herbicide atrazine (Forson and Storfer, 2006a). However, in tiger salamanders, intermediate and high levels of atrazine decreased leukocyte counts and increased infection rates at intermediate levels (Forson and Storfer, 2006b). Thus, atrazine could result in increased infection and disease emergence in some salamander populations. Additional work has shown that chlorpyrifos, another heavily used pesticide, can increase ATV susceptibility in tiger salamanders relative to either stressor alone (Kerby and Storfer, 2009). When combined with atrazine, chlorpyrifos had an additive effect, increasing susceptibility to ATV infection and mortality (Kerby et al., 2010). Carbaryl, the most commonly used household pesticide in North America, also showed an additive effect when combined with predation and increased mortality by 50% (Kerby, unpublished). In addition to pesticides, one field study showed that aluminum was correlated with ranavirus outbreaks in Acadia National Park, Maine, USA (Gahl and Calhoun, 2010).

Aside from toxins, anthropogenically disturbed environments can increase the likelihood of ranavirus epizootics. In a survey of ponds containing green frogs (*Rana clamitans*) in Ontario, Canada, industrial activity, human habitation, and degree of human influence (e.g., distance to nearest road) all had significant effects on ranavirus presence, whereas other (natural) variables showed no correlation (St-Amour et al., 2008). Two studies suggest that cattle grazing can also influence the likelihood and possible severity of ranavirus outbreaks (Gray et al., 2007; Greer et al. 2008). In a study of green frogs in TN, ponds in cattle grazing areas were 3.9 times more likely than those in nongrazed areas to experience ranavirus epizootics (Gray et al., 2007). Greer et al. 2008 showed that in the Kaibab Plateau of Northern Arizona increased cattle access led to reduced pond vegetation, increased salamander densities in vegetated pond segments, and consequently increased viral transmission and prevalence.

In addition, ranaviruses are a significant concern to conservation biologists given their widespread geographic range, their propensity to infect multiple host species, and the combination of anthropogenic transport and habitat disturbance that enhances their prevalence in host populations. Listing ranaviruses as an OIE notifiable pathogen is a major step in stemming the global transport and trade of infected amphibians (Schloegel et al., 2010). In addition, researchers themselves may be a source of infection and should disinfect equipment between field sites to block transmission. Likewise, moratoriums on the transport of amphibians as fishing bait should be considered throughout North America. Although ranaviruses have caused apparent declines in some amphibian populations, others appear to fluctuate, and some may have become adapted to the virus. Further research should investigate variations in resistance and perhaps provide key (genomic or other) information about those populations or species that have adapted to ranavirus infections and developed disease resistance.

ACKNOWLEDGMENTS

This work was supported by NSF award no. IOS-07-42711 to VGC. We thank James Jancovich (Arizona State University) for the phylogenetic analysis of iridovirid species/strains depicted in Figure 9.3.

We have recently shown that ranavirus vIF-2α blocked cytotoxicity mediated by human or zebrafish PKR in a heterologous yeast system. Like K3L, vIF-2α acted as a pseudosubstrate inhibitor and blocked the PKR-mediated phosphorylation of eIF-2α, but not the autophosphorylation of PKR (Rothenberg et al., BioMed Central, in press).

REFERENCES

Agarkova, I. V., Dunigan, D. D., and Van Etten, J. L. (2006). Virion-associated restriction endonucleases of chloroviruses. *J. Virol.* 80, 8114–8123.

Alcami, A. and Koszinowski, U. H. (2000). Viral mechanisms of immune evasion. *Immunol. Today* 6, 365–372.

Andres, G., Garcia-Escudero, R., Simon-Mateo, C., and Vineula, E. (1998). African swine fever virus is enveloped by a two-membraned collapsed cisterna derived from the endoplasmic reticulum. *J. Virol.* 72, 8988–9001.

Ariel, E., Kielgast, J., Svart, H. E., Larsen, K., Tapiovaara, H., Jensen, B. B., and Holopainen, R. (2009). Ranavirus in wild edible frogs *Pelophylax kl. esculentus* in Denmark. *Dis. Aquat. Organ.* 85, 7–14.

Besch, R., Poeck, H., Hohenauer, T., Senft, D., Hacker, G., Berking, C., Hornung, V., Endres, S., Ruzicka, T., Rothenfusser, S., and Hartmann, G. (2009). Proapoptotic signaling induced by RIG-I and MDA-5 results in type I interferon-independent apoptosis in human melanoma cells. *J. Clin. Invest.* 119, 2399–2411.

Biacchesi, S., LeBerre, M., Lamoureux, A., Louise, Y., Lauret, E., Boudinot, P., and Bremont, M. (2009). Mitochondrial antiviral signaling protein plays a major role in induction of the fish innate immune response against RNA and DNA viruses. *J. Virol.* 83, 7815–7827.

Bird, S., Zou, J., and Secombes, C. J. (2006). Advances in fish cytokine biology give clues to the evolution of a complex network. *Curr. Pharm. Des.* 12, 3051–3069.

Blaustein, A. R., Romansic, J. M., Kiesecker, J. M., and Hatch, A. C. (2003). Ultraviolet radiation, toxic chemicals and amphibian population declines. *Divers. Distrib.* 9, 123–140.

Bolker, B. M., de Castro, F., Storfer, A., Mech, S. G., Harvey, E., and Collins, J. P. (2008). Disease as a selective force precluding widespread cannibalism: a case study of an iridovirus of tiger salamanders, *Ambystoma tigrinum. Evol. Ecol. Res.* 10, 105–128.

Bollinger, T. K., Mao, J., Schock, D., Brigham, R. M., and Chinchar, V. G. (1999). Pathology, isolation and molecular characterization of an iridovirus from tiger salamanders in Saskatchewan. *J. Wildlife Dis.* 35, 413–429.

Bos, D. H. and DeWoody, J. A. (2005). Molecular characterization of major histocompatibility complex II alleles in wild tiger salamander (*Ambystoma tigrinum*). *Immunogenetics* 57, 775.

Bouchier-Hayes, L. and Martin, S. J. (2002). CARD games in apoptosis and immunity. *EMBO Rep.* 3, 616–621.

Brunner, J. L. and Sesterhenn T. (2001). Disinfection of Ambystoma tigrinum virus (ATV). Froglog 48-2

Brunner, J. L., Richards, K., and Collins, J. P. (2005). Dose and host characteristics influence virulence of ranavirus infections. *Oecologia* 144, 399–406.

Brunner, J. L., Schock, D. M., and Collins, J. P. (2007). Transmission dynamics of the amphibian ranavirus *Ambystoma tigrinum* virus. *Dis. Aquat. Organ.* 77, 87–95.

Brunner, J. L., Schock, D. M., Davidson, E. W., and Collins, J. P. (2004). Intraspecific reservoirs: complex life history and the persistence of a lethal ranavirus. *Ecology* 85, 560–566.

Buller, R. M. and Pallumbo, G. J. (1991). Poxvirus pathogenesis. *Microbiol. Rev.* 55, 80–122.

Burkett-Cadena, N. D., Graham, S. P., Hassan, H. K., Guyer, C., Eubanks, M. D., Katholi, C. R., and Unnasch, T. R. (2008). Blood feeding patterns of potential arbovirus vectors of the genus *Culex* targeting ectothermic hosts. *Am. J. Trop. Med. Hyg.* 79, 809–815.

Burton, E. C., Miller, D. L., Styer, E. L., and Gray, M. J. (2008). Amphibian ocular malformation associated with frog virus 3. *Vet. J.* 177, 442–444.

Carey, C., Cohen, N., and Rollins-Smith, L. (1999). Amphibian declines: an immunological perspective. *Dev. Comp. Immunol.* 23, 459–472.

Charlemagne, J. and Tournefier, A. (1977). Humoral response to *Salmonella typhimurium* antigens in normal and thymectomized urodele amphibian *Pleurodeles waltlii* Michar. *Eur. J. Immunol.* 7, 500–502.

Charlemagne, J. and Tournefier, A. (1998). Immunology of amphibians. In: Pastoret, P., et al. (eds.), *Handbook of Vertebrate Immunology*. Academic Press, pp. 63–72.

Chinchar, V. G. (2002). Ranaviruses (family Iridoviridae): emerging cold-blooded killers. *Arch. Virol.* 147, 447–470.

Chinchar, V. G., Goorha, R., and Granoff, A. (1984). Early proteins are required for the formation of frog virus 3 assembly sites. *Virology* 135, 148–156.

Chinchar, V. G. and Granoff, A. (1984). Isolation and characterization of a frog virus 3 variant resistant to phosphonoacetate: genetic evidence for a virus-specific DNA polymerase. *Virology* 138, 357–361.

Chinchar, V. G., Granoff, A. (1986). Temperature-sensitive mutants of frog virus 3: biochemical and genetic characterization. *J. Virol.* 58, 192–202.

Chinchar, V. G., Hyatt, A., Miyazaki, T., and Williams, T. (2009). Family Iridoviridae: poor viral relations no longer. *Curr. Top. Microbiol. Immunol.* 328, 123–170.

Clark, H. F., Brennan, J. C., Zeigel, R. F., and Karzon, D. T. (1968). Isolation and characterization of viruses from the kidneys of *Rana pipiens* with renal adenocarcinoma before and after passage in the red eft (*Triturus viridescens*). *J. Virol.* 2, 629–640.

Collins, J. P., Brunner, J. L., Jancovich, J. K., and Schock, D. M. (2004). A model host–pathogen system for studying infectious disease dynamics in amphibians: tiger salamanders (*Ambystoma tigrinum)* and *Ambystoma tigrinum* virus. *Herpetol. J.* 14, 195–200.

Collins, J. P., Jones, T. R., and Berna, H. J. (1988). Conserving genetically distinctive populations: the case of the Huachuca tiger salamanders (Ambystoma tigrinum stebbinsi Lowe) In: Szaro, R.C., Severson, K.C.,and Patton, D.R. (eds.), *Management of Amphibians, Reptiles, and Small Mammals in North America*. GTR-RM-166, US Department of Agriculture Forest Service, Rocky Mountain Forest and Range Experiment Station, Fort Collins, C.O., pp. 45–53.

Collins, J. P. and Storfer, A. (2003). Global amphibian declines: sorting the hypotheses. *Divers. Distrib.* 9, 89–98.

Collins, J. P., Zerba, K. E., and Sredl, M. J. (1993). Shaping intraspecific variation: development, ecology and the evolution of morphology and life history variation in tiger salamanders. *Genetica* 89, 167–183.

Colonna, M. (2007). All roads lead to CARD9. *Nat. Immunol.* 8, 554–555.

Cotter, J. D., Storfer, A., Page, R. B., Beachy, C. K., and Voss, S. R. (2008). Transcriptional response of Mexican axolotls to Ambystoma tigrinum virus (ATV) infection. *BMC Genom.* 9, 493. (doi:10.1186/1471-2165-9-493).

Cullen, B. R., Owens, L., and Whittington, R. J. (1995). Experimental infection Australian anurans (*Limnodynastes terraereginae* and *Litoria latopalmata*) with Bohle iridovirus. *Dis. Aquat. Organ.* 23, 83–92.

Cunningham, A. A., Hyatt, A. D., Russell, P., and Bennett, P. M. (2007). Experimental transmission of a ranavirus disease of common toads (*Bufo bufo*) to common frogs (*Rana temporaria*). *Epidemiol. Infect.* 135, 1213–1216.

Cunningham, A. A., Langton, T. E. S., Bennett, P. M., Lewin, J. F., Drury, S. E. N., Gough, R. E., and MacGregor, S. K. (1996). Pathological and microbiological findings from incidents of unusual mortality of the common frog (*Rana temporaria*). *Philos. Trans. R. Soc. Lond. B* 351, 1539–1557.

Cunningham, A. A., Tems, C. A., and Russell, P. H. (2008). Immunohistochemical demonstration of ranavirus antigen in the tissues of infected frogs (*Rana temporaria*) with systemic haemorrhagic or cutaneous ulcerative disease. *J. Comp. Pathol.* 138, 3–11.

Cupp, E. W., Zhang, D., Yue, X., Cupp, M. S., Guyer, C., Sprenger, T. R., and Unnasch, T. R. (2004). Identification of reptilian and amphibian blood meals from mosquitoes in an Eastern equine encephalitis virus focus in Central Alabama. *Am. J. Trop. Med. Hyg.* 71, 272–276.

Daszak, P., Cunningham, A. A., and Hyatt, A. D. (2000). Wildlife ecology—emerging infectious diseases of wildlife: threats to biodiversity and human health. *Science* 287, 443–449.

Davison, A. J., Eberle, R., Ehlers, B., Hayward, G. S., McGeoch, D. J., Minson, A. C., Pellett, P. E., Roizman, B., Studdert, M. J., and Thiry, E. (2009). The order Herpesvirales. *Arch. Virol.* 154, 171–177.

Densmore, C. L. and Green, D. E. (2007). Diseases of amphibians. *ILAR J.* 48, 235–240.

Docherty, D. E., Meteyer, C. U., Wang, J., Mao, J., Case, S., and Chinchar, V. G. (2003). Diagnostic and molecular evaluation of three iridovirus-associated salamander mortality events. *J. Wildlife Dis.* 39, 556–566.

Duffus, A. L., Pauli, B. D., Wozney, K., Brunetti, C. R., and Berrill, M. (2008). Frog virus 3-like infections in aquatic amphibian communities. *J. Wildlife Dis.* 44, 109–120.

DuPasquier, L. (2001). The immune system of invertebrates and vertebrates. *Comp. Biochem. Physiol. Part B* 129, 1–15.

Eaton, H. E., Metcalf, J., Penny, E., Tcherepanov, V., Upton, C., and Brunetti, C. R. (2007). Comparative genomic analysis of the family Iridoviridae: re-annotating and defining the core set of iridovirus genes. *Virol. J.* 4, 11. (doi:10.1186/1743-422X-4-11).

Epifano, C., Krijnse-Locker, J., Salas, M. L., Rodriguez, J. M., and Salas, J. (2006a). The African swine fever virus nonstructural protein pB602L is required for formation of the icosahedral capsid of the virus particle. *J. Virol.* 80, 12260–12270.

Epifano, C., Krijnse-Locker, J., Salas, M. L., Salas, J., and Rodriguez, J. M. (2006b). Generation of filamentous instead of icosahedral particles by repression of African swine fever virus structural proteins pB438L. *J. Virol.* 80, 11456–11466.

Ewald, P. W. (1994). *Evolution of Infectious Diseases*. Oxford University Press, New York, *298* pp.

Fauquet, C.M., Mayo, M.S.A., Maniloff, J., Desselberger, U.,and Ball, L.A. (eds.). (2005). *Virus Taxonomy: Classification and Nomenclature of Viruses. Eighth Report of the International Committee on Taxonomy of Viruses*. Elsevier/Academic Press, San Diego.

Federici, B. A., Bideshi, D. K., Tan, Y., Spears, T., and Bigot, Y. (2009). Ascoviruses: superb manipulators of apoptosis for viral replication and transmission. *Curr. Top. Microbiol. Immunol.* 328, 171–196.

Fellah, J. S., Andre, S., Kerourn, F., Guerci, F., Bleaux, C., and Charlemagne, J. (2002). Structure, diversity, and expression of the TCR delta chains in the Mexican axolotl. *Eur. J. Immunol.* 32, 1349–1358.

Felsenstein, J. (1985). Confidence limits on phylogenies: an approach using the bootstrap. *Evolution* 39, 783–791.

Fenner, B. J., Goh, W., and Kwang, J. (2006). Sequestration and protection of dsRNA by the betanodavirus B2 protein. *J. Virol.* 80, 6822–6833.

Findlay, B. B. and McFadden, G. (2006). Anti-immunology: evasion of the host immune system by bacterial and viral pathogens. *Cell* 124, 767–782.

Fitzpatrick, B. M. and Shaffer, H. B. (2007). Hybrid vigor between native and introduced salamanders raises new challenges for conservation. *Proc. Natl. Acad. Sci. U. S. A.* 104, 15793–15798.

Forson, D. D. and Storfer, A. (2006a). Atrazine increases ranavirus susceptibility in the tiger salamander, *Ambystoma tigrinum*. *Ecol. Appl.* 16, 2325–2332.

Forson, D. and Storfer, A. (2006b). Effects of atrazine and iridovirus infection on survival and life-history traits of the long-toed salamander (*Ambystoma macrodactylum*). *Environ. Toxicol. Chem.* 25, 168–173.

Gahl, M. K. and Calhoun, A. J. K. (2008). Landscape setting and risk of *Ranavirus* mortality events. *Biol. Conserv.* 141, 2679–2689.

Gahl, M. K. and Calhoun, A. J. K. (2010). The role of multiple stressors in ranavirus-caused amphibian mortalities in Acadia National Park wetlands. *Can. J. Zool.* 88, 108–121.

Gantress, J., Maniero, G. D., Cohen, N., and Robert, J. (2003). Development and characterization of a model system to study amphibian immune responses to iridoviruses. *Virology* 311, 254–262.

Goorha, R. (1982). Frog virus 3 DNA replication occurs in two stages. *J. Virol.* 43, 519–528.

Goorha, R., Granoff, A., Willis, D. B., and Murti, K. G. (1984). The role of DNA methylation in virus replication: inhibition of frog virus 3 replication by 5-azacytidine. *Virology* 138, 94–102.

Granoff, A., Came, P. E., and Breeze, D. C. (1966). Viruses and renal carcinoma of *Rana pipiens*: I. The isolation and properties of virus from normal and tumor tissues. *Virology* 29, 133–148.

Granoff, A., Came, P. E., and Rafferty, K. A. (1965). The isolation and properties of viruses from *Rana pipiens*: their possible relationship to the renal adenocarcinoma of the leopard frog. *Ann. N. Y. Acad. Sci.* 126, 237–255.

Gray, M. J., Miller, D. L., and Hoverman, J. T. (2009). Ecology and pathology of amphibian ranaviruses. *Dis. Aquat. Organ.* 87, 243–266.

Gray, M. J., Miller, D. L., Schmutzer, A. C., and Baldwin, C. A. (2007). Frog virus 3 prevalence in tadpole populations inhabiting cattle-access and non-access wetlands in Tennessee, USA. *Dis. Aquat. Organ.* 77, 97–103.

Green, D. E., Converse, K. A., and Schrader, A. K. (2002). Epizootiology of sixty-four amphibian morbidity and mortality events in the USA, 1996–2001. *Ann. N. Y. Acad. Sci.* 969, 323–339.

Greer, A. L., Berrill, M., and Wilson, P. J. (2005). Five amphibian mortality events associated with ranavirus infection in south central Ontario, Canada. *Dis. Aquat. Organ.* 67, 9–14.

Greer, A. L. and Collins, J. P. (2008). Habitat fragmentation as a result of biotic and abiotic factors controls pathogen transmission throughout a host population. *J. Anim. Ecol.* 77, 364–399.

Greer, A. L., Briggs, C. I., and Collins, J. P. (2008). Testing a key assumption of host–pathogen theory: density and disease transmission. *Oikos* 117, 1667–1673.

Greer, A. L., Brunner, J. L., and Collins, J. P. (2009). Spatial and temporal patterns of *Ambystoma tigrinum* virus (ATV) prevalence in tiger salamanders *Ambystoma tigrinum nebulosum*. *Dis. Aquat. Organ.* 85, 1–6.

Guselnikov, S. V., Reshetnikova, E. S., Najakshin, A. M., Mechetina, L. V., Robert, J., and Taranin, A. (2010). The amphibians *Xenopus laevis* and *Silurana tropicalis* possess a family of activating KIR-related immunoglobulin-like receptors. *Dev. Comp. Immunol.* 34, 308–315.

Hara, H., Ishihara, C., Takeuchi, A., Imanishi, T., Xue, L., et al. (2007). The adaptor protein CARD9 is essential for the activation of myeloid cells through ITAM-associated and Toll-like receptors. *Nat. Immunol.* 8, 619–629.

Harp, E. M. and Petranka, J. W. (2006). Ranavirus in wood frogs (*Rana sylvatica*): potential sources of transmission within and between ponds. *J. Wildlife Dis.* 42, 307–318.

Harty, R. N., Pitha, P. M., and Okumura, A. (2009). Antiviral activity of innate immune protein ISG15. *J. Innate Immun.* 1, 397–404.

Hausmann, S., Marq, J. B., Tapparel, C., Kolakofsky, D., and Garcin, D. (2008). RIG-I and dsRNA-induced IFNβ activation. *PLoS One* 3(12), e3965.

Hoelzer, K., Shackelton, L. A., and Parrish, C. R. (2008). Presence and role of cytosine methylation in DNA viruses of animals. *Nucleic Acids Res.* 36, 2825–2837.

Hofmann, Bucher P., and Tschopp, J. (1997). The CARD domain: a new apoptotic signaling motif. *Trends Biol. Sci.* 22, 155–156.

Houdayer, M. and Fougereau, M. (1972). Phylogeny of immunoglobulins: immune reaction of the axolotl, *Ambystoma mexicanum.* Kinetics of the immune response and characterization of antibodies. *Ann. Inst. Pasteur (Paris)* 123, 3–28.

Hoverman, J. T., Gray, M. J., and Miller, D. L. (2010). Anuran susceptibilities to ranaviruses: role of species identity, exposure route, and a novel virus isolate. *Dis. Aquat. Organ.* 89, 97–107.

Huelsenbeck, J. P., Rannala, B., and Yang, Z. (1997). Statistical tests of host-parasite cospeciation. *Evolution* 51, 410–419.

Huang, Y., Huang, X., Liu, H., Gong, J., Ouyang, Z., Cui, H., Cao, J., Zhao, Y., Wang, X., Jiang, Y., and Qin, Q. (2009). Complete sequence determination of a novel reptile iridovirus isolated from soft-shelled turtle and evolutionary analysis of Iridoviridae. *BMC Genom.* 10, 224. (doi:10.1186/1471-2164-10-224).

Hueffer, K., Parker, J. S. J., Weichert, W. S., Geisel, R. E., Sgro, J.-Y., and Parrish, C. R. (2003). The natural host range shift and subsequent evolution of canine parvovirus resulted from virus-specific binding to the canine transferrin receptor. *J. Virol.* 77, 1718–1726.

Hughes, A. L. and Yeager, M. (1997). Molecular evolution of the vertebrate immune system. *Bioessays* 19, 777–786.

Hussain, M., Abrahim, A. M., and Asgari, S. (2010). An ascovirus-encoded RNase III autoregulates its expression and suppresses RNA interference-mediated gene silencing. *J. Virol.* 84, 3624–3630.

Hyatt, A. D., Gould, A. R., Zupanovic, Z., Cunningham, A. A., Hengstberger, S., Whittington, R. J., Kattenbelt, J., and Coupar, B. E. (2000). Comparative studies of piscine and amphibian iridoviruses. *Arch. Virol.* 145, 301–331.

Jancovich, J. K., Bremont, M., Touchman, J. W., and Jacobs, B. L. (2010). Evidence for multiple recent host species shifts among the ranavirus (family Iridoviridae). *J. Virol.* 84, 2636–2647.

Jancovich, J. K., Chinchar, V. G., Hyatt, A., Miyazaki, T., Williams, T., and Zhang, Q. Y. (in press). Family Iridoviridae. *9th Report of the International Committee on Taxonomy Viruses.*

Jancovich, J. K., Davidson, E. W., Morado, J. F., Jacobs, B. L., and Collins, J. P. (1997). Isolation of a lethal virus from the endangered tiger salamander *Ambystoma tigrinum stebbinsi. Dis. Aquat. Organ.* 31, 161–167.

Jancovich, J. K., Davidson, E. W., Parameswaran, N., Mao, J., et al. (2005). Evidence for emergence of an amphibian iridoviral disease because of human-enhanced spread. *Mol. Ecol.* 14, 213–224.

Jancovich, J. K., Davidson, E. W., Seiler, A., Jacobs, B. L., and Collins, J. P. (2001). Transmission of the *Ambystoma tigrinum* virus to alternative hosts. *Dis. Aquat. Organ.* 46, 159–163.

Jancovich, J. K., Mao, J. H., Chinchar, V. G., Wyatt, C., et al. (2003). Genomic sequence of a ranavirus (family Iridoviridae) associated with salamander mortalities in North America. *Virology* 316, 90–103.

Johnston, J. B. and McFadden, G. (2003). Poxvirus immunomodulatory strategies: current perspectives. *J. Virol.* 77, 6093–6100.

Johnston, S. C. and Ward, B. M. (2009). Vaccinia virus protein F12 associates with intracellular enveloped virions through interaction with A36. *J. Virol.* 83, 1708–1717.

Kasahara, M., Suzuki, T., and DuPasquier, L. (2004). On the origins of the adaptive immune system: novel insights from invertebrates and cold-blooded vertebrates. *Trends Immunol.* 25, 105–111.

Kats, L. B. and Ferrer, R. P. (2003). Alien predators and amphibian declines: review of two decades of science and the translation to conservation. *Divers. Distrib.* 9, 99–110.

Kawai, T. and Akira, S. (2009). The roles of TLRs, RLRs, and NLRs in pathogen recognition. *Int. Immunol.* 21(4), 317–337.

Kawai, T. and Akira, S. (2010). The role of pattern-recognition receptors in innate immunity: update on Toll-like receptors. *Nat. Immunol.* 112, 373–384.

Kerby, J. L., Richards-Hrdlicka, K. L., Storfer, A., and Skelly, D. K. (2010). An examination of amphibian sensitivity to environmental contaminants: are amphibians poor canaries?, *Ecol. Lett.* 13, 60–67.

Kerby, J. L. and Storfer, A. (2009). Combined effects of atrazine and chlorpyrifos on susceptibility of the tiger salamander to *Ambystoma tigrinum* virus. *Ecohealth* 6, 91–98.

Kerscher, O., Felberbaum, R., and Hochstrasser, M. (2006). Modification of proteins by ubiquitin and ubiquitin-like proteins *Ann. Rev. Cell. Dev. Biol.* 22, 159–180.

Kim, Y.-S., Ke, F., Lei, X.-Y., Zhu, R., and Zhang, Q.-Y. (2010). Viral envelope protein 53R gene highly specific silencing and iridovirus resistance in fish cells by AmiRNA. *PLoS One* 5(4), e10308.

Klenk, K. and Komar, N. (2003). Poor replication of West Nile virus (New York 1999 strain) in three reptilian and one amphibian species. *Am. J. Trop. Med. Hyg.* 69, 260–262.

Kostyukov, M. A., Alekseev, A. N., Bulchev, V. P., and Gordeeva, Z. E. (1986). Experimentally proven infection of *Culex pipiens* L. mosquitoes with West Nile fever virus via the Lake Pallas *Rana ridibunda* frog and its transmission via bites. *Med. Parazitol. (Mosk.)* 6, 76–78.

Kreuze, J. F., Savenkov, E. I., Cuellar, W., Li, X., and Valkonen, J. P. T. (2005). Viral class 1 RNase III involved in suppression of RNA silencing. *J. Virol.* 79, 7227–7238.

Krug, A., Luker, G. D., Barchet, W., Leib, D. A., Akira, S., and Colonna, M. (2004). Herpes simplex virus type 1 activates murine natural interferon-producing cells through toll-like receptor 9. *Blood* 103, 1433–1437.

Langland, J. O. and Jacobs, B. L. (2002). The role of the PKR inhibitory genes E3L and K3L in determining vaccinia virus host range. *Virology* 299, 133–141.

Lannoo, M. (ed.) (2005). Amphibian Declines: The Conservation Status of United States Species. University of California Press.

Lantermann, M., Schwantes, A., Sliva, K., Sutter, G., and Schnierle, B. S. (2007). Vaccinia virus dsRNA-binding protein E3 does not interfere with siRNA-mediated gene silencing in mammalian cells. *Virus Res.* 126, 1–8.

Lei, Y., Moore, C. B., Liesman, R. M., O'Connor, B. P., Bergstrahl, D. T., Chen, Z. J., Pickles, R. J., and Ting, J. P. Y. (2009). MAVS-mediated apoptosis and its inhibition by viral proteins. *PLos One* 4(5), e5466.

Levy, D. E. and Marie, I. J. (2004). RIGing an antiviral defense—it's in the CARDs. *Nat. Immunol.* 5, 699–701.

Li, W. X., Li, H., Lu, R., Li, F., Dus, M., Atkinson, P., Brydon, E. W. A., Johnson, K. L., Garcia-Sastre, Ball, L. A., Palese, P., and Ding, S. W. (2004). Interferon antagonist proteins of influenza and vaccinia viruses are suppressors of RNA silencing. *Proc. Natl. Acad. Sci. U. S. A.* 101, 1350–1355.

Loeb, K. R. and Haas, A. L. (1992). The interferon inducible 15 kDa ubiquitin homolog conjugates to intracellar proteins. *J. Biol. Chem.* 267, 7806–7813.

Majji, S., LaPatra, S., Long, S. M., Sample, R., Bryan, L., Sinning, A., and Chinchar, V. G. (2006). Rana catesbeiana virus Z (RCV-Z): a novel pathogenic ranavirus. *Dis. Aquat. Organ.* 73, 1–11.

Majji, S., Thodima, V., Sample, R., Whitley, D., Deng, Y., Mao, J., and Chinchar, V. G. (2009). Transcriptome analysis of frog virus 3, the type species of the genus *Ranavirus,* family Iridoviridae. *Virology* 391, 293–303.

Maniero, G. D., Morales, H. D., Gantress, J., and Robert, J. (2006). Generation of a long-lasting, protective, and neutralizing antibody response to the ranavirus FV3 by the frog *Xenopus. Dev. Comp. Immunol.* 30, 649–657.

Mao, J., Green, D. E., Fellers, G., and Chinchar, V. G. (1999). Molecular characterization of iridoviruses isolated from sympatric amphibians and fish. *Virus Res.* 63, 45–62.

Mao, J., Hedrick, R. P., and Chinchar, V. G. (1997). Molecular characterization, sequence analysis, and taxonomic position of newly isolated fish iridoviruses. *Virology* 229, 212–220.

Marr, S., Morales, H., Bottaro, A., Cooper, M., Flajnik, M., and Robert, J. (2007). Localization and differential expression of activation-induced cytidine deaminase in the amphibian *Xenopus* upon antigen stimulation and during early development. *J. Immunol.* 179, 6783–6789.

Mazzoni, R., de Mesquita, A. J., Fleury, L. F. F., de Brito, W. M. E. D., Nunes, I. A., Robert, J., Morales, H., Coelho, A. S. G., Barthasson, D. L., Galli, L., and Catroxo, M. H. B. (2009). Mass mortality associated with frog virus 3-like ranavirus infection in farmed tadpoles *Rana catesbeiana* from Brazil. *Dis. Aquat. Organ.* 86, 181–191.

Miller, D. L., Rajeev, S., Gray, M. J., and Baldwin, C. A. (2007). Frog virus 3 infection, cultured American bullfrogs. *Emerg. Infect. Dis.* 13, 342–343.

Mohamed, M. R., Rahman, M. M., Rice, A., Moyer, R. W., Warden, S. J., and McFadden, G. (2009). Cowpox virus expresses a novel ankyrin repeat NF-κB inhibitor that controls inflammatory cell influx into virus-infected tissues and is critical for virus pathogenesis. *J. Virol.* 83, 9223–9236.

Moody, N. J. G. and Owens, L. (1994). Experimental demonstration of pathogenicity of a frog virus, Bohle iridovirus, for a fish species, barramundi *Lates calcarifer*. *Dis. Aquat. Organ.* 18, 95–102.

Moore, J. B. and Smith, G. L. (1992). Steroid hormone synthesis by a vaccine enzyme: a new type of virus virulence factor. *EMBO J.* 11, 1973–1980.

Morales, H. D., Abramawitz, L., Gertz, J., Sowa, J., Vogel, A., and Robert, J. (2010). Innate immune responses and permissiveness to ranavirus infection of peritoneal leukocytes in the frog *Xenopus laevis*. *J. Virol.* 84, 4912–4922.

Morales, H. D. and Robert, J. (2007). Characterization of primary and memory CD8 T-cell responses against ranavirus (FV3) in *Xenopus laevis*. *J. Virol.* 81, 2240–2248.

Panchanathan, V., Chaudhri, G., and Karapiah, G. (2005). Interferon function is not required for recovery from a secondary poxvirus infection. *Proc. Natl. Acad. Sci. U. S. A.* 102, 12921–12926.

Parris, M. J., Storfer, A., Collins, J. P., and Davidson, E. W. (2005). Life history responses to pathogens in tiger salamander (*Ambystoma tigrinum*) larvae. *J. Herpetol.* 39, 366–372.

Pasmans, F., Blahak, S., Martel, A., Pantchev, N., and Zwart, P. (2008). Ranavirus-associated mass mortality in imported red tailed knobby newts (*Tylototriton kweichowensis*): a case report. *Vet. J.* 176, 257–259.

Pearman, P. B. and Garner, T. W. J. (2005). Susceptibility of Italian agile frog populations to an emerging strain of ranavirus parallels population genetic diversity. *Ecol. Lett.* 8, 401–408.

Pearman, P. B., Garner, T. W., Straub, M., and Greber, U. F. (2004). Response of the Italian agile frog (*Rana latastei*) to a ranavirus, frog virus 3: a model for viral emergence in naive populations. *J. Wildlife Dis.* 40, 660–669.

Pfennig, D. W., Loeb, M. L. G., and Collins, J. P. (1991). Pathogens as a factor limiting the spread of cannibalism in tiger salamanders. *Oecologia* 88, 161–166.

Picco, A. M., Brunner, J. L., and Collins, J. P. (2007). Susceptibility of the endangered California tiger salamander, *Ambystoma californiense,* to *Ranavirus* infection. *J. Wildlife Dis.* 43, 286–290.

Picco, A. M. and Collins, J. P. (2008). Amphibian commerce as a likely source of pathogen pollution. *Conserv. Biol.* 22, 1582–1589.

Qi, Z., Nie, P., Secombes, C. J., and Zou, J. (2010). Intron-containing type I and type III IFN coexist in amphibians: refuting the concept that a retroposition event gave rise to type I IFNs. *J. Immunol.* 184, 5038–5046.

Rathinam, V. A. K., Jiang, Z., Waggoner, S. N., Sharma, S., Cole, L. E., Waggoner, L., Vanaja, S. K., Monks, B. G., Ganesan, S., Latz, E., Hornung, V., Vogel, S. N., Szomolanyi-Tsuda, E., and Fitzgerald, K. A. (2010). The AIM2 inflammasome is essential for host defense against cytosolic bacteria and DNA viruses. *Nat. Immunol.* 11, 395–402.

Reading, P. C., Moore, J. B., and Smith, G. L. (2003). Steroid hormone synthesis by vaccine virus suppresses the inflammatory response to infection. *J. Exp. Med.* 197, 1269–1278.

Rhen, T. and Cidlowski, J. A. (2005). Anti-inflammatory actions of glucocorticoids—new mechanisms for old drugs. *N. Engl. J. Med.* 353, 1711–1723.

Richman, A. D., Herrera, G., Reynoso, V. H., Mendez, G., and Zambrano, L. (2007). Evidence for balancing selection at the *DAB* locus in axolotl, *Ambystoma mexicanum. Int. J. Immunogenet.* 34, 475–478.

Ridenhour, B. J. and Storfer, A. T. (2008). Geographically variable selection in *Ambystoma tigrinum* virus (Iridoviridae*)* throughout the western USA. *J. Evol. Biol.* 21, 1151–1159.

Robert, J. (2010). Emerging ranaviral infectious diseases and amphibian decline. *Diversity* 2, 314–330. (doi:10/3390/d1030314).

Robert, J., Abramowitz, L., Gantress, J., and Morales, H. D. (2007). *Xenopus laevis*: a possible vector of ranavirus infection? *J. Wildlife Dis.* 43, 645–652.

Robert, J., Morales, H., Buck, W., Cohen, N., Marr, S., and Gantress, J. (2005). Adaptive immunity and histopathology in frog virus 3-infected *Xenopus. Virology* 332, 667–675.

Robert, J. and Ohta, Y. (2009). Comparative and developmental study of the immune system in *Xenopus. Dev. Dynam.* 238, 1249–1270.

Rojas, S., Richards, K., Jancovich, J. K., and Davidson, E. W. (2005). Influence of temperature on *Ranavirus* infection in larval salamanders *Ambystoma tigrinum. Dis. Aquat. Organ.* 63, 95–100.

Rollins-Smith, L. (1998). Metamorphosis and the amphibian immune system. *Immunol. Rev.* 166, 221–230.

Sadler, A. J. and Williams, B. R. G. (2008). Interferon-inducible antiviral effectors. *Nat. Rev. Immunol.* 8, 559–568.

Saitou, N. and Nei, M. (1987). The neighbor-joining method: a new method for reconstructing phylogenetic trees. *Mol. Biol. Evol.* 4, 406–425.

Sammut, B., Du Pasquier, L., Ducoroy, P., Laurens, V., Marcuz, A., and Tournefier, A. (1999). Axolotl MHC architecture and polymorphism. *Eur. J. Immunol.* 29, 2897–2907.

Sample, R. C. (2010). Elucidation of frog virus 3 gene function and pathways of virion formation. PhD dissertation. University of Mississippi Medical Center.

Sample, R., Bryan, L., Long, S., Majji, S., Hoskins, G., Sinning, A., Olivier, J., and Chinchar, V. G. (2007). Inhibition of iridovirus protein synthesis and virus replication by antisense morpholino oligonucleotides targeted to the major capsid protein, the 18 kDa immediate-early protein, and a viral homolog of RNA polymerase II. *Virology* 358, 311–320.

Schaerlinger, B. and Frippiat, J.-P. (2008). IgX antibodies in the urodele amphibian *Ambystoma mexicanum*. *Dev. Comp. Immunol.* 32, 908–915.

Schetter, C., Grunemann, B., Holker, I., and Doerfler, W. (1993). Patterns of frog virus 3 DNA methylation and DNA methyltransferase activity in nuclei of infected cells. *J. Virol.* 67, 6973–6978.

Schloegel, L. M., Daszak, P., Cunningham, A. A., Speare, R., and Hill, B. (2010). Two amphibian diseases, chytridiomycosis and ranaviral disease, are now globally notificable to the World Organization for Animal Health (OIE): an assessment. *Dis. Aquat. Organ.* (doi:10.3354/dao02140).

Schloegel, L. M., Picco, A. M., Kilpatrick, A. M., Davies, A. J., Hyatt, A. D., and Daszak, P. (2009). Magnitude of the US trade in amphibians and presence of *Batrachochytrium dendrobatidis* and ranavirus infection in imported North American bullfrogs (*Rana catesbeiana*). *Biol. Conserv.* 142, 1420–1426.

Schock, D. M. and Bollinger, T. K. (2005). An apparent decline of northern leopard frogs (*Rana pipiens*) on the Rafferty Dam mitigation lands near Estevan, Saskatchewan. *Blue Jay* 63, 144–154.

Schock, D. M., Bollinger, T. K., Chinchar, V. G., Jancovich, J. K., and Collins, J. P. (2008). Experimental evidence that amphibian ranaviruses are multi-host pathogens. *Copeia* 1, 133–143.

Sen, G. C. and Sarkar, S. N. (2005). Hitching RIG to action. *Nat. Immunol.* 6, 1074–1076.

Shackelton, L. A., Parrish, C. R., Truyen, U., and Holmes, E. C. (2005). High rate of viral evolution associated with the emergence of carnivore parvovirus. *Proc. Natl. Acad. Sci. U. S. A.* 102, 379–384.

St-Amour, V., Garner, T. W. J., Schulte-Hostedde, A. I., and Lesbarrères, D. (2010). Effects of two amphibian pathogens on the developmental stability of green frogs. *Conserv. Biol.* 24, 788–794.

St-Amour, V., Wong, W. M., Garner, T. W. J., and Lesbarrères, D. (2008). Anthropogenic influence on prevalence of 2 amphibian pathogens. *Emerg. Infect. Dis.* 14, 1175–1176.

Storfer, A., Alfaro, M. E., Ridenhour, B. J., Jancovich, J. K., Mech, S. G., Parris, M. J., and Collins, J. P. (2007). Phylogenetic concordance analysis shows an emerging pathogen is novel and endemic. *Ecol. Lett.* 10, 1075–1083.

Storfer, A., Mech, S. G., Reudink, M. W., Ziemba, R. E., Warren, J. L., and Collins, J. P. (2004). Introgression by non-native species in the endangered tiger salamander, *Ambystoma tigrinum stebbinsi*. *Copeia* 2004, 783–796.

Stuart, S. N., Chanson, J. S., Cox, N. A., Young, B. E., Rodrigues, A. S. L., Fischman, D. L., and Waller, R. W. (2004). Status and trends of amphibian declines and extinctions worldwide. *Science* 306, 1783–1786.

Sun, Wei, Huang, Y., Zhao, Z., Gui, J., and Zhang, Q. (2006). Characterization of the Rana grylio virus 3β-hydroxysteroid dehydrogenase and its novel role in suppressing virus-induced cytopathic effect. *Biochem. Biophys. Res. Commun.* 351, 44–50.

Takezaki, N., Rzhetsky, A., and Nei, M. (1995). Phylogenetic test of the molecular clock and linearized trees. *Mol. Biol. Evol.* 12, 823–833.

Tamura, K., Dudley, J., Nei, M., and Kumar, S. (2007). MEGA4: molecular evolutionary genetics analysis (MEGA) software version 4.0. *Mol. Biol. Evol.* 24, 1596–1599.

Tan, W. G. H., Barkman, T. J., Chinchar, V. G., and Essani, K. (2004). Comparative genome analysis of frog virus 3, type species of the genus *Ranavirus* (family Iridoviridae). *Virology* 323, 70–84.

Teacher, A. G. F., Garner, T. W. J., and Nichols, R. A. (2009a). Evidence for directional selection at a novel major histocompatibility class I marker in wild common frogs (*Rana temporaria*) exposed to a viral pathogen (*Ranavirus*). *PLoS One* 4, e4616. (doi:10.1371/journal.pone.0004616).

Teacher, A. G. F., Garner, T. W. J., and Nichols, R. A. (2009b). Population genetic patterns suggest a behavioral change in wild common frogs (*Rana temporaria*) following disease outbreaks (*Ranavirus*). *Mol. Ecol.* 18, 3163–3172.

Tortorella, D., Gewurz, B. E., Furman, M. H., Schust, D. J., and Ploegh, H. L. (2000). Viral subversion of the immune system. *Ann. Rev. Immunol.* 18, 861–926.

Tournefier, A. (1975). Incomplete antibodies and immunoglobulin characterization in adult urodeles, *Pleurodeles waltlii* Micah and *Triturus alpestris* Laur. *Immunology* 29, 209–217.

Tournefier, A. and Charlemagne, J. (1975). Antibodies against *Salmonella* and SRBC in urodele amphibians: synthesis and characterization. *Adv. Exp. Med. Biol.* 64, 161–171.

Tournefier, A., Laurens, V., Chapusat, C., Ducoroy, P., Padros, M. R., Saladori, F., and Sammut, B. (1998). Structure of MHC class I and class II cDNAs and possible immunodeficiency linked to class II expression in Mexican axolotl. *Immunol. Rev.* 166, 259–277.

Tulman, E. R., Delhon, G. A., Ku, B. K., and Rock, D. L. (2009). African swine fever virus. *Curr. Top. Microbiol. Immunol.* 328, 43–88.

Une, Y., Sakuma, A., Matsueda, H., Nakai, K., and Murakami, M. (2009). Ranavirus outbreak in North American bullfrogs (*Rana catesbeiana*), Japan, 2008. *Emerg. Infect. Dis.* 15, 1146–1147.

Wake, D. B. and Vredenburg, V. T. (2008). Are we in the midst of the sixth mass extinction? A view from the world of amphibians. *Proc. Natl. Acad. Sci. U. S. A.* 105, 11466–11473.

Ward, B. M., Weisberg, A. S., and Moss, B. (2003). Mapping and functional analysis of internal deletions within the cytoplasmic domains of the vaccinia virus A33R and A36R entry proteins. *J. Virol.* 77, 4113–4126.

Weinmann, N., Papp, T., de Matos, A. P. A., Teifke, J. P., and Marschang, R. E. (2007). Experimental infection of crickets (*Gryllus bimaculatus*) with an invertebrate iridovirus isolated from a high-casqued chameleon (*Chamaeleo hoehnelii*). *J. Vet. Diagn. Invest.* 19, 674–679.

Weng, S. P., He, J. G., Wang, X. H., Lu, L., Deng, M., and Chan, S. M. (2002). Outbreaks of an iridovirus disease in cultured tiger frog, *Rana tigrina rugulosa,* in southern China. *J. Fish Dis.* 25, 423–427.

Werts, C., Girardin, S. E., and Philpott, D. J. (2006). TIR, CARD, PYRIN: three domains for an anti-microbial triad. *Cell Death Differ.* 13, 798–815.

Whitney, E., Jamback, H., Means, R. G., and Watthews, T. H. (1968). Arthropod-borne virus survey in St. Lawrence Country, New York. *Am. J. Trop. Med. Hyg.* 17, 645–650.

Whitley, D. S., Yu, K., Sample, R. C., Sinning, A., Henegar, J., Norcross, E., and Chinchar, V. G. (2010). Frog virus 3 ORF 53R, a putative myristoylated membrane protein, is essential for virus replication *in vitro*. *Virology* 405, 448–456.

Williams, T. (1996). The iridoviruses. *Adv. Virus Res.* 46, 345–412.

Williams, T., Barbosa-Solomieu, V., and Chinchar, V. G. (2005). A decade of advances in iridovirus research. *Adv. Virus Res.* 65, 173–249.

Willis, D. B., Goorha, R., and Granoff, A. (1984). DNA methyltransferase induced by frog virus 3. *J. Virol.* 49, 86–91.

Willis, D. B. and Granoff, A. (1980). Frog virus 3 DNA is heavily methylated at CpG sequences. *Virology* 107, 250–257.

Willis, D. B. and Granoff, A. (1985). Trans-activation of an immediate-early frog virus 3 promoter by a virion protein. *J. Virol.* 56, 495–501.

Wolf, K., Bullock, G., Dunbar, C., and Quimby, M. (1968). Tadpole edema virus: viscerotropic pathogen for anuran amphibians. *J. Infect. Dis.* 118, 253–262.

Xie, J., Lu, J., Deng, M., Weng, S., Zhu, J., Wu, Y., Gan, L., Chan, S. M., and He, J. (2005). Inhibition of reporter gene and iridovirus-tiger frog virus in fish cell by RNA interference. *Virology* 338, 43–52.

Yoneyama, M., Kikuchi, M., Natsukawa, T., Shinobu, N., Imaizumi, T., Miyagishi, M., Taira, K., Akira, S., and Fujita, T. (2004). The RNA helicase RIG-I has an essential function in double-stranded RNA-induced innate antiviral responses. *Nat. Immunol.* 5, 730–737.

Zenke, K. and Kim, K. H. (2008). Functional characterization of the RNase III gene of rock bream iridovirus. *Arch. Virol.* 153, 1651–1656.

Zhang, Q. Y., Xiao, f., Li, Z. Q., Gui, G. F., Mao, J., and Chinchar, V. G. (2001). Characterization of an iridovirus from the cultured pig frog (*Rana grylio*) with lethal syndrome. *Dis. Aquat. Organ.* 48, 27–36.

Zou, J., Tafalla, C., Truckle, J., and Secombes, C. J. (2007). Identification of a second group of type I IFNs in fish sheds light on IFN evolution in vertebrates. *J. Immunol.* 179, 3859–3871.

Zuckerandl, E. and Pauling, L. (1965). Evolutionary divergence and convergence in proteins. In: Bryson, V. and Vogel, H.J. (eds.), *Evolving Genes and Proteins*. Academic Press, New York, pp. 97–166.

CHAPTER 10

VIRUSES OF INSECTS

DECLAN C. SCHROEDER
Marine Biological Association of the UK, Plymouth, UK

CONTENTS

10.1 INTRODUCTION

The phylogenetic relationship between and within major insect taxa is the subject of much debate with molecular systematics often contradicting morphological interpretations. The current view is that insects ("Hexapoda") are neither a sister group of crustaceans (crabs, shrimp, etc.) nor myriapods (centipedes, millipedes, etc.). Instead, insects are now seen to be nested within the crustaceans (Regier et al., 2010). Here, I will seek to provide a brief description of the range of viruses whose ultimate hosts are insects. Viruses that infect the European honeybee, *Apis mellifera*, is specifically discussed as recent unexplained worldwide colony collapses have reopened the debate of the role viruses play in the potential demise of the honeybee. This chapter will neither, therefore, review viruses of all crustacea (subphylum Pancrustacea) nor will it examine in detail the interactions of all the numerous viruses known to infect the million or so species strong class Insecta. The viruses that infect aquatic crustaceans are covered in Volume 2, Chapter 7, while the multitude of viruses vectored by insects to their ultimate plant hosts are addressed in Volume 1, Chapter 11. Those viruses that infect vertebrate hosts and get vectored by insects are addressed in Volume 2, Chapter 11 (terrestrial mammals), Chapter 13 (human hosts), and Chapter 14 (avian hosts).

Virus diseases of insects were at first etiologically misdiagnosed to be of protozoan, bacterial, and fungal origin. The first documented disease of insects, caused by what is now known to be a virus, was jaundice of the silkworm (Steinhaus, 1949). Records as far back as the early sixteenth century refer to what is commonly known as polyhedrosis caused by an occluded-type baculovirus (Table 10.1). The sixth edition of the Bergey's Manual of Determinative Bacteriology by Holmes (1948) represents the first attempt to classify insect viruses. Notably, in the 1940s only two groups/genera of viruses were known to infect insects; those that infect Lepidoptera (we know today as

Studies in Viral Ecology: Animal Host Systems: Volume 2, First Edition. Edited by Christon J. Hurst.

TABLE 10.1 Viruses of Insects

Subphylum[a]	Superclass[a]	Class	Order	Common Species Names (Number or Species)	Nature of Virus	Virus Family[b] (Genus)	Genera
Pancrustacea	Hexapoda	Insecta	Lepidoptera	Moths, butterflies, and so on (>180,000)	dsDNA	Baculoviridae	*Alphabaculovirus* and *Betabaculovirus*
						Iridoviridae	*Iridovirus*
						Poxviridae	*Betaentomopoxvirus*
						Ascoviridae	*Ascovirus*
						Unassigned	*Nudivirus*
					ssDNA	Parvoviridae	*Densovirus*, *Iteravirus*, *Pefudensovirus*
					dsRNA	Reoviridae	*Cypovirus*
					ssRNA	Dicistroviridae	*Cripavirus*
						Unassigned	*Iflavirus*
						Tetraviridae	*Betatetravirus* and *Omegatetravirus*
						Nodaviridae	*Alphanodavirus*
					RT-virus	Metaviridae	*Metavirus*, *Semotivirus*, and *Errantivirus*
			Diptera	"True flies"—mosquito's, gnats, and so on (>240,000)	dsDNA	Baculoviridae	*Deltabaculovirus*
						Iridoviridae	*Iridovirus* and *Chloriridovirus*
						Poxviridae	*Gammaentomopoxvirus*
					ssDNA	Parvoviridae	*Densovirus*, *Brevi-*, and *Pefudensovirus*
					dsRNA	Reoviridae	*Cypovirus*
						Birnaviridae	*Entomobirnavirus*
					ssRNA	Rhabdoviridae	Unclassified
						Dicistroviridae	*Cripavirus*
						Nodaviridae	*Alphanodavirus*
					RT-virus	Pseudoviridae	*Hemivirus*
						Metaviridae	*Metavirus*, *Semotivirus*, and *Errantivirus*

Hymenoptera	Ants, bees, and so on (>130,000)	dsDNA	Baculoviridae	*Gammabaculovirus*
			Iridoviridae	*Iridovirus*
			Polydnaviridae	*Bracovirus* and *Ichnovirus*
		dsRNA	Reoviridae	*Cypovirus*
		ssRNA	Dicistroviridae	*Cripavirus*
			unassigned	Unassigned (CBPV)
Coleoptera	Beetles (>350,000)	dsDNA	Iridoviridae	*Iridovirus*
			Poxviridae	*Alphaentomopoxvirus*
			Unassigned	*Nudivirus*
		ssRNA	Nodaviridae	*Alphanodavirus*
Hemiptera	"True bugs"—aphids, cicadas, and so on (>80,000)	dsDNA	Iridoviridae	*Iridovirus*
		ssDNA	Parvoviridae	*Pefudensovirus*
		ssRNA	Dicistroviridae	*Cripavirus*
Orthoptera	Grasshoppers, crickets, and so on	dsDNA	Iridoviridae	*Iridovirus*
			Poxviridae	*Betaentomopoxvirus*
			Unassigned	*Nudivirus*
		ssDNA	Parvoviridae	*Pefudensovirus*
		ssRNA	Dicistroviridae	*Cripavirus*
Odonata	Dragonflies, damselflies, and so on	ssDNA	Parvoviridae	*Pefudensovirus*
Blattaria	Cockroaches (>4,500)	dsDNA	Poxviridae	*Alphaentomopoxvirus*
		ssDNA	Parvoviridae	*Pefudensovirus*

[a]Classification based on Regier et al. (2010).
[b]Based mainly on 8th Report of ICTV (Fauquet et al., 2005) and more recent publications.

baculoviruses) and those that cause sacbrood disease in honeybees (belonging to the picorna-like viruses) (Holmes, 1948). Over the past 60 years or so, many more virus species have since been identified, with many being assigned to families (Table 10.2). As it was in the sixteenth century, the baculoviruses and picorna-like viruses (cripaviruses and iflaviruses) are arguably still the best characterized.

10.2 DIVERSITY OF INSECT VIRUSES

Insects have to overcome an array of viruses at different points in their life cycle. These viruses come in the form of many shapes, sizes, types, and sizes of genomes (Table 10.2), thereby posing many ecological questions on their hosts as to how to survive the varied array of accompanying infection strategies.

10.2.1 Double-Stranded DNA Viruses

10.2.1.1 Ascoviridae Ascoviruses retard growth and development of lepidopteron larvae. These infections are often fatal. Some species have a broad tissue tropism, while others are restricted to a small number of cells within a specific tissue type. Opaque hemolymph (due to presence of refractile virion vesicles) is typical of ascoviral disease (Federici and Govindarajan, 1990).

10.2.1.2 Baculoviridae Baculoviruses can be broadly separated into two major groups, nucleopolyhedroviruses (NPVs) and granuloviruses (GVs), based on the virion phenotype as seen in the gut epithelium (midgut) when consumed by susceptible larvae. In the former, the virion is occluded (enclosed) in a crystalline protein matrix that is polyhedral in shape (polyhedrosis), while in the latter, an ovicylindrical shape containing generally a single virion. Members of the Baculoviridae have been divided into four genera (Jehle et al., 2006), with those affecting Lepidoptera having been divided into Alpha- and Betabaculoviruses encompassing the NPVs and GVs, respectively. Those infecting Hymenoptera and Diptera are named Gamma- and Deltabaculoviruses, respectively (Table 10.1).

After consumption by larvae, the occlusions are solubilized and the virions are released within the midgut epithelium. Secondary infection proceeds in the hemocoel. Depending on the viral species, progeny virions are either restricted to the gut epithelium or transmitted to other internal organs and tissues. In Lepidoptera, fat bodies are the primary location of occluded virus production. Ultimately, occlusion bodies are released into the environment upon liquefaction of the host. The typical infection process takes around 7 days and the released occluded virions represent a stable form of the virus. Nucleopolyhedroviruses are known to infect at least three orders of insects (Table 10.1) and one order of Malacostraca (Volume 2, Chapter 7). Granuloviruses, however, are known to only infect lepidopteron species. A more comprehensive review on baculoviruses can be found in Rohrmann's book entitled "Baculovirus Molecular Biology" (Rohrmann, 2008).

10.2.1.3 Unassigned Nudivirus Various nonoccluded, enveloped, rod-shaped viruses previously assigned to the family Baculoviridae infect lepidopteron, coleopteron, and orthopteron insects (Table 10.1). These viruses are transmitted when larvae feed on contaminated food and their sites of replication are midgut and fat body tissues. Some species are, however, highly infectious to the reproductive tissues of its host. Nudiviruses are also known to infect decapods from the marine environment (Wang and Jehle, 2009).

10.2.1.4 Iridoviridae A diverse group of insect iridoviruses that have been reported to infect at least six orders of insects (Table 10.1). Five genera are currently recognized with two of those, *Iridovirus* and *Chloriridovirus*, known to infect insects and terrestrial isopods (Williams, 2008). Transmission mechanisms for these insect viruses are still poorly understood. However, laboratory-based experiments

TABLE 10.2 Physiochemical Properties of Insect Viruses

Nature of Virus	Virus Family (Genus)	Virion Type	Virion Size (nm)	Virus Genome Size (kbp)
dsDNA	Baculoviridae (*Alpha-*, *Beta-*, *Gamma-*, and *Deltabaculovirus*)	Enveloped, occluded, rod shaped, and ovicylindrical	(30–60) × (250–300)	80–180
	Iridoviridae (*Iridovirus* and *Chloriridovirus*)	Nonenveloped, icosahedral	120–300	135–303
	Poxviridae (*Alpha-*, *Beta-*, *Gammaentomopoxvirus*)	Enveloped, occluded, ovoid, oval, and brick shaped	350 × 250, 450 × 250, 320 × 230 × 110	225–380
	Ascoviridae (*Ascovirus*)	Enveloped, bacilliform, ovoid, allentoid	130 × 200–400	120–180
	Polydnaviridae (*Bracovirus* and *Ichnovirus*)	Enveloped, cyclindrical	(30–40) × (8–330)	150–250
	Unassigned (*Nudivirus*)	Enveloped, rod shaped	100 × 200	80–230
ssDNA	Parvoviridae (*Densovirus*, *Iteravirus*, *Brevidensovirus*, *Pefudensovirus*)	Nonenveloped, icosahedral	18–24	4–6
dsRNA	Reoviridae (*Cypovirus*)	Nonenveloped, icosohedral, may be occluded	58–70	24
	Birnaviridae (*Entomobirnavirus*)	Nonenveloped, icosahedral	30–40	4.5–6.0
ssRNA	Dicistroviridae (*Cripavirus*)	Nonenveloped, icosahedral	27–30	7–10
	Unassigned (*Iflavirus*)	Nonenveloped, icosahedral	27–30	9–10
	Rhabdoviridae (unclassified)	Enveloped, bullet shaped	(45–100) × (100–430)	11–15
	Tetraviridae (*Beta-* and *Omegatetravirus*)	Nonenveloped, icosahedral	40	6.5–7.7
	Nodaviridae (*Alphanodavirus*)	Nonenveloped, icosahedral	32–33	4.5
RT-virus	Pseudoviridae (*Hemivirus*)	Nonenveloped, icosahedral	20–30	5–9
	Metaviridae (*Metavirus*, *Semotivirus*, and *Errantivirus*)	Enveloped, ovoid	50	4–10

reveal that transmission can occur through manual injection of virions into the insects (not DNA alone), cohabitation with infected individuals, feeding, or wounding. Acute infections usually result in an iridescent symptom. This discoloration is caused by paracrystalline arrays of particles in infected tissues.

10.2.1.5 *Polydnaviridae* Polydnaviruses are unique to insects and replicate exclusively in the nuclei of the calyx cells located in the female reproductive tract of adult hymenopterans. Two genera are recognized, *Bracovirus* and *Ichnovirus*, detected in braconid and ichneumonid wasps. During oviposition, many endoparasitic wasps inject viral particles into their insect hosts enabling these parasitoids to evade or directly suppress their hosts' immune system, especially avoiding encapsulation by hemocytes. Structurally, bracovirus virions resemble nudivirus and baculovirus virions, and ichnovirus virions resemble those of ascoviruses. Whereas nudiviruses, baculoviruses, and ascoviruses take responsibility for replicating their own DNA to produce progeny virions, polydnavirus DNA is integrated into and replicated from the wasp genome that then directs virion synthesis (Federici and Bigot, 2003).

10.2.1.6 *Poxviridae* Arguably the most famous poxvirus is the *variola virus* that obliterated the Aztec Empire in 1510. Introduced into the New World by Cortez, this virus is believed to be responsible for over 3.5 million deaths over a 2-year period. The family Poxviridae is subdivided into two subfamilies, Chordopoxvirinae and Entomopoxvirinae, composed of eight vertebrate genera and three insect genera, respectively. Insect poxviruses have been detected in Coleoptera (Alpha-), Blattaria (Alpha-), Lepidoptera (Beta-), Orthoptera (Beta-), Diptera (Gamma-), and possibly Hymenoptera (unclassified) (Radek and Fabel, 2000). These distinctions are mainly based on virion morphology.

Poxviruses have a biphasic life cycle within insects where both occluded (OBs) and nonoccluded (cell-released) virus particles are produced. It is speculated that the role of the crystalline matrix protein associated with OBs (spheroids) is to protect virions during horizontal transmission (Arif, 1995). The replication cycle of entomopoxviruses normally begins with the oral ingestion of OBs. After liberation of the occluded viruses in the midgut of the infected insect, the viral envelope fuses with the microvillus membrane of the gut epithelial cells and releases the viral core into the cytoplasm (Granados, 1973). After a period of latency, the virus progeny is assembled in the cell cytoplasm of hemocytes and adipose tissue cells. Mature virion particles may migrate either toward the cytoplasmic membrane for exocytosis or toward a site where occlusion into the OBs occurs (Kurstak and Garzon, 1977).

10.2.2 Single-Stranded DNA Viruses

10.2.2.1 *Parvoviridae* Two subfamilies, Parvovirinae and Densovirinae, composed of five vertebrate and four arthropod-infecting genera, respectively, make up the family Parvoviridae (Mukha et al., 2006). Densoviruses replicate in most tissues of larvae, nymphs, and adult members of the orders Blattaria, Diptera, Hemiptera, Lepidoptera, Odonata, and Orthoptera within the class Insecta (Table 10.1); and order Decapoda, class Malacostraca (Volume 2, Chapter 7).

10.2.3 Double-Stranded RNA Viruses

10.2.3.1 *Birnaviridae* Three genera constitute the family Birnaviridae with *Entomobirnavirus* thus far found to infect only *Drosophila melanogaster*. The other two genera, *Avibirnavirus* and *Aquabirnavirus*, infect birds and aquatic animals in freshwater, brackish and marine environments, respectively. *Drosophila* X virus causes extensive lysis of mesodermal and epidermal cells within 26 h (Teninges et al., 1979).

10.2.3.2 *Reoviridae* The family Reoviridae contain members in 12 genera that infect animals, plants, and fungi. The genus *Cypovirus* have been isolated mainly from insects in the orders Lepidoptera, Diptera, and Hymenoptera (Table 10.1). The cypoviruses replicate within the cytoplasm of infected insect cells and typically produce inclusion bodies (polyhedra) that are composed primarily of a single viral protein (polyhedrin), within which the virus particles can become either singly or (more usually) multiply embedded. Cypoviruses usually infect the midgut cells, particularly of the insect larval stages and can produce chronic rather than fatal disease (Green et al., 2007).

10.2.4 Single-Stranded RNA Viruses

10.2.4.1 Tetraviridae All virus species were exclusively isolated from Lepidoptera species, principally from Saturniid, Limacodid, and Noctuid moths. Two genera are described, *Betatetravirus* and *Omegatetravirus*, with the distinction between these based on the structure of their genomes. These viruses replicate in the midgut epithelial cells and cause cytolysis, leading to larval death within 6–10 days. Oral transmission has been demonstrated experimentally. At high densities, horizontal transmission is a major route of infection. Symptoms vary from inapparent to chronic, which include retarded growth and abnormal pupae and adults (Christian et al., 2001).

10.2.4.2 Nodaviridae The family Nodaviridae contains two genera, *Alphanodavirus* and *Betanodavirus*, which predominantly infect insects and fish, respectively. Although original hosts of alphanodaviruses are limited to insects, their RNAs can replicate in insect, vertebrate, plant, and even yeast cells (Liu et al., 2006). Moreover, certain species are capable of multiplying in and killing both an insect and a vertebrate host. Replication takes place in the cytoplasm of infected cells, in close association with the mitochondria of the host muscle cells. In their insect hosts, alphanodaviruses typically cause stunting, paralysis, and death.

10.2.4.3 Picornavirales (Including the Dicistroviridae and Unclassified Iflavirus) Members of the newly proposed order Picornavirales include viruses infecting animals (family Picornaviridae), plants (families Sequiviridae and Comoviridae and unassigned genera *Sadwavirus* and *Cheravirus*), algae (family Marnaviridae), and insects (family Dicistroviridae, unassigned genus *Iflavirus*, and unassigned chronic bee paralysis virus, CBPV) (Baker and Schroeder, 2008; Christian et al., 2005; Olivier et al., 2008). The European honeybee *A. mellifera* is the host to a wide range of picorna-like viruses. Due to the environmental and economic importance of insect pollinators, the precise role of these viruses in worldwide honeybee colony collapses has been the subject of much research (see next section). Nonetheless, insects in the orders Lepidoptera, Orthoptera, Diptera, and Hymenoptera (Table 10.1) are all known hosts of these viruses.

It is still under debate as to whether these viruses follow a persistent, latent, inapparent, or progressive infection strategy. Persistent, often also referred to as chronic, infections imply that the rate of infection within a host is in balance with the reproduction rate of the infected cell type or host itself. This is achieved through a combination of changing virus replication and host immune responses. Latent infections are when the virus lays dormant within the host (replication inactive) until activation by defined stimuli. Progressive infections are caused by viruses that enter the host cell and replicate undetected for many cellular generations over many years before manifesting overt or acute symptoms. These three infections strategies all evade the host immune system, which results in the inability of the host to fully expel the virus, with the latter scenario often being lethal. Inapparent, often also referred to as covert, infections are indicative of a highly evolved virus and natural host relationship. Moreover, these infections are distinct, in that the natural host can eventually clear itself from this short-term infection. Members of the family Dicistroviridae follow a classic chronic acute type infection strategy, since relatively low loads can rapidly translate into overt symptoms of deformed pupae or paralysis and ultimately death for the adult honeybee, depending on the species of virus and mode of transmission. While members of the unassigned genus *Iflavirus* are generally considered as less virulent, they are known to cause overt symptoms such as larvae liquefaction (liquid-filled sacs) and wing deformities in developing honeybees. Recent findings suggest that one such virus, deformed wing virus (DWV), may be a major factor in overwintering colony losses (Highfield et al., 2009).

Many of the picorna-like viruses either are vectored by or associated with other organisms such as the microsporidian, *Nosema apis*, or the Varroa mite, *Varroa destructor*. However, transmission is not restricted to vectoring. They can be transmitted orally, via food, cleaning, and feeding activities. The presence of these viruses in queen, drone, and subsequent brood suggest vertical and venereal transmission routes. All of these in combination with many other factors complicate the true role these viruses play in the life cycle of the honeybee (Genersch, 2010).

10.2.4.4 Rhabdoviridae The best-known rhabdovirus is the rabies virus, where Pasteur in the 1880s found that passaging this virus through alternate hosts resulted in the production of attenuated strains suitable to immunize patients. The family Rhabdoviridae are a diverse group of viruses subdivided into six genera; all being important pathogens of humans, livestock, and plants that are often vectored by insects. However, *Sigma virus* (SIGMAV) occurs naturally in *D. melanogaster* and is maintained in fly populations through vertical transmission via germ cells. SIGMAV does not appear to adversely affect *Drosophila* sp. in their natural environment; however, SIGMAV-infected flies remain irreversibly paralyzed and die after CO_2 anesthetization (Tsai et al., 2008).

10.2.5 Reverse Transcribing (RT) Viruses

10.2.5.1 Metaviridae The genomes of insects are rich in retroelements that structurally are similar to the RT viruses of vertebrates. These genetic elements are able to jump from one place in the genome to new locations and indeed from one genome to another. Different classes of retroelements may have an incredible impact upon the survival and overall fitness of an insect population. Members of the family Metaviridae, are often referred to as the Ty3-*gypsy* family, are common in insect orders Lepidoptera and Diptera (Terzian et al., 2001).

10.3 ECOLOGY OF HONEYBEE VIRUSES

Beekeeping is synonymous with honey production but often its contribution to providing all-year-round supply of bees for pollination is often not recognized. In fact, the honeybee and its contribution to pollination are today entirely dependent on beekeepers. The wide distribution of beekeepers across a large geographical area ensures that honeybees play a significant role in crop pollination, including the pollination of commercially important crops such as fruit, which require a long growth season and thus flower early. Certain agricultural crops such as rape, borage, apples, and so on, gain particularly from the early presence of honeybees. Any loss of honeybee colonies will have a severe impact on what we eat and how we live (Genersch, 2010).

Viruses infecting honeybees have been isolated and characterized over the past 50 years, with there being over 18 known viruses described in the literature (Allen and Ball, 1996). The picorna-like single-stranded positive sense RNA viruses dominate this group and are consequently, the best studied to date. Honeybee viruses of this group include the deformed wing virus, acute bee paralysis virus (ABPV), chronic bee paralysis virus, sacbrood virus (SBV), black queen cell virus (BQCV), and the Kashmir bee virus (KBV). They are commonly described to cause inapparent, symptomless infections in their hosts and as a consequence, often go undetected. Certain triggers, such as immune suppression of the honeybee caused by parasites, such as by the mite *V. destructor* (Varroa), are considered to induce viral replication with viral symptoms consequently being observed (Yue and Genersch, 2005). Symptoms of virus infection include crumpled, deformed wings seen in bees infected with DWV (Figure 10.1); trembling, flightless bees seen in bees infected with CBPV; and the accumulation of fluid in brood cells seen in bees infected with SBV. As well as suppressing honeybee immunity, Varroa is also considered important in vectoring viruses, passing them between honeybees (Bowen-Walker et al., 1999).

The wider implications of virus infection and Varroa infestation are thought to be a reduction in honeybee populations, a decline in pollination and a reduction in honey production. A study in the United States observed a correlation between an variant of ABPV (Israeli acute paralysis virus, IAPV) and "colony collapse disorder (CCD)" (Cox-Foster et al., 2007). It is, however, important to note that in a comprehensive study by vanEngelsdorp et al. (2009), which looked at more than 200 variables (including viruses), found that none of these on their own could distinguish CCD from control colonies. Moreover, they identified no single risk factor either consistently or sufficiently abundant in CCD colonies to suggest a single causal agent (vanEngelsdorp et al., 2009). DWV, on the other hand, is thought to have an intricate relationship with Varroa, with numerous studies showing the link between mites parasitizing honeybee pupae and the resultant adults developing malformed wings. Taken together, the expectation is that DWV-associated colony collapse would typically occur in the presence of large Varroa infestation carrying high levels of DWV and with a high proportion of deformed honeybees.

Highfield et al. (2009) showed that despite the presence of multiple virus infections over

FIGURE 10.1 Snapshot of a comb within a husbanded honeybee colony. Arrow: honeybee with deformed wings. (Image courtesy of S. J. Martin.) (*See the color version of this figure in Color Plates section.*)

a year, a significant correlation was observed only between DWV viral load and overwintering colony losses. As stated earlier, the long-held view has been that DWV is relatively harmless to the overall health status of honeybee colonies unless it is in association with severe Varroa mite infestations. These important findings suggest that although DWV can potentially act independently of Varroa mites to bring about colony losses, those colonies that have died demonstrated typical Varroa loads. Therefore, DWV may be a major factor in unexplained overwintering colony losses (Highfield et al., 2009).

10.4 SUMMARY

The correlation between our understanding of a disease and the socioeconomic importance of particular organisms is once again exemplified here. It comes as no surprise that historically the two best-studied examples of insect virology comes from the two most economically important insects, the silkworm and the honeybee. Similarly, our limited understanding of the impact of insect viruses on the class Insecta are limited to a few hundred observations (if that) within a million strong species group. Moreover, research is often limited to the harm and thus control of insects in transmitting viral diseases within and between edible crops. Thus, it is important that research continues to explore new avenues to better our understanding the role viruses play in shaping the ecology of insects, especially the honeybee. Not to do so would have long lasting effects on the environment we live.

REFERENCES

Allen, M. and Ball, B. (1996). The incidence and world distribution of honey bee viruses. *Bee World* 77, 141–162.

Arif, B. M. (1995). Recent advances in the molecular biology of entomopoxviruses. *J. Gen. Virol.* 76, 1–13.

Baker, A. and Schroeder, D. (2008). The use of RNA-dependent RNA polymerase for the taxonomic assignment of Picorna-like viruses (order Picornavirales) infecting *Apis mellifera* L. populations. *Virol. J.* 5, 10.

Bowen-Walker, P. L., Martin, S. J., and Gunn, A. (1999). The transmission of deformed wing virus between honeybees (*Apis mellifera* L.) by the Ectoparasitic Mite Varroa jacobsoni Oud. *J. Invertebr. Pathol.* 73, 101–106.

Christian, P. D., Dorrian, S. J., Gordon, K. H. J., and Hanzlik, T. N. (2001). Pathology and properties of the tetravirus *Helicoverpa armigera* stunt virus. *Biol. Control* 20, 65–75.

Christian, P., Fauquet, C., Gorbalenya, A., King, A., Knowles, N., Le Gall, O., and Stanway, G. (2005). Picornavirales: a proposed order of positive sense RNA viruses. ICTV Poster Session.

Cox-Foster, D. L., Conlan, S., Holmes, E. C., Palacios, G., Evans, J. D., Moran, N. A., Quan, P., Briese, T., Hornig, M., Geiser, D. M., Martinson, V., vanEngelsdorp, D., Kalkstein, A. L., Drysdale, A., Hui, J., Zhai, J., Cui, L., Hutchison, S. K., Simons, J. F., Egholm, M., Pettis, J. S., and Lipkin, W. I. (2007). A metagenomic survey of microbes in honey bee colony collapse disorder. *Science* 318, 283–287.

Fauquet, C. M., Mayo, M. A., Maniloff, J., Desselberger, U., and Ball, L. A. (2005). *Virus Taxonomy: Eighth Report of the International Committee on Taxonomy of Viruses*. Elsevier/Academic Press, San Diego/London.

Federici, B. A. and Bigot, Y. (2003). Origin and evolution of polydnaviruses by symbiogenesis of insect DNA viruses in endoparasitic wasps. *J. Insect Physiol.* 49, 419–432.

Federici, B. A. and Govindarajan, R. (1990). Comparative histopathology of three ascovirus isolates in larval noctuids. *J. Invertebr. Pathol.* 56, 300–311.

Genersch, E. (2010). Honey bee pathology: current threats to honey bees and beekeeping. *Appl. Microbiol. Biotechnol.* 87, 87–97.

Granados, R. R. (1973). Entry of an insect poxvirus by fusion of the virus envelope with the host cell membrane. *Virology* 52, 305–309.

Green, T. B., White, S., Rao, S., Mertens, P. P. C., Adler, P. H., and Becnel, J. J. (2007). Biological and molecular studies of a cypovirus from the black fly *Simulium ubiquitum* (Diptera: Simuliidae). *J. Invertebr. Pathol.* 95, 26–32.

Highfield, A. C., El Nagar, A., Mackinder, L. C. M., Noel, L. M. L. J., Hall, M. J., Martin, S. J., and Schroeder, D. C. (2009). Deformed wing virus implicated in overwintering honeybee colony losses. *Appl. Environ. Microbiol.* 75, 7212–7220.

Holmes, F. (1948). Order Virales, the filterable viruses. In: *In Bergey's Manual of Determinative Bacteriology*, 6th edition. Williams & Wilkins, Baltimore, pp. 1225–1228.

Jehle, J. A., Blissard, G. W., Bonning, B. C., Cory, J. S., Herniou, E. A., Rohrmann, G. F., Theilmann, D. A., Thiem, S. M., and Vlak, J. M. (2006). On the classification and nomenclature of baculoviruses: a proposal for revision. *Arch. Virol.* 151, 1257– 1266.

Kurstak, E. and Garzon, S. (1977). Entomopoxviruses (poxviruses of invertebrates). In: Maramosch, K. (ed.), *The Altas of Insect and Plant Viruses*. Academic Press, New York, pp. 29–39.

Liu, C., Zhang, J., Yi, F., Wang, J., Wang, X., Jiang, H., Xu, J., and Hu, Y. (2006). Isolation and RNA1 nucleotide sequence determination of a new insect nodavirus from *Pieris rapae* larvae in Wuhan city, China. *Virus Res.* 120, 28–35.

Mukha, D. V., Chumachenko, A. G., Dykstra, M. J., Kurtti, T. J., and Schal, C. (2006). Characterization of a new densovirus infecting the German cockroach, *Blattella germanica*. *J. Gen. Virol.* 87, 1567–1575.

Olivier, V., Blanchard, P., Chaouch, S., Lallemand, P., Schurr, F., Celle, O., Dubois, E., Tordo, N., Thiéry, R., Houlgatte, R., and Ribière, M. (2008). Molecular characterisation and phylogenetic analysis of Chronic bee paralysis virus, a honey bee virus. *Virus Res.* 132, 59–68.

Radek, R. and Fabel, P. (2000). A new entomopoxvirus from a cockroach: light and electron microscopy. *J. Invertebr. Pathol.* 75, 19–27.

Regier, J. C., Shultz, J. W., Zwick, A., Hussey, A., Ball, B., Wetzer, R., Martin, J. W., and Cunningham, C. W. (2010). Arthropod relationships revealed by phylogenomic analysis of nuclear protein-coding sequences. *Nature* 463, 1079– 1083.

Rohrmann, G. F. (2008). *Baculovirus Molecular Biology*. National Library of Medicine: NCBI, Bethesada, MD.

Steinhaus, E. A. (1949). Nomenclature and classification of insect viruses. *Bacteriol. Rev.* 13, 203–223.

Teninges, D., Ohanessian, A., Richard-Molard, C., and Contamine, D. (1979). Isolation and biological properties of Drosophila X virus. *J. Gen. Virol.* 42, 241–254.

Terzian, C., Pelisson, A., and Bucheton, A. (2001). Evolution and phylogeny of insect endogenous retroviruses. *BMC Evol. Biol.* 1, 3.

Tsai, C. W., McGraw, E. A., Ammar, E. D., Dietzgen, R. G., and Hogenhout, S. A. (2008). *Drosophila melanogaster* mounts a unique immune response to the Rhabdovirus sigma virus. *Appl. Environ. Microbiol.* 74, 3251–3256.

vanEngelsdorp, D., Evans, J. D., Saegerman, C., Mullin, C., Haubruge, E., Nguyen, B. K., Frazier, M., Frazier, J., Cox-Foster, D., Chen, Y., Underwood, R., Tarpy, D. R., and Pettis, J. S. (2009). Colony collapse disorder: a descriptive study. *PLoS ONE* 4, e6481.

Wang, Y. and Jehle, J. A. (2009). Nudiviruses and other large, double-stranded circular DNA viruses of invertebrates: new insights on an old topic. *J. Invertebr. Pathol.* 101, 187–193.

Williams, T. (2008). Natural invertebrate hosts of iridoviruses (Iridoviridae). *Neoptrop. Entomol.* 37, 615–632.

Yue, C. and Genersch, E. (2005). RT-PCR analysis of deformed wing virus in honeybees (*Apis mellifera*) and mites (*Varroa destructor*). *J. Gen. Virol.* 86, 3419–3424.

CHAPTER 11

VIRUSES OF TERRESTRIAL MAMMALS

LAURA D. KRAMER[1,2] and NORMA P. TAVAKOLI[1,2]
[1]Wadsworth Center, New York State Department of Health, Albany, NY
[2]Department of Biomedical Sciences, School of Public Health, State University of New York, Albany, NY

CONTENTS

11.1 INTRODUCTION

11.1.1 Approach

A wide range of viruses cause disease in humans and other mammals leading to substantial economic losses and significant medical impact. However, the consequences of viral infection vary widely, from asymptomatic to mild or serious illness, and in some cases death. Zoonotic viruses are a major source of infection in humans and can lead to recurring as well as novel outbreaks. Examples include *Influenzavirus* reassortants that directly infect humans and cause large numbers of infections across the world; the severe acute respiratory syndrome (SARS) *Coronavirus* that during the 2003 outbreak infected humans, had a case fatality rate in Canada of 12.4% (Fung and Yu, 2003), and led to huge economic losses even though its spread was contained. Approximately 5000 different viruses have been described to date. However, this is a small fraction of the true number of viruses in existence. Assuming 50,000 vertebrates, each with 20 endemic viruses, there are likely 1,000,000 vertebrate viruses. Thus, 99.8% of vertebrate viruses remain to be discovered (Morse, 1993).

Zoonotic pathogens are agents that naturally infect nonhuman animals, but can be transmitted directly between these vertebrates and man or indirectly from vectors to man. Some zoonotic agents have the potential to cause extensive outbreaks. In some cases the source of infection is domestic and synanthropic animals

Studies in Viral Ecology: Animal Host Systems: Volume 2, First Edition. Edited by Christon J. Hurst.

(e.g., cat scratch disease and urban rabies), while in others the cycle occurs outside human habitats and is associated with feral and wild animals (e.g., arboviruses and wildlife rabies). Some zoonotic pathogens can circulate in both sylvatic and urban cycles (e.g., *chikungunya virus* and *yellow fever virus* (YFV)). Zoonoses account for the majority (60.3%) of emerging infectious disease (EID) events (Jones et al., 2008; Patz et al., 2004; Taylor et al., 2001). Furthermore, 71.8% of zoonotic EID events were determined to be caused by pathogens with a wildlife origin (Jones et al., 2008). Two examples of such emerging zoonotic pathogens are *Nipah virus* (NIPV) that emerged in Perak, Malaysia, and SARS *Coronavirus* in Guangdong Province, China. It is notable that viral and prion pathogens comprise approximately 25–44% of EID events, depending on the study (Cleaveland et al., 2001; Jones et al., 2008; Woolhouse et al., 2005). The species of animals harboring emerging agents infectious for humans are numerous, but have been reported to be predominantly ungulates, carnivores, and rodents, in that order, for viral pathogens (Cleaveland et al., 2001).

The enormous number of viruses detected in terrestrial mammals makes it prohibitive to cover all of them fully in a single chapter. Therefore, we have taken the approach of providing a general listing of this large group of viruses (Table 11.1), briefly describing virus classification and the main characteristics that distinguish the orders of viruses infecting terrestrial mammals, and focusing the chapter more intensively on those viruses that contribute to cross-species infections or spillover events. We will review taxonomic classes of viruses, but will also examine viruses of terrestrial mammals in ecological groupings, that is, viruses of wildlife, livestock, and companion animals. Although the classical viral taxonomy is extremely important, this is covered more extensively in Chapter 2.

Pathogen spillover occurs when epidemics in a host population are driven by transmission from a novel population other than the maintenance or reservoir species. The maintenance or reservoir host is the main vertebrate host in the

TABLE 11.1 Listing of Viral Families Affecting Vertebrates

Virus Family	Viral Genome
Adenoviridae	DNA, double-stranded
Alloherpesviridae	DNA, double-stranded
Anelloviridae	DNA, single-stranded
Arenaviridae	RNA, single-stranded (– sense)
Arteriviridae	RNA, single-stranded (+ sense)
Asfarviridae	DNA, double-stranded
Astroviridae	RNA, single-stranded (+ sense)
Birnaviridae	RNA, double-stranded
Bornaviridae	RNA, single-stranded (– sense)
Bunyaviridae	RNA, single-stranded (– sense)
Caliciviridae	RNA, single-stranded (+ sense)
Circoviridae	DNA, single-stranded
Coronaviridae	RNA, single-stranded (+ sense)
Filoviridae	RNA, single-stranded (– sense)
Flaviviridae	RNA, single-stranded (+ sense)
Hepadnaviridae	DNA, partially double-stranded
Hepeviridae	RNA, single stranded (+ sense)
Herpesviridae	DNA, double-stranded
Iridoviridae	DNA, double-stranded
Nodaviridae	RNA, single-stranded (+ sense)
Orthomyxoviridae	RNA, single-stranded (– sense)
Papillomaviridae	DNA, partially double-stranded
Paramyxoviridae	RNA, single-stranded (– sense)
Parvoviridae	DNA, single-stranded
Picobirnaviridae	RNA, double-stranded
Picornaviridae	RNA, single-stranded (+ sense)
Polyomaviridae	DNA, partially double-stranded
Poxviridae	DNA, double-stranded
Reoviridae	RNA, double stranded
Retroviridae	RNA, single-stranded (+ sense)

TABLE 11.1 *(Continued)*

Virus Family	Viral Genome
Rhabdoviridae	RNA, single-stranded (– sense)
Togaviridae	RNA, single-stranded (+ sense)

Source: This information has been modified from the ICTV Document "Master Species List 2009."

enzootic cycle. This host harbors the pathogen and serves as a source of infection. Transmission between this primary host and a novel host population (recipient host) occurs when habitats overlap. These interspecies transmission events are generally not perpetuated in the newly involved recipient species, which had not been part of the enzootic transmission cycle. The likelihood that a virus will become endemic in the new host species depends on a complex set of interactions (Kuiken et al., 2006). Ecological, anthropogenic, and evolutionary factors affect the probability that zoonotic viruses maintained by wildlife reservoir hosts will emerge and infect new host species. The discovery of *Hendra virus* (HENV) and *Nipah virus* in Australasia, SARS *Coronavirus* in China, HPAI H5N1 *Influenzavirus* in Southeast Asia, and the spread and persistence of epidemic *West Nile virus* (WNV) in the United States have pointed to the importance of cross-species transmission. Childs et al. (2007) have described a four-step process by which zoonotic viruses are transmitted to and infect other species. Two of these, interspecies contact and cross-species virus transmission (spillover), are essential and sufficient to cause epidemic emergence. Sustained transmission and virus adaptation within the spillover host are transitions not required for virus emergence, but determine the magnitude and scope of subsequent disease outbreaks (Childs, 2004). Pathogen jumps across species appear to be common, but very little is known about the process leading to establishment of the virus in new host species. A list of pathogens that have jumped species is provided in Table 11.2, modified from Woolhouse et al. (2005).

The virus–host interaction is rather complex and is influenced by the genetics of both the virus and the host, immune status of the hosts, population density, and more. The complexity

TABLE 11.2 Viral Pathogens That Have Jumped Species

Viruses	Original Host	New Host	Year Reported
Rinderpest virus	Eurasian cattle	African ruminants	Late 1800s
Myxoma virus	Brush/Brazilian rabbit	European rabbit	1950s
Ebolavirus	Unknown	Humans	1977
Feline panleukopenia virus/ canine parvovirus	Cats	Dogs	1978
Simian immunodeficiency virus/ human immunodeficiency virus 1	Primates	Humans	1983
Simian immunodeficiency virus/ human immunodeficiency virus 2	Primates	Humans	1986
Canine/phocine distemper virus	Canids	Seals	1988
Hendra virus	Bats	Horses and humans	1994
Australian bat lyssavirus	Bats	Humans	1996
Influenza A virus (H5N1)	Chicken	Humans	1997
Nipah virus	Bats	Pigs and humans	1999
SARS-Coronavirus	Palm civets	Humans	2003
Monkeypox virus	Prairie dogs	Humans	2003

Source: Woolhouse et al. (2005).

of the interaction between the virus and host is further exacerbated by the rapid evolution of some viruses through mutation, recombination, or reassortment (Table 11.3). In the case of arthropod-borne pathogens, complexity is further increased by the vectorial capacity of the vector, which includes the vector's population density, longevity, feeding behavior, vector genetics, and more (Figure 11.1). Vertebrate and arthropod hosts are important determinants of pathogen transmission in other ways as well. Their reservoir competence (i.e., susceptibility to infection, duration of infection, level of infectiousness) (Komar et al., 2003) is critical to their ability to act as amplifying hosts, as is their abundance and their contact with vectors (Figure 11.1). An amplifying host is one in which a pathogen replicates to a high enough titer that a vector such as a mosquito that feeds on it will probably become infectious. Additional important contributing factors include the vertebrate immune status, reproductive rate, that is, how frequently new susceptible hosts appear, host population density, and presence of other hosts, that is, biodiversity.

11.1.2 Classification

The classification of viruses is based on genome type (single-stranded or double-stranded RNA or DNA, positive- or negative-sense genome), morphology, replication strategy, and the type of disease they cause. Classification guidelines have been established by the International Committee on Taxonomy of Viruses (ICTV) and follow the standard sequence: order, family, subfamily, genus, species, and subspecies. The six orders that have currently been established are Herpesvirales, Mononegavirales, Nidovirales, Picornavirales, Tymovirales, and Caudovirales. Caudovirales are bacteriophages and Tymovirales are plant viruses and therefore these groups will not be discussed further in this chapter. In addition, there are genera for which no family has as yet been established. These "floating" genera include numerous viruses that infect terrestrial vertebrates.

11.1.2.1 Order Herpesvirales The order Herpesvirales contains large double-stranded DNA viruses. Within this order, there are

TABLE 11.3 Genetic Changes and Recent Evidence of Viral Evolution

Mechanism	Example
Point mutation	Avian influenza virus (H5N2), minimally pathogenic strain change into lethal epidemic strain by just a few point mutations
Point mutation	Canine parvovirus, host range mutation of feline panleukopenia virus leading to the emergence of a new global pathogen
Intramolecular recombination	Western equine encephalitis virus, originated by recombination between eastern equine encephalitis virus and a Sindbis-like alphavirus that no longer exists
Genetic reassortment (shift)	Pandemic human influenza A subtypes H1N1 (1918), H2N2 (1957), and H3N2 (1968), each originated by gaining genes from avian viruses, in each case after passage through swine
Intramolecular recombination and mutation	Polio vaccine viruses, reversion of attenuated viruses to virulence following vaccination
Probable mutation in individual host	Feline infectious peritonitis virus, evolution of lethal virus from temperate feline coronovirus
Possible mutation	Lelystad virus (porcine reproductive and respiratory syndrome virus) may have emerged via host range mutation of a rodent arterivirus
Possible mutation	Equine morbillivirus, the cause of acute respiratory distress syndrome in horses and humans in Australia, may have required a host range mutation to jump from its reservoir in bats

Source: Murphy et al. (1999).

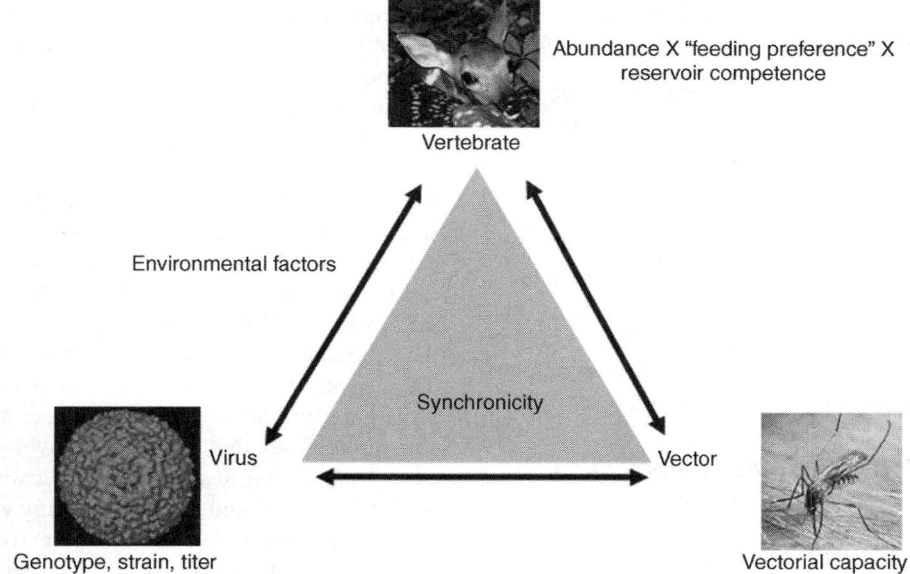

FIGURE 11.1 Factors affecting transmission and establishment of arboviruses. (*See the color version of this figure in Color Plate section.*)

3 families, 3 subfamilies, 17 genera, and 90 species (Davison, 2010). The subfamily Alphaherpesvirinae includes the genera *Simplexvirus* (e.g., herpes simplex virus 1) and *Varicellovirus* (e.g., varicella zoster virus). The subfamily Betaherpesvirinae includes the genus *Cytomegalovirus* (e.g., human herpesvirus 5) and *Roseolovirus* (e.g., human herpesvirus 6). The subfamily Gammaherpesvirinae includes the genus *Lymphocryptovirus* (e.g., Epstein–Barr virus).

11.1.2.2 Order Mononegavirales The order Mononegavirales includes nonsegmented negative-strand single-stranded RNA plant and animal viruses. This order is comprised of four families: Bornaviridae (e.g., Borna disease virus), Rhabdoviridae (e.g., rabies virus), Filoviridae (e.g., Ebola virus), and Paramyxoviridae (e.g., measles virus).

11.1.2.3 Order Nidovirales The order Nidovirales includes single-stranded positive-sense RNA viruses that infect vertebrates. This order consists of three families: Arteriviridae, Coronaviridae, and Roniviridae. Viruses belonging to the family Arteriviridae are thus far known to primarily cause disease in animals (e.g., *Arterivirus*). Coronaviridae, in general, infect animals and birds and are characterized by crown-shaped projections on the surface of the virion. The order Roniviridae contains a single genus, *Okavirus*, which predominantly infects crustaceans, and thus will not be discussed further in this chapter on viruses of terrestrial mammals.

11.1.2.4 Order Picornavirales The order Picornavirales contains small positive-strand RNA viruses that variously infect plants, insects, and vertebrate hosts. Within this order, the family Picornaviridae consists of 12 distinct genera and includes many important pathogens of humans and animals (e.g., enteroviruses and foot-and-mouth disease virus (FMDV)).

11.1.2.5 Floating Virus Genera Many viral genera are yet to be assigned to higher classification levels by the ICTV, and thus they are referred to as floating genera.

11.1.3 Structure and Genome Organization

11.1.3.1 Order Herpesvirales The order Herpesvirales shares an overall morphology in which an icosahedral capsid contains a large nonsegmented double-stranded linear DNA genome (Pellett et al., 1985). The capsid is surrounded by a tegument protein and subsequently by a lipid envelope that contains a number of membrane-associated proteins. The entire particle forms the virion. The genome of herpesviruses encodes 100–200 genes and has both direct and indirect terminal repeats.

11.1.3.2 Order Mononegavirales The members of the order Mononegavirales possess a nonsegmented negative-sense RNA genome. The genomic RNA and RNA-dependent RNA polymerase (RDRP) form a helical nucleocapsid that is surrounded by an envelope. The genome contains 6–10 genes, although a greater number of gene products are processed through RNA editing.

11.1.3.3 Order Nidovirales The order Nidovirales consists of viruses that have positive-sense single-stranded RNA genomes. The group contains the virus *murine hepatitis virus* (MHV) that, with a genome of 31.5 kb, has the largest known nonsegmented RNA genome. Because they are of positive sense, the members of this group of viruses do not have a polymerase function included in their virion as the genome can be read directly as mRNA when it first enters the host cell. Within the cytoplasm of the host cell, the virus uses some of the host cell proteins to replicate and express its genes. The structural proteins are encoded at the 3′-end of the genome, while the nonstructural proteins including the proteinases are located at the 5′-end (de Vries et al., 1997). All members of this order are enveloped.

The *Coronavirus* virion is roughly spherical, approximately 100–120 nm in diameter and have 20 nm long spikes. The *Torovirus* virion, however, is pleomorphic and 120–140 nm in length with surface projections. Arterivirions are smaller (50–70 nm) and lack large surface projections. The nucleocapsid structures of the viruses in the three families are very different as the coronaviruses have a loosely wound helix (Macnaughton et al., 1978), the toroviruses a compact tubular structure (Weiss and Horzinek,), and the arteriviruses an isometric structure (Horzinek, 1981). This group of viruses expresses the structural proteins separately from the nonstructural ones. A set of subgenomic mRNAs at the 3′-region of the genome encode the structural proteins (Brian and Baric, 2005). The 5′-end of the genome encodes one main proteinase and between one and three accessory proteinases that are mainly involved in expressing the replicase gene. Replication of the RNA genome and its correct timing within the viral life cycle depends on the activation and inactivation of specific viral proteins by the viral proteinases. The function of many other proteins identified on the genomes of Nidovirales is as yet unknown.

11.1.3.4 Order Picornavirales As their name “pico-” indicates, the picornaviruses grouped within the order Picornavirales are small RNA viruses. The viral particle is approximately 27–30 nm in diameter. Picornaviruses are nonenveloped and have single-stranded positive-sense RNA genomes encased in an icosahedral capsid (Stanway, 1990). The RNA has protein on the 5′-end that acts as a primer for transcription by RNA polymerase. The viral genome is approximately 7.8 kb in length and has a 5′- and 3′-untranslated region that is involved in translation, negative-strand synthesis, and perhaps virulence. The genes encoding the structural proteins are at the 5′-end of the genome, whereas those encoding the nonstructural proteins are at the 3′-end.

11.1.3.5 Other Notable Double-Stranded DNA Viruses: The Poxviruses and Adenoviruses The poxviruses belong to the family Poxviridae and are large, complex viruses that have an unusual morphology. They are slightly pleomorphic and range from ovoid to brick shape. Poxviruses consist of a surface membrane, a core, and lateral bodies and are generally enveloped. The virion is approximately 200 nm in diameter and 300 nm in length. The complete genome of poxviruses is 130–375 kb long and consists of linear double-stranded DNA with terminal inverted repeats. Adenoviruses are the largest nonenveloped viruses. Their icosahedral viral particles are 90–100 nm in length and are composed of a nucleocapsid and a double-stranded DNA genome. The adenoviral capsid is associated with spikes that allow the virus to attach to the host cell. The genome of adenoviruses is 26–45 kb in length and on both 5′-ends is associated with a 55 kDa protein that acts as primers in viral replication.

11.2 NONARTHROPOD-BORNE VIRUSES

The nonarthropod-borne viruses, as their name suggests, are transmitted by means other than an arthropod host. These viruses are generally transmitted from human to human, among animals, or from animals to humans. Their modes of transmission include fecal–oral route (*Adenovirus*), aerosol, urine, feces, saliva, or nasal discharge (e.g., *Influenzavirus A*), mechanical route (e.g., *Myxoma virus*), direct contact by biting (e.g., *Rabies virus*), and venereal (e.g., HIV-1). The terrestrial mammals that act as hosts for these viruses vary widely and include humans and domesticated and wild animals.

11.2.1 Nonarthropod-Borne Viruses of Wildlife

Viruses can have a catastrophic effect on wildlife regardless of whether the vertebrate host is the maintenance host or an incidental one. *Canine parvovirus* (CPV) 2, *Canine distemper virus* (CDV), *Fowlpox virus*, *Nipah virus*, *Fibroma virus*, and *Monkeypox virus* are but a few of the many examples of viruses that cause severe disease in their hosts. Some wildlife viruses pose a threat to man and domesticated animals, and conversely, endangered species of animals are threatened by crossover of viruses from domestic animals that come into contact with them as a result of spillover events. Therefore, the potential for spillover of viruses from their natural maintenance cycles needs to be taken into consideration in land management and wildlife conservation decisions. This need was highlighted by several significant events. The black-footed ferret nearly reached extinction in the western United States as prairie dogs, the mainstay of the ferrets' diet, lost habitat through agricultural development of prairies, followed by decimation of the populations of prairie dogs by plague. Then, in 1985, *Canine distemper virus*, which generally infects domestic dogs, further threatened the black-footed ferret population. Spillover of canine distemper infecting domestic dogs also killed one-third of the Serengeti lions in 1974. Other examples of spillover of viruses leading to emerging disease threats include HIV AIDS, *Ebolavirus*, *Hantavirus*, *Nipah virus*, and SARS *Coronavirus*. In each case, it was anthropogenic change or human behavior that led to situations that altered contact among susceptible hosts leading to overlap of habitats and consequent pathogen transmission. The spread and incidence of cross-species transmission appear to be increasing due to globalization and growing human populations, especially where these impinge on previously undisturbed habitats. It is clear that wildlife conservation and public health are intertwined and both must be considered in formulating policy for disease control and optimization of ecosystem health (Cleaveland et al., 2007).

11.2.1.1 DNA Viruses

11.2.1.1.1 Herpesviruses Herpesviruses are ubiquitous in nature and most animal species examined have been found to harbor one or more diverse herpesviruses. They generally are spread by direct physical contact and the most common form of transmission for many herpesviruses is sexual contact. Host terrestrial mammals include humans, dogs, cats, nonhuman primates (macaques, chimpanzee, gorilla, marmoset, and orangutan), cattle, water buffalo, sheep, elephants, horses, guinea pigs, mice, and rats. These viruses are species specific and cross-species infections have not been observed. Human herpesviruses infect people of all ages and lifestyles. In humans, diseases caused by herpesviruses vary and include primary and recurrent epithelial lesions, keratitis, encephalitis, hepatitis, retinitis, pneumonia, and infectious mononucleosis. Herpesviruses have the ability to establish and maintain a latent state in their host and reactivate following cellular stress. During latency, a small subset of viral genes are expressed and the viral genome is maintained in the host cell nucleus. Following reactivation, the virus can replicate, cause disease, and also be transmitted.

Certain herpes infections are fatal in mammals other than humans. For example, *Equid herpesvirus 1* (EHV-1) is highly contagious and has caused significant and costly problems in the equine industry. The virus is enzootic throughout the world and in 2007 met the criteria to be considered an emerging infectious disease (USDA, 2008). EHV-1 causes abortions and neurological disease in horses and has given rise to large-scale outbreaks in recent years (Del et al., 2000; Taouji et al., 2002). Intensive annual vaccinations are in progress to help prevent and control the disease. Elephants, particularly young ones, are highly susceptible to an elephantine herpesvirus (*Elephantid herpesvirus 1*). In 2007, a 6.5-year-old elephant died of a previously undiscovered herpesvirus after briefly showing evidence of illness. Because the virus was new, it was not medically recognized (Garner et al., 2009).

Herpesviruses can be transmitted among primates (*Herpes taraminus*), from human to primate (*H. hominis*) or from primate to human (*H. simiae*). *H. taraminus* is perhaps uniquely transmitted by oral or fecal contact, and is fatal for marmosets and owl monkeys. *H. hominis* can cause encephalitis and conjunctivitis and is occasionally fatal in tamarins and marmosets. *H. simiae* is transmitted by bites and scratches and can be fatal for humans and apes.

11.2.1.1.2 Poxviruses Poxviruses, members of the family Poxviridae, have been a scourge of the human race for many centuries. However, the use of effective vaccines led to the official eradication in 1979 of smallpox, the most notable disease caused by these groups of viruses. Because vaccination against smallpox was subsequently suspended, a large proportion of the world's human population is now immunologically näve to orthopox viruses as a group, and therefore the possibility of the emergence or reemergence of variola, monkeypox, camelpox, taterapox, and other orthopox diseases has increased. Adequate supplies of vaccinia virus, which can be used as vaccines, should therefore be maintained and available to combat future smallpox outbreaks.

Monkeypox disease (family Poxviridae; genus *Orthopoxvirus*) poses one of the greatest global disease threats to wild carnivores, including lions, African wild dogs, and several types of seal. It was first isolated from sick animals in a colony of captive cynomolgus monkeys in Copenhagen in 1958, and then identified in 1970 in a child in the Democratic Republic of Congo (DRC) (Heymann et al., 1998). Monkeys, squirrels (*Funisciurus anerythrus* and *Heliosciurus rufobrachium*) in the areas surrounding human settlements (Khodakevich et al., 1987), and rodents appear to be the predominant reservoirs for monkeypox virus (Jezek et al., 1983), which is endemic in rainforests in central and western Africa. Outbreaks of this disease have occurred in the DRC. The disease is very similar to smallpox, although the signs and symptoms may be milder and the case fatality rate is approximately 10% (Meyer et al., 2002). *Monkeypox virus* was introduced into the United States in 2003 in Wisconsin through an infected rainforest rat that was imported from Gambia for the pet trade. The virus was then transmitted to prairie dogs for sale as pets, leading to an outbreak of disease. Between May and July 2003, over 70 monkeypox infections were reported in individuals from multiple states in the United States (CDC, 2003; Guarner et al., 2004).

11.2.1.2 RNA Viruses *Canine distemper virus* (family Paramyxoviridae; genus *Morbillivirus*) is a negative-strand RNA virus with a broader host range than most morbilliviruses. It is mostly associated with domesticated dogs and ferrets as well as raccoons, coyotes, foxes, and skunks and appears to be expanding its host range (Harder and Osterhaus, 1997). The contact between the infected wild canids and domestic dogs facilitates the spread of the virus. Some animals, for example, raccoons, acquire the disease most often when their populations become large or concentrated. Outbreaks appear to run in cycles of 5–7 years. The virus appears to be present in the environment, is highly contagious, and is transmitted via airborne particles and is shed through bodily secretions, mainly respiratory secretions, and droppings from an infected animal. The disease does not affect humans. Canine distemper was once the leading cause of death in puppies worldwide. However, due to vaccination efforts, the incidence of the disease has decreased significantly. The disease primarily manifests with gastrointestinal and respiratory symptoms, including fever, loss of appetite, discharge from nose and eyes, a rough coat of hair, emaciated appearance, conjunctivitis, diarrhea, pneumonia, rhinitis, and vomiting. Neurological symptoms such as disorientation or wandering aimlessly are also observed. The neurological symptoms can mimic rabies; the animal may smack its lips, convulse, and appear to be chewing gum. Most of the infected dogs suffer from encephalomyelitis and die from neurological complications such as seizures, paralysis, and loss of mental and motor skills. In 2010 there was a notable outbreak of canine distemper in dogs in California and in raccoons in Washington State. It is strongly recommended that pet owners vaccinate their dogs against this disease as there is no effective treatment.

The virus has a wide host range evidenced by the fact that in the early 1990s distemper outbreaks caused by CDV-like morbilliviruses were reported in lions, tigers, and jaguars in zoos and safari parks in the United States (Appel et al., 1994). In 1994, 1000 lions died in Tanzania, one-third of the Serengeti lion population, from distemper carried by dogs, as determined by the close phylogenetic relationship between CDV isolates from lions and domestic dogs. The epidemic spread north to lions in the Maasai Mara National Reserve of Kenya and infected hyenas, bat-eared foxes, and leopards (Roelke-Parker et al., 1996). An additional CDV infection occurred in 2001 leading to similar losses in the lion population. Both outbreaks were preceded by extreme drought conditions that led to debilitated populations of Cape buffalo, a major prey species of lions. The buffalo suffered heavy tick infestations and became even more common in the lions' diet, resulting in unusually high levels of tick-borne blood parasites in the lions. The *Canine distemper virus* suppressed the lions' immunity, which allowed the elevated levels of blood parasites to reach fatally high levels, leading to mass die-offs of lions. Pathogen evolution at known functional sites in the CDV genome has been demonstrated to be an important mechanism in the success of cross-species transmission (McCarthy et al., 2007).

Control strategies include vaccination of neighboring domestic and feral dog populations, which prevents circulation of CDV in these animals and thereby limits transmission to wildlife. However, problems with this approach are both logistical and biological.

11.2.1.3 Cross-Species Transmission of Bat Viruses One of the oldest viruses known to man is rabies (family Rhabdoviridae; genus *Lyssavirus*). This negative-strand RNA virus was first described in pre-Mosaic Eshmuna Code of Babylon, twenty-third century B.C. (Rupprecht et al., 1995). Of all the viruses carried by bats, it is the only one directly transmitted by bite, while the other bat viruses are spread through excreta and body fluids. The *Lyssavirus* genus currently segregates into two phylogroups containing seven genotypes, but is constantly evolving as evidenced by the fact that new lyssaviruses are isolated from bats.

The genotypes that have been discovered differ in neurotropism, pathogenesis, induction of apoptosis, and immunogenicity. Lyssaviruses appear to have evolved in chiropters (bats) before the emergence of carnivoran rabies, and that emergence very likely occurred through host-switching from bats to carnivores. Examples of occasional rabies spillover events have been noted between different species of bats and from bats to domestic animals and terrestrial wildlife (Shankar et al., 2005). Bat rabies is a growing concern for both public and animal health. If infection with divergent lyssavirus genotypes is not prevented by available vaccine strains, then there is a need to broaden the spectrum of vaccines.

Rabies virus infects all warm-blooded animals and is present in animal populations in most parts of the world. In Western Europe and Australia, rabies is mainly prevalent in the bat population (Kuzzmin and Rupprecht, 2007). In Asia, most of the Americas, and Africa, the primary hosts are dogs (Hanlon et al., 2007) and the viral infection then is transmitted to humans by bites of the affected dogs. In the United States, feral raccoons and skunks are also responsible for transmitting rabies to humans. Other animals that transmit the disease include monkeys, foxes, coyotes, wolves, mongoose, groundhogs, bears, cats, domestic farm animals, and cattle (Hanlon et al., 2007). The disease is generally transmitted by the bite of an infected animal and causes acute encephalitis that is usually fatal, if left untreated.

The early symptoms of rabies include headache, malaise, fever, vomiting, and loss of appetite. The disease progresses to a neurological phase with symptoms that include acute pain, violent movements, confusion, depression, and hydrophobia. Finally, the patient may experience periods of hyperactivity and lethargy, seizures, and paralysis eventually leading to coma and multiple organ failure (Jackson, 2007). In animals, the symptoms are similar and cause viciousness and "mad dog" behavior. In some cases, animals do not present the hyperactive stage but instead appear in a stupor. In the latter stage of the disease, damage to motor neurons leads to loss of coordination, paralysis, drooling, and the inability to swallow (Hanlon et al., 2007). Respiratory failure and, less frequently, cardiac arrest are the main causes of death.

Other viruses in the *Lyssavirus* genus include *Australian bat lyssavirus* that has been found in a variety of bat species in Australia and *Mokola virus* that has been found in shrews in Nigeria and numerous other mammalian species in sub-Saharan Africa, including rodents and domestic cats in Zimbabwe.

Bats are also the natural reservoir hosts of other pathogens, including *Nipah virus* and *Hendra virus* (family Paramyxoviridae; genus *Henipavirus*) (Hyatt et al., 2004). The natural reservoir for *Nipah virus* is flying foxes (genus *Pteropus*) in Australia and that of *Hendra virus* possibly is bats of the genus *Pteropus* in Malaysia. The increase in overlap between bat habitats and pig farms in Malaysia has led to the contact of pigs with urine and feces of infected bats, thus creating an environment for transmission of *Nipah virus* from flying foxes to pigs (Chua et al., 2002). Human encroachment into bat territory has allowed an increase in contact between bats and humans, illustrating cross-species transmission and leading to emergence of henipaviruses.

Ebolavirus (order Mononegavirales; family Filoviridae) causes short-lived but high-risk epidemics in tropical areas of Africa and has been detected in three species of fruit bats (suborder Megachiroptera): *Hypsignathus monstrosus* (hammer-headed fruit bats), *Epomops franqueti* (singing fruit bats), and *Myonycteris torquata* (little collared fruit bats) in Central Africa. Great apes and forest duikers feed on fruits contaminated with infected bat saliva, suggesting a possible chain of transmission events leading to *Ebolavirus* spillover to these incidental hosts. While the bats are chronically and asymptomatically infected, the incidental hosts are highly susceptible to lethal disease. Humans presumably encounter Ebola by consumption of or other physical contact with the infected vertebrates.

Coronaviruses (order Nidovirales; family Coronaviridae) infect a wide range of mammals and birds and occur worldwide. Although most diseases associated with these viruses are mild, sometimes they can cause more severe symptoms in humans, such as the infection of the respiratory tract known as severe acute respiratory syndrome. Three species of horseshoe bats (*Rhinolophus* spp.) are now recognized as the natural reservoir hosts of SARS *coronavirus* (Li et al., 2005). In the 2003 SARS outbreak, palm civets used as human food source were believed to have acted as the amplification hosts for the virus, but there is evidence suggesting that bats could have been the primary or introductory host (Lau et al., 2005). Since bats act as reservoirs of a number of different viruses, continuous surveillance of these flying mammals is required to prevent and control outbreaks of emerging zoonotic diseases.

11.2.2 Nonarthropod-Borne Viruses of Livestock

Livestock are susceptible to infection with a broad scope of viruses leading to disease and these diseases can result in large economic losses to the farmer or rancher. These viruses belong to the same diverse families as do wildlife viruses and include such pathogens as *Alcelaphine herpesvirus 1* (AHV-1), *Ovine herpesvirus 2* (OHV-2), *Foot-and-mouth disease virus*, *Bovine viral diarrhea virus* (BVDV) 1 and 2, *Equine arteritis virus* (EAV), influenza viruses, lentiviruses, *porcine circovirus*, *Bluetongue virus*, and the vesicular stomatitis viruses among others. Midges, cockroaches, and grasshoppers have been demonstrated in the laboratory to transmit some of these viruses after brief periods of incubation, and occasionally these and other related viruses may depend on arthropod transmission for perpetuation, for example, *Bluetongue virus* and midges of the genus *Culicoides*. An excellent reference for viral diseases in cattle is *Viral Diseases of Cattle*, 2nd edition, by Robert F. Kahrs (Iowa State University Press, Ames, IA, 2001. This book provides a thorough description of the viral etiology, disease symptoms, treatment, and, where applicable, immunization.

11.2.2.1 Alcelaphine Herpesvirus 1 and Ovine Herpesvirus 2 Alcelaphine herpesvirus 1 and Ovine herpesvirus 2 (family Herpesviridae; genus *Rhadinovirus*) belong to a group of ruminant gamma herpesviruses. AHV-1 causes inapparent infection in its reservoir host wildebeest as does the OHV-2 in sheep. However, both viruses jump species and are usually fatal in cattle, antelope, deer, and buffalo. In Africa, over one million wildebeest migrate over 1800 miles every year. Many of the 400,000 wildebeest calves born each year are infected with malignant catarrhal fever (MCF), and consequently the migration presents a threat to local farmers as this virus infects livestock causing major concern. To avoid MCF, the farmers move their cattle to poorer upland grazing where they consequently become exposed to other serious diseases.

11.2.2.2 Foot-and-Mouth Disease Virus *Foot-and-mouth disease virus* is a picornavirus and the prototypic member of the *Aphthovirus* genus in the family Picornaviridae (Grubman and Baxt, 2004). The virus is the causative agent of foot-and-mouth disease, a severe, highly contagious disease of cattle and swine. The disease also affects sheep, deer, goats, wild bovids, and other cloven-hoofed ruminants as well as over 70 species of wild animals (Fenner et al., 1993) such as elephants and hedgehogs. Humans are generally not at risk although in rare cases they can be infected by contact with infected animals. FMD is believed to be the most economically devastating livestock disease in the world and therefore a significant threat to the agricultural industry.

FMDV is spread by the interaction between domestic and wildlife populations, humans and any material that comes into contact with an infected animal, and enters the body through inhalation. Signs of the disease include fever, salivation, and lameness due to vesicles forming in the mouth and feet. The disease also causes abortions, loss of milk, myocarditis, and even death (Grubman and Baxt, 2004); the latter two outcomes occur chiefly in newborn animals.

FMDV is endemic in large areas of Asia, Africa, and South America. Since the virus can easily be transmitted to domestic livestock, the possibility of an outbreak in other parts of the world cannot be ruled out. Between 1998 and 2001, the Pan Asia strain of this virus was responsible for a massive outbreak in Asia, parts of Europe, and Africa and in 2000 and 2001 spread to countries that had previously been free of the virus including Korea and the Netherlands (Knowles et al., 2005). During the first six months of 2009, 122 outbreaks of FMD spanning Asia, Africa, and the Middle East have been reported to the World Organization for Animal Health (OIE) (Arzt et al., 2010). In addition, the virus has been reintroduced into Taiwan and occurs sporadically in at least nine South American countries (Sumption et al., 2008). These examples emphasize the importance of effective surveillance and control measures to recognize outbreaks and prevent spread of disease.

In Africa, the African buffalo plays an important role as the natural host of FMDV but other wildlife such as impala and kudu may also be involved (Alexandersen and Mowat, 2005). The proximity of infected deer, camels, and alpacas to livestock could create a potential risk of FMDV transmission, especially if all of these animal species live in crowded conditions where interanimal contact becomes more likely. The disease can cause severe lesions in the mouth and feet of Asian elephants, which also appear to be susceptible to FMD.

11.2.2.3 Bovine Viral Diarrhea Virus 1 and 2 There are two viral diseases in cattle, which account for much of the financial loss experienced by farmers annually, caused by *bovine viral diarrhea virus* and *bovine respiratory syncytial virus* (BRSV). BVDV-1 and BVDV-2 are RNA viruses and members of the *Pestivirus* genus and the family Flaviviridae. The two types of BVDV mainly infect cattle and have also been found in pigs, sheep, and goats. The clinical signs of infection can vary from subclinical to severe signs that kill the animal (Baker, 1995). Bovine virus diarrhea is characterized by fever, anorexia, respiratory disease, and occasionally diarrhea. Severe acute BVDV infection can lead to hemorrhagic symptoms, discharge from nose and mouth, bloody diarrhea, and pyrexia. Other outcomes of infection are abortions, pneumonia, and immunosuppression in cattle. In general, animals recover from the infection but death from the disease does occur in severe chronic cases. The most common consequences of the disease are respiratory and reproductive disorders that can have profound economical consequences on the cattle industry. Milk production is reduced significantly in dairy herds that are infected with the virus. Delayed growth, increased susceptibility to other diseases, increased mortality of young calves, and early culling to prevent spread of disease are further causes of significant economic loss (Houe, 2003).

BVDV can be transmitted congenitally. In addition, it is transmitted through excretions and secretions, including nasal discharge, milk, and semen. Infection in bulls can lead to venereal transmission and can spread rapidly to multiple cows due to artificial insemination practices. Persistently infected animals are a continuous source of infection as they shed very high levels of virions each day. Immediate removal of persistently infected cattle from the herd, as well as vaccination, are strongly recommended in order to prevent the disease and control the spread of the virus.

11.2.2.4 Bovine Respiratory Syncytial Virus Bovine respiratory syncytial virus is a major cause of lower respiratory disease in calves (Baker and Frey, 1985). It is estimated that the U.S. cattle industry losses due to disease caused by the virus is approximately \$1 billion (Duncan and Potgieter, 1993). Ruminant RSV isolates include ovine, bovine, and caprine. These viruses are members of the genus *Pneumovirus* in the family Paramyxoviridae and are enveloped, single-stranded, nonsegmented, negative-sense RNA viruses (McIntosh and Chanock, 1990). The virus causes pneumonia in calves with symptoms including fever, dullness, loss of appetite, and coughing, often combined with nasal discharge. The disease also occurs in stressed adult cattle and abortions may occur in pregnant animals. Vaccinating calves for RSV, decreasing stress, and good ventilation are essential for preventing pneumonia in calves.

11.2.2.5 Arteriviruses *Equine arteritis virus* is the type species of the genus *Arteriviruses* in the order Nidovirales and the family Arteriviridae (Cavanagh, 1997). The virus causes equine viral arteritis that is characterized by an influenza-like illness in adult horses, abortion in pregnant mares, and interstitial pneumonia in young foals (Timoney and McCollum, 1993). The symptoms of the disease include fever, depression, edema, conjunctivitis, and nasal discharges. EAV also infects donkeys, zebras, and South American camelids and is distributed among populations of Equidae throughout the world. The virus is transmitted by aerosolization and venereal transmission (Timoney and McCollum, 1993).

Porcine reproductive and respiratory syndrome virus (PRRSV), which also belongs to the *Arterivirus* genus, causes porcine reproductive and respiratory syndrome (PRRS) disease in pigs. This is an economically important disease of pig livestock as it can lead to reproductive failure in breeding stock and respiratory distress in young pigs. The disease is highly contagious with symptoms including petechiae, high fever, depression, anorexia, cough, shivering, diarrhea, and disorders of the respiratory tract (Tian et al., 2007). In 2006, an epidemic of the virus in China infected approximately 2,000,000 pigs and killed in the order of 400,000 pigs, thus causing great economic losses (Tian et al., 2007). There have also been recent outbreaks in Vietnam, Sweden, and South Africa.

11.2.2.6 Influenza Virus Swine flu is an acute respiratory disease of pigs caused by a member of the genus *Influenzavirus A* (family Orthomyxoviridae). Transmission of this virus occurs by aerosols and direct contact between animals and therefore the risk of transmission is increased when animals are in proximity. The symptoms of infection in pigs include fever, lethargy, difficulty in breathing, sneezing, eye redness or inflammation, and decreased appetite (Kothalawala et al., 2006). The associated weight loss and poor growth can lead to economic loss for farmers even though at 1–4% the resulting mortality in pigs is fairly low (Merck, 2008). Transmission of swine influenza from pigs to humans is not common although human individuals with frequent exposure to pigs have an increased risk of infection. The 2009 pandemic H1N1 influenza strain marks the first pandemic of influenza specifically known to be of swine origin (van der Meer et al., 2010). The virus was first reported in humans in Mexico and the United States, and approximately 2 months later was identified on a swine farm in Canada (CDC, 2009b; Fraser et al., 2009; Weingartl et al., 2010). The virus has since then spread to humans on all continents and also infected ferrets, dogs, and a cheetah (USDA, 2009). With few exceptions, the animals suffered mild flu-like symptoms and recovered fully. Certain varieties of influenza virus have infected other terrestrial mammals. For example, the avian Flu H5N1 has been reported in cats, leopards, and tigers (Keawcharoen et al., 2004; Leschnik et al., 2007).

11.2.3 Nonarthropod-Borne Viruses of Companion Animals

We will only briefly touch on viruses infecting companion animals such as dogs and cats since this topic has been thoroughly reviewed recently (Patel and Heldens, 2009). Because the focus of this chapter is more on spillover events and pathogens that jump across species, we will address the jump by contemporary strains of equine influenza virus H3N8 (Orthomyxoviridae) to racing dogs in 2004. Investigations indicated that there was a single interspecies transfer of virus, with adaptation in the new host. Influenza virus subtype H3N8 is one of the most important pathogens of horse populations and has been detected throughout the world (van Maanen and Cullinane, 2002). The virus causes equine influenza that is highly contagious in horses, but generally, after a period of flu-like symptoms, the horses make a full recovery. In 2004, transmission of equine influenza to dogs led to a dog-flu outbreak in Florida and subsequently the virus spread among dogs across the United States (Crawford et al., 2005; Dubovi and Njaa, 2008). The symptoms of canine and equine influenza are similar and include fever, nasal discharge, anorexia, and lethargy (Dubovi and Njaa, 2008; van Maanen and Cullinane, 2002). Ferrets are susceptible to most strains of influenza that infect humans and can also pass on the infection to humans. Ferrets have long been used as animal models for disease caused by influenza virus.

Parvoviruses are small, single-stranded DNA viruses belonging to the family Parvoviridae. They are responsible for acute gastroenteritis and leukopenia in young domesticated carnivores and also wild carnivores. Feline panleukopenia virus (FPLV) and canine parvovirus infect cats and dogs, respectively. Disease caused by each of the viruses has similar symptoms: diarrhea, vomiting, appetite loss, fever, and lethargy. Parvoviruses are in general restricted in their host range. However, new mutations in the canine parvovirus have allowed it to infect cats also (Ikeda et al., 2002). The new variants of the canine parvovirus can replicate in a wide variety of cat populations, including leopard cats, and antibodies to the virus have been detected in large felids (e.g., lions, tigers, leopards, wildcats, civets, otters, and even bears). Such interspecies transmissions could help facilitate viral recombination resulting in the emergence of new antigenic subtypes that are better able to infect a wide variety of hosts.

11.2.3.1 Adenoviruses Adenoviruses are primarily spread via respiratory droplets; however, they can also be spread by the fecal–oral route. Since they are fairly resistant to harsh conditions, adenoviruses are capable of existing outside the host for long periods of time. They cause disease in a wide variety of species in different ecological environments.

In humans, adenovirus infections mostly result in upper respiratory tract disease, conjunctivitis, tonsillitis, ear infection, and croup. In young children, adenoviral infections can lead to severe bronchiolitis or pneumonia and in some cases coughing fits that resemble whooping cough. There have also been reports of adenovirus infections causing meningitis or encephalitis (Soeur et al., 1991; Straussberg et al., 2001). Besides infecting humans, adenoviruses also infect dogs, horses, cattle, sheep, and goats. In dogs, canine adenovirus type 1 causes canine hepatitis and eye infections, while both canine adenovirus type 1 and 2 can cause respiratory infections.

11.2.3.2 Reoviruses Reoviruses are primarily respiratory and enteric viruses; however, they may also cause central nervous system infections. Members of this viral family, the Reoviridae, have been isolated from humans, macaque monkeys, baboons, cattle, pigs, horses, cats, dogs, and mice and infections have been described in small mammals, cattle, sheep, horses, pigs, dogs, and cats (Tyler et al., 1989). Interestingly, cases of respiratory illness of reoviral origin have been found in nonhuman primates, including chimpanzees, marmosets, African monkeys, and macaques. Rare cases of meningitis, hepatitis, and hepatic biliary atresia caused by reoviruses have also been reported in primates (Rosenberg et al., 1983; Sabin, 1959).

11.2.3.3 Lentiviruses Five serotypes of lentiviruses (genus *Lentivirus*), which belong to the family Retroviridae, are recognized in relationship to the respective vertebrate host of the virus. These include primate, bovine, equine, ovine/caprine, and feline lentiviruses. Some have broad host ranges, and as an example simian immunodeficiency virus (SIV) is found in over 40 species of primates. SIV infections are generally nonpathogenic in their hosts. However, SIV from chimpanzees and sooty mangabeys are thought to have crossed the species barrier and resulted in HIV-1 and HIV-2, respectively, where they are highly pathogenic, causing immunodeficiency (Gao et al., 1999; Keele et al., 2006; Reeves and Doms, 2002). Feline immunodeficiency virus (FIV) is the causative agent of AIDS in cats and has been found to be endemic in the domestic cat population worldwide (Zislin, 2005). The virus is also endemic in some larger cats, such as the African lion, but these species do not necessarily exhibit symptoms. In cats, FIV affects the immune system in a manner similar to that of HIV affecting humans. During the acute stage of the disease caused by FIV, cats suffer from fever, depression, and generalized lymphadenopathy. During a subsequent subclinical stage, symptoms disappear but the infected cats remain viremic. Finally, during the chronic stage, these cats can suffer from conjunctivitis, enteritis, dermatitis, and gingivitis and they succumb to tumors and chronic and opportunistic infections due to their suppressed immune system (Bendinelli et al., 1995; Pedersen et al., 1989).

11.2.3.4 Arenaviruses Arenaviruses (family Arenaviridae) are associated with rodents and in general chronically infected rodents do not show signs of disease. Each type of arenavirus is usually associated with one particular rodent host species in which it is maintained. For example, the type species, lymphocytic choriomeningitis virus (LCMV) is maintained in the house mouse. The arenaviruses are divided into two groups: the New World or Tacaribe complex and the Old World or lymphocytic choriomeningitis (LCM)/Lassa complex. Humans are incidental hosts for LCMV and can become infected when they come into contact with excretions of an infected rodent. Pet rodents such as hamsters also represent possible reservoirs for LCMV.

Lassa virus causes Lassa fever and is endemic in West Africa. The disease is characterized by high fever, severe myalgia, vomiting, diarrhea, conjunctivitis, coagulopathy, hemorrhagic skin rash, occasional visceral hemorrhage, and necrosis of liver and spleen. Neurological problems, including encephalitis, have also been described as symptoms for this disease. The vectors of Lassa virus are multimammate rats (*Mastomys natalensis*), in which infection is asymptomatic and persistent.

11.3 ARTHROPOD-BORNE VIRUSES

The arboviruses (arthropod-borne viruses) are a broadly defined group of unrelated RNA viruses with the common feature that they are transmitted by arthropod species, most commonly mosquitoes, ticks, and sand flies (Table 11.4). The African swine fever virus (ASFV), the causative agent of African swine fever (ASF), represents the single known exception to this rule because it is a large double-stranded DNA virus and it is also the only member of the family Asfarviridae. The multitude of factors influencing activity of this group of viruses include such vertebrate host factors as population composition and density, competence (infectability and infectiousness), and immunity; vector factors including species composition and competence and longevity; and virus factors including strain and genotype (Figure 11.1). The genetics of the vector, virus, and vertebrate have a critical impact on the intensity of virus transmission. Additional layers of complexity are the influence of environmental factors such as rainfall and temperature and the need for behavioral synchronicity leading to contact between the vector and vertebrate at a time when one of them is infectious and the other susceptible. Arthropod-borne viruses frequently cause disease in humans when they spill over after amplification in the enzootic cycle (amplification cycle between a vertebrate host and arthropod vector).

11.3.1 Asfarviridae

The African swine fever virus is the only DNA virus that is known to date to be transmitted by

TABLE 11.4 Important Arboviruses Causing Human Disease

Family/Virus	Vector	Vertebrate Host	Ecology	Human Disease	Geographic Distribution	Epidemics
Togaviridae						
Chikungunya virus	Mosquitoes	Humans, primates	U, S, R	SFI	Africa, Asia	Yes
Ross River virus[a]	Mosquitoes	Humans, marsupials	R, S, U	SFI	Australia, South Pacific	Yes
Mayaro virus[a]	Mosquitoes	Birds	R, S, U	SFI	South America	Yes
O'nyong-nyong virus[a]	Mosquitoes	?	R	SFI	Africa	Yes
Sindbis virus	Mosquitoes	Birds	R	SFI	Asia, Africa, Australia, Europe, Americas	Yes
Barmah forest virus[a]	Mosquitoes	Marsupials, wild birds	R	SFI	Australia	Yes
Eastern equine encephalitis virus	Mosquitoes	Birds	R	SFI, ME	Americas	Yes
Western equine encephalitis virus	Mosquitoes	Birds, rabbits	R	SFI, ME	Americas	Yes
Venezuelan equine encephalitis virus[a]	Mosquitoes	Rodents	R	SFI, ME	Americas	Yes
Flaviviridae						
Dengue virus 1–4[a]	Mosquitoes	Humans, primates	U, S, R	SFI, HF	Worldwide in tropics	Yes
Yellow fever virus[a]	Mosquitoes	Humans, primates	R, S, U	SFI, HF	Africa, South America	Yes
Japanese encephalitis virus[a]	Mosquitoes	Birds, pigs	R, S, U	SFI, ME	Asia, Pacific	Yes
Murray Valley encephalitis virus	Mosquitoes	Birds	R	SFI, ME	Australia	Yes
Rocio virus	Mosquitoes	Birds	R	SFI, ME	South America	Yes
St. Louis encephalitis virus	Mosquitoes	Birds	R, S, U	SFI, ME	Americas	Yes
West Nile virus[a]	Mosquitoes	Birds	R, S, U	SFI, ME	Africa, Asia, Europe, United States	Yes
Kyasanur forest disease virus[a]	Ticks	Primates, rodents, camels	R	SFI, HF, ME	India, Saudi Arabia	Yes
Omsk hemorrhagic fever virus	Ticks	Rodents	R	SFI, HF	Asia	No
Tick-borne encephalitis virus	Ticks	Birds, rodents	R, S	SFI, ME	Europe, Asia, North America	No

Bunyaviridae						
Sand fly fever Naples virus[a]	Sand flies	Humans, possibly others	R	SFI	Europe, Africa, Asia	Yes
Rift Valley fever virus[a]	Mosquitoes	Wildlife, livestock	R	SFI, HF, ME	Africa, Middle East	Yes
La Crosse virus	Mosquitoes	Rodents	R, S	SFI, ME	North America	No
California encephalitis virus	Mosquitoes	Rodents	R	SFI, ME	North America, Europe, Asia	Yes
Crimean-Congo hemorrhagic fever	Ticks	Rodents	R	SFI, HF	Europe, Asia, Africa	Yes
Oropouche virus[a]	Midges	Primates, sloths, possibly wild birds	<u>R</u>, S, U	SFI	Central and South America	Yes

Source: Gubler, 2002.

U, urban; S, suburban, R, rural; SFI, systemic febrile illness; ME, meningoencephalitis; HF, hemorrhagic fever. Underline designates the most important ecology.

[a]Arboviruses that produce significant human viremia levels.

arthropod vectors. The virus belongs to the family Asfarviridae (Penrith and Vosloo, 2009). ASFV infects domestic pigs, European wild boar, warthogs, and bushpigs and causes a lethal hemorrhagic disease. In the wild, the virus exists in a cycle of transmission between wild porcines and argasid ticks. The virus is endemic in sub-Saharan Africa and outbreaks have been reported in Europe (Sardinia, Portugal, Spain, and Belgium), the Caribbean Islands, and Asia (Georgia and Russia) (Costard et al., 2009). The 2007 outbreak in Georgia resulted in the death due to disease and preventative culling of over 80,000 pigs (Rowlands et al., 2008). By late 2007 in Russia, the disease had spread from domesticated swine to wild boars (Rowlands et al., 2008). The acute form of African swine fever is generally characterized by decreased appetite, high fever, reddening of skin, pronounced hemorrhage of the lymph nodes and occasionally enlargement of the spleen, vomiting, and diarrhea. The acute form of this disease can cause abortions in sows and has a high mortality rate. The chronic form is characterized by joint inflammation, respiratory symptoms, and pericarditis and has low mortality rates. There is no treatment for ASF and therefore to control the disease, herds of affected and exposed pigs have to be depopulated.

11.3.2 Togaviridae

The family Togaviridae contains two genera, *Alphavirus* and *Rubivirus*. These viruses have enveloped, spherical (65–70 nm diameter) virions that contain a single-stranded, positive-sense RNA genome 10,000–12,000 nucleotides in length. The 5′-end has a methylated cap and the 3′-terminus has a polyadenylated tail. There are 27 alphaviruses that have been isolated from mammals, including humans and horses, rodents, fish, and birds. Transmission between vertebrate species occurs mainly via mosquitoes. These alphaviruses that cause disease in man and their hosts, vectors, and distributions are listed in Table 11.4. These viruses cause a spectrum of disease from fever and rash to encephalitis. Horses can also be infected and represent dead-end hosts (i.e., viremia levels too low to infect mosquitoes) for such arboviruses as eastern equine encephalitis virus (EEEV), but can be amplifying hosts (concentration of virus in the blood sufficiently high to infect vector mosquitoes) in the case of other arboviruses such as Venezuelan equine encephalitis virus (VEEV).

11.3.3 Flaviviridae

The family Flaviviridae consists of three genera: *Flavivirus*, *Hepacivirus*, and *Pestivirus*. Members of the family are spherical, enveloped virions, and approximately 50 nm in diameter (Rice, 1996). The virion contains a host-derived lipid bilayer surrounding a nucleocapsid core consisting of the viral RNA complexed with multiple copies of the capsid protein (Mukhopadhyay et al., 2003). The viral genome is linear, positive-sense, single-stranded RNA of approximately 11 kb in length. The hepaciviruses consist of a single type species, hepatitis C virus. The pestiviruses are viruses of *Bovidae* and *Suidae* and include the causative agents of the diseases bovine virus diarrhea/mucosal disease and classical swine fever (see 11.2.2). There are approximately 70 known viruses in the genus *Flavivirus*. Medically important member viruses cause encephalitis and are maintained in their enzootic transmission cycles by *Culex* species mosquitoes and avian hosts or they cause hemorrhagic fever in humans and are maintained by *Aedes* species mosquitoes and primates (Gould et al., 2003).

The Japanese encephalitis serocomplex of the genus *Flavivirus* is an antigenic grouping of eight species and two strains/subtypes: Japanese encephalitis virus (JEV), Murray Valley encephalitis virus, St. Louis encephalitis virus, West Nile virus, Kunjin virus, Alfuy virus, Cacipacore virus, Yaounde virus, Koutango virus, and Ustusu virus (Thiel et al., 2005). JEV, after which the serocomplex is named, causes an estimated 45,000 human cases of illness that result in 10,000 deaths annually.

Pigs and ardeid wading birds are the significant vertebrate hosts in the normal amplification cycle, with the former critical for pre-epizootic amplification (Soman et al., 1977). Many mammals demonstrate evidence of antibody to this virus but are not important amplification hosts. Clinical disease is uncommon in pigs, but JEV infection may lead to fetal abortion and stillbirth in infected sows and aspermia in boars (Takashima et al., 1988). Humans and horses are incidental hosts for JEV, but may develop fatal encephalitis.

11.3.3.1 West Nile Virus

West Nile virus is another medically important flavivirus in the JEV serocomplex that has reemerged as a serious public health threat since the mid-1990s. It was first isolated from a febrile woman in the West Nile province of Uganda in 1937 (Smithburn et al., 1940). Neurological disease (ND) associated with human infection with this virus was first apparent during outbreaks in elderly patients in Israel in the 1950s (Zeller and Schuffenecker, 2004). The virus is actively transmitted in Africa, Australia, Eurasia, and the Americas, making it the most widely distributed arbovirus in the world. The known strains of this virus fall into one of two major lineages based on sequence homology, and are designated lineages 1 and 2 (Berthet et al., 1997), with a new third lineage initially detected in Europe in 1997 and then subsequently in 1999 (Bakonyi et al., 2005, 2006). Two additional lineages, one in Russia (Lvov et al., 2004) and one in India (Bondre et al., 2007), were identified. Each lineage is comprised of several genotypes, with a particular genotype tending to dominate in a given region (Hammam et al., 1965; Lanciotti et al., 2002). Severe ND in humans has been associated only with lineage 1 strains (Jia et al., 1999). Strains within lineage 2, while not associated with overt human outbreaks with ND, vary phenotypically with respect to virulence in animal models (Samuel and Diamond, 2006) and have caused large outbreaks of West Nile fever.

Until 1994, outbreaks of West Nile disease were sporadic and occurred primarily in the Mediterranean region, Africa, and Eastern Europe (Hayes, 2001). After 1994, however, reports of more frequent and severe outbreaks often associated with neuroinvasive disease became common, especially in the Mediterranean Basin (Marfin and Gubler, 2001). A severe outbreak in birds occurred in Israel in 1998 with marked mortality in migrating white storks and domestic geese (Malkinson et al., 2002). Romania and Russia reported severe outbreaks in humans in 1996 and 1999, respectively (Dauphin et al., 2004; MacKenzie et al., 2004). Then in 1999, a single point introduction in the New York City area led to the current epidemic of WNV in North America (Ebel et al., 2001; Lanciotti et al., 1999), followed by a dramatic range expansion that currently encompasses most of the contiguous United States, Canada, Mexico, Central America, the Caribbean, and South America (Bernard et al., 2001b; Bernard and Kramer, 2001; Blitvich et al., 2003; Cruz et al., 2005; Dupuis et al., 2003; Kramer and Bernard, 2001; Berrocal et al., 2006; Bosch et al., 2007).

The strain of WNV introduced into the United States in 1999 had 99.7% homology in both nucleic acid and amino acid sequence with a 1998 Israeli goose isolate (Lanciotti et al., 1999). While WNV in the Americas remains a relatively homogeneous virus population (Beasley et al., 2003; Davis et al., 2003; Ebel et al., 2001, 2004), a single genotype that differs slightly but significantly from the introduced genotype has displaced previously existing strains in the United States (Ebel et al., 2004; Davis et al., 2005). Experimental studies suggest that the dominance of this new viral genotype was facilitated by interactions between the virus and certain *Culex* mosquito vectors (Ebel et al., 2004).

11.3.3.2 Ecology of West Nile Virus

West Nile virus is a zoonotic pathogen, maintained in an enzootic cycle between ornithophilic mosquitoes and avian hosts. This virus has been detected in approximately 326 native and captive avian species, although the crow family Corvidae and especially the American crows (*Corvus brachyrhynchos*) are exquisitely susceptible to the disease. As an example, 50% of American crows that were experimentally infected with a North American WNV strain from lineage 1 died (LD_{50}) following inoculation of <10 plaque-forming units (PFU) in African Green monkey kidney cell culture (Bernard et al., 2001a). Some populations of this avian species have decreased by 90% following the introduction of WNV into the United States (Hochachka et al., 2004). It is not clear why WNV has proven to be so virulent for birds in North America, but one possible culprit is the change of a single amino acid residue in the NS3 helicase (Brault et al., 2007). However, while WNV circulating in the United States was demonstrated to be more virulent for American crows than a Kenyan isolate (Brault et al., 2004), no difference in pathogenicity between the two strains was evident in house sparrows (Langevin et al., 2005). Therefore, alternative explanations for the increased virulence of North American WNV must be considered. One possibility is that birds in the United States have had relatively little exposure to related flaviviruses, with the exception of St. Louis encephalitis virus. Further studies need to be conducted to elucidate the reason for the virulence differences.

Most vertebrate species appear susceptible to infection by WNV; however, morbidity and mortality vary greatly (Komar et al., 2003). Incidental hosts, including humans and equids, are generally considered "dead-end" hosts because they do not mount a sufficiently high viremia to subsequently infect mosquitoes. Enhanced virulence of North American WNV strains isolated from birds and mosquitoes compared to Old World lineage 1 strains is at least partly mediated by envelope protein glycosylation (Beasley et al., 2005). Vertebrates are most commonly infected with WNV by the bite of infected mosquitoes, but bird-to-bird transmission has been demonstrated, as well as the infection of crows and cats following their consumption of infected carcasses (Komar et al., 2003; McLean et al., 2001).

Domestic (Komar, 2003; McLean et al., 2002), captive (Ludwig et al., 2002), and wild mammals (Gomez et al., 2008a; McLean et al., 2002) become infected with WNV, but their role in the transmission cycle is probably less important because of the generally low levels of viremia achieved in these mammals. Nonetheless, evidence is accumulating that some wild mammals may contribute to WNV amplification (Root et al., 2006). Domestic dogs and cats seem minimally affected by infection with WNV. Disease in dogs has not been documented; however, a serosurvey in New York City of dogs in the 1999 epidemic area indicated that dogs were frequently infected. There is a single published report of WNV isolated from a dog in southern Africa (Botswana) in 1982, and a dead cat in 1999 in New York City, NY. High WNV seroprevalence has been detected in three common peridomestic wild mammal species (the Virginia opossum, *Didelphis virginiana*, the northern raccoon, *Procyon lotor*, and the eastern gray squirrel, *Sciurus carolinensis*) and lower seroprevalence in the eastern chipmunk, *Tamias striatus* in the eastern United States (Gomez et al., 2008a) and mesopredators (mid-size predators such as raccoons, skunks, and foxes) in western United States (Bentler et al., 2007). Low-level viremias and lack of mortality have been observed in WNV-infected eastern gray squirrels in the eastern United States. There was no evidence of morbidity or mortality in eastern gray squirrels infected experimentally with WNV (Gomez et al., 2008b), although mean viremias of $10^{5.1}$ and $10^{4.8}$ PFU mL^{-1} blood were observed on days 3 and 4 postinfection (DPI), respectively. These results suggest that 2.1% of *Culex pipiens* feeding on squirrels during 1–5 DPI would become infectious. Thus, *S. carolinensis* are unlikely to be important amplifying hosts and may instead dampen the intensity of transmission in most host communities. California gray squirrels (*S. griseus*) are used as indicators of spillover of the virus from the amplification cycle in California in the western United States. Horses are dead-end hosts for WNV, that is, they do not mount a sufficiently high viremia to infect feeding mosquitoes (Bunning et al., 2002), but they may experience severe morbidity and death if infected (Trock et al., 2001). Clinical signs of WNV in equines include weakness, a wide stance, stumbling, toe dragging, leaning, and eventually paralysis. More than 28,600 clinically ill mammals, mostly horses but also including miscellaneous other mammals, with a 35% mortality rate among clinically ill horses (including death from disease and euthanasia) have been reported from 1999 to 2009. A formalin-inactivated whole-virus vaccine directed against WNV has been demonstrated to have excellent efficacy in horses and the numbers of equine cases dropped precipitously after the vaccine's release in the United States in 2001 (Hayes and Gubler, 2005).

Culex mosquitoes are the predominant vectors of WNV, with the most important mosquito species varying regionally and globally. Approximately 75 mosquito species have been found infected with WNV worldwide (Kramer et al., 2007b). Mosquito species and populations vary in vector competence for this virus, that is, they vary in their ability to become infected following an infectious blood meal, their ability to support replication of this virus, and their ability to transmit virus upon taking subsequent blood meals. Longevity of individual female mosquitoes and their vertebrate host feeding preferences are also critical factors (Kramer and Ebel, 2003). *C. quinquefasciatus*, an important vector in urban areas in the southern United States, feeds broadly on both avian and mammalian hosts, including man (Niebylski and Meek, 1992). The predominantly ornithophilic *C. pipiens*, *C. nigripalpus*, and *C. tarsalis*, three other important vector species in the United States, have been demonstrated to shift their feeding from birds to mammals in the late summer and early fall (Edman and Taylor, 1968; Kilpatrick et al., 2006; Nelson et al., 1976), thereby acting as bridge vectors and facilitating spillover of the virus from the enzootic bird–mosquito cycle to infection of equine and human hosts. Host switching may be due to migratory departure of a preferred host, lack of tolerance for mosquito bites by avian hosts as the mosquito population increases, or genetics of the mosquito species in the United States (Kilpatrick et al., 2006; Spielman, 2001). Other mosquito species that feed frequently on mammals, for example, *C. salinarius* Coquillett, probably also become involved as bridge vectors in areas where they are common, exposing animals not involved in the normal enzootic transmission cycle of virus. *C. pipiens* account for >80% of the total risk of human WNV infections in the New York region, over the transmission season. The combined risk associated with four other important mosquito species evaluated, *Aedes japonicus, A. vexans, A. trivittatus*, and *C. salinarius*, represented one-quarter the threat posed by *C. pipiens* and *C. restuans* (Kilpatrick et al., 2005).

WNV survives adverse seasons in temperate environments, most likely in the diapausing adult *Culex* female in hibernacula. However, persistent WNV infection has been observed in experimentally infected birds, hamsters, and mice (Komar et al., 2003; Tesh et al., 2005; Appler et al., 2010); therefore, recrudescence of infectious virus in birds cannot be ruled out as a possible means of perpetuation of the virus. Infected migratory birds may also play an important role in reintroduction of virus after harsh winters or dry seasons when mosquitoes are not active. It is known that at least in Europe and Asia, the virus is reintroduced by migrating birds (Malkinson and Banet, 2002) and possibly may overwinter in adult mosquitoes.

Alternative nonvector-borne modes of WNV transmission have been noted, that is, transfusion and transplantation transmission in humans, ingestion of infected breast milk, and intrauterine transmission (Kramer et al., 2007a). Fecal–oral and peroral transmission have been observed in experimental animals. WNV has been detected in the urine of humans (Tonry et al., 2005) more than 6 years after infection (Murray et al., 2010). Experimentally infected hamsters have been demonstrated to shed WNV virus in their urine for over 52 days and they can become infected via the oral route

(Tesh et al., 2005). Cats and other vertebrates also become infected by WNV via the oral route following feeding on dead infected animals (Austgen et al., 2004; Komar et al., 2003).

11.3.3.3 West Nile Disease From 1999 to 2009, approximately 29,683 cases of WN disease have been reported in the United States, of which 12,126 (40.9%) were ND with 1164 fatalities (3.9% of all; 9.6% of ND) (www.cdc.gov/ncidod/dvbid/westnile). The 2002/2003 epidemics in the United States represent the largest outbreaks of ND ever reported in the Western Hemisphere, making WNV the dominant vector-borne viral pathogen in North America. Approximately 1.4 million total infections were estimated to have occurred through 2004 (Davis et al., 2006; Petersen and Hayes, 2004). There have been no reported cases of WN disease in tropical America with a few exceptions, for example, Puerto Rico. The reasons for this are not clear but hypotheses include the presence of cross-protective flaviviruses (Yamshchikov et al., 2005), decreased virulence of the circulating virus, and greater biodiversity in the region. Similarly, there have been no overt cases in the United Kingdom, even with evidence of serological conversions in sentinel chickens (Buckley et al., 2006). The lack of human cases in northern Europe, as compared to southern Europe, may be attributed to the genetics of the predominant vector, *C. pipiens* (Fonseca et al., 2004), virus strain differences, climatic factors, among other possible explanations.

Infection most commonly takes place following bite of an infected mosquito and leads to a broad spectrum of disease in humans, ranging from fever to severe ND, including meningitis, encephalitis, and acute flaccid paralysis. This is typical of most *Culex*-transmitted flavivirus infections such as WNV (Gould et al., 2003). Severe disease is more common in older individuals and immunocompromised hosts are at greatest risk. Transplant patients are at extreme risk for ND following blood transfusion, donor transmission, or community exposure (Hoekstra, 2005). Fever cases had been considered mild and unremarkable before the largest outbreaks occurred in the United States in 2002 and 2003, but it is now clear that the fever cases can also be serious and may lead to death. Long-term neurological and functional sequelae are common, with more severe complications following ND (Sejvar, 2007).

11.3.3.4 Other Flaviviruses Dengue virus and yellow fever virus (genus *Flavivirus*; family Flaviviridae) differ from the JEV serogroup viruses in that humans are not dead-end hosts for these two virus species, but rather the human is a critical component of the virus' urban transmission cycle. The predominant urban vector of these two flaviviruses, *A. aegypti*, is anthropophilic and takes a blood meal nearly every day, leading to rapid spread of the virus through a host community. However, both viruses have a sylvatic cycle in tropical rainforests where the virus cycles between nonhuman primates and forest species of mosquitoes, predominantly *Aedes* and *Haemagogus* species. Infected forest mosquitoes may occasionally feed on humans entering the forests. Most *yellow fever virus* transmission to humans takes place in an intermediate savannah cycle in semihumid habitats in Africa where villages are located. Here semidomestic mosquitoes feed on both humans and monkeys. Large-scale outbreaks of yellow fever occur in urban areas heavily populated with humans and infested with *Aedes* mosquitoes. In Africa, the number of yellow fever cases has increased over the past two decades due to declining population immunity to infection, deforestation, urbanization, population movements, and climate change. In the Americas, unlike Africa, urban outbreaks of yellow fever have not been observed for many decades even though *A. aegypti* is present. The reasons for this are not clear, but may in part reflect social changes such as the use of window screens in houses and air conditioning and mosquito control practices. Dengue continues to increase in scope leading to hyperendemicity in most parts of the world where this virus exists. The potential for dengue virus "spillover" into urban populations from sylvatic cycles is debated among researchers (Vasilakis, 2010), and the evidence for further maintenance of these sylvatic cycles may be dwindling due to constant reduction in size of natural forests (Rodhain, 1991).

11.3.4 Bunyaviridae

The family Bunyaviridae is comprised of more than 300 named viruses. Membership in the family is primarily based on structural features (specifically a trisegmented RNA genome of negative sense) and a basically similar protein coding strategy within each viral genomic segment. Viruses in the family Bunyaviridae have tripartite single-stranded RNA genomes consisting of three size classes designated as large (L), medium (M), and small (S) RNA segments. Each RNA segment exhibits a pseudocircular structure due to its complementary ends. The L segment encodes the RNA-dependent RNA polymerase, necessary for viral RNA replication and mRNA synthesis. The M segment encodes the viral glycoproteins G1 and G2, which project from the viral surface and aid the virus in attaching to and entering the host cell. The S segment encodes the nucleocapsid protein (N). The L and M segments are of negative sense. For the genera of *Phlebovirus* and *Tospovirus*, the S segment is of ambisense, that is, genes at the 5′-end of the RNA strand are of positive sense and contain an open reading frame, and the 3′-end contains negative-sense coding. The S segment encodes the viral nucleoprotein (N) in the negative sense and a nonstructural (NSs) protein in ambisense (Nguyen and Haenni, 2003). The total length of the viral genome size ranges from 10.5 to 22.7 kb (ICTVdB).

Membership in this viral family is also based on antigenic interrelatedness among the viruses. The family is divided into five genera (*Orthobunyavirus*, *Nairovirus*, *Phlebovirus*, *Hantavirus*, and *Tospovirus*). All but the hantaviruses are transmitted by arthropods—predominantly mosquitoes, sand flies, and ticks—and all infect animals with the exception of the tospoviruses, which are plant viruses transmitted by thrips. Humans are an incidental host for all but the tospoviruses, which do not infect humans, and possibly sand fly-transmitted viruses, where humans may be amplification hosts.

11.3.4.1 *Orthobunyavirus*

Viruses in the genus *Orthobunyavirus* have been divided into 16 serogroups. They are found in tropical, temperate, and Arctic regions where they are closely associated with their mosquito vectors and a wide variety of vertebrates. Transovarial transmission is thought to be the mechanism by which the viruses are maintained in the mosquito population, particularly in the Arctic. *Lacrosse virus* (LACV) is maintained and amplified in nature by several distinct transmission mechanisms in deciduous forest habitats. The predominant horizontal transmission/amplification cycle involves *A. triseriatus* mosquitoes serving as the principal vector of LACV and chipmunks (*T. striatus*) and squirrels (*S. carolinensis*) as the principal vertebrate hosts. The virus is maintained over the winter by transovarial transmission in mosquito eggs, meaning that if the female mosquito is infected, she may lay eggs that carry the virus, and the F1 adults may be able to transmit the virus perorally. This transovarial transmission route may be considered an amplification mechanism since during each gonadotropic cycle, multiple infected progeny result. LACV may also be horizontally transmitted venereally from male to female *A. triseriatus* (Thompson and Beaty, 1978).

The Simbu serogroup of the *Orthobunyavirus* genus was first described by Casals (1957) and currently contains 25 related viruses, which have been isolated from all continents, except Europe (Calisher, 1996). These viruses are found in temperate and tropical zones of the world. Most of the Simbu serogroup viruses have been isolated from arthropods that were presumed to be their vectors and from vertebrate hosts. One member of this serogroup, Akabane virus, is primarily transmitted by biting midges of the *Culicoides* species.

Outbreaks of this disease resulting in congenital malformations in ruminants have occurred in Japan, Australia, Israel, Turkey, Korea, and Taiwan. Genetic analyses of Akabane isolates in Japan have suggested that these viruses frequently invade that country in infected *Culicoides* from overseas tropical and subtropical areas (Kono et al., 2008). *Akabane* virus strains were recently divided into four genetically distinct groups (I–IV) and one subgroup (Ia and Ib). Neurovirulence of these strains belonging to genogroup Ia is greater than that of strains in genogroup II (Kono et al., 2008). *Akabane* virus has presented a serious threat to the livestock industry for many decades. It can cause disease in cattle, sheep, and goats. The disease is characterized by the development of lesions in the fetus of susceptible animals. The virus does not produce any signs of disease in young or adult animals. However, problems arise when infection occurs by this virus for the first time during pregnancy and if the virus then passes through the placenta to the fetus, potentially causing severe disease of the fetal central nervous system. The fetus can be infected at almost any stage of gestation and the nature of the problems produced depends upon the stage of pregnancy at which the infection takes place. Infections in early gestation generally result in abortion. Calves that were infected *in utero* and survived gestation will generally show symptoms of one or two syndromes: arthrogryposis (multiple joint contractures) and hydranencephaly (loss of cerebral hemispheres and their replacement by sacs filled with cerebrospinal fluid). Most abnormalities in sheep and goats develop following infection between 28 and 56 days of gestation; few, if any, abnormalities are observed after infection at other times.

The second member of the *Orthobunyavirus* genus, *Oropouche* virus (OROV), is of particular importance as it has been responsible for several large outbreaks of a dengue-like illness in human populations in South America. Transmission of OROV is thought to involve two distinct cycles: an epidemic urban cycle with humans as the primary vertebrate host and a silent jungle cycle. The epidemic vector of OROV is thought to be the biting midge *Culicoides paraensis* (Goeldi). In the presumed sylvatic cycle, primates, sloths, and birds have been proposed as vertebrate reservoirs, but the suspected arthropod vector has not been identified and conclusive evidence of a jungle cycle is lacking. Since the first isolation of OROV from the blood of a febrile forest worker in 1955 in Trinidad, the virus has caused at least 27 epidemics in rural and urban communities of Brazil, Panama, and Peru. The earliest documented epidemic of OROV in 1961 in Belem, Brazil, infected more than 11,000 people.

11.3.4.2 Phlebovirus There are at least 45 viruses in the *Phlebovirus* genus of the family Bunyaviridae. These viruses, with the exception of *Rift Valley fever virus* (RVFV), generally do not replicate in mosquitoes. Many members of the genus *Phlebovirus* are transovarially transmitted by their arthropod hosts. RVFV can cause severe disease in domestic animals and man. The virus was first isolated in 1930 during an epizootic of fatal hepatic necrosis and abortion in sheep on a farm in the Rift Valley in Kenya (Daubney et al., 1933). In 1977, activity associated with this virus was detected in Egypt, where there was a massive outbreak with 25–50% of sheep and cattle infected in some areas along the Nile and 200,000 human cases with $\geq$600 deaths. Recurrent epidemics of this disease then followed in 1993, 1997, and 2003. There have been other severe outbreaks of RVFV in the past 20 years, including an epidemic in Senegal–Mauritania in 1987 with an estimated 89,000 victims and 220 deaths. Northeastern Kenya experienced an outbreak of RVFV during 1997–1998, affecting an estimated 27,500 individuals with 170 hemorrhagic fever-associated deaths (Woods et al., 2002). RVFV is also found in Madagascar that experienced epidemics of disease caused by this virus during 1990–1991 and 2008. Only a few deaths occurred in the 1990–1991 outbreak; however, the 2008 outbreak resulted in 476 cases and

19 deaths. RVFV spread to the Arabian peninsula in 2000 (Saudi Arabia and Yemen), where there were an estimated 2000 cases, including acute hepatic necrosis, hepatitis, delayed-onset encephalitis, and retinitis, resulting in at least 245 deaths with a case fatality rate of 12%. RVFV activity has also been noted in the Comoros Islands and the virus can be found in all parts of Africa, where sporadic epidemics of associated disease occur in man resulting in significant morbidity and mortality.

Rift Valley fever epizootics occur under the appropriate climatological conditions (e.g., wet seasons and resulting flooding required for *Aedes* mosquito breeding). RVFV is transmitted mainly by mosquitoes, although it can also be transmitted by sand flies. Although this virus may appear at times to be absent in an arid region, it is capable of maintaining a presence by lying dormant in the eggs of infected mosquitoes. Aerosol transmission had also been documented where a man became infected after coming into contact with carcasses of RVFV-diseased animals. This virus can also be transmitted by bodily fluids of infected animals, especially to slaughterhouse workers and animal handlers. Most of the associated infections are symptomatic and usually present as a mild nonspecific febrile illness, but a small proportion ($\sim$1%) result in the development of hemorrhagic fever, retinal vasculitis, and encephalitis. Because it can infect many different vertebrates and many different mosquitoes, this virus presents a great threat to public and veterinary health.

RVFV is maintained in an enzootic cycle among wildlife, such as African buffaloes, and a wide variety of mosquito species ($>$30 species, 6 genera including *Aedes*, *Anopheles*, *Culex*, and *Mansonia*, among others, and by other arthropods including sand flies). Transmission is enzootic during most years, when the associated virus persists during dry season/interepizootic periods through vertical transmission in *Aedes* mosquito eggs. But during wet years and especially following droughts, the virus may become epizootic or epidemic. Flooding results in mass hatching of infected *Aedes* eggs, which subsequently produce adult *Culex* mosquitoes, some of which may be infected with RVFV and will be able to transmit this infection to ruminants leading to an outbreak. Large numbers of mosquitoes and possibly multiple viral lineages may become active after such flooding events. However, because infection rates among *Aedes* adults reared from field-collected larvae are typically low, horizontal amplification in mammals is required for RVFV maintenance. Further support for peroral transmission from *Aedes* species mosquitoes to mammals is evident in the large variety of wildlife naturally infected with RVFV. Herdsmen bring their livestock to the water to drink and to grasslands to eat, and thereby multiple susceptible hosts wander in proximity with infectious mammal-feeding *Aedes* mosquitoes that emerge in these same habitats.

11.3.4.3 Nairovirus The *Nairovirus* genus includes seven species of virus, among which is Crimean Congo hemorrhagic fever virus (CCHFV). All the Nairoviruses are transmitted by argasid or ixodid ticks and are largely zoonotic pathogens; three of these viruses have been implicated as causes of human disease: *Dugbe virus*, *Nairobi sheep disease virus*, and CCHFV. Nairobi sheep disease is predominantly a disease of sheep and goats, resulting in hemorrhagic gastroenteritis with very high morbidity and mortality. CCHFV is maintained in ticks where transovarial transmission has been demonstrated. Certain mammalian hosts, such as hedgehogs, hares, and domestic animals, support amplification of this virus. Humans can become infected by this virus as a result of tick bite, accidental parenteral exposure, or nosocomial infection; associated human mortality rates of 10–40% have been reported.

11.3.4.4 Hantavirus Hantaviruses are the only Bunyaviridae that are not transmitted by arthropod vectors, but rather by urine, saliva, or feces of infected rodents. In the United States, deer mice, cotton rats, rice rats, and white-footed mice carry these viruses. Inhalation of aerosolized virus from rodent excreta, consumption of virus-contaminated food, and, less frequently, bites from infected animals can transmit hantaviruses to humans. Many of these viruses cause a spectrum of disorders that are called hemorrhagic fever with renal syndrome and are characterized by renal failure and hemorrhagic manifestations. *Sin Nombre virus*, which is carried by deer mice, is the most common cause of hantavirus cardiopulmonary syndrome (HCPS) in North America (CDC, 2009a). Early symptoms of HCPS include fatigue, fever, and muscle ache that may or may not be accompanied by headaches, dizziness, chills, and abdominal problems such as nausea, vomiting, diarrhea, and abdominal pain. Later symptoms of this disease can include coughing and shortness of breath followed by pulmonary edema. Thrombocytopenia and elevated white blood cells are typical laboratory findings associated with this disease, and respiratory distress requiring mechanical ventilation is also common (Khan et al., 1996). Environmental and ecological changes such as increase in rainfall can lead to an increase in the population of potentially infected rodents and more frequent human–rodent contact, which can then lead to an increase in the numbers of human cases (Parmenter et al., 1993). In 1993, there was an outbreak of the disease in the southwestern region of the United States. Of the initial 18 cases, 14 patients (78%) died, emphasizing the severity of disease (CDC, 1993). Each year approximately 20–40 cases of HCPS are reported in the United States, mostly in the southwest (CDC, 2009a). There is no antiviral treatment for HCPS and therefore the recommendation is to eliminate rodent infestations of human residences to facilitate avoidance of contact with infected rodents.

11.4 SUMMARY

There are a large number of viruses that infect terrestrial mammals. A good reference for further reading on infections in terrestrial wildlife is *Infectious Diseases of Wild Mammals* by Elizabeth S. Williams and Ian K. Barker (Blackwell Publishing, Iowa, 2001). Some viruses of terrestrial mammals cause disease in their primary host and some only cause disease when they jump species. The complexity of the interaction between the virus and host is exacerbated by the rapid evolution of some viruses through mutation, recombination, or reassortment. Changes in the virus, the biology of the host such as its lifespan and reproductive rate, and the environmental factors such as temperature can alter the intensity of transmission in nature. In some cases, even a very few mutations in the viral genome can alter the host range, virulence, or pathogenicity of a virus, thus changing its ecological niche and leading to unexpected outbreaks.

Many of the viruses of terrestrial mammals described in this chapter are important pathogens for humans. For example, an *Influenzavirus* outbreak in 1918–1919 killed approximately 20 million people and influenza reassortants remain a future threat as evidenced by the 2009 H1N1 outbreak. Seasonal influenza kills approximately 250,000–500,000 individuals annually (WHO, 2009). It is estimated that in 2008 over 33 million people were infected with HIV and 2 million people have died of AIDS. There is still no satisfactory treatment or vaccine for HIV. Other viruses are also emerging: *Ebola virus*, *Hantavirus*, *Dengue virus*, and *Lassa fever virus* are causing increasing incidence of hemorrhagic fever in humans. In addition, viruses that cause disease in livestock such as cattle, horses, and pigs can be economically devastating (e.g., *foot-and-mouth disease virus*, the bovine viral diarrhea viruses, and *bovine respiratory syncytial virus*). Many of these viruses of livestock are capable of crossing host species barriers and thus creating a greater threat to the human population. In addition, as viruses are inadvertently introduced into new environments and adapt to new terrestrial mammal hosts, unanticipated disease outbreaks may occur.

ACKNOWLEDGMENT

The authors wish to thank Dr. Elizabeth Kauffman for her valuable assistance in preparing this chapter.

REFERENCES

Alexandersen, S. and Mowat, N. (2005). Foot and mouth disease: host range and pathogenesis. In: Mahy, B. W. J. (ed.), *Foot and Mouth Disease Virus*. Springer, Heidelberg, Germany, pp. 9–42.

Appel, M. J., Yates, R. A., Foley, G. L., Bernstein, J. J., Santinelli, S., Spelman, L. H., Miller, L. D., Arp, L. H., Anderson, M., and Barr, M. (1994). Canine distemper epizootic in lions, tigers, and leopards in North America. *J. Vet. Diagn. Invest.* 6, 277–288.

Appler, K. K., Brown, A. N., Stewart, B. S., Behr, M. J., Demarest, V. L., Wong, S. J., and Bernard, K. A. (2010). Persistence of West Nile virus in the central nervous system and periphery of mice. *PLoS One* 5, e10649.

Arzt, J., White, W. R., Thomsen, B. V., and Brown, C. C. (2010). Agricultural diseases on the move early in the third millennium. *Vet. Pathol.* 47, 15–27.

Austgen, L. E., Bowen, R. A., Bunning, M. L., Davis, B. S., Mitchell, C. J., and Chang, G.-J. J. (2004). Experimental infection of cats and dogs with West Nile virus. *Emerg. Infect. Dis.* 10, 82–86.

Baker, J. C. (1995). The clinical manifestations of bovine viral diarrhea infection. *Vet. Clin. North. Am. Food. Anim. Pract.* 11, 425–445.

Baker, J.C and Frey, M. L. (1985). Bovine respiratory syncytial virus. *Vet. Clin. North Am. Food Anim. Pract.* 1, 259–275.

Bakonyi, T., Hubalek, Z., Rudolf, I., and Nowotny, N. (2005). Novel flavivirus or new lineage of West Nile virus, central Europe. *Emerg. Infect. Dis.* 11, 225–231.

Bakonyi, T., Ivanics, E., Erdelyi, K., Ursu, K., Ferenczi, E., Weissenbock, H., and Nowotny, N. (2006). Lineage 1 and 2 strains of encephalitic West Nile virus, central Europe. *Emerg. Infect. Dis.* 12, 618–623.

Beasley, D. W., Davis, C. T., Guzman, H., Vanlandingham, D. L., Travassos da Rosa, A. P. A., Parsons, R. E., Higgs, S., Tesh, R. B., and Barrett, A. D. T. (2003). Limited evolution of West Nile virus has occurred during its southwesterly spread in the United States. *Virology* 309, 190–195.

Beasley, D. W., Whiteman, M. C., Zhang, S., Huang, C. Y., Schneider, B. S., Smith, D. R., Gromowski, G. D., Higgs, S., Kinney, R. M., and Barrett, A. D. (2005). Envelope protein glycosylation status influences mouse neuroinvasion phenotype of genetic lineage 1 West Nile virus strains. *J. Virol.* 79, 8339–8347.

Bendinelli, M., Pistello, M., Lombardi, S., Poli, A., Garzelli, C., Matteucci, D., Ceccherini-Nelli, L., Malvaldi, G., and Tozzini, F. (1995). Feline immunodeficiency virus: an interesting model for AIDS studies and an important cat pathogen. *Clin. Microbiol. Rev.* 8, 87–112.

Bentler, K. T., Hall, J. S., Root, J. J., Klenk, K., Schmit, B., Blackwell, B. F., Ramey, P. C., and Clark, L. (2007). Serologic evidence of West Nile virus exposure in North American mesopredators. *Am. J. Trop. Med. Hyg.* 76, 173–179.

Bernard, K. A., Dupuis, A. P. II, Ebel, G. D., Kauffman, E. B., Jones, S. A., and Kramer, L. D. (2001a). Comparison of West Nile virus infection in naturally infected wild birds. In: Proceedings of the 20th Annual Meeting of the American Society for Virology, p. 78.

Bernard, K. A. and Kramer, L. D. (2001). West Nile virus activity in the United States, 2001. *Viral Immunol.* 14, 319–338.

Bernard, K. A., Maffei, J. G., Jones, S. A., Kauffman, E. B., Ebel, G. D., Dupuis, A. P. II, Ngo, K. A., Nicholas, D. C., Young, D. M., Shi, P.-Y., Kulasekera, V. L., Eidson, M., White, D. J., Stone, W. B., and Kramer, L. D. (2001b). West Nile virus infection in birds and mosquitoes, New York State, 2000. *Emerg. Infect. Dis.* 7, 679–685.

Berrocal, L., Peña, J., González, M., and Mattar, S. (2006). West Nile virus; ecology and epidemiology of an emerging pathogen in Colombia. *Rev. Salud Publica (Bogota)* 8, 218–228.

Berthet, F. X., Zeller, H. G., Drouet, M. T., Rauzier, J., Digoutte, J. P., and Deubel, V. (1997). Extensive nucleotide changes and deletions within the envelope glycoprotein gene of Euro-African West Nile viruses. *J. Gen. Virol.* 78(Part 9), 2293–2297.

Blitvich, B. J., Fernandez-Salas, I., Contreras-Cordero, J. F., Marlenee, N. L., Gonzalez-Rojas, J. I.,

Komar, N., Gubler, D. J., Calisher, C. H., and Beaty, B. J. (2003). Serological evidence of West Nile virus infection in horses, Coahuila State, Mexico. *Emerg. Infect. Dis.* 9, 853–856.

Bondre, V. P., Jadi, R. S., Mishra, A. C., Yergolkar, P. N., and Arankalle, V. A. (2007). West Nile virus isolates from India: evidence for a distinct genetic lineage. *J. Gen. Virol.* 88, 875–884.

Bosch, I., Herrera, F., Navarro, J. C., Lentino, M., Dupuis, A., Maffei, J., Jones, M., Fernández, E., Pérez, N., Pérez-Emán, J., Guimarães, A. E., Barrera, R., Valero, N., Ruiz, J., Velásquez, G., Martinez, J., Comach, G., Komar, N., Spielman, A., and Kramer, L. (2007). West Nile virus, Venezuela. *Emerg. Infect. Dis.* 13, 651–653.

Brault, A. C., Huang, C. Y., Langevin, S. A., Kinney, R. M., Bowen, R. A., Ramey, W. N., Panella, N. A., Holmes, E. C., Powers, A. M., and Miller, B. R. (2007). A single positively selected West Nile viral mutation confers increased virogenesis in American crows. *Nat. Genet.* 39, 1162–1166.

Brault, A. C., Langevin, S. A., Bowen, R. A., Panella, N. A., Biggerstaff, B. J., Miller, B. R., and Nicholas, K. (2004). Differential virulence of West Nile strains for American crows. *Emerg. Infect. Dis.* 10, 2161–2168.

Brian, DA and Baric, R. S. (2005). Coronavirus genome structure and replication. *Curr. Top. Microbiol. Immunol.* 287, 1–30.

Buckley, A., Dawson, A., and Gould, E. A. (2006). Detection of seroconversion to West Nile virus, Usutu virus and Sindbis virus in UK sentinel chickens. *Virol. J.* 3, 71–76.

Bunning, M. L., Bowen, R. A., Cropp, C. B., Sullivan, K. G., Davis, B. S., Komar, N., Godsey, M. S., Baker, D., Hettler, D. L., Holmes, D. A., Biggerstaff, B. J., and Mitchell, C. J. (2002). Experimental infection of horses with West Nile virus. *Emerg. Infect. Dis.* 8, 380–386.

Calisher, C. H. (1996). History, classification, and taxonomy of viruses in the family Bunyaviridae. In: Elliott, R. M. (ed.), *The Bunyaviridae*. Plenum Press, New York, pp. 1–17.

Casals, J. (1957). The versatile parasites: I. The arthropod-borne group of animal viruses. *Trans. N. Y. Acad. Sci.* 19, 219–235.

Cavanagh, D. (1997). Nidovirales: a new order comprising Coronaviridae and Arteriviridae. *Arch. Virol.* 142, 629–633.

CDC. (1993). Update: hantavirus disease: southwestern United States, 1993. *MMWR Morb. Mortal Wkly. Rep.* 42, 570–572.

CDC. (2003). Update: multistate outbreak of monkeypox–Illinois, Indiana, Kansas, Missouri, Ohio, and Wisconsin. *MMWR Morb. Mortal Wkly. Rep.* 52, 642–646.

CDC. (2009a). Hantavirus pulmonary syndrome in five pediatric patients: four states, 2009. *MMWR Morb. Mortal Wkly. Rep.* 58, 1409–1412.

CDC. (2009b). Swine influenza A (H1N1) infection in two children: Southern California, March–April 2009. *MMWR Morb. Mortal Wkly. Rep.* 58, 400–402.

Childs, J. E. (2004). Zoonotic viruses of wildlife: hither from yon. *Arch. Virol. Suppl.* 18, 1–11.

Childs, J. E., Richt, J. A., and MacKenzie, J. S. (2007). Introduction: conceptualizing and partitioning the emergence process of zoonotic viruses from wildlife to humans. *Curr. Top. Microbiol. Immunol.* 315, 1–31.

Chua, K. B., Chua, B. H., and Wang, C. W. (2002). Anthropogenic deforestation, El Nino and the emergence of Nipah virus in Malaysia. Malays. *J. Pathol.* 24, 15–21.

Cleaveland, S., Haydon, D. T., and Taylor, L. (2007). Overviews of pathogen emergence: which pathogens emerge, when and why? *Curr. Top. Microbiol. Immunol.* 315, 85–111.

Cleaveland, S., Laurenson, M. K., and Taylor, L. H. (2001). Diseases of humans and their domestic mammals: pathogen characteristics, host range and the risk of emergence. *Philos. Trans. R. Soc. Lond. B Biol. Sci.* 356, 991–999.

Costard, S., Wieland, B., de Glanville, W., Jori, F., Rowlands, R., Vosloo, W., Roger, F., Pfeiffer, D. U., and Dixon, L. K. (2009). African swine fever: how can global spread be prevented? *Philos. Trans. R. Soc. Lond. B Biol. Sci.* 364, 2683–2696.

Crawford, P. C., Dubovi, E. J., Castleman, W. L., Stephenson, I., Gibbs, E. P., Chen, L., Smith, C., Hill, R. C., Ferro, P., Pompey, J., Bright, R. A., Medina, M. J., Johnson, C. M., Olsen, C. W., Cox, N. J., Klimov, A. I., Katz, J. M., and Donis, R. O. (2005). Transmission of equine influenza virus to dogs. *Science* 310, 482–485.

Cruz, L., Cardenas, V. M., Abarca, M., Rodriguez, T., Reyna, R. F., Serpas, M. V., Fontaine, R. E., Beasley, D. W., da Rosa, A. P., Weaver, S. C., Tesh, R. B., Powers, A. M., and Suarez-Rangel, G.

(2005). Serological evidence of West Nile virus activity in El Salvador. *Am. J. Trop. Med. Hyg.* 72, 612–615.

Daubney, R., Hudson, J. R., and Gamham, P. C. (1933). Enzootic hepatitis of Rift Valley fever: an undescribed virus in sheep, cattle and man from East Africa. *East Afr. Med. J.* 10, 2–19.

Dauphin, G., Zientara, S., Zeller, H., and Murgue, B. (2004). West Nile: worldwide current situation in animals and humans. *Comp. Immunol. Microbiol. Infect. Dis.* 27, 343–355.

Davis, C. T., Beasley, D. W., Guzman, H., Raj, R., D'Anton, M., Novak, R. J., Unnasch, T. R., Tesh, R. B., and Barrett, A. D. (2003). Genetic variation among temporally and geographically distinct West Nile virus isolates, United States, 2001, 2002. *Emerg. Infect. Dis.* 9, 1423–1429.

Davis, L. E., DeBiasi, R., Goade, D. E., Haaland, K. Y., Harrington, J. A., Harnar, J. B., Pergam, S. A., King, M. K., DeMasters, B. K., and Tyler, K. L. (2006). West Nile virus neuroinvasive disease. *Ann. Neurol.* 60, 286–300.

Davis, C. T., Ebel, G. D., Lanciotti, R. S., Brault, A. C., Guzman, H., Siirin, M., Lambert, A., Parsons, R. E., Beasley, D. W., Novak, R. J., Elizondo-Quiroga, D., Green, E. N., Young, D. S., Stark, L. M., Drebot, M. A., Artsob, H., Tesh, R. B., Kramer, L. D., and Barrett, A. D. (2005). Phylogenetic analysis of North American West Nile virus isolates, 2001–2004: evidence for the emergence of a dominant genotype. *Virology* 342, 252–265.

Davison, A. J. (2010). Herpesvirus systematics. *Vet. Microbiol.* 143, 52–69.

de Vries, A. A., Horzinek, M. C., Rottier, P. J. M., and De Groot, R. J. (1997). The genome organization of the Nidovirales: similarities and differences between arteri-, toro-, and coronaviruses. *Semin. Virol.* 8, 33–47.

Del, P. F., Wilkins, P. A., Timoney, P. J., Kadushin, J., Vogelbacker, H., Lee, J. W., Berkowitz, S. J., and La Perle, K. M. (2000). Fatal nonneurological EHV-1 infection in a yearling filly. *Vet. Pathol.* 37, 672–676.

Dubovi, E. J. and Njaa, B. L. (2008). Canine influenza. *Vet. Clin. North Am. Small Anim. Pract.* 38, 827–835.

Duncan, R. B. Jr. and Potgieter, L. N. (1993). Antigenic diversity of respiratory syncytial viruses and its implication for immunoprophylaxis in ruminants. *Vet. Microbiol.* 37, 319–341.

Dupuis, A. P., Marra, P. P., and Kramer, L. D. (2003). Serologic evidence of West Nile virus transmission, Jamaica, West Indies *Emerg. Infect. Dis.* 9, 860–863.

Ebel, G. D., Carricaburu, J., Young, D., Bernard, K. A., and Kramer, L. D. (2004). Genetic and phenotypic variation of West Nile virus in New York, 2000–2003. *Am. J. Trop. Med. Hyg.* 71, 493–500.

Ebel, G. D., Dupuis, A. P. II, Ngo, K. A., Nicholas, D. C., Kauffman, E. B., Jones, S. A., Young, D. M., Maffei, J. G., Shi, P.-Y., Bernard, K. A., and Kramer, L. D. (2001). Partial genetic characterization of West Nile virus strains, New York State. *Emerg. Infect. Dis.* 7, 650–653.

Edman, J. D. and Taylor, D. J. (1968). *Culex nigripalpus*: seasonal shift in the bird–mammal feeding ratio in a mosquito vector of human encephalitis. *Science* 161, 67–68.

Fenner, F. J., Gibbs, P. J., Murphy, F. A., Rott, R., Studdert, M. J., and White, D. O. (1993). Picornaviridae. In: Veterinary virology. Academic Press, New York, pp. 403–423.

Fonseca, D. M., Keyghobadi, N., Malcolm, C. A., Mehmet, C., Schaffner, F., Mogi, M., Fleischer, R. C., and Wilkerson, R. C. (2004). Emerging vectors in the *Culex pipiens* complex. *Science* 303, 1535–1538.

Fraser, C., Donnelly, C. A., Cauchemez, S., Hanage, W. P., Van, K., Hollingsworth, T. D., Griffin, J., Baggaley, R. F., Jenkins, H. E., Lyons, E. J., Jombart, T., Hinsley, W. R., Grassly, N. C., Balloux, F., Ghani, A. C., Ferguson, N. M., Rambaut, A., Pybus, O. G., Lopez-Gatell, H., Apluche-Aranda, C. M., Chapela, I. B., Zavala, E. P., Guevara, D. M., Checchi, F., Garcia, E., Hugonnet, S., and Roth, C. (2009). Pandemic potential of a strain of influenza A (H1N1): early findings. *Science* 324, 1557–1561.

Fung, W. K. and Yu, P. L. H. (2003). SARS case-fatality rates. *Can. Med. Assoc. J.* 169, 277–278.

Gao, F., Bailes, E., Robertson, D. L., Chen, Y., Rodenburg, C. M., Michael, S. F., Cummins, L. B., Arthur, L. O., Peeters, M., Shaw, G. M., Sharp, P. M., and Hahn, B. H. (1999). Origin of HIV-1 in the chimpanzee *Pan troglodytes troglodytes*. *Nature* 397, 436–441.

Garner, M. M., Helmick, K., Ochsenreiter, J., Richman, L. K., Latimer, E., Wise, A. G., Maes, R. K., Kiupel, M., Nordhausen, R. W., Zong, J. C., and Hayward, G. S. (2009). Clinico-pathologic features of fatal disease attributed to new variants of endotheliotropic herpesviruses in two Asian elephants (*Elephas maximus*). *Vet. Pathol.* 46, 97–104.

Gomez, A., Kramer, L. D., Dupuis, A. P., Kilpatrick, A. M., Davis, L. J., Jones, M. J., Daszak, P., and Aguirre, A. A. (2008b). Experimental infection of eastern gray squirrels (*Sciurus carolinensis*) with West Nile virus. *Am. J. Trop. Med. Hyg.* 79, 447–451.

Gomez, A., Kilpatrick, A. M., Kramer, L. D., Dupuis, A. P., Maffei, J. G., Goetz, S. J., Marra, P. P., Daszak, P., and Aguirre, A. A. (2008a). Land use and West Nile virus seroprevalence in wild mammals. *Emerg. Infect. Dis.* 14, 962–965.

Gould, E. A., Lamballerie, X., Zanotto, P. M., and Holmes, E. C. (2003). Origins, evolution, and vector/host coadaptations within the genus *Flavivirus*. *Adv. Virus Res.* 59, 277–314.

Grubman, M. J. and Baxt, B. (2004). Foot-and-mouth disease. *Clin. Microbiol. Rev.* 17, 465–493.

Guarner, J., Johnson, B. J., Paddock, C. D., Shieh, W. J., Goldsmith, C. S., Reynolds, M. G., Damon, I. K., Regnery, R. L., and Zaki, S. R. (2004). Monkeypox transmission and pathogenesis in prairie dogs. *Emerg. Infect. Dis.* 10, 426–431.

Gubler, D. J. (2002). The global emergence/resurgence of arboviral diseases as public health problems. *Arch. Med. Res.* 33, 330–342.

Hammam, H. M., Clarke, D. H., and Price, W. H. (1965). Antigenic variations of West Nile virus in relation to geography. *Am. J. Epidemiol.* 82, 40–55.

Hanlon, A. C., Niezgoda, M., and Rupprecht, C. E. (2007). Rabies in terrestrial animals. In: Jackson, A. C. and Wunner, W. H. (eds), *Rabies*. Elsevier Academic Press, London, pp. 201–258.

Harder, T. C. and Osterhaus, A. D. (1997). Canine distemper virus: a morbillivirus in search of new hosts? *Trends Microbiol.* 5, 120–124.

Hayes, C. G. (2001). West Nile virus: Uganda, 1937, to New York City, 1999. *Ann. N. Y. Acad. Sci.* 951, 25–37.

Hayes, E. B. and Gubler, D. J. (2005). West Nile virus: epidemiology and clinical features of an emerging epidemic in the United States. *Annu. Rev. Med.* 57, 181–194.

Heymann, D. L., Szczeniowski, M., and Esteves, K. (1998). Re-emergence of monkeypox in Africa: a review of the past six years. *Br. Med. Bull.* 54, 693–702.

Hochachka, W. M., Dhondt, A. A., McKowan, K. A., and Kramer, L. D. (2004). Impact of West Nile virus on American crows in the northeastern United States, and its relevance to existing monitoring programs. *EcoHealth* 1, 60–68.

Hoekstra, C. (2005). West Nile virus: a challenge for transplant programs. *Prog. Transplant.* 15, 397–400.

Horzinek, M. C. (1981). *Non-Arthropod-Borne Togaviruses*. Academic Press, London.

Houe, H. (2003). Economic impact of BVDV infection in dairies. *Biologicals* 31, 137–143.

Hyatt, A. D., Daszak, P., Cunningham, A. A., Field, H., and Gould, A. R. (2004). Henipaviruses: gaps in the knowledge of emergence. *EcoHealth* 1, 25–38.

Ikeda, Y., Nakamura, K., Miyazawa, T., Takahashi, E., and Mochizuki, M. (2002). Feline host range of canine parvovirus: recent emergence of new antigenic types in cats. *Emerg. Infect. Dis.* 8, 341–346.

Jackson, A. C. (2007). Human disease. In: Jackson, A. C. and Wunner, W. H. (eds), *Rabies*. Elsevier Academic Press, London, pp. 309–340.

Jezek, Z., Gromyko, A. I., and Szczeniowski, M. V. (1983). Human monkeypox. *J. Hyg. Epidemiol. Microbiol. Immunol.* 27, 13–28.

Jia, X. Y., Briese, T., Jordan, I., Rambaut, A., MacKenzie, J. S., Hall, R. A., Scherret, J. H., and Lipkin, I. (1999). Genetic analysis of West Nile New York 1999 encephalitis virus. *Lancet* 354, 1971–1972.

Jones, K. E., Patel, N. G., Levy, M. A., Storeygard, A., Balk, D., Gittleman, J. L., and Daszak, P. (2008). Global trends in emerging infectious diseases. *Nature* 451, 990–993.

Keawcharoen, J., Oraveerakul, K., Kuiken, T., Fouchier, R. A., Amonsin, A., Payungporn, S., Noppornpanth, S., Wattanodorn, S., Theamboonniers, A., Tantilertcharoen, R., Pattanarangsan, R., Arya, N., Ratanakorn, P., Osterhaus, D. M., and Poovorawan, Y. (2004). Avian influenza H5N1 in tigers and leopards. *Emerg. Infect. Dis.* 10, 2189–2191.

Keele, B. F., Van, H. F., Li, Y., Bailes, E., Takehisa, J., Santiago, M. L., Bibollet-Ruche, F., Chen, Y., Wain, L. V., Liegeois, F., Loul, S., Ngole, E. M., Bienvenue, Y., Delaporte, E., Brookfield, J. F., Sharp, P. M., Shaw, G. M., Peeters, M., and Hahn, B. H. (2006). Chimpanzee reservoirs of pandemic and nonpandemic HIV-1. *Science* 313, 523–526.

Khan, A. S., Khabbaz, R. F., Armstrong, L. R., Holman, R. C., Bauer, S. P., Graber, J., Strine, T., Miller, G., Reef, S., Tappero, J., Rollin, P. E., Nichol, S. T., Zaki, S. R., Bryan, R. T., Chapman, L. E., Peters, C. J., and Ksiazek, T. G. (1996). Hantavirus pulmonary syndrome: the first 100 US cases. *J. Infect. Dis.* 173, 1297–1303.

Khodakevich, L., Szczeniowski, M., Manbu, mD., Jezek, Z., Marennikova, S., Nakano, J., and Messinger, D. (1987). The role of squirrels in sustaining monkeypox virus transmission. *Trop. Geogr. Med.* 39, 115–122.

Kilpatrick, A. M., Kramer, L. D., Campbell, S. R., Alleyne, E. O., Dobson, A. P., and Daszak, P. (2005). West Nile virus risk assessment and the bridge vector paradigm. *Emerg. Infect. Dis.* 11, 425–429.

Kilpatrick, A. M., Kramer, L. D., Jones, M. J., Marra, P. P., and Daszak, P. (2006). West Nile virus epidemics in North America are driven by shifts in mosquito feeding behavior. *PLoS Biol.* 4, e82.

Knowles, N. J., Samuel, A. R., Davies, P. R., Midgley, R. J., and Valarcher, J. F. (2005). Pandemic strain of foot-and-mouth disease virus serotype O. *Emerg. Infect. Dis.* 11, 1887–1893.

Komar, N. (2003). West Nile virus: epidemiology and ecology in North America. *Adv. Virus Res.* 61, 185–234.

Komar, N., Langevin, S., Hinten, S., Nemeth, N., Edwards, E., Hettler, D., Davis, B., Bowen, R., and Bunning, M. (2003). Experimental infection of North American birds with the New York 1999 strain of West Nile virus. *Emerg. Infect. Dis.* 9, 311–322.

Kono, R., Hirata, M., Kaji, M., Goto, Y., Ikeda, S., Yanase, T., Kato, T., Tanaka, S., Tsutsui, T., Imada, T., and Yamakawa, M. (2008). Bovine epizootic encephalomyelitis caused by Akabane virus in southern Japan. *BMC Vet. Res.* 4, 20.

Kothalawala, H., Toussaint, M. J., and Gruys, E. (2006). An overview of swine influenza. *Vet. Q.* 28, 46–53.

Kramer, L. D. and Bernard, K. A. (2001). West Nile virus in the western hemisphere. *Curr. Opin. Infect. Dis.* 14, 519–525.

Kramer, L. D. and Ebel, G. D. (2003). Dynamics of flavivirus infection in mosquitoes. *Adv. Virus Res.* 60, 187–232.

Kramer, L. D., Li, J., and Shi, P.-Y. (2007a). West Nile virus. *Lancet Neurol.* 6, 171–181.

Kramer, L. D., Styer, L. M., and Ebel, G. D. (2007b). A global perspective on the epidemiology of West Nile virus. *Annu. Rev. Entomol.* 53, 61–81.

Kuiken, T., Holmes, E. C., McCauley, J., Rimmelzwaan, G. F., Williams, C. S., and Grenfell, B. T. (2006). Host species barriers to influenza virus infections. *Science* 312(5772), 394–397.

Kuzzmin, I. V. and Rupprecht, C. E. (2007). Bat rabies. In: Jackson, A. C. and Wunner, W. H. (eds), *Rabies*. Elsevier Academic Press, London, pp. 259–308.

Lanciotti, R. S., Ebel, G. D., Deubel, V., Kerst, A. J., Murri, S., Meyer, R., Bowen, M., McKinney, N., Morrill, W. E., Crabtree, M. B., Kramer, L. D., and Roehrig, J. T. (2002). Complete genome sequences and phylogenetic analysis of West Nile virus strains isolated from the United States, Europe, and the Middle East. *Virology* 298, 96–105.

Lanciotti, R. S., Roehrig, J. T., Deubel, V., Smith, J., Parker, M., Steele, K., Crise, B., Volpe, K. E., Crabtree, M. B., Scherret, J. H., Hall, R. A., MacKenzie, J. S., Cropp, C. B., Panigrahy, B., Ostlund, E., Schmitt, B., Malkinson, M., Banet, C., Weissman, J., Komar, N., Savage, H. M., Stone, W., McNamara, T., and Gubler, D. J. (1999). Origin of the West Nile virus responsible for an outbreak of encephalitis in the northeastern United States. *Science* 286, 2333–2337.

Langevin, S. A., Brault, A. C., Panella, N. A., Bowen, R. A., and Komar, N. (2005). Variation in virulence of West Nile virus strains for house sparrows (*Passer domesticus*). *Am. J. Trop. Med. Hyg.* 72, 99–102.

Lau, S. K., Woo, P. C., Li, K. S., Huang, Y., Tsoi, H. W., Wong, B. H., Wong, S. S., Leung, S. Y., Chan, K. H., and Yuen, K. Y. (2005). Severe acute respiratory syndrome coronavirus-like virus in Chinese horseshoe bats. *Proc. Natl. Acad. Sci. U. S. A.* 102, 14040–14045.

Leschnik, M., Weikel, J., Mostl, K., Revilla-Fernandez, S., Wodak, E., Bago, Z., Vanek, E., Benetka, V., Hess, M., and Thalhammer, J. G. (2007). Subclinical infection with avian influenza A (H5N1) virus in cats. *Emerg. Infect. Dis.* 13, 243–247.

Li, W., Shi, Z., Yu, M., Ren, W., Smith, C., Epstein, J. H., Wang, H., Crameri, G., Hu, Z., Zhang, H., Zhang, J., McEachern, J., Field, H., Daszak, P., Eaton, B. T., Zhang, S., and Wang, L. F. (2005). Bats are natural reservoirs of SARS-like coronaviruses. *Science* 310, 676–679.

Ludwig, G. V., Calle, P. P., Mangiafico, J. A., Raphael, B. L., Danner, D. K., Hile, J. A., Clippinger, T. L., Smith, J. F., Cook, R. A., and McNamara, T. (2002). An outbreak of West Nile virus in a New York City captive wildlife population. *Am. J. Trop. Med. Hyg.* 67, 67–75.

Lvov, D. K., Butenko, A. M., Gromashevsky, V. L., Kovtunov, A. I., Prilipov, A. G., Kinney, R., Aristova, V. A., Dzharkenov, A. F., Samokhvalov, E. I., Savage, H. M., Shchelkanov, M. Y., Galkina, I. V., Deryabin, P. G., Gubler, D. J., Kulikova, L. N., Alkhovsky, S. K., Moskvina, T. M., Zlobina, L. V., Sadykova, G. K., Shatalov, A. G., Lvov, D. N., Usachev, V. E., and Voronina, A. G. (2004). West Nile virus and other zoonotic viruses in Russia: examples of emerging–reemerging situations. *Arch. Virol. Suppl.* 18, 85–96.

MacKenzie, J. S., Gubler, D. J., and Petersen, L. R. (2004). Emerging flaviviruses: the spread and resurgence of Japanese encephalitis, West Nile and dengue viruses. *Nat. Med.* 10, S98–S109.

Macnaughton, M. R., Davies, H. A., and Nermut, M.V. (1978). Ribonucleoprotein-like structures from coronavirus particles. *Gen. Virol.* 39, 545–549.

Malkinson, M. and Banet, C. (2002). The role of birds in the ecology of West Nile virus in Europe and Africa. *Curr. Top. Microbiol. Immunol.* 267, 309–322.

Malkinson, M., Banet, C., Weisman, Y., Pokamunski, S., King, R., Drouet, M. T., and Deubel, V. (2002). Introduction of West Nile virus in the Middle East by migrating white storks. *Emerg. Infect. Dis.* 8, 392–397.

Marfin, A. A. and Gubler, D. J. (2001). West Nile encephalitis: an emerging disease in the United States. *Clin. Infect. Dis.* 33, 1713–1719.

McCarthy, A. J., Shaw, M. A., and Goodman, S. J. (2007). Pathogen evolution and disease emergence in carnivores. *Proc. Biol. Sci.* 274, 3165–3174.

McIntosh, K. and Chanock, R. M. (1990). Respiratory syncytial virus. In: Knipe, D. M. (ed.), *Virology*. Raven Press, New York, pp. 1045–1072.

McLean, R. G., Ubico, S. R., Bourne, D., and Komar, N. (2002). West Nile virus in livestock and wildlife. *Curr. Top. Microbiol. Immunol.* 267, 271–308.

McLean, R. G., Ubico, S. R., Docherty, D. E., Hansen, W. R., Sileo, L., and McNamara, T. S. (2001). West Nile virus transmission and ecology in birds. *Ann. N. Y. Acad. Sci.* 951, 54–57.

Merck. (2008). *Swine influenza*. Merck Publishing. www.merckvetmanualcom/mvm/indexjsp?cfile=htm/bc/121407htm.

Meyer, H., Perrichot, M., Stemmler, M., Emmerich, P., Schmitz, H., Varaine, F., Shungu, R., Tshioko, F., and Formenty, P. (2002). Outbreaks of disease suspected of being due to human monkeypox virus infection in the Democratic Republic of Congo in 2001. *J. Clin. Microbiol.* 40, 2919–2921.

Morse, S. (1993). *Emerging Viruses*. Oxford University, New York.

Mukhopadhyay, S., Kim, B. S., Chipman, P. R., Rossmann, M. G., and Kuhn, R. J. (2003). Structure of West Nile virus. *Science* 302, 248.

Murphy, F. A., Gibbs, E. P., Horzinek, M. C., and Studdert, M. J. (1999). Viral genetics and evolution. In: Murphy, F. A., Gibbs, E. P., Horzinek, M. C., and Studdert, M. J. (eds), *Veterinary Virology*, 3rd edition. Academic Press, New York, pp. 61–80.

Murray, K., Walker, C., Herrington, E., Lewis, J. A., McCormick, J., Beasley, D. W., Tesh, R. B., and Fisher-Hoch, S. (2010). Persistent infection with West Nile virus years after initial infection. *J. Infect. Dis.* 201, 2–4.

Nelson, R. L., Tempelis, C. H., Reeves, W. C., and Milby, M. M. (1976). Relation of mosquito density to bird: mammal feeding ratios of *Culex tarsalis* in stable traps. *Am. J. Trop. Med. Hyg.* 25, 644–654.

Nguyen, M. and Haenni, A. L. (2003). Expression strategies of ambisense viruses. *Virus Res.* 93, 141–150.

Niebylski, M. L. and Meek, C. L. (1992). Blood-feeding of *Culex* mosquitoes in an urban environment. *J. Am. Mosq. Control Assoc.* 8, 173–177.

Parmenter, R. R., Brunt, J. W., Moore, D. I., and Ernest, S. M. (1993). The hantavirus epidemic in the southwest: rodent population dynamics and the implications for transmission of hantavirus-associated adult respiratory distress syndrome (HARDS) in the four corners region. Sevilleta LTER No. 41, 1–45.

Patel, J. R. and Heldens, J. G. (2009). Review of companion animal viral diseases and immunoprophylaxis. *Vaccine* 27, 491–504.

Patz, J. A., Daszak, P., Tabor, G. M., Aguirre, A. A., Pearl, M., Epstein, J., Wolfe, N. D., Kilpatrick, A. M., Foufopoulos, J., Molyneux, D., and Bradley, D. J. (2004). Unhealthy landscapes: policy recommendations on land use change and infectious disease emergence. *Environ. Health Perspect.* 112, 1092–1098.

Pedersen, N. C., Yamamoto, J. K., Ishida, T., and Hansen, H. (1989). Feline immunodeficiency virus infection. *Vet. Immunol. Immunopathol.* 21, 111–129.

Pellett, P. E., McKnight, J. L., Jenkins, F. J., and Roizman, B. (1985). Nucleotide sequence and predicted amino acid sequence of a protein encoded in a small herpes simplex virus DNA fragment capable of trans-inducing alpha genes. *Proc. Natl. Acad. Sci. U. S. A.* 82, 5870–5874.

Penrith, M. L. and Vosloo, W. (2009). Review of African swine fever: transmission, spread and control. *J. S. Afr. Vet. Assoc.* 80, 58–62.

Petersen, L. R. and Hayes, E. B. (2004). Westward ho? The spread of West Nile virus. *N. Engl. J. Med.* 351, 2257–2259.

Reeves, J. D. and Doms, R. W. (2002). Human immunodeficiency virus type 2. *J. Gen. Virol.* 83, 1253–1265.

Rice, C. M. (1996). Flaviviridae: the viruses and their replication. In: Fields, B. N., Knipe, D. M., and Howley, P. M. (eds), *Fields Virology*, 3rd edition. Lippincott-Raven Publishers, Philadelphia, PA, pp. 931–960.

Rodhain, F. (1991). The role of monkeys in the biology of dengue and yellow fever. *Comp. Immunol. Microbiol. Infect. Dis.* 14, 9–19.

Roelke-Parker, M. E., Munson, L., Packer, C., Kock, R., Cleaveland, S., Carpenter, M., O'Brien, S. J., Pospischil, A., Hofmann-Lehmann, R., Lutz, H., Mwamengele, G. L., Mgasa, M. N., Machange, G. A., Summers, B. A., and Appel, M. J. (1996). A canine distemper virus epidemic in Serengeti lions (*Panthera leo*). *Nature* 379, 441–445.

Root, J. J., Oesterle, P. T., Nemeth, N. M., Klenk, K., Gould, D. H., McLean, R. G., Clark, L., and Hall, J. S. (2006). Experimental infection of fox squirrels (*Sciurus niger*) with West Nile virus. *Am. J. Trop. Med. Hyg.* 75, 697–701.

Rosenberg, D. P., Morecki, R., Lollini, L. O., Glaser, J., and Cornelius, C. E. (1983). Extrahepatic biliary atresia in a rhesus monkey (*Macaca mulatta*). *Hepatology* 3, 577–580.

Rowlands, R. J., Michaud, V., Heath, L., Hutchings, G., Oura, C., Vosloo, W., Dwarka, R., Onashvili, T., Albina, E., and Dixon, L. K. (2008). African swine fever virus isolate, Georgia, 2007. *Emerg. Infect. Dis.* 14, 1870–1874.

Rupprecht, C. E., Smith, J. S., Fekadu, M., and Childs, J. E. (1995). The ascension of wildlife rabies: a cause for public health concern or intervention? *Emerg. Infect. Dis.* 1, 107–114.

Sabin, A. B. (1959). Reoviruses: a new group of respiratory and enteric viruses formerly classified as ECHO type 10 is described. *Science* 130, 1387–1389.

Samuel, M. A. and Diamond, M. S. (2006). Pathogenesis of West Nile virus infection: a balance between virulence, innate and adaptive immunity, and viral evasion. *J. Virol.* 80, 9349–9360.

Sejvar, J. J. (2007). The long-term outcomes of human West Nile virus infection. *Clin. Infect. Dis.* 44, 1617–1624.

Shankar, V., Orciari, L. A., De, M. C., Kuzmin, I. V., Pape, W. J., O'Shea, T. J., and Rupprecht, C. E. (2005). Genetic divergence of rabies viruses from bat species of Colorado, USA. *Vector Borne Zoonotic Dis.* 5, 330–341.

Smithburn, K. C., Hughes, T. P., Burke, A. W., and Paul, J. H. (1940). A neurotropic virus isolated from the blood of a native of Uganda. *Am. J. Trop. Med. Hyg.* 20, 471–473.

Soeur, M., Wouters, A., de Saint-Georges, A., Content, J., and Depierreux, M. (1991). Meningoencephalitis and meningitis due to an adenovirus type 5 in two immunocompetent adults. *Acta Neurol. Belg.* 91, 141–150.

Soman, R. S., Rodrigues, F. M., Guttikar, S. N., and Guru, P. Y. (1977). Experimental viraemia and transmission of Japanese encephalitis virus by mosquitoes in ardeid birds. *Indian J. Med. Res.* 66, 709–718.

Spielman, A. (2001). Structure and seasonality of nearctic *Culex pipiens* populations. *Ann. N. Y. Acad. Sci.* 951, 220–234.

Stanway, G. (1990). Structure, function and evolution of picornaviruses. *J. Gen. Virol.* 71(Part 11), 2483–2501.

Straussberg, R., Harel, L., Levy, Y., and Amir, J. (2001). A syndrome of transient encephalopathy associated with adenovirus infection. *Pediatrics* 107, E69.

Sumption, K., Rweyemamu, M., and Wint, W. (2008). Incidence and distribution of foot-and-mouth disease in Asia, Africa and South America; combining expert opinion, official disease information and livestock populations to assist risk assessment. *Transbound. Emerg. Dis.* 55, 5–13.

Takashima, I., Watanabe, T., Ouchi, N., and Hashimoto, N. (1988). Ecological studies of Japanese encephalitis virus in Hokkaido: interepidemic outbreaks of swine abortion and evidence for the virus to overwinter locally. *Am. J. Trop. Med. Hyg.* 38, 420–427.

Taouji, S., Collobert, C., Gicquel, B., Sailleau, C., Brisseau, N., Moussu, C., Breuil, M. F., Pronost, S., Borchers, K., and Zientara, S. (2002). Detection and isolation of equine herpesviruses 1 and 4 from horses in Normandy: an autopsy study of tissue distribution in relation to vaccination status. *J. Vet. Med. B Infect. Dis. Vet. Public Health* 49, 394–399.

Taylor, L. H., Latham, S. M., and Woolhouse, M. E. (2001). Risk factors for human disease emergence. *Philos. Trans. R. Soc. Lond. B Biol. Sci.* 356, 983–989.

Tesh, R. B., Siirin, M., Guzman, H., Travassos da Rosa, A. P., Wu, X., Duan, T., Lei, H., Nunes, M. R., and Xiao, S. Y. (2005). Persistent West Nile virus infection in the golden hamster: studies on its mechanism and possible implications for other flavivirus infections. *J. Infect. Dis.* 192, 287–295.

Thiel, H. J., Collett, M. S., Gould, E. A., Heinz, F. X., Houghton, M., Meyer, G., Purcell, R. H., and Rice, C. M., (2005). Flaviviridae. In: Fauquet, C. M., Mayo, M. A., Maniloff, J., Desselberger, U., and Ball, L. A. (eds), *Virus Taxonomy: Classification and Nomenclature of Viruses: Eighth Report of the International Committee on the Taxonomy of Viruses*. Elsevier, San Diego, pp. 981–998.

Thompson, W. H. and Beaty, B. J. (1978). Venereal transmission of La Crosse virus from male to female *Aedes triseriatus*. *Am. J. Trop. Med. Hyg.* 27, 187–196.

Tian, K., Yu, X., Zhao, T., Feng, Y., Cao, Z., Wang, C., Hu, Y., Chen, X., Hu, D., Tian, X., Liu, D., Zhang, S., Deng, X., Ding, Y., Yang, L., Zhang, Y., Xiao, H., Qiao, M., Wang, B., Hou, L., Wang, X., Yang, X., Kang, L., Sun, M., Jin, P., Wang, S., Kitamura, Y., Yan, J., and Gao, G. F. (2007). Emergence of fatal PRRSV variants: unparalleled outbreaks of atypical PRRS in China and molecular dissection of the unique hallmark. *PLoS One* 2, e526.

Timoney, P. J. and McCollum, W. H. (1993). Equine viral arteritis. *Vet. Clin. North Am. Equine Pract.* 9, 294–309.

Tonry, J. H., Xiao, S. Y., Siirin, M., Chen, H., da Rosa, A. P., and Tesh, R. B. (2005). Persistent shedding of West Nile virus in urine of experimentally infected hamsters. *Am. J. Trop. Med. Hyg.* 72, 320–324.

Trock, S. C., Meade, B. J., Glaser, A. L., Ostlund, E. N., Lanciotti, R. S., Cropp, B. C., Kulasekera, V., Kramer, L. D., and Komar, N. (2001). West Nile virus outbreak among horses in New York State, 1999 and 2000. *Emerg. Infect. Dis.* 7, 745–747.

Tyler, K. L., Virgin, H. W., Bassel-Duby, R., and Fields, B. N. (1989). Antibody inhibits defined stages in the pathogenesis of reovirus serotype 3 infection of the central nervous system. *J. Exp. Med.* 170, 887–900.

USDA. (2008) Equine herpesvirus myeloencephalopathy: mitigation experiences, lessons learned, and future needs. USDA-APHIS-V. S. CEAH., Fort Collins, CO.

USDA. (2009). 2009 Pandemic H1N1 influenza presumptive and confirmed results. U. S. Department of Agriculture.

van der Meer, F. J., Orsel, K., and Barkema, H. W. (2010). The new influenza A H1N1 virus: balancing on the interface of humans and animals. *Can. Vet. J.* 51, 56–62.

van Maanen, C. and Cullinane, A. (2002). Equine influenza virus infections: an update. *Vet. Q.* 24, 79–94.

Vasilakis, N., Cardosa, J., Diallo, M., Sall, A. A., Holmes, E. C., Hanley, K. A., Weaver, S. C., Mota, J., and Rico-Hesse, R. (2010). Sylvatic dengue viruses share the pathogenic potential of urban/endemic dengue viruses. *J. Virol.* 84, 3726–3728.

Weingartl, H. M., Berhane, Y., Hisanaga, T., Neufeld, J., Kehler, H., Emburry-Hyatt, C., Hooper-McGreevy, K., Kasloff, S., Dalman, B., Bystrom, J., Alexandersen, S., Li, Y., and Pasick, J. (2010). Genetic and pathobiologic characterization of pandemic H1N1 2009 influenza viruses from a naturally infected swine herd. *J. Virol.* 84, 2245–2256.

Weiss, M. and Horzinek, M. C. (1987). The proposed family Toroviridae: agents of enteric infections. *Arch. Virol.* 92, 1–15.

WHO. (2009). Influenza (Seasonal). Fact sheet N° 211. World Health Organization. http://www.who int/mediacentre/factsheets/fs211/en/index html.

Woods, C. W., Karpati, A. M., Grein, T., McCarthy, N., Gaturuku, P., Muchiri, E., Dunster, L.,

Henderson, A., Khan, A. S., Swanepoel, R., Bonmarin, I., Martin, L., Mann, P., Smoak, B. L., Ryan, M., Ksiazek, T. G., Arthur, R. R., Ndikuyeze, A., Agata, N. N., and Peters, C. J. (2002). An outbreak of Rift Valley fever in Northeastern Kenya, 1997–98. *Emerg. Infect. Dis.* 8, 138–144.

Woolhouse, M. E.J. Haydon, D. T., and Antia, R. (2005). Emerging pathogens: the epidemiology and evolution of species jumps. *Trends Ecol. Evol.* 20, 238–244.

Yamshchikov, G., Borisevich, V., Kwok, C. W., Nistler, R., Kohlmeier, J., Seregin, A., Chaporgina, E., Benedict, S., and Yamshchikov, V. (2005). The suitability of yellow fever and Japanese encephalitis vaccines for immunization against West Nile virus. *Vaccine* 23, 4785– 4792.

Zeller, H. G. and Schuffenecker, I. (2004). West Nile virus: an overview of its spread in Europe and the Mediterranean Basin in contrast to its spread in the Americas. *Eur. J. Clin. Microbiol. Infect. Dis.* 23, 147–156.

Zislin, A. (2005). Feline immunodeficiency virus vaccine: a rational paradigm for clinical decision-making. *Biologicals* 33, 219–220.

CHAPTER 12

VIRUSES OF CETACEANS*

MARIE-FRANÇOISE VAN BRESSEM[1,2] and JUAN ANTONIO RAGA[3]
[1]Cetacean Conservation Medicine Group (CMED/CEPEC), Bogota, Colombia
[2]Centro Peruano de Estudios Cetológicos (CEPEC), Museo de Delfines, Pucusana, Lima 20, Peru
[3]Marine Zoology Unit, Cavanilles Institute of Biodiversity and Evolutionary Biology, University of Valencia, Valencia, Spain

CONTENTS

12.1 INTRODUCTION

Because of the recognition of morbillivirus epidemics in marine mammals in the 1980s, virus research substantially increased in cetaceans and pinnipeds during the last decades (Osterhaus and Vedder, 1988; Grachev et al., 1989; Domingo et al., 1990; Van Bressem et al., 1993; Forsyth et al., 1998). Several previously unknown DNA and RNA viruses were isolated and sequenced (Mahy et al., 1988; Visser et al., 1993; Bracht et al., 2006; Rehtanz et al., 2006; Van Bressem et al., 2007a). Their epidemiology and ecology have been investigated providing further insight into the causes of massive cetacean die-offs as well as into the negative impact of viral diseases upon populations (Duignan et al., 1996; Van Bressem et al., 1999, 2009a; Härkönen et al., 2006; Harris et al., 2008). These studies also indicated that the emergence and severity of morbillivirus, poxvirus, and calicivirus diseases were, in some cases, related to chemical and biological contamination of the environment (Ross et al., 1996; Ross, 2002; Smith and Boyt, 1990; Van Bressem et al., 2009a, 2009b). For example, environmental changes may affect the "normal" course of a poxvirus infection in odontocetes, eventually leading to septicemia and death, or modify the classical

* This chapter is dedicated to the memory of our dear friend and colleague Dr Thomas Barrett who led the path of research in the field of marine mammal morbilliviruses and was always there to help.

Studies in Viral Ecology: Animal Host Systems: Volume 2, First Edition. Edited by Christon J. Hurst.

endemic pattern (Van Bressem et al., 2009b; Duignan and Van Bressem, 2010). Caliciviruses and influenza A viruses that are commonly detected in whales and dolphins could infect humans during whaling operations, handling and consumption of cetacean meat, and during petting activities in marine parks (Van Bressem et al., 2009a). Molecular studies of cetacean viruses have enlightened some mechanisms of viral evolution proving their use beyond the field of cetology (Van Bressem et al., 2007a; Rector et al., 2008; Gottschling et al., 2011). In this chapter, we will review cetacean viruses that have been characterized by antigenic and molecular techniques and for which ecological data are available.

12.2 DNA VIRUSES

12.2.1 Poxviruses

Poxviruses are large, double-stranded DNA viruses that mostly replicate in the cytoplasm and belong to the family Poxviridae. This family is divided into two subfamilies based on the hosts they infect: Chordopoxvirinae (vertebrates) and Entomopoxvirinae (invertebrates). Until recently, the International Committee on Taxonomy of Viruses recognized eight genera among the Chordopoxvirinae: *Orthopoxvirus*, *Parapoxvirus*, *Leporipoxvirus*, *Capripoxvirus*, *Suipoxvirus*, *Avipoxvirus*, *Molluscipoxvirus*, and *Yatapoxvirus* (Buller et al., 2005). Other genera are awaiting acceptance. Several poxviruses may cause serious and potentially lethal diseases in humans and animals (variola virus, monkey poxvirus virus, lumpy skin disease virus, etc.). They may cause a localized, self-limited infection with little spread from the original site of inoculation or provoke a fulminant, systemic infection characterized by a generalized rash and a high mortality rate (variola and ectromelia viruses) (Buller and Palumbo, 1991). Poxviruses use all portals of entry for infecting the host. The skin is a predilected route followed by the respiratory and digestive tracts. Immunity following a primary viral infection or vaccination with a live, avirulent virus is long-lasting in the case of many but not all poxviruses (Fenner et al., 1993; Buller and Palumbo, 1991). Among the genera of Chordopoxvirinae, the genus *Orthopoxvirus* is notable for its inclusion of viruses that may infect a broad range of species and cross family as well as order barriers (e.g., monkeypox virus, cowpox virus, and vaccinia virus). These viruses are closely related antigenically and genetically, and an extensive cross-neutralization and cross-protection occur between them (Moss, 1996).

The viruses causing tattoo skin disease (TSD) in cetaceans belong to a new genus of *Chordopoxvirinae* and are divided into two subgroups: cetacean poxvirus (CPV)-1 in odontocetes and CPV-2 in mysticetes. The CPV-1 subgroup is further divided into poxviruses of Delphinidae and Phocoenidae (Bracht et al., 2006; Pearce et al., 2008). CPVs have a common, most immediate ancestor with terrestrial poxviruses of the genus *Orthopoxvirus* (Bracht et al., 2006). Serological studies in Peruvian small cetaceans where these viruses are endemic demonstrated the presence of cowpox virus neutralizing antibodies in the sera of all long-beaked common dolphins (*Delphinus capensis*) ($n = 6$) and bottlenose dolphins (*Tursiops truncatus*) ($n = 8$) examined as well as in 63.0% of 27 dusky dolphins (*Lagenorhynchus obscurus*) and in 82.4% of 17 Burmeister's porpoises (*Phocoena spinipinnis*) caught off Peru in 1993–1995 (Van Bressem et al., 2006) further suggesting that cetacean poxviruses are related to orthopoxviruses.

TSD is characterized by very typical, irregular, gray, black, or yellowish, stippled cutaneous lesions that may occur on any body part but show a preferential corporal distribution depending on the species (Figure 12.1). Individual tattoo lesions may persist for months or years and recurrence is possible (Van Bressem and Van Waerebeek, 1996; Van

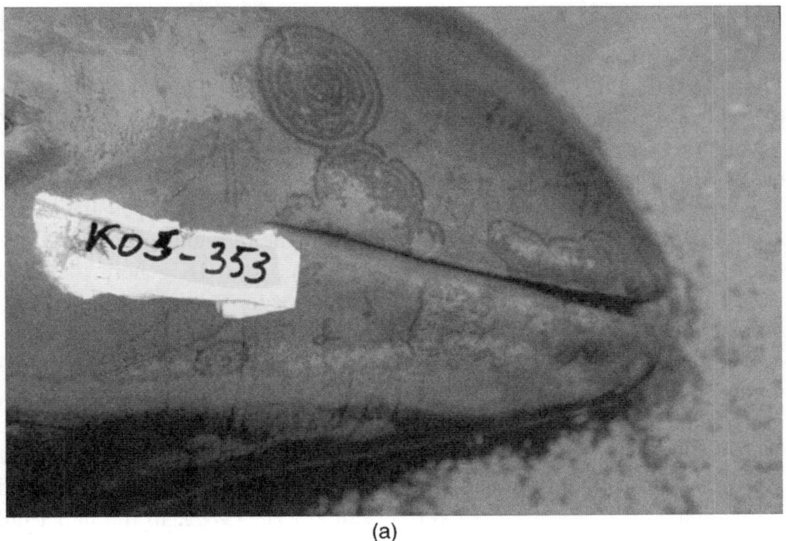

(a)

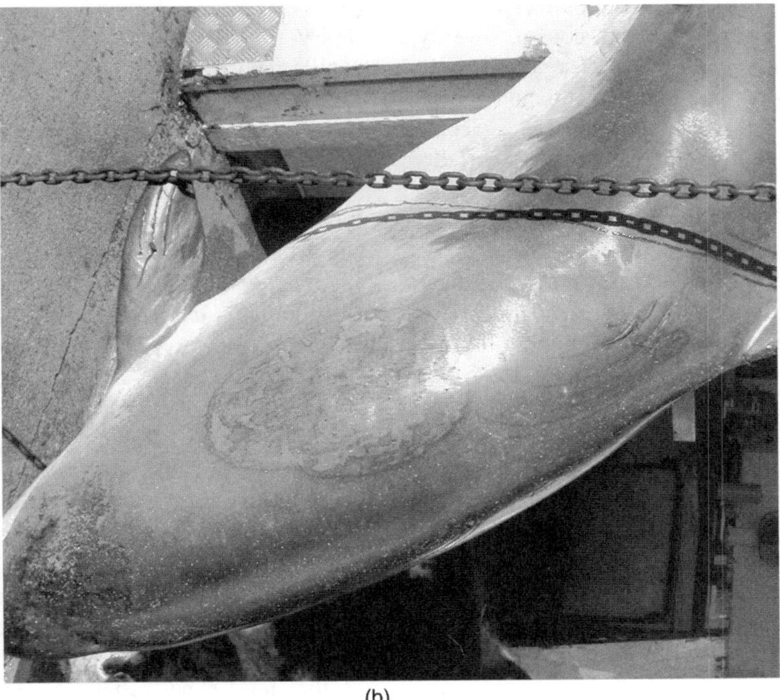

(b)

FIGURE 12.1 (a) Tattoos on the head of an immature, male Burmeister's porpoise *P. spinipinnis* caught in central Peru in July 1994. (b) Giant tattoo on the flank and back of an adult female harbor porpoise *P. phocoena* from the Northeast Atlantic, killed by a bottlenose dolphin *T. truncatus* in 2005 along the coast of the British Isles (courtesy of Drs. Davison, Monies, and Jepson).

Bressem et al., 2003). TSD has been reported in Delphinidae, Phocoenidae, Ziphiidae, Balaenopteridae and Balaenidae from Europe, the Americas, the Middle East, Australia, and New Zealand (Van Bressem et al., 2009b; Duignan and Van Bressem, 2010; Baldwin et al., 2010). A recent study in 17 cetacean species (1392 individuals) from 3 oceans and contiguous seas showed a common pattern for endemic TSD: a significant increase in prevalence in juveniles compared to calves, presumably due to juveniles that had lost maternal humoral immunity, and a significantly higher prevalence in juveniles than in adults, possibly because a high percentage of adults had acquired active immunity following infection. This epidemiological pattern was found inverted in the samples of poor health odontocetes, possibly the result of a depressed immune system (Van Bressem et al., 2009b). The role of environmental factors in the course, severity, and epidemiology of the disease is unknown. Immunosuppressive contaminants and severe, chronic stress levels among sociable cetaceans in areas with persistently elevated mortality rates from by-catch may contribute to higher prevalences of TSD in adults, cases of progressive TSD as evidenced by very large tattoo lesions (Figure1b), as well as to recurrence (Van Bressem et al., 2009b). When endemic, TSD does not appear to induce a high mortality rate (Van Bressem and Van Waerebeek, 1996; Van Bressem et al., 2003). However, it may cause lethal disease in neonates and calves without protective immunity and thus could interfere with host population dynamics (Van Bressem et al., 1999, 2009a, 2009b). Superinfection by other viruses, fungi, or bacteria may exacerbate TSD and lead to systemic disease and death (Smith et al., 1983, Van Bressem et al., 2003, 2007b; Duignan and Van Bressem, 2010).

12.2.2 Papillomaviruses

Papillomaviruses (PVs) are small, non-enveloped, double-stranded DNA viruses (family Papillomaviridae). They are epitheliotropic pathogens that may induce proliferation of the stratified squamous epithelia of the skin and mucosae and cause lesions known as warts, papillomas, and condylomas in mammals and other vertebrates (Howley and Lowy, 2001). High-risk PVs may induce invasive carcinomas (Lowy and Howley, 2001). Until now, at least 189 different PV types have been found in human beings and animals. They have been classified into 18 genera identified by Greek letters according to phylogenetic criteria (de Villiers et al., 2004, Bernard et al., 2010).

Until recently, papillomaviruses associated with genital lesions had only been found in humans and other primates, although papillomas and warts of the genital tract had been observed in several other mammalian species (Sundberg, 1987). In the late 1990s and early 2000 years, cetacean papillomaviruses, including *P. spinipinnis* papillomavirus type 1 (PsPV-1), *P. phocoena* PVs type 1, 2, and 3 (PphV-1, PphPV-2, and PphPV-3), *T. truncatus* papillomavirus type 1, 2, and 3 (TtPV-1, TtPV-2 and TtPV-3), and *D. delphis* PV type 1 (DdPV-1) were isolated from genital lesions in small cetaceans (Figure 12.2) from South America, Europe and the United States (Van Bressem et al., 2007a; Rector et al., 2008, Gottschling et al., 2011). The *P. spinipinnis* and *Delphinidae* PVs belong to the genera *Omikronpapillomavirus* and *Upsilonpapillomavirus*, respectively (Van Bressem et al., 2007a; Bernard et al., 2010; Gottschling et al., 2011). They are characterized by the absence of the E7 oncogene (Rehtanz et al., 2006; Van Bressem et al., 2007a; Rector et al., 2008; Gottschling et al., 2011). Comparison of phylogenetic reconstructions indicated that Omikron + Upsilon-PVs constitute a monophyletic group with PphPV-3 + Alpha + Omega + Dyodelta-PVs in the early genes analysis, but are closely related to Xi + Phi-PVs in the late genes analysis (Van Bressem et al., 2007a; Gottschling et al., 2011). These data strongly suggest that with the exception of PphPV-3, all currently known cetacean papillomaviruses are recombinant viruses (Van Bressem et al., 2007a; Gottschling et al., 2011). PphPV-3 may be a relative of a donor that

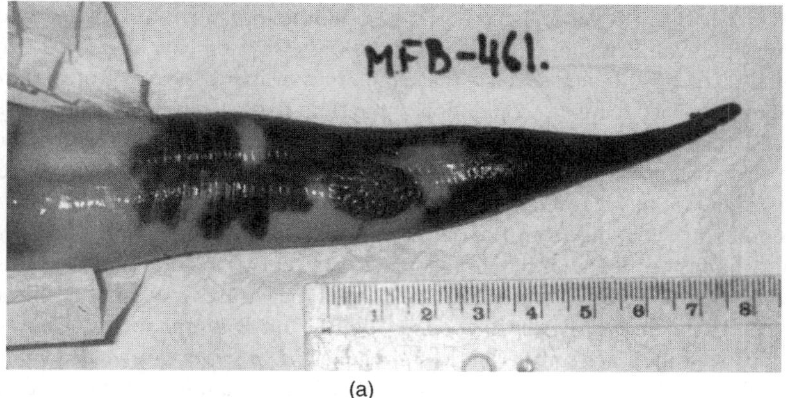

(a)

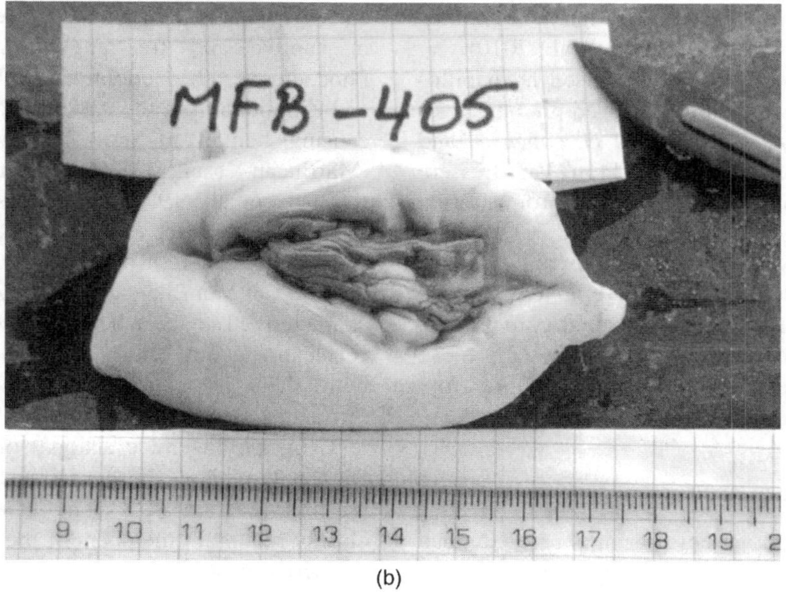

(b)

FIGURE 12.2 (a) Papilloma on the penis of an adult Burmeister's porpoise *P. spinipinnis*. (b) Papillomas in the genital slit of an immature female dusky dolphin *L. obscurus* caught in gillnets off central Peru in 1993.

contributed early genes to the other cetacean papillomaviruses (Gottschling et al., 2011). The recombination of an ancestor of PsPV-1 and an ancestor of TtPV-2 may have generated the common ancestor of TtPV-1 and TtPV-3 (Rector et al., 2008). The presence of a variant of TtPV-3 in a genital wart of an Atlantic white-sided dolphin (*Lagenorhynchus acutus*) suggests interspecies transmission, possibly during heterospecific mating (Gottschling et al., 2011).

Prevalence of genital warts was 48.5% (CI 33.0–64.0%) in 33 *P. spinipinnis* caught in central Peru during 1993–1995 (Van Bressem et al., 1996). Males were almost three times more often infected than females (Van Bressem et al., 1996). The reasons for the higher prevalence in males are unknown, but hormonal and behavioral differences as well as environmental factors may play a role (Van Bressem et al., 1996, 2007a). *P. spinipinnis* may become infected early in life through vertical and horizontal transmission (Van Bressem et al., 1996, 2007a). Genital warts of sufficient severity that may impede, or at least hamper, copulation affected 2 of 20 male *P. spinipinnis* examined. PVs in some circumstances (especially if nonrandomly distributed) may exert an indirect impact on population dynamics (Van Bressem et al., 1999). A high prevalence of papillomavirus antibodies was detected in free-ranging *T. truncatus* from the Indian River Lagoon, Florida, and the estuarine waters near Charleston, South Carolina, and in captive *T. truncatus* from Hawaii and Portugal. In these specimens, the probability of seroconversion in immatures increased with age (Rehtanz et al., 2010).

Genital warts possibly caused by papillomaviruses were also observed in 66.7% of 78 *L. obscurus*, 50% of 10 *D. capensis*, and 33% of 9 *T. truncatus* from Peru (Van Bressem et al., 1996), in 3 *T. truncatus* from Florida (Bossart et al., 2005) and 28.7% of 251 *T. truncatus* from Cuba (Cruz et al., 2006), in a Guiana dolphin (*Sotalia guianensis*) from Brazil (M. Marcondes, personal communication to M.F.B., May 2007), and in 9.7% of 31 sperm whales (*Physeter macrocephalus*) from Iceland (Lambertsen et al., 1987).

12.2.3 Herpesviruses

Herpesviridae are enveloped, double-stranded DNA viruses with relatively large complex genomes and replicate in the nucleus of mammals, birds, and reptiles. They are able to establish and maintain a latent state in their host and reactivate following a variety of psychological or physical stressors (Sainz et al., 2001; Freeman et al., 2010). Although most members of the herpesviridae are of relatively low virulence in their respective hosts, some lack strict host specificity, and cross-species transmission to an inadvertent host can be associated with severe and fatal disease (Westmoreland and Mansfield, 2008). Herpesviruses are further divided into three subfamilies (Alpha-, Beta-, and Gammaherpesvirinae) on the basis of biological characteristics and genomic organization (Pellet and Roizman, 2007; Davison et al., 2009). The alphaherpesviruses are neurotropic (infect nervous system tissue), have a short reproductive cycle ($\sim$18 h) with efficient cell destruction and present a variable host range (Davison et al., 2006). They include four genera: the *Simplexvirus*, *Varicellovirus*, *Mardivirus*, and *Iltovirus*, with the latest two genera found in birds. The gammaherpesviruses are lymphotropic and specific for T- or B-lymphocytes. They include the genera *Lymphocryptovirus* and *Rhadinovirus* that infect mammals (Davison et al., 2006).

Alphaherpesviruses, gammaherpesviruses, and herpes-like viruses have been detected in captive and free-ranging Delphinidae, Phocoenidae, Monodontidae, Physeteridae, and Ziphiidae from the Americas and Europe, causing benign to serious, lethal diseases (Martineau et al., 1988; Barr et al., 1989; Kennedy et al., 1992a; Van Bressem et al., 1994; Blanchard et al., 2001; Manire et al., 2006; Smolarek Benson et al., 2006; Esperón et al., 2008; Van Elk et al., 2009). All partially characterized cetacean herpesviruses are likely cetacean-specific and probably co-evolved with their hosts for thousands of years (Smolarek Benson et al., 2006).

Using phylogenetic analysis of the partial DNA polymerase genes amplified by polymerase chain reaction (PCR), gammaherpesviruses closely related to members of the genus *Rhadinovirus* were detected in lesions of the oral and genital mucosae sampled in 2001–2005 in Atlantic *T. truncatus*, a Risso's dolphin (*Grampus griseus*), a Blainville's beaked whale (*Mesoplodon densirostris*), and a dwarf sperm whale (*Kogia sima*) beached along the U.S. Atlantic coast or kept in captivity in the U.S. The five viruses isolated in *T. truncatus* slightly differed between themselves according to the population/community where they circulated, with nucleotide and amino acid identities ranging from 91.6% to 98.2% and 93.8% to 98.8%, respectively (Smolarek Benson et al., 2006; Van Elk et al., 2009). The virus from the *G. griseus* was closely related to the *T. truncatus* viruses with nucleotide and amino acid identities ranging from 87.9% to 89.8% and 91.3% to 94.2%, respectively. The rates of identity with the viruses from *M. densirostris* and *K. sima* were lower (Smolarek Benson et al., 2006). These data indicated that different species and strains of gammaherpesviruses infect the odontocete families Delphinidae, Ziphiidae and Physeteridae. Using the same molecular techniques, Van Elk et al. (2009) demonstrated that a virus identical to the gammaherpesvirus detected in two captive *T. truncatus* and a free-ranging *T. truncatus* stranded on Islamorada Key, Florida (Smolarek Benson, 2005) was endemic in *T. truncatus* kept in captivity in the Netherlands in 2007–2008. The virus was probably introduced into the captive population by one or two infected *T. truncatus* born in Florida waters, caught in the wild, and sent to the marine mammal park (Van Elk et al., 2009). These results emphasize the risk of introducing new infectious agents through dolphin trade worldwide. It should be stressed that the genital lesions from which the gammaherpesviruses were isolated, amplified, and sequenced were, to our opinion, very similar to the genital warts caused by papillomaviruses in dolphins and porpoises from Peru and the British Isles. Though the presence of PVs was investigated in one study (Van Elk et al., 2009), the primers may not have been adequate to amplify a *T. truncatus* PV. Two other notable studies did not examine the presence of PVs in the genital lesions (Saliki et al., 2006; Smolarek Benson et al., 2006). Gammaherpesviruses are characterized by *in vitro* and *in vivo* infection of lymphoblastoid cells, may affect host immunocompetence, and may be associated with an asymptomatic carrier state or with generalized and oncogenic diseases. However, to our knowledge, none are known to cause genital papilloma-like lesions. The gammaherpesviruses may have been present in the index dolphins 5 and 16 (Van Elk et al., 2009) before they were captured and their presence in the warts may be linked to lymphocyte infection. Papillomaviruses circulate in captive and free-ranging *T. truncatus* from Europe and the United States, including Florida waters from where the two *T. truncatus* kept in captivity in the Netherlands originated (Rector et al., 2008; Rehtanz et al., 2010) and could also be present in the dolphins from the Dutch marine mammal park. The role of gammaherpesviruses in the aetiology of genital warts deserves further studies.

Four previously unrecognized alphaherpesviruses closely related to simplexviruses were detected by PCR and phylogenetic analysis of the partially cloned DNA polymerase and terminase genes in samples of cutaneous lesions and necrotic organs from four *T. truncatus* from the Atlantic coast of the United States (Smolarek Benson et al., 2006; Blanchard et al., 2001; Manire et al., 2006). These viruses were identical or very similar and could cause an apparently benign skin disease or a generalized, fatal infection (Smolarek Benson et al., 2006). Using similar molecular techniques another simplexvirus possibly related to human herpesvirus-1 was detected in the brain of an adult male *T. truncatus* that stranded along the shores of Tenerife (Canary Islands) in 2001 (Esperón et al., 2008). A virus antigenically related to alphaherpesviruses was detected by immunohistochemistry and EM in cerebral cortical neurons of a female harbor porpoise (*Phocoena phocoena*) stranded along the coast of Sweden in 1988 with encephalitis (Kennedy et al., 1992a). Herpes-like virus particles were also observed by EM in epithelial cells from skin lesions (Figure 12.3) sampled in two *L. obscurus* caught in Peru and in stranded and captive beluga whales *Delphinapterus leucas* from the St. Lawrence estuary and the Churchill River (Martineau et al., 1988; Barr et al., 1989; Van Bressem et al., 1994). In most affected specimens, the skin disorders did not seem to be more than mildly pathogenic (Barr et al., 1988; Van Bressem et al., 1994; Manire et al., 2006). Serum antibodies against a virus closely related to Alphaherpesvirinae were detected by a virus neutralization and a blocking ELISA for bovine herpesvirus type 1 test in, respectively, 46% and 58% of 13 belugas found dead along the shores of the St. Lawrence estuary in 1995–1997 (Mikaelian et al., 1999).

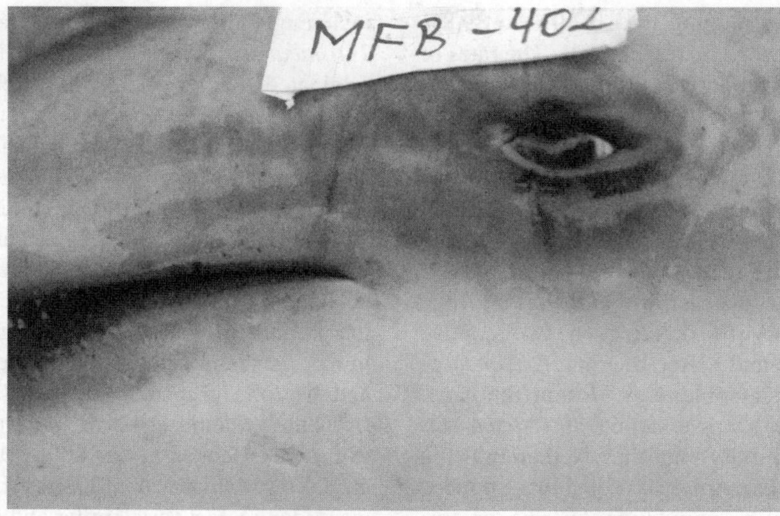

FIGURE 12.3 Herpes-like lesions (black dots) on the head of an immature, female dusky dolphin *L. obscurus* caught off central Peru in 1993.

Taking all these data into account, it is likely that cetacean alphaherpesviruses commonly circulate in several populations from the Pacific and Atlantic Oceans, including the Churchill River and St. Lawrence estuary. Their impact on these populations is unknown and should be further investigated as lethal infections occur.

12.3 RNA VIRUSES

12.3.1 Morbilliviruses

The genus *Morbillivirus* belongs to the family Paramyxoviridae of the subfamily Paramyxovirinae. They are unsegmented, linear negative-sense, single-stranded RNA viruses. All members of this genus may cause serious and potentially lethal diseases in their hosts. Transmission occurs mostly after the inhalation of aerosolized virus shed by infected individuals. Morbilliviruses are extremely infectious and are likely to infect most of the immunologically naïve individuals in a population. Herd formation and migration increase the probability of transmission (Black, 1991; Pomeroy et al., 2008). Surviving individuals have a lasting immunity. Morbilliviruses require large populations of individuals (e.g., 300,000 for measles virus in humans) to be maintained endemically (Black, 1991). Until recently, the genus *Morbillivirus* included human measles virus (MV), bovine rinderpest virus (RPV), peste-des-petits ruminants virus (PPRV), and canine distemper virus (CDV). In 1988, a newly recognized virus called phocine distemper virus (PDV) caused mass mortalities in harbor seals (*Phoca vitulina*) and gray seals (*Halichoerus grypus*) from Northern Europe. At about the same time, another previously unknown virus, the porpoise morbillivirus (PMV), was isolated from *P. phocoena* found dead along the coasts of the British Islands and the Netherlands in 1988–1990 (Kennedy et al., 1988; McCullough et al., 1991; Visser et al., 1993). In 1990–1992, a third marine mammal morbillivirus, the dolphin morbillivirus (DMV), decimated the population of Mediterranean striped dolphins (*Stenella coeruleoalba*) (Van Bressem et al., 1991; Aguilar and Raga, 1993). The pilot whale morbillivirus (PWMV) was detected by PCR in a long-finned pilot whale (*Globicephala melas*) from New Jersey (Taubenberger et al., 2000).

PMV and DMV are antigenically and genetically very similar and are thought to be strains of the same virus for which the name cetacean morbillivirus (CeMV) was suggested (Barrett et al., 1993; Bolt et al., 1994; Banyard et al., 2008). They may infect different families of odontocetes, as indicated by the finding of a porpoise morbillivirus in the Delphinidae *T. truncatus* (Taubenberger et al., 1996). They are more closely related to the ruminant morbilliviruses and measles virus than to the carnivore viruses (Barrett et al., 1993; Visser et al., 1993; Blixenkrone-Möller et al., 1994, 1996; Bolt et al., 1994). Molecular and antigenic data suggest that DMV is closest to the putative morbillivirus ancestor (Blixenkrone-Möller et al., 1994, 1996), implying that it may be the "archevirus" of the genus and may have infected cetaceans for hundreds of thousands or even millions of years. The wide geographic and species distribution of CeMV, the occurrence of potentially very large numbers of hosts among cetacean species from polar waters to tropical seas for some 2–5 million years (Barnes et al., 1985), and the gregarious behavior and migratory habits of many whales and dolphins also argue in favor of this hypothesis (Van Bressem et al., 1999). However, recent splits decomposition analysis suggested that DMV may have originated from recombination events between CDV and PPRV (McCarthy and Goodman, 2010). This hypothesis is in conflict with antigenic and genetic data that have repeatedly showed that DMV is only distantly related to CDV and remains to be further investigated.

CeMV causes lethal disease characterized by pneumonia, nonsuppurative meningoencephalitis, and prominent lymphoid cell depletion in naïve individuals of Delphinidae and Phocoenidae (Domingo et al., 1992; Duignan et al., 1992). During the 1990–1992 epidemic, most affected *S. coeruleoalba* were in poor body condition and several had ulcerative stomatitis (Domingo et al., 1992; Duignan et al., 1992). CeMV is endemic in several species of cetaceans worldwide (Duignan et al., 1995a, 1995b; Van Bressem et al., 2001). In these populations, the virus may cause mortalities among young, naïve individuals that have lost maternal immunity and play a natural regulation role. Pilot whales (*Globicephala* spp.) and other gregarious species are thought to be reservoirs of infection, act as vectors, and spread the virus to other species with which they associate (Duignan et al., 1995a; Van Bressem et al., 1998, 2009a). In the absence of population immunity, CeMV may trigger devastating epidemics that may recur.

PMV and DMV caused die-offs in inshore and estuarine *T. truncatus* from different areas of the U.S. Atlantic coast in 1982, 1987–1988, and 1993–1994 (Lipscomb et al., 1994; Krafft et al. 1995; Duignan et al., 1996; Taubenberger et al., 1996; Rosel et al., 2009). Serological data indicated that the outbreak of mortality in the Indian Banana River, Florida, in January–May 1982 was likely due to morbillivirus infection (Hersh et al., 1990; Duignan et al., 1996). Forty-three carcasses were recovered during that year among a community estimated at 211 individuals (Hersh et al., 1990). From June 1987 till May 1988, epizootics of PMV and DMV ravaged communities of inshore *T. truncatus*, starting from New Jersey and eventually reaching Florida (Lipscomb et al., 1994; Taubenberger et al., 1996; Rosel et al., 2009). This epidemic was associated with the stranding of at least 645 *T. truncatus* (McLellan et al., 2002). A comparative interquartile analysis showed that the stranding pattern was anomalous during that outbreak with "migrating" mortality northward in the summer and southward in the winter (McLellan et al., 2002). The presence of DMV in a *S. coeruleoalba* that died during this die-off demonstrated that multiple species were affected (Taubenberger et al., 1996). Contacts between inshore and estuarine *T. truncatus* and offshore species (offshore *T. truncatus*, *S. coeruleoalba*, and short-beaked common dolphin (*Delphinus delphis*)) where CeMV is endemic (Duignan et al., 1996) may be at the origin of the epizootics. The geography of the outbreak and the observed strain variation may reflect the movement and migrations of these species and the virus they harbor. Seasonal overlap in the inshore/estuarine *T. truncatus* populations at certain times of the year may have favored propagation of the disease (Rosel et al., 2009). Recent genetic data implied that the 1987–1988 outbreak is related to a signature of decreased population size for the *T. truncatus* present in summer from Northern Virginia to New Jersey (Rosel et al., 2009). In 1993–1994, PMV caused a third outbreak of mortality in a separated population of *T. truncatus* from the Gulf of Mexico, spanning from Florida (Panama City) to Texas (Lipscomb et al., 1996; Taubenberger et al., 1996; Rosel et al., 2009). A total of 171 specimens were retrieved from the entire Texas coast in March and April 1994 (Worthy, 1998).

In the Mediterranean, the 1990–1992 DMV epidemic apparently started along the central coast of Spain in July 1990 affecting predominantly *S. coeruleoalba* (Figure 12.4a), currently the most abundant odontocete in this ocean province (Aguilar, 2000; Aguilar and Raga, 1993). It expended to France, Morocco, Italy, and Greece in 1991 and apparently ended in 1992 (Aguilar and Raga, 1993; Van Bressem et al., 1993; Cebrian, 1995). Morbillivirus infection was demonstrated in blood and tissue samples from *S. coeruleoalba* stranded in Spain, Italy, and Greece in 1990–1992 (Van Bressem et al., 1993; Visser et al., 1993). Sexually mature individuals suffered the highest mortality. However, dependent calves also represented a significant portion of the toll possibly indirectly because of the death of their mothers (Calzada et al., 1994). Although no precise mortality rates could be estimated for this die-off, probably thousands of animals perished (Aguilar and Raga, 1993; Forcada et al., 1994). As a relative measure of the impact, the mean school size in the epizootic's core regions significantly decreased to less than 30% of the preoutbreak number (Aguilar and Raga, 1993; Forcada et al., 1994). DMV apparently did not persist as an endemic infection in the Mediterranean striped dolphins after the epidemic terminated (Van Bressem et al., 2001), presumably because the abundance (117,880, CI = 68,379–148,000) in the western Mediterranean Sea (Forcada et al., 1994) was too low to support endemic infection. In 2001–2003, the density (0.49 dolphin per km^2) of striped dolphins estimated in the Gulf of Valencia was again close to the maximum reported for this species in the western Mediterranean (Gómez de Segura et al., 2006). Between October 2006 and April 2007, at least 27 *G. melas* stranded along the southern Spanish Mediterranean coast and the Balearic Islands (Fernández et al., 2008). In early July 2007, dead or moribund *S. coeruleoalba* and *G. melas* were found in the Gulf of Valencia (Figure 12.4b) (Raga et al., 2008). Morbillivirus lesions and antigen were detected in all nine examined *G. melas* and in 13 of 17 *S. coeruleoalba*. A DMV strain closely related to the virus isolated during the 1990–1992 epidemic was detected in 7 of 10 *S. coeruleoalba* and in 9 *G. melas* by reverse transcription (RT)-PCR (Fernández et al., 2008; Raga et al., 2008). In summer–autumn 2007, over 200 *S. coeruleoalba* were found dead along the beaches of Spain. Juveniles were more frequently affected than adults in the 2006–2007 outbreak, likely because older dolphins were still protected by the immunity developed during the 1990–1992 epidemic (Raga et al., 2008; Van Bressem et al., 2009a). The virus apparently reached the French Mediterranean coast in August and Italy's Ligurian Sea coast in August–November 2007 (Garibaldi et al., 2008). The high striped dolphin population density likely favored the propagation of the epidemic (Raga et al., 2008). Taking into account that both the 1990–1992 and 2006–2007 DMV epidemics started close to or in the Gibraltar Strait and that DMV was circulating in the North Sea in January 2007 (Wohlsein et al., 2007), it was suggested that DMV-infected cetaceans, possibly *G. melas*, entered the Strait of Gibraltar and transmitted the infection to *S. coeruleoalba* with which they occasionally form mixed groups (Van Bressem et al., 2009a). Environmental factors (fisheries interactions, inbreeding, migration, high contaminant loads, higher sea surface temperatures, and limited prey

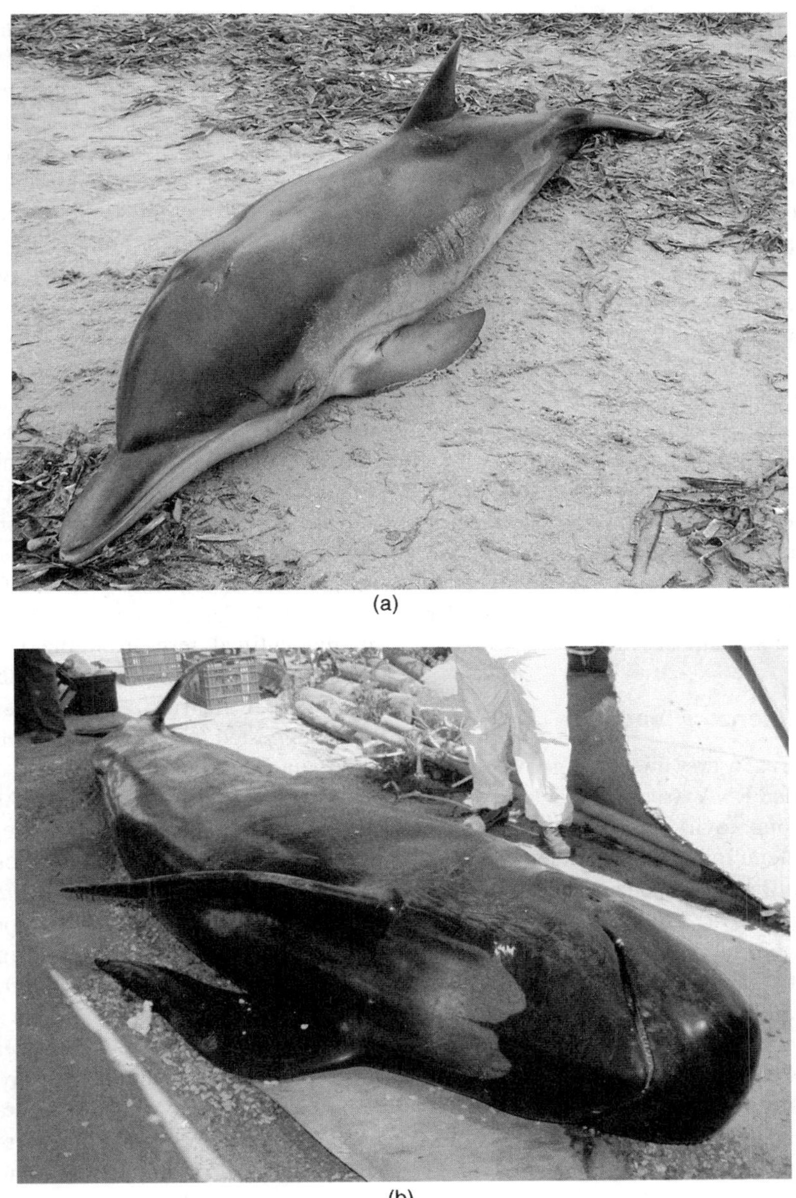

(a)

(b)

FIGURE 12.4 (a) Immature, male, striped dolphin *S. coeruleoalba* stranded along the coast of Alicante, Spain in 2007. (b) Immature, male, long-finned pilot whale (*G. melas*) stranded along the coast of Alicante, Spain in 2007.

availability) may have synergistically interacted to increase the severity of the disease and may favor recurrent epidemics with a profound, accumulative impact on the population dynamics of Mediterranean *S. coeruleoalba* (Van Bressem et al., 2009a).

CeMV also caused the death of several *P. phocoena* and a white-beaked dolphin

(*Lagenorhynchus albirostris*) stranded along the coasts of northern Europe in 1988–1990 (Kennedy et al., 1988, 1991, 1992b; Visser et al., 1993; Osterhaus et al., 1995). Surveys carried out on serum samples from cetaceans stranded or by-caught during 1989–1999 along the coasts of the British Isles suggested that populations of *P. phocoena* and *D. delphis* from the Northeast Atlantic and North Sea are losing their immunity to the virus and may be at risk from new introductions (Van Bressem et al., 1998, 2001). The reintroduction of CeMV into these populations could cause increased mortalities. Altogether these data suggest that partially isolated, small populations and communities are more at risk of suffering morbillivirus epidemics that may further reduce their numbers. The presence of cetacean morbillivirus strains throughout the world oceans is an additional risk to porpoises, dolphins and whales, particularly to those under pressure from human activities.

12.3.2 Influenza A Virus

Influenza viruses are linear, negative-sense, single-stranded RNA viruses that belong to the family Orthomyxoviridae. They are divided into three genera: Influenza A, Influenza B, and Influenza C viruses. Influenzavirus A includes only one species, the Influenza A virus which causes influenza in birds and mammals (Kawaoka et al., 2006). Influenza A viruses are further classified into subtypes based on the viral surface proteins hemagglutinin (HA or H) and neuraminidase (NA or N). Antigenic drift results in point mutations in the HA and NA glycoproteins (Webby and Webster, 2001). Wild birds are apparently the reservoir of influenza A virus and generally only develop an asymptomatic disease. There is phylogenetic evidence that all mammalian virus lineages originally derive from avian influenza viruses after initial cross-species transmission of the viruses from birds to mammals (Reperant et al., 2009). Some strains of influenza A virus may cause severe disease with respiratory symptoms in domestic poultry, horses, pigs, carnivores, cetaceans, and humans.

Two influenza A viruses, A/whale/Maine/1/84 (H13N9) and A/whale/Maine/2/84 (H13N2), were isolated from the lung and hilar node of a sick *G. melas* herded ashore near Portland, ME, in October 1984 (Hinshaw et al., 1986). A third strain of influenza A virus designated A/whale/P0/19/76 (H1N3) was isolated from the lungs and liver of "striped" whales (Balaenopteridae, possibly Antarctic minke whales *Balaenoptera bonaerensis*) caught in the South Pacific in 1975–1976 (Lvov et al., 1978; Murphy and Webster, 1996). The pilot whale was extremely emaciated and swam with difficulty. Necropsy revealed hemorrhagic lungs, a small and friable liver, and a greatly enlarged hilar node (Hinshaw et al., 1986). The antigenic, genetic, and biological properties of the isolates indicated that the *G. melas* and the rorqual viruses are closely related to influenza A viruses isolated from gulls *Larus* sp. and an unidentified tern from the Caspian sea, respectively (Hinshaw et al., 1986; Mandler et al., 1990; Callan et al., 1995). These observations suggested that the viruses were independently introduced into these species from avian sources (Mandler et al., 1990). Associations between seabirds and cetaceans are very common and, in the North Atlantic, some species such as *B. acutorostrata* and *G. melas* are found associated with more bird species and more regularly than others (Figure 12.5) (Evans, 1982). The Arctic tern *Sterna paradisaea* has also been seen foraging in association with *B. bonaerensis* in the open ocean north of the pack-ice zone (Higgins and Davies, 1996). Such behaviors may favor virus transmission between cetacean and wild birds. Transmission from whales and dolphins to humans could occur during whaling and hunting as well as during rescue operations and autopsy.

Serological studies in cetaceans from the Arctic, the Antarctic, and North Pacific have further suggested a sporadic occurrence of the infection, a periodical introduction from infected birds and the absence of enzootic virus maintenance in these species (Nielsen et al., 2001; Ohishi et al., 2006). Thus, only 5% and 6% of the 140 common minke whales

FIGURE 12.5 Minke whale (*B. acutorostrata*) feeding among mixed seabird flocks, including mainly kittiwakes (*Rissa tridactyla*) and some herring gulls (*Larus argentatus*) (image and bird identification courtesy of Dr. P.G.H. Evans).

(*Balaenoptera acutorostrata*) and 34 Dall's porpoises (*Phocoenoides dalli*) caught in the North Pacific in 2000–2001 had specific antibodies against the virus, respectively (Ohishi et al., 2006). The highest titers were detected in *B. acutorostrata* (320 till 2560). Among the 34 belugas taken in fisheries off Baffin Island in 1990–1991, 14.7% had antibodies (dilution 1:5) to influenza A but none of the 62 specimens killed in 1992–1994 were positive (Nielsen et al., 2001). Antibodies were not found in 104 *B. bonaerensis* caught by Japanese whalers in 2000 and 2001 in Antarctic waters (Ohishi et al., 2006).

12.3.3 Caliciviruses

The caliciviruses are small non-enveloped viruses with a single-stranded, positive-sense, polyadenylated RNA genome and belong to the family Caliciviridae. This family is divided into four genera: *Vesivirus*, *Lagovirus*, *Noravirus*, and *Sapovirus* (Cubitt et al., 2006). Marine caliciviruses are classified into the genus *Vesivirus*. They form a distinct lineage within this genus and are strains of the virus species vesicular exanthema of swine virus (VESV) that originated in the North Pacific (Smith and Boyt, 1990; Smith et al., 1998). During 1932–1952, VSV caused outbreaks of vesicular disease in swine herd from California through the ingestion of raw garbage feed contaminated with marine mammal and fish products (Smith et al., 1998). The disease subsequently extended to all major swine-growing areas after garbage containing infected raw pork was fed to pigs (Smith et al., 1998). VESV was officially eradicated in swine in 1956. However, a virus identical to VESV was isolated from a rectal swab of an aborting California sea lion (*Zalophus californianus*) from San Miguel Island in 1972 (Smith et al., 1973). Because it could not be called VESV since the disease had been officially eradicated, it was called San Miguel Sea Lion virus (SMSV) (Smith et al., 1998). Other strains of SMSV were subsequently isolated from pinnipeds, cetaceans, fishes, and parasites from the North Pacific indicating that the marine environment is the primary reservoir for VESV (Smith et al., 1998).

Several caliciviruses have been detected in cetaceans. Most, if not all, are serotypes of the SMSV and VESV. Cetacean calicivirus (CCV Tur-1) was isolated from vesicular skin lesions that developed on a tattoo skin lesions and old scars in two Atlantic *T. truncatus*. The vesicles quickly eroded, leaving shallow ulcers in one of the dolphins (Smith et al., 1983). SMSV-9 serotype, initially recovered from a

Z. californianus, was subsequently isolated from a blowhole swab in a *T. truncatus* kept in captivity at the Navy Operational Support Center, Hawaii (Smith and Boyt, 1990; www.kindplanet.org/smith/isolations.htm). Neutralizing antibodies to several SMSV and VESV serotypes were detected by serology in gray whales (*Eschrichtius robustus*), fin whales (*Balaenoptera physalus*), sei whales (*Balaenoptera borealis*), and *P. macrocephalus* from the North Pacific (Smith and Latham, 1978; Smith and Boyt, 1990), in bowhead whales (*Balaena mysticetus*) caught in Barrow, Alaska (Smith et al., 1987; O'Hara et al., 1998), and in presumably captive *T. truncatus* (Smith and Boyt, 1990). Interestingly, marine vesiviruses could not be isolated in cell cultures or detected by RT-PCR in miscellaneous tissue samples collected during 2004–2007 in Atlantic *T. truncatus*, a spinner dolphin (*Stenella longirostris*), *G. griseus*, rough-toothed dolphins (*Steno bredanensis*), a *P. macrocephalus* and pygmy sperm whales (*Kogia breviceps*) stranded or kept in captivity along the Atlantic coast (including the Gulf of Mexico) of the United States (McClenahan, 2008). All the sera sampled during 1998–2008 in captive and stranded *T. truncatus*, pygmy killer whales (*Feresa attenuata*), *P. phocoena*, *D. delphis*, *G. griseus*, *S. bredanensis*, pantropical spotted dolphin (*Stenella attenuata*), and *K. breviceps* from the same area were negative for the presence of marine vesivirus antibodies by ELISA (McClenahan, 2008). These data suggest that these viruses do not circulate in cetaceans from the U.S. Atlantic coast (McClenahan, 2008).

Besides the skin lesions observed in Atlantic *T. truncatus* affected by CCV Tur-1, very little is known of the symptoms caused by vesiviruses in cetaceans. In pinnipeds, infection by these viruses was associated with abortion and cutaneous vesicular lesions (reviewed in Smith and Boyt, 1990). SMSV and VESV serotypes have a wide range of phylogenetically unrelated host species, ranging from a liver fluke *Zalophotrema* sp. to marine and terrestrial mammals including humans. They may persist in nonlytic cycles in many reservoir hosts (Smith et al., 1980a; Gelberg et al., 1982; Smith and Boyt, 1990). California sea lions and one of their prey, the opaleye fish *Girella nigricans*, may be primarily involved in the maintenance of the viruses in the North Pacific Ocean (Smith et al., 1980a; Smith and Boyt, 1990). Transmission of the virus to other marine mammals is probably linked to contacts between species, migratory pathways, and vectors (Smith et al., 1980a, 1980b; Smith and Boyt, 1990). Metazoan parasites like the liver fluke *Zalophotrema* sp. and the lungworm *Parafilaroides decorus* may act as mechanical vectors (Smith et al., 1980a, 1980b). The capture and worldwide dissemination of marine animals for display purpose may favor virus spread (Smith and Boyt, 1990). Direct transmission may occur while terrestrial animals are foraging on marine mammal carcasses, a concept that may offer the virus new routes of dissemination (Figure 12.6).

12.3.4 Retroviruses

Retroviruses are enveloped, positive-sense, single-stranded RNA viruses and are members of the family Retroviridae. They are grouped into seven genera: alpha-, beta-, gamma-, delta-, epsilon-, lenti-, and spumaretroviruses. Of these, the first six belong to the subfamily Orthoretrovirinae, while the last is part of the subfamily Spumavirinae. Retroviruses use a unique replication scheme in which a long, single-stranded RNA genome is converted by reverse transcription into a double-stranded DNA molecule that is inserted into and becomes a permanent resident of the host genome (Goff, 2001). Retroviruses can be transmitted horizontally or vertically in various ways as exogenous viruses (infectious particles in extracellular space). If a retrovirus happens to infect a germline cell, all its descendants will carry the resulting provirus as part of their genome. Such endogenous retroviruses are transmitted from parents to offspring like Mendelian genes. Most of these endogenous retroviruses are truncated and inactivated by

FIGURE 12.6 Dogs feeding on dolphin remains on a beach of central Peru in 1993.

point mutations and deletions accumulated in the course of evolution, but there are some examples of functional endogenous retroviruses in various animal species (Burmeister, 2001; Weiss, 2006). Several retroviruses, with human deficiency viruses as the most notorious, may cause serious, potentially lethal diseases, including malignant tumors and immune suppression, in a large range of vertebrate species (Burmeister, 2001). Endogenous retroviruses may also increase susceptibility to viral and bacterial pathogens (Bhadra et al., 2006).

An endogenous retrovirus named killer whale endogenous retrovirus (KWERV) that presented all the classical features of a gammaretrovirus was detected in the brain of an adult female killer whale (*Orcinus orca*) stranded in Puget Sound in 2007 using a PCR primer system for detecting oncoretroviruses (Burmeister et al., 2001; LaMere et al., 2009; St. Leger and LaMere, personal communication to M.F.B., January 2010). Sequences of KWERV were subsequently detected in DNA from peripheral blood mononuclear cells and tissue samples from 13 other killer whales (11 captive, 2 dead free-ranging) as well as in sperm cells from a captive *O. orca* using a variety of specific primers. In addition, liver tissue samples from the following stranded and captive Delphinidae were PCR positive for all the *gag*, *pol*, and *env* KWERV genes: a *D. delphis*, a Commerson's dolphin (*Cephalorhynchus commersonii*), two *T. truncatus*, a false killer whale (*Pseudorca crassidens*), a *G. griseus*, a *S. bredanensis*, and a Pacific white-sided dolphin (*Lagenorhynchus obliquidens*). *Gag* products from all the delphinids were nearly identical (95–98%) to the original KWERV nucleotide sequence. Distinct endogenous retroviruses that were positive for KWERV *gag* but not for *pol* and *env* were detected in stranded Physeteridae (*K. simus* and *K. breviceps*) and Phocoenidae (*P. phocoena*). This suggested that related exogenous retroviruses infected several odontocete species and endogenized independently at different times (LaMere et al., 2009). KWERV-related viruses were not detected in liver samples from two captive belugas and a stranded fin whale (origin not provided). The absence of evidence of productive infection *in vivo* in *O. orca* suggested that KWERV is unlikely to cause disease in this species (LaMere et al., 2009). Finally, a possible betaretrovirus was detected by PCR using primers targeting two highly conserved domains within retrovirus genes in genomic DNA from tissue samples of a *G. griseus* of which origin and biological data are unknown (Gifford et al., 2005).

The presence of endogenous retroviruses in cetaceans indicates that exogenous retroviruses also likely occur in this order. As exogenous beta- and gammaviruses may cause leukemia and other malignant tumors in mammals, it would be worthwhile to further explore the role of these viruses in cetacean cancers.

12.4 CONCLUSIONS

An increasing number of virus families, genera, and types have been detected in odontocetes and mysticetes during the past 20 years. Some viruses may trigger lethal epizootics with high mortality rates, while others only cause benign lesions or may be asymptomatic. Some are order- or family-specific, while others have a broader host range infecting animals from different classes and phyla (Figure 12.7). Infections by species-specific cetacean viruses are likely to have occurred for thousands of years with some equilibrium between host populations and pathogens as in other species (Begon et al., 1996). Viruses infecting cetaceans use different strategies to perpetuate themselves (Figure 12.7). Morbilliviruses depend on the population threshold in order for them to be maintained endemically. Papillomaviruses are sexually transmitted, while herpesviridae establish and maintain a latent state in their host after a primary infection and reactivate following specific stimuli. Endogenous retroviruses the endogenous retroviruses are integrated in the animal genome. Influenza A virus and caliciviruses have their reservoir in miscellaneous unrelated species. Extrinsic anthropogenic factors, including biological, acoustic, and chemical pollution, climate change, fisheries, and heavy boat traffic, may in some cases (morbilliviruses, poxviruses, herpesviruses, and possibly papillomaviruses), disturb the virus–host

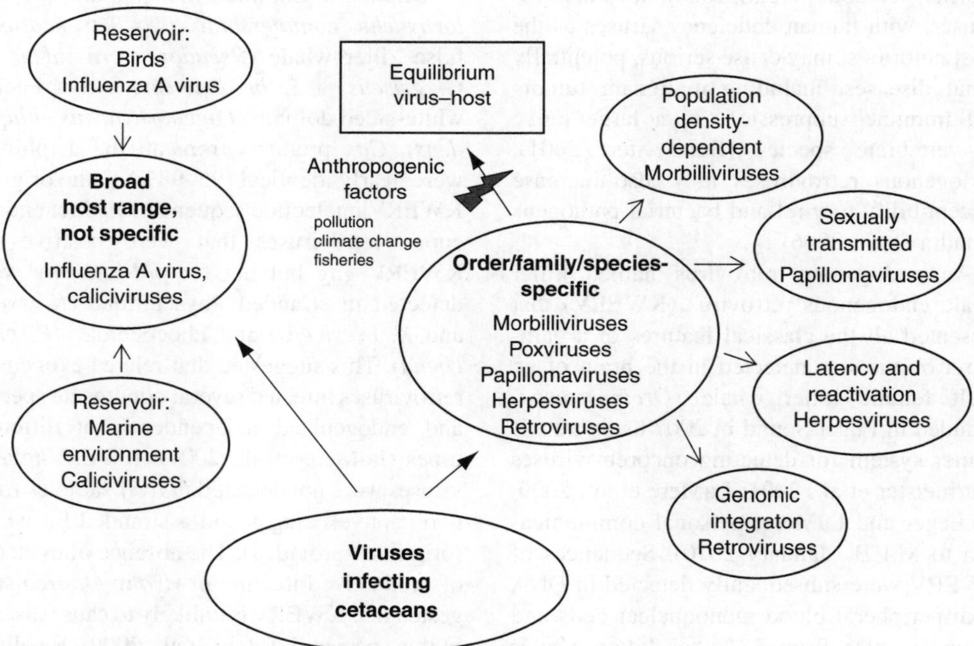

FIGURE 12.7 Pattern of infection and strategies used by viruses infecting cetaceans. Anthropogenic factors may alter the equilibrium between the host and the virus.

equilibrium. Such disruptions include reducing the number of animals in a population and hence reducing the likelihood for successful establishment of enzootic infections, lowering the population immune response, depressing food supplies, and increasing stress (Figure 12.7) (Van Bressem et al., 1999; Fair and Becker, 2000; Ross, 2002; Burek et al., 2008, Acevedo-Whitehouse and Duffus, 2009). At least four virus families (Paramyxoviridae, Poxviridae, Herpesviridae, and Papillomaviridae) have the potential to exert a negative impact on the population dynamics of cetaceans by increasing natural mortality and by negatively affecting reproduction. While morbilliviruses are the more conspicuous of these four virus families because of the massive die-offs that they may cause, papillomaviruses, poxviruses, and herpesviruses deserve further attention.

We recommend that virological studies are systematically included in all advanced biological research programs on cetaceans to account for the potential impact of these microparasites when constructing population dynamics models and to provide biological data on the specimen cetaceans from which the viruses originate in order to improve our knowledge of epidemiology and ecology of these microorganisms.

ACKNOWLEDGMENTS

We thank Drs. Koen Van Waerebeek and Gérard Orth for fruitful discussions and Drs. Jepson, Davison, Monies, and Evans for lending us images used in this chapter. M.F.B. and J.A.R. were supported by a grant of the Cetacean Society International (CSI) and the Service of Conservation of Biodiversity, Conselleria de Medi Ambient, Aigua, Territori i Habitage, Generalitat Valenciana, Spain, respectively.

REFERENCES

Acevedo-Whitehouse, K. and Duffus, A. L. (2009). Effects of environmental change on wildlife health. *Philos. Trans. R. Soc. Lond. B* 364, 3429–3438.

Aguilar, A. (2000). Population biology, conservation threats, and status of Mediterranean striped dolphins (*Stenella coeruleoalba*). *J. Cetacean Res. Manag.* 2, 17–26.

Aguilar, A. and Raga, J. A. (1993). The striped dolphin epizootic in the Mediterranean Sea. *Ambio* 22, 524–528.

Baldwin, R., Collins, T., Minton, G., Findlay, K., Corkeron, P., Willson, A., and Van Bressem, M.-F. (2010). Arabian Sea humpback whales: canaries for the Northern Indian Ocean? Paper SC/62/SH20 presented to the IWC Scientific Committee, June 2010, Agadir, Morocco.

Banyard, A. C., Grant, R. J., Romero, C. H., and Barrett, T. (2008). Sequence of the nucleocapsid gene and genome and antigenome promoters for an isolate of porpoise morbillivirus. *Virus Res.* 132, 213–219.

Barnes, L. G., Domning, D. P., and Ray, C. E. (1985). Status of studies on fossil marine mammals. *Mar. Mamm. Sci.* 1, 15–53.

Barr, B., Dunn, J. L., Daniel, M. D., and Banford, A. (1989). Herpes-like viral dermatitis in a beluga whale (*Delphinapterus leucas*). *J. Wildl. Dis.* 25, 608–611.

Barrett, T., Visser, I. K., Mamaev, L., Goatley, L., Van Bressem, M. F., and Osterhaust A. D. (1993). Dolphin and porpoise morbilliviruses are genetically distinct from phocine distemper virus. *Virology* 193, 1010–1012.

Begon, M., Harper, J. L., and Townsend, C. R. (1996). *Ecology*, 3rd edition. Blackwell Science, Oxford.

Bernard, H. U., Burk, R. D., Chen, Z., van Doorslaer, K., zur Hausen, H., and de Villiers, E. M. (2010). Classification of papillomaviruses (PVs) based on 189 PV types and proposal of taxonomic amendments. *Virology* 401, 70–79.

Bhadra, S., Lozano, M. M., Payne, S. M., and Dudley, J. P. (2006). Endogenous MMTV proviruses induce susceptibility to both viral and bacterial pathogens. *PLoS Pathog.* 12, 1134–1143.

Black, F. (1991). Epidemiology of paramyxoviridae. In: Kingsburry, D. W. (ed.), *The Paramyxoviruses*. Plenum Press, New York, pp. 509–536.

Blanchard, T. W., Santiago, N. T., Lipscomb, T. P., Garber, R. L., Mcfee, W. E., and Knowles, S.

(2001). Two novel alphaherpesviruses associated with fatal disseminated infections in Atlantic bottlenose dolphins. *J. Wildl. Dis.* 37, 297–305.

Blixenkrone-Möller, M., Bolt, G., Gottschalk, E., and Kenter, M. (1994). Comparative analysis of the gene encoding the nucleocapsid protein of dolphin morbillivirus reveals its distant evolutionary relationship to measles virus and ruminant morbilliviruses. *J. Gen. Virol.* 75, 2829–2834.

Blixenkrone-Möller, M., Bolt, G., Jensen, T. D., Harder, T., and Svansson, V. (1996). Comparative analysis of the attachment protein gene (H) of dolphin morbillivirus. *Virus Res.* 40, 47–55.

Bolt, G., Blixenkrone-Möller, M., Gottschalk, E., Wishaupt, R. G. A., Welsh, M. J., Earle, P. J. A., and Rima, B. K. (1994). Nucleotide and deduced amino acid sequences of the matrix (M) and fusion (F) protein genes of cetacean morbilliviruses isolated from a porpoise and a dolphin. *Virus Res.* 34, 291–304.

Bossart, G. D., Ghim, S. J., Rehtanz, M., Goldstein, J., Varela, R., Ewing, R. Y., Fair, P. A.; Lenzi, R. J. B., Hicks, C. L., Schneider, L. S., McKinnie, C. J., Reif, J. S., Sanchez, R., Lopez, A., Novoa, S., Bernal, J., Goretti, M., Rodriguez, M., Defran, R. H., and Jenson, A. B. (2005). Orogenital neoplasia in Atlantic bottlenose dolphins (*Tursiops truncatus*). *Aquat. Mamm.* 31, 473–480.

Bracht, A. J., Brudek, R. L., Ewing, R. Y., Manire, C. A., Burek, A., Rosa, C., Beckmen, K. B., Maruniak, J. E., and Romero, C. H. (2006). Genetic identification of novel poxviruses of cetaceans and pinnipeds. *Arch. Virol.* 151, 423–438.

Buller, R. M., Arif, B. M., Black, D. N., Dumbell, K. R., Esposito, J. J., Lefkowitz, E. J., McFadden, G., Moss, B., Mercer, A. A., Moyer, R. W., Skinner, M. A., and Tripathy, D. N. (2005). Family Poxviridae. In: Fauquet, C. M., Mayo, M. A., Maniloff, J., Desselberger, U., and Ball, L. A. (eds), *Virus Taxonomy: Classification and Nomenclature of Viruses. The Eighth Report of the International Committee on Taxonomy of Viruses*. Elsevier/Academic Press, London, pp. 117–133.

Buller, R. M. and Palumbo, G. J. (1991). Poxvirus pathogenesis. *Microbiol. Rev.* 55, 80–122.

Burek, K. A., Gulland, F. M., and O'Hara, T. M. (2008). Effects of climate change on Arctic marine mammal health. *Ecol. Appl.* 18, S126–S134.

Burmeister, T. (2001). Oncogenic retroviruses in animals and humans. *Rev. Med. Virol.* 11, 369–380.

Callan, R. J., Early, G., Kida, H., and Hinshaw, V. S. (1995). The appearance of H3 influenza viruses in seals. *J. Gen. Virol.* 6, 199–203.

Calzada, N., Lockyer, C. H., and Aguilar, A. (1994). Age and sex composition of the striped dolphin die-off in the western Mediterranean. *Mar. Mamm. Sci.* 10, 299–310.

Cruz, D., Guevara, C., Blanco, M., Sánchez, L., and Chamizo, E. G. (2006). Papilomatosis en genitales de delfines (*Tursiops truncatus*) de zonas costeras del archipelago cubano. Sixth International Congress of Veterinary Sciences, April 13, 2006, La Habana, Cuba.

Cebrian, D. (1995) The striped dolphin *Stenella coeruleoalba* epizootic in Greece, 1991–1992. *Biological Conservation*, **74**: 143–145.

Cubitt, D., Bradley, D. W., Carter, M. J., Chiba, S., Estes, M. K., Saif, L. J., Schaffer, F. L., Smith, A. W., Studdert, M. J., and Thiel, H. J. (2006). Family Caliciviridae. In: Büchen-Osmond, C. (ed.), *ICTVdB: The Universal Virus Database*, version 4, Columbia University, New York. www.ncbi.nlm.nih.gov/ICTVdb/ICTVdB/index.htm

Davison, A. J., Eberle, R., Ehlers, B., Hayward, G. S., McGeoch, D. J., Minson, A. C., Pellett. Ph. E. Roizman, B., Studdert, M. J., and Thiry, E. (2009). The order Herpesvirales. *Arch. Virol.* 154, 171–177.

Davison, A. J., Eberle, R., Hayward, G. S., McGeoch, D. J., Minson, A. C., Pellett, P. E., Roizman, B., Studdert, M. J., and Thiry, E. (2006). Family Herpesviridae. In: Büchen-Osmond, C. (ed.), *ICTVdB: The Universal Virus Database*, version 4, Columbia University, New York. www.ncbi.nlm.nih.gov/ICTVdb/ICTVdB/index.htm

de Villiers, E. M., Fauquet, C., Broker, T. R., Bernard, H. U., and zur Hausen, H. (2004). Classification of papillomaviruses. *Virology* 324, 17–27.

Domingo, M., Ferrer, L., Pumarola, M., Marco, A., Plana, J., Kennedy, S., McAlisky, M., and Rima, B. K. (1990). Morbillivirus in dolphins. *Nature* 348, 21.

Domingo, M., Visa, J., Pumarola, M., Marco, A., Ferrer, L., Rabanal, R., and Kennedy, S. (1992). Pathologic and immunocytochemical studies of morbillivirus infection in striped dolphins (*Stenella coeruleoalba*). *Vet. Pathol.* 29, 1–10.

Duignan, P. J., Geraci, J. R., Raga, J. A., and Calzada, N. (1992). Pathology of morbillivirus infection in striped dolphins (*Stenella coeruleoalba*) from Valencia and Murcia, *Spain. Can. J. Vet. Res.* 56, 242–248.

Duignan, P. J., House, C., Geraci, J. R., Duffy, N., Rima, B. K., Walsh, M. T., Early, G., St. Aubin, D. J., Sadove, S., Koopman, H., and Rhinehart, H. (1995b). Morbillivirus infection in cetaceans of the western Atlantic. *Vet. Microbiol.* 44, 241–249.

Duignan, P. J., House, C., Geraci, J. R., Early, G., Copland, H., Walsh, M. T., Bossart, G. D., Cray, C., Sadove, S., St. Aubin, D. J., and Moore, M. (1995a) Morbillivirus infection in two species of pilot whales (*Globicephala* sp.) from the western Atlantic. *Mar. Mamm. Sci.* 11, 150–162.

Duignan, P. J., House, C., Odell, D. K., Wells, R. S., Hansen, W., Walsh, M. T., St. Aubin, D. J., Rima, B. K., and Geraci, J. R. (1996). Morbillivirus in bottlenose dolphins: evidence for recurrent epizootics in the western Atlantic and Gulf of Mexico. *Mar. Mamm. Sci.* 12, 499–515.

Duignan, P. J. and Van Bressem, M. F. (2010). Cetacean tattoo skin disease (TSD): not just a fashion statement. 59th Wildlife Disease Association Annual Meeting, May 30–June 4, 2010, Iguazú, Argentina (abstract).

Esperón, F., Fernández, A., and Sánchez-Vizcáno, J. M. (2008). Herpes simplex-like infection in a bottlenose dolphin stranded in the Canary Islands. Dis. Aquat. Org. 81, 73–76.

Evans, P. G. H. (1982). Associations between seabirds and cetaceans: a review. *Mamm. Rev.* 12, 187–206.

Fair, P. A. and Becker, P. R. (2000). Review of stress in marine mammals. *J. Aquat. Ecosyst. Stress Recovery* 7, 335–354.

Fenner, F. J., Gibbs, E. P. G., Murphy, F. A., Rott, R., Studdert, M. J., and White, D. O. (1993). *Veterinary Virology*, 2nd edition. Academic Press, San Diego.

Fernández, A., Esperón, F., Herraéz, P., Espinosa de los Monteros, A., Clavel, C., Bernabé, A., Sanchez-Vizcaino, M., Verborgh Ph, DeStephanis, R., Toledano, F., and Bayon, A. (2008). Morbillivirus and pilot whale deaths, Mediterranean Sea. *Emerg. Infect. Dis.* 14, 792–794.

Forcada, J., Aguilar, A., Hammond, P. S., Pastor, X., and Aguilar, R. (1994). Distribution and numbers of striped dolphins in the western Mediterranean Sea after the 1990 epizootic outbreak. *Mar. Mamm. Sci.* 10, 137–150.

Forsyth, M. A., Kennedy, S., Wilson, S., Eybatov, T., and Barrett, T. (1998). Canine distemper virus in a Caspian seal. *Vet. Rec.* 143, 662–664.

Freeman, M. L., Sheridan, B. S., Bonneau, R. H., and Hendricks, R. L. (2010). Psychological stress compromises CD8+ T cell control of latent herpes simplex virus type 1 infections. *J. Immunol.* 179, 322–328.

Garibaldi, F., Mignone, W., Caroggio, P., Ballardini, M., Podestà, M., Bozzetta, E., Casalone, C., Marsilio, F., Di Francesco, C. E., Proietto, U., Colangelo, P., Scaravelli, D., and Di Guardo, G. (2008). Serological evidence of *Morbillivirus* infection in striped dolphins (*Stenella coeruleoalba*) found stranded on the Ligurian Sea coast of Italy. Proceedings of the 22nd European Cetacean Society Conference, March 10-12, 2008, Egmond aan Zee, The Netherlands, pp. 192–193.

Gelberg, H. B., Dieterich, R. A., and Lewis, R. M. (1982). Vesicular exanthema of swine and San Miguel sea lion virus: experimental and field studies in otarid seals, feeding trials in swine. *Vet. Pathol.* 19, 413–423.

Gifford, R., Kabat, P., Martin, J., Lynch, C., and Tristem, M. (2005). Evolution and distribution of class II-related endogenous retroviruses. *J. Virol.* 79, 6478–6486.

Goff, S. P. (2001). Retroviridae: the retroviruses and their replication. In: Knipe, D. M. and Howley, P. M. (eds), *Fields Virology*. Lippincott Williams and Wilkins, Philadelphia, pp. 1999–2069.

Gómez de Segura, A., Crespo, E. A., Pedraza, S. N., Hammond, P. S., and Raga, J. A. (2006). Abundance of small cetaceans in the waters of the central Spanish Mediterranean. *Mar. Biol.* 150, 149–160.

Gottschling, M., Bravo, I. G., Schulz, E., Bracho, M. A., Deaville, R., Jepson, P. D., Van Bressem, M.-F., Stockfleth, E., and Nindl, I. (2011). Modular organizations of novel cetacean papillomaviruses. *Mol. Phyl. Evol.* (in press).

Grachev, M. A., Kumarev, V. P., Mammev, V. P., Zorin, V. L., Baranova, L. V., Denikina, N. N., Belicov, S. I., Petrov, E. A., Kolsnik, V. S., Kolsnik, R. S., Beim, A. M., Kudelin, V. N., Nagieva, F. G., and Sidorovo, V. N. (1989). Distemper virus in Baikal seals. *Nature* 338, 209.

Härkönen, T., Dietz, R., Reijnders, P., Teilmann, J., Harding, K., Hall, A., Brasseur, S., Siebert, U., Goodman, S. J., Jepson, P. D., Dau Rasmussen, T., and Thompson, P. (2006). The 1988 and 2002 phocine distemper virus epidemics in European harbour seals. *Dis. Aquat. Org.* 68, 115–130.

Harris, C. M., Travis, J. M., and Harwood, J. (2008). Evaluating the influence of epidemiological parameters and host ecology on the spread of phocine distemper virus through populations of harbour seals. *PLoS ONE* 3, 1–6.

Hersh, S. L., Odell, D. K., and Asper, E. D. (1990). Bottlenose dolphin mortality patterns in the Indian/Banana river system in Florida. In: Leatherwood, S. P. and Reeves, R. R. (eds), *The Bottlenose Dolphin*. Academic Press, San Diego, pp. 155–164.

Higgins, P. J. and Davies, S. J. J. F. (1996). *The Handbook of Australian, New Zealand and Antarctic Birds*, Vol. 3 Oxford University Press, Melbourne.

Hinshaw, V. S., Bean, W. J. Geraci, J., Fiorelli, P., Early, G., and Webster, R. G. (1986). Characterization of two influenza A viruses from a pilot whale. *J. Virol.*, 58, 655–656.

Howley, P. M. and Lowy, D. R. (2001). Papillomaviruses and their replication. In: Knipe, D. M. and Howley, P. M. (eds), *Fields Virology*, 4th edition, Vol. 2 Lippincott Williams & Wilkins, Philadelphia, pp. 2197–2229.

Kawaoka, Y., Cox, N. J., Haller, O., Hongo, S., Kaverin, N., Klenk, H. D., Lamb, R. A., McCauley, J., Palese, P., Rimstad, E., and Webster, R. G. (2006). Family Orthomyxoviridae. In: Büchen-Osmond, C. (ed.) *ICTVdB: The Universal Virus Database*, version 4. Columbia University, New York. www.ncbi.nlm.nih.gov/ICTVdb/Ictv/fs_index.htm

Kennedy, S., Kuiken, T., Ross, H. M., McAliskey, M., Moffett, D., McNiven, M., and Carole, M. (1992b). Morbillivirus infection in two common porpoises (*Phocoena phocoena*) from the coasts of England and Scotland. *Vet. Rec.* 131, 286–290.

Kennedy, S., Lindstedt, I. J. M. C. Aliskey, M. M., McConnell, S. A., and McCullough, S. J. (1992a). Herpesviral encephalitis in a harbour porpoise (*Phocoena phocoena*). *J. Zoo Wildl. Med.* 23, 374–379.

Kennedy, S., Smyth, J. A., Cush, P. F., McAliskey, M., McCullough, S. J., and Rima, B. K. (1991). Histological and immunocytochemical studies of distemper in harbour porpoises. *Vet. Pathol.* 28, 1–7.

Kennedy, S., Smyth, J. A., Cush, P. F., McCullough, S. J., Allan, G. M., and McQuaid, S. (1988). Viral distemper now found in porpoises. *Nature* 336, 21.

Krafft, A., Lichy, J. H., Lipscomb, T. P., Klaunberg, B. A., Kennedy, S., and Taubenberger, J. K. (1995). Postmortem diagnosis of morbillivirus infection in bottlenose dolphins (*Tursiops truncatus*) in the Atlantic and Gulf of Mexico epizootics by polymerase chain reaction-based assay. *J. Wildl. Dis.* 31, 410–415.

Lambertsen, R. H., Kohn, B. A., Sundberg, J. P., and Buergelt, C. D. (1987). Genital papillomatosis in sperm whale bulls. *J. Wildl. Dis.* 23, 361–367.

LaMere, S. A., St. Leger, J. A., Schrenzel, M. D., Anthony, S. J., Rideout, B. A., and Salomon, D. R. (2009). Molecular characterization of a novel gammaretrovirus in killer whales (*Orcinus orca*). *J. Virol.*, 83, 12956–12967.

Lipscomb, T. P., Kennedy, S., Moffett, D., Krafft, A., Klaunberg, B. A., Lichy, J. H., Regan, G. T., Worthy, G. A., and Taubenberger, J. K. (1996). Morbilliviral epizootic in bottlenose dolphins of the Gulf of Mexico. *J. Vet. Diagn. Invest.* 8, 283–290.

Lipscomb, T. P., Schulman, F. Y., Moffett, D., and Kennedy, S. (1994). Morbilliviral disease in Atlantic bottlenose dolphins (*Tursiops truncatus*) from the 1987-1988 epizootic. *J. Wildl. Dis.* 30, 567–571.

Lowy, D. R. and Howley, P. M. (2001). Papillomaviruses. In: Knipe, D. M. and Howley, P. M. (eds), *Fields Virology*, 4th edition, Vol. 2, Lippincott Williams & Wilkins, Philadelphia, pp. 2231–2264.

Lvov, D. K., Zdanov, V. M., Sazonov, A. A., Braude, N. A., Vladimrtceva, E. A., Agafonova, L. V., Skljanskaja, E. I., Kaverin, N. V., Reznik, V. I., Pysina, T. V., Oserovic, A. M., Berzin, A. A., Mjasnikova, I. A., Podcernjaeva, R. Y., Klimenko, S. M., Andrejev, V. P., and Yakhno, M. A. (1978). Comparison of influenza viruses isolated from man and from whales. *Bull. World Health Org.* 56, 923–930.

Mahy, B. W. J., Barrett, T., Evans, S., Anderson, E. C., and Bostock, C. J. (1988). Characterization of a seal morbillivirus. *Nature* 336, 115.

Mandler, J., Gorman, O. T., Ludwig, S., Schroeder, E., Fitch, W. M., Webster, R. G., and Scholtissek, C. (1990). Derivation of the nucleoproteins (NP) of influenza A viruses isolated from marine mammals. *Virology* 176, 255–261.

Manire, C. A., Smolarek, K. A., Romero, C. H., Kinsel, M. J., Clauss, T. M., and Byrd, L. (2006). Proliferative dermatitis associated with a novel alphaherpesvirus in an Atlantic bottlenose dolphin (*Tursiops truncatus*). *J. Zoo Wildl. Med.* 37, 174–181.

Martineau, D., Lagace, A., Beland, P., Higgins, R., Armstrong, D., and Shugart, L. R. (1988). Pathology of stranded beluga whales (*Delphinapterus leucas*) from the St. Lawrence estuary, Quebec. *Can. J. Comp. Pathol.* 98, 287–311.

McCarthy, A. J. and Goodman, S. J. (2010). Reassessing conflicting evolutionary histories of the Paramyxoviridae and the origins of respiroviruses with Bayesian multigene phylogenies. *Infect. Genet. Evol.* 10, 97–107.

McClenahan, S. D. (2008). Characterization of two novel marine caliciviruses: molecular and serological approaches for improved diagnostics. PhD thesis, University of Florida, p. 220.

McCullough, S. J., McNeilly, F., Allan, G. M., Kennedy, S., Smyth, J. A., Cosby, S. L., McQuaid, S., and Rima, B. K. (1991) Isolation and characterisation of a porpoise morbillivirus. *Arch. Virol.* 118, 247–252.

McLellan, W., Friedlaender, A., Mead, J., Potter, C., and Pabst, D. A. (2002). Analysing 25 years of bottlenose dolphin (*Tursiops truncatus*) strandings along the Atlantic coast of the USA: do historic records support the coastal migratory stock hypothesis. *J. Cetacean Res. Manag.* 4, 297–304.

Mikaelian, I., Tremblay, M. P., Montpetit, C., Tessaro, S. V., Cho, H. J., House, C., Measures, L., and Martineau, D. (1999). Seroprevalence of selected viral infections in a population of beluga whales (*Delphinapterus leucas*) in Canada. *Vet. Rec.* 144, 50–51.

Moss, B (1996). Poxviridae: the viruses and their replication. In: Fields, B. N., Knipe, D. M., Howley, P. M. et al. (eds), *Fields Virology*, 3rd edition. Lippincott-Raven Publishers, Philadelphia, pp. 2637–2671.

Murphy, B. R. and Webster, R. G. (1996). Orthomyxoviruses. In: Fields, B. N., Knipe, D. M., Howley, P. et al. (eds), *Fields Virology*, 3rd edition. Lippincott-Raven Publishers, Philadelphia, pp. 1397–1445.

Nielsen, O., Clavijo, A., and Boughen, J. A. (2001). Serologic evidence of influenza A infection in marine mammals of arctic Canada. *J. Wildl. Dis.* 37, 820–825.

O'Hara, T. M., House, C., House, J. A., Suydam, R. S., and George, J. C. (1998). Viral serologic survey of bowhead whales in Alaska. *J. Wildl. Dis.* 34, 39–46.

Ohishi, K., Maruyama, T., Ninomiya, A., Kida, H., Zenitani, R., Bando, T., Fujise, Y., Nakamatsu, K., Miyazaki, N., and Boltunov, A. N. (2006). Serologic investigation of influenza A virus infection in cetaceans from the western North Pacific and the southern oceans. *Mar. Mamm. Sci.* 22, 214–221.

Osterhaus, A. D. M. E., De Swart, R. L., Vos, H. W., Ross, P. S., Kenter, M. J. H., and Barrett, T. (1995). Morbillivirus infections of aquatic mammals: newly identified members of the genus. *Vet. Microbiol.* 44, 219–227.

Osterhaus, A. D. M. E. and Vedder, E. J. (1988). Identification of virus causing recent seal deaths. *Nature* 335, 20.

Pearce, G., Blacklaws, B. A., Gajda, A. M., Jepson, P., Deaville, R., and Van Bressem M.-F. (2008). Molecular identification and phylogenetic relationships in poxviruses from cetacean skin lesions. The 22nd Annual Conference of the European Cetacean Society, March 10-12, 2008, Egmond aan Zee, The Netherlands (abstract).

Pellet, P. E. and Roizman, B. (2007). The family Herpesviridae: a brief introduction. In: Knipe, D. M., Howley, P., Griffin, D. E. et al. (eds), *Fields Virology*, 5th edition. Lippincott Williams & Wilkins, New York, pp. 2479–2499.

Pomeroy, L. W., Bjørnstad, O. N., and Holmes, E. C. (2008). The evolutionary and epidemiological dynamics of the paramyxoviridae. *J. Mol. Evol.* 66, 98–106.

Raga, J. A., Banyard, A., Domingo, M., Corteyn, M., Van Bressem M.-F., Fernández, M., Aznar, F. J., and Barrett, T. (2008). Dolphin morbillivirus epizootic resurges in the Mediterranean. *Emerg. Infect. Dis.* 14, 471–473.

Rector, A., Stevens, H., Lacave, G., Lemey, P., Mostmans, S., Salbany, A., Vos, M., Van Doorslaer, K., Ghim, S. J., Rehtanz, M., Bossart, G. D., Jenson, A. B., and Van Ranst, M. (2008). Genomic characterization of novel dolphin papillomaviruses provides indications for recombination within the Papillomaviridae. *Virology* 378, 151–161.

Rehtanz, M., Ghim, S. J., McFee, W., Doescher, B., Lacave, G., Fair, P. A., Reif, J. S., Bossart, G. D., and Jenson, A. B. (2010). Papillomavirus antibody prevalence in free-ranging and captive bottlenose dolphins (*Tursiops truncatus*). *J. Wildl. Dis.* 46, 136–145.

Rehtanz, M., Ghim, S. J., Rector, A., Van Ranst, M., Fair, P., Bossart, G. D., and Jenson, A. B. (2006). Isolation and characterization of the first American bottlenose dolphin papillomavirus: *Tursiops truncatus* papillomavirus type 2. *J. Gen. Virol.* 87, 3559–3565.

Reperant, L. A., Rimmelzwaan, G. F., and Kuiken, T. (2009). Avian influenza viruses in mammals. *Revue Scientifique et Technique de l'Office International des Epizooties* 28, 137–159.

Rosel, P. E., Hansen, L., and Hohn, A. A. (2009). Restricted dispersal in a continuously distributed marine species: common bottlenose dolphins *Tursiops truncatus* in coastal waters of the western North Atlantic. *Mol. Ecol.* 18, 5030–5045.

Ross, P. S. R. (2002). The role of immunotoxic environmental contaminants in facilitating the emergence of infectious diseases in marine mammals. *Hum. Ecol. Risk Assess.* 8, 277–292.

Ross, P. S. R., De Swart, R. L., Van Loveren, H., Osterhaus, A. D. M. E., and Vos, J. G. (1996). The immunotoxicity of environmental contaminants to marine wildlife: a review. *Annu. Rev. Fish Dis.* 6, 151–165.

Sainz, B., Loutsch, J. M., Marquart, M. E., and Hill, J. M. (2001). Stress-associated immunomodulation and herpes simplex virus infections. *Med. Hypotheses* 56, 348–356.

Saliki, J. T., Cooper, E. J., Rotstein, D. S., Caseltine, S. L., Pabst, D. A., McLellan, W. A., Govett, P., Harms, C., Smolarek, K. A., and Romero, C. H. (2006). A novel gammaherpesvirus associated with genital lesions in a Blainville's beaked whale (*Mesoplodon densirostris*). *J. Wildl. Dis.* 42, 142–148.

Smith, A. W., Akers, T. G., Madin, S. H., and Vedros, N. A. (1973). San Miguel Sea Lion Virus isolation, preliminary characterization and relationship to vesicular exanthema of swine. *Nature* 244, 108–109.

Smith, A. W. and Boyt, P. M. (1990). Calicivirus of ocean origin: a review. *J. Zoo Wildl. Med.* 21, 3–23.

Smith, A. W. and Latham, A. B. (1978). Prevalence of vesicular exanthema of swine antibodies among feral mammals associated with the southern California coastal zones. *Am. J. Vet. Res.* 39, 291–296.

Smith, A. W., Skilling, D. E., Bernirschke, K., Albert, F. T., and Barlough, J. E. (1987). Serology and virology of the bowhead whale (*Balaena mysticetus* L.). *J. Wildl. Dis.* 23, 92–98.

Smith, A. W., Skilling, D. E., and Brown, R. J. (1980b) Preliminary investigation of a possible lung worm (*Parafilaroides decorus*), fish (*Girella nigricans*) and marine mammal (*Callorhinus ursinus*) cycle for San Miguel sea lion virus type 5. *Am. J. Vet. Res.* 41, 1846–1850.

Smith, A. W., Skilling, D. E., Cherry, N., Mead, J. H., and Matson, D. O. (1998). Calicivirus emergence from ocean reservoirs: zoonotic and interspecies movements. *Emerg. Infect. Dis.* 4, 13–19.

Smith, A. W., Skilling, D. E., Dardiri, A. H., and Latham, A. B. (1980a). Calicivirus pathogenic for swine: a new serotype isolated from opaleye *Girella nigricans,* an ocean fish. *Science* 209, 940–941.

Smith, A. W., Skilling, D. E., and Ridgway, S. (1983). Calicivirus-induced vesicular disease in cetaceans and probable interspecies transmission. *J. Am. Vet. Med. Assoc.* 83, 1223–1225.

Smolarek Benson, K. A. (2005). Molecular identification and genetic characterization of cetacean herpesviruses and porpoise morbillivirus. MS thesis, University of Florida, p. 118.

Smolarek Benson, K. A., Manire, C. A., Ewing, R. Y., Saliki, J. T., Townsend, F. I., Ehlers, B., and Romero, C. H. (2006). Identification of novel alpha- and gammaherpesviruses from cutaneous and mucosal lesions of dolphins and whales. *J. Virol. Methods* 136, 261–266.

Sundberg, J. P. (1987). Papillomavirus infection in animals. In: Syrjanen, K., Gissman, L., and

Koss, L. G. (eds), *Papillomaviruses and Human Disease*. Springer, London, pp. 41–103.

Taubenberger, J. K., Tsai, M., Atkin, T. J., Fanning, T. G., Krafft, A. E., Moeller, R. B., Kodsi, S. E., Mense, M. G., and Lipscomb, T. P. (2000). Molecular genetic evidence of a novel morbillivirus in a long-finned pilot whale (*Globicephalus* (sic) *melas*). *Emerg. Infect. Dis.* 6, 42–45.

Taubenberger, J. K., Tsai, M., Krafft, A. E., Lichy, J. H., Reid, A. H., Schulman, F. Y., and Lipscomb, T. P. (1996). Two morbilliviruses implicated in bottlenose dolphin epizootics. *Emerg. Infect. Dis.* 2, 213–216.

Van Bressem M.-F., Cassonnet, P., Rector, A., Desaintes, C., Van Waerebeek, K., Alfaro-Shigueto, J., Van Ranst, M., Orth, G. (2007a). Genital warts in Burmeister's porpoises: characterization of *Phocoena spinipinnis* papillomavirus type 1 (PsPV-1) and evidence for a second, distantly related PsPV. *J. Gen. Virol.* 88, 1928–1933.

Van Bressem M.-F., Gaspar, R., and Aznar, J. (2003). Epidemiology of tattoo skin disease in bottlenose dolphins (*Tursiops truncatus*) from the Sado estuary, Portugal. *Dis. Aquat. Org.* 56, 171–179.

Van Bressem M.-F., Jepson, P., and Barrett, T. (1998) Further insight on the epidemiology of cetacean morbillivirus in the northeastern Atlantic. *Mar. Mamm. Sci.* 14, 605–613.

Van Bressem M.-F., Raga, J. A., Di Guardo, G., Jepson, P. D., Duignan, P., Siebert, U., Barrett, T., Santos, M. C. O., Moreno, I. B., Siciliano, S., Aguilar, A., and Van Waerebeek, K. (2009a). Emerging infectious diseases in cetaceans worldwide and the role of environmental stressors. *Dis. Aquat. Org.* 86, 143–157.

Van Bressem M.-F. and Van Waerebeek, K. (1996). Epidemiology of poxvirus in small cetaceans from the Eastern South Pacific. *Mar. Mamm. Sci.* 12, 371–382.

Van Bressem M.-F., Van Waerebeek, K., Aznar, J., Raga, J. A., Jepson, P. D., Duignan, P. J., Deaville, R., Flach, L., Viddi, F., Baker, J. R., Di Beneditto, A. P., Echegaray, M., Genov, T., Reyes, J. C., Felix, F., Gaspar, R., Ramos, R., Peddemors, V., Sanino, G. P., and Siebert, U. (2009b). Epidemiological pattern of tattoo skin disease: a general health indicator for cetaceans? *Dis. Aquat. Org.* 85, 225–237.

Van Bressem M.-F., Van Waerebeek, K., and Bennett, M. (2006). Orthopoxvirus neutralising antibodies in small cetaceans from the Southeast Pacific. *Lat. Am. J. Aquat. Mamm.* 5, 49–54.

Van Bressem, M. F., Van Waerebeek, K., Garcia-Godos, A., Dekegel, D., and Pastoret, P. P. (1994). Herpes-like virus in dusky dolphins *Lagenorhynchus obscurus,* from coastal Peru. 1994. *Mar. Mamm. Sci.* 10, 354–359.

Van Bressem M.-F., Van Waerebeek, K., Piérard, G., and Desaintes, C. (1996). Genital and lingual warts in small cetaceans from coastal Peru. *Dis. Aquat. Org.* 26, 1–10.

Van Bressem M.-F., Van Waerebeek, K., and Raga, J. A. (1999). A review of virus infections of cetaceans and the potential impact of morbilliviruses, poxviruses and papillomaviruses on host population dynamics. *Dis. Aquat. Org.* 38, 53–65.

Van Bressem M.-F., Van Waerebeek, K., Reyes, J. C., Félix, F., Echegaray, M., Siciliano, S., Di Beneditto, A. P., Flach, L., Viddi, F., Avila, I. C., Bolaños, J., Castineira, E., Montes, D., Crespo, E., Flores, P. A. C., Haase, B., Mendonça de Souza, S. M. F., Laeta, M., and Fragoso, A. B. (2007b). A preliminary overview of skin and skeletal diseases and traumata in small cetaceans from South American waters. *Lat. Am. J. Aquat. Mamm.* 6, 7–42.

Van Bressem, M.-F., Visser, I. K. G., De Swart, R. L., Örvell, C., Stanzani, L., Androukaki, E., Siakavara, K., and Osterhaus, A. D. M. E. (1993). Dolphin morbillivirus in different parts of the Mediterranean Sea. *Arch. Virol.* 129, 235–242.

Van Bressem M.-F., Visser, I. K. G., Van De Bildt, M. W. G., Teppema, J. S., Raga, J. A., and Osterhaus, A. D. M. E. (1991). Morbillivirus infection in Mediterranean striped dolphins (*Stenella coeruleoalba*). *Vet. Rec.* 129, 471–472.

Van Bressem, M., Waerebeek, K. V., Jepson, P. D., Raga, J. A., Duignan, P. J., Nielsen, O., Di Beneditto, A. P., Siciliano, S., Ramos, R., Kant, W., Peddemors, V., Kinoshita, R., Ross, P. S., López-Fernandez, A., Evans, K., Crespo, E., and Barrett, T. (2001). An insight into the epidemiology of dolphin morbillivirus worldwide. *Vet. Microbiol.* 81, 287–304.

Van Elk, C. E., Van de Bildt, M. W. G., de Jong, A. A. W., Osterhaus, A. D. M. E., and Kuiken, T. (2009). Genital herpesvirus in bottlenose dolphins (*Tursiops truncatus*): cultivation, epidemiology, and associated pathology. *J. Wildl. Dis.* 45, 895–906.

Visser, I. K. G., Van Bressem M.-F., De Swart, R. L., van de Bildt, M. W. G., Vos, H. W., Van der Heijden, R. W. J., Saliki, J. T., Örvell, C., Kitching, P., Kuiken, T., Barrett, T., and Osterhaus, A. D. M. E. (1993). Characterization of morbilliviruses isolated from dolphins and porpoises in Europe. *J. Gen. Virol.* 74, 631–641.

Webby, R. J. and Webster, R. G. (2001). Emergence of influenza A viruses. *Philos. Trans. R. Soc. B* 356, 1817–1828.

Weiss, R. A. (2006). The discovery of endogenous retroviruses. *Retrovirology* 3, 67.

Westmoreland, S. V. and Mansfield, K. G. (2008). Comparative pathobiology of Kaposi sarcoma associated herpesvirus and related primate rhadinoviruses. *Comp. Med.* 58, 31–42.

Wohlsein, P., Puff, C., Kreutzer, M., Siebert, U., and Baumgärtner, W. (2007). Distemper in a dolphin. *Emerg. Infect. Dis.* 13, 1959–1961.

Worthy, G. A. J. (1998). Patterns of bottlenose dolphin, *Tursiops truncatus,* strandings in Texas. In: Zimmennan, R. (ed.), *Characteristics and Causes of Texas Marine Strandings*, NOAA Technical Report NMFS 143. U.S. Department of Commerce, Seattle, p. 85.

CHAPTER 13

THE RELATIONSHIP BETWEEN HUMANS, THEIR VIRUSES AND PRIONS*

CHRISTON J. HURST[1,2]
[1]Departments of Biology and Music, Xavier University, Cincinnati, OH
[2]Engineering Faculty, Universidad del Valle, Ciudad Universitaria Meléndez, Santiago de Cali, Valle, Colombia

CONTENTS

*This chapter represents a revision of "Relationship between humans and their viruses", which appeared as chapter 14 of the book *Viral Ecology*, edited by Christon J. Hurst, published in 2000 by Academic Press. All of the artwork contained in this chapter appears courtesy of Christon J. Hurst.

Studies in Viral Ecology: Animal Host Systems: Volume 2, First Edition. Edited by Christon J. Hurst.

13.1 INTRODUCTION

Many of the viruses which can infect humans should not be considered as viruses of humans, but rather are zoonotic. Zoonotic viruses are those viruses of animals which can cross boundaries such that they occasionally infect humans. Some examples of diseases induced in humans by zoonotic viruses are: dengue, Ebola fever, the equine encephalitids (i.e., eastern, Saint Louis, Venezuelan, and western), hantavirus pneumonia, Lassa fever, Marburg fever, rabies, and yellow fever. Additionally, it should be noted that the zoonotic category includes most, if not all, of the human illnesses induced either by arboviruses (viruses whose transmission is vectored by arthropods) or by the hemorrhagic fever viruses. With respect to the zoonotic viruses humans are, at best, alternate hosts. Humans do in fact usually represent dead-end hosts for these zoonotic viruses, meaning that subsequent transmission of the viruses either to new humans or back to the virus' natural host is not sustained.

There is a subgroup of zoonotic viruses which, although principally remaining viruses of animals, seem to have adapted themselves to use humans as natural hosts. This adaptation is indicated by the fact that these viruses have demonstrated an ability to sustain a chain of transmission among humans. Examples of zoonotic viruses which have shown this ability to adapt themselves to become viruses of humans are those members of the family Flaviviridae (genus *Flavivirus*) which induce the human diseases known as dengue and yellow fever.

All of the above mentioned zoonotic viruses contrast with the viral agents which clearly are known by their nature to be viruses of humans. Examples of viruses of humans are those that induce the diseases known as acquired immunodeficiency syndrome, fever blisters, measles, mumps, polio, rubella, smallpox, T-cell leukemia, T-cell lymphoma, and type A influenza. This chapter is intended primarily to address those viruses which are to be considered as being viruses of humans. Those viruses of terrestrial mammals which are considered to be zoonotic are addressed in further detail by Kramer & Tavakoli in chapter 11 of this book.

Every virus species needs to have a successful overall approach for sustaining its existence. That overall approach must enable the virus to attain its two principle goals, namely, that the virus be able to reproduce itself within a host and that the virus then be transmitted onward to a new host. Those mechanisms which any given virus species employs for achieving its sustainment have been, of course, developed through a process which involved an initiation of events by random chance followed by an evolutionary selection. The most successful overall approaches may be those that subsequently evolve into the types of relationships between a virus and its host species which will allow the virus to persist without eliminating the host population. This latter point is very important because the virus, in turn, may become extinct if it kills off the host population. It is for this reason that excessive virulence will be detrimental to the virus, and an interesting side point may be that if an individual host cannot successfully surmount the infection, then death of that individual host may be seen as an altruistic defense mechanism for the host population as a whole. This latter point might be viewed as explaining the reason why the dramatic deaths associated with Lassa and Ebola hemorrhagic fevers are caused by the host's immune response.

13.2 ACHIEVING THE GOAL OF VIRAL REPRODUCTION

Achieving self reproduction is the first of the virus' principle goals. The involved processes can be divided into three aspects. I will define the first as being an overall 'strategy of the infection course', representing the general approach used by the virus as it establishes the interactive nature of its association within the body of a hosting individual. This aspect includes the likely duration of a relationship which may last the lifetime of the host, and largely is characterized by the extent to which infectious progeny viral particles may be produced during the course of the viral-host association. The second aspect will be called the 'strategy of viral replication', involving sequential issues of where the virus begins its march through the host's body and the physical trajectory which the virus follows until the time that the virus exits that body in hopes of encountering of a new host. The third aspect will be called 'strategies for evading host defensive mechanisms' and describes those approaches, if any are known, which the virus uses to avoid the host's defensive mechanisms. The host generally fights back all of the way, with the progress of the interaction being in fact an involved and highly interactive process! But, please remember that this book is being written to describe the perspective of the virus. If the question can be raised as to whether or not a virus is living, then the answer might well be that from the viral perspective living things exist only to serve as hosts to support viral replication.

13.2.1 Strategy of the Infection Course

As mentioned in chapter 1, the goal of establishing an effective course of infection is an aspect of viral reproduction which can be attained in many ways. Those strategies which viruses use can be summarized into six basic patterns and these often are the evolutionary result of interactions between the viral and host species.

13.2.1.1 Productive Infections Five out of the six course patterns which viral infections have been observed to follow are considered to be productive. These five involve the host acquiring that particular viral species in the form of an infectious viral particle (a virion), following which progeny viral particles subsequently are produced within that host and those progeny can be transmitted to infect other potential host individuals. Productive, in this usage of the word, thereby means that the progeny viral particles produced during the course of an infection will be sufficiently great in number to effectively serve for transmitting the infection to a new host with some reasonable probability. The productive approach involves interactions between the virus and host which determine the course of the infection.

a. "*Short term - initial*". During this type of infection pattern, virions are only produced during a short time course which begins within a few hours of when the infection initiates and the duration of that production lasts from between days to a few months. The viral infection usually then ends completely. The human host either may or may not survive beyond the course of this short infection. Host survival depends upon the type of virus involved, the extent to which the involved virus and humans have had time to coevolve as species, and whether or not the ancestral humans of that particular subgroup of the human host population previously had contact with the causative virus. Coevolution usually will tend to make the outcome of this pattern of viral infection sufficiently mild as to be associated with a fairly low incidence of mortality in an otherwise healthy population of human hosts. Some examples of this pattern would be the infections caused by the human caliciviruses (Caliciviridae family), human influenzaviruses (Orthomyxoviridae family), human enteroviruses (Picornaviridae family), and the human rotaviruses (Reoviridae family).

b. "*Recurrent*". This pattern often involves a very pronounced initial production of virions and accompanying symptoms, after which the virus persists quiescently within the body of the host as only viral nucleic acid accompanied by a minimal generation of highly specialized viral proteins. There will be either little or no evidence of virions produced by the host during these periods of quiescence, and generally the viral infection cannot be transmitted onward to a new host during the quiescent periods. There will be recurrences, characterized by an increase or temporarily renewed generation of virions and a resulting reappearance of symptoms that may be severe but usually are not life threatening. The viral infection can be transmitted to a new host during these recurrences. This cyclical pattern of quiescence periods and recurrence may last for years, and indeed often spans the remaining natural lifetime of the individual host. Some examples of this pattern would be the infections caused by the human herpesviruses (Herpesviridae family) and human papillomaviruses (Papillomaviridae family).

c. "*Increasing to end-stage*". This is a prolonged course of infection that may be nearly asymptomatic at the early stage, and the initial level of virion production may be minimal. Following this slow, almost innocuous start, there is a gradual but inevitable progression of the infection. During this time, a worsening in the health condition of the host corresponds to an increasingly higher level of virion production and there also is a corresponding increase in the likelihood of viral transmission to a new host. This pattern of infection eventually ends in death of the host, a process which may take from 10 to 40 years before it reaches that inevitable conclusion. That finality often is precipitated by progressive viral destruction of the host's immunological defense systems which then results in lethal secondary infections. An example of this pattern would be the infections caused by the human immunodeficiency viruses and the human T-lymphotrophic viruses, all belonging to the Retroviridae family.

d. "*Persistent - episodic*". This pattern represents a prolonged nonfatal infection that may persist for the remainder of the hosts natural lifetime. There is a continuous production of virions within the host, but the infection only episodically results in symptoms. The viral genome does not become quiescent and the host remains infectious throughout the course of this associative interaction. Members of the family Picobirnaviridae often produce this pattern of productive infection.

e. "*Persistent but inapparent*". This pattern represents a prolonged nonfatal infection that seemingly never results in overt symptoms of illness attributable to that particular virus. Viral infections that follow this pattern are persistently productive and the viral genome does not become quiescent. The host often remains infectious for the remainder of their natural lifetime. Some examples of viruses which produce this pattern would be members of the family Anelloviridae, and in certain notably rare instances, infection by Human immunodeficiency virus 2 which is a member of the genus *Lentivirus* of the family Retroviridae.

13.2.1.2 Non-Productive Infections

The sixth basic pattern of viral infection is considered essentially to be nonproductive. A non-productive infection is one in which the production of infectious virus particles either never occurs or is so rare and limited that the virus must transmit itself through other means, usually done by transferring a copy of only the virus' nucleic acid genome. In these instances the viral infection normally is acquired by a direct transfer of the virus' genetic material from the human parents to their developing fetuses, with this transfer occurring via the egg and sperm cells. There may never be apparent health effects associated with this type of infection. An example of this pattern would be the infections caused by the endogenous retroviruses (Retroviridae family), whose genomes are incorporated into the chromosomal material of every cell in the human body (Villareal, 1997). The non-productive pattern of infection seems to suggest the highest degree of coevolution between a virus and its host, since a non-productive virus has no means of transmitting itself to a new host without some very active, albeit possibly unwitting, participation on the part of the present host.

13.2.2 Strategy of Viral Replication

This section addresses the questions of where and how the virus begins its march through the individual host's body, and how the virus then continues the course of that attack, leading ultimately to the concept of viral reproduction strategies at the host population level.

13.2.2.1 Cellular Metabolic Level Discussion of the strategy of viral replication within the body of a host organism begins at the most basic level, which is the attachment of the virus to a particular molecule present on the surface the virus' host cells. Such a molecule is termed to be the virus' receptor, and will be some cellular protein or lipid component naturally produced by those cells. The virus' choice of receptor is a product of viral evolution. After binding to its receptor, the virus gains entrance to the interior of the cell and viral replication begins. Those viruses whose genome is composed of DNA generally focus the center of their replication in the nucleus. Contrastingly, those viruses whose genome is composed of RNA generally focus their center of replication in the cytoplasm. During the course of its replication, the virus must decide which cellular systems and machinery it will use. Some large viruses carry the genomic coding capacity for many of their own enzymes, others may rely almost completely upon the enzymatic machinery possessed by the host cell. Many viruses, such as those belonging to the genus *Enterovirus* of the family Picornaviridae, are said to be highly cytopathogenic, meaning that they usually quickly kill the host cell as a product of infecting that cell. Other viruses, such as that which occupies the genus *Rubivirus* of the family Togaviridae, may establish a prolonged severe crippling of the cell rather than quickly killing it outright. A further discussion of these issues can be found in chapter 3 by Debi Nayak.

13.2.2.2 Tissue and Organ Tropism Level Viruses greatly differ with respect to the tissues which they tend to target for infection. This then leads, on a larger scale, to an identification of those organs which the viruses are affecting. This selective targeting is referred to as being a tropism. Viral tropisms can be divided into those which are considered primary versus those considered to be secondary. Primary tropisms will be associated with the production of those viral particles that subsequently contribute to transmission of the viral infection to a new host. As such, the primary tropisms tend to be related to those sites (termed to be portals) through which the virus either enters or exits the body of the host. Secondary tropisms may represent accidents. Some of these accidents may come about as a result of the molecule which a virus uses as its receptor existing on the cells of tissues which are unrelated to those that the virus must employ in order to achieve its transmission. Nevertheless, secondary tropisms may contribute greatly to the types and severity of the illnesses associated with infection of humans by any particular virus (see Figure 13.1).

13.2.2.3 Host Population Level When considered at the host population level, the strategy of viral replication includes the ease or likelihood with which a virus is transmitted to new hosts, plus the severity of infection and accompanying likelihood of death (including the age-related likelihood of death) for any given host individual.

13.2.3 Strategies for Evading Host Defensive Mechanisms

Through the course of evolution, many viruses have developed mechanisms for either countering or evading the human immune and nonimmune defenses as means for aiding the virus' probability of success.

13.2.3.1 Avoiding the Host's Immune Defenses The human immune system includes both humoral (antibody mediated) and cellular components. The cellular components can include granulomatous reactions, which play a role in defense against protozoans albeit their possible role in anti-viral defenses seems incompletely explored. Those mechanisms which viruses use either to avoid or minimize attack by the immune system can be divided into the four groups listed below. The use of these types of mechanisms seems particularly crucial in association with those viral infections which persist within an individual human host for extremely long periods of time. Some such

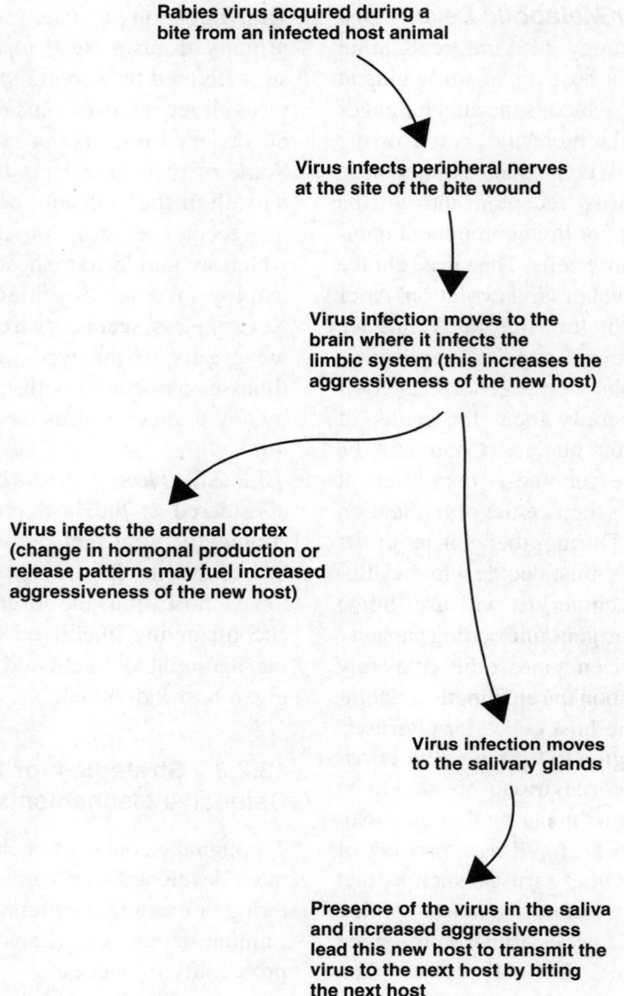

FIGURE 13.1 This figure shows the viral ecology of Rabies Virus (genus *Lyssavirus*, family Rhabdoviridae) in association with a natural host. Transmission of this virus between hosts occurs when an infected animal bites an uninfected animal, with the virus being transferred by saliva into the bite wound. The subsequent movement of the viral infection into the nervous system and the salivary glands of the newly bitten host animal is considered to represent primary tropisms, as infection at these sites is directly related to movement of the virus into the body of this current host and subsequent transfer of the virus to the next host. Infection of the adrenal cortex is considered to represent a secondary tropism, since those viruses produced in the adrenal cortex will not be transferred to any subsequent host animal. However, infection of the adrenal cortex may play a role in the virus' ecology by augmenting the aggressiveness of this newly infected host and thereby increasing the likelihood that this animal will then bite other potential host animals.

interactions often encompass decades in a dance that must seem to last forever.

a. *Antigenic mimicry*. The produced antigens are similar to those of the host, as with the prions. This can benefit the virus but also may lead to viral-triggered autoimmune responses that can be very detrimental to the hosting organism.

b. *Rapid viral mutation*. This mechanism includes both antigenic drifting and shifting.

Some viral types demonstrate rapid viral mutation during the course of an infection, as occurs with the human immunodeficiency viruses of the family Retroviridae. Other virus types, such as the influenza viruses of the family Orthomyxoviridae, demonstrate rapid viral mutation between reinfections of the same host.

c. *Low antigenicity*. Some viruses inherently seem to provoke little, if any, immune response. Often this occurs because the virus persists in a latent state within host cells, during which time either little or no viral antigenic material is produced. Examples include the endogenous retroviruses of the family Retroviridae and the human herpesviruses of the family Herpesviridae.

d. *Infect the immune cells*! The most direct attack may be the most effective. Exceptionally notorious examples of this approach are the genus *Rubivirus* of the family Togaviridae and the genus *Lentivirus* of the family Retroviridae. Aside from the above groupings, some viruses such as the Norwalk virus of the family Caliciviridae seem to be antigenic but provoke an immune response which is minimally effective.

13.2.3.2 Avoiding the Host's Non-Immune Defenses

The body has non-immune defense mechanisms which help to protect against viral infections. These mechanisms are associated with the portals through which viruses can enter the body of the host. Examples of non-immune defenses include the degradative enzymes secreted as a part of pancreatic fluid, saliva, and tears. Numerous viruses are resistant to these and, in particular, many gastroenteritis viruses such as the rotaviruses of the family Reoviridae and the astroviruses of the family Astroviridae have evolved such an effective resistance to attack by proteolytic enzymes that those viruses virtually need partial proteolysis to facilitate their infectivity. Various glands associated with mucosal tissues secrete antimicrobial compounds into the mucus which those tissues produce. Some mucosal tissues also possess cilia whose movement helps to expell both the mucus and any foreign materials, including pathogens, that become entrapped within the mucus. The influenza viruses of the family Orthomyxoviridae are known for their ability to paralyze the activity of the mucosal cilia located within the respiratory tract. Another prominent example of a non-immune defense is the stomach acid produced to aid digestion of organic compounds. One of the defining characteristics for the Human enterovirus species A - D of the family Picornaviridae is their resistance to acidic exposure, which increases their ability to reach target cells within the intestines and facilitates their transmission by the fecal-oral route. Prions that are transmissible by ingestion of an infected host are resistant to both low pH exposure and proteolytic enzymes.

13.3 ACHIEVING THE GOAL OF VIRAL TRANSMISSION BETWEEN HOSTS

The task of achieving viral transmission between hosting individuals involves two aspects. The first of these aspects is the type of infectious material in which a virus will leave its present host. The second aspect involves the route by which that virus subsequently can encounter its proximate host.

13.3.1 Type of Infectious Bodily Material in Which Virus is Released from the Host

The types of bodily materials in which viruses can be released include substances that exit during the course of normal body functions. Among these substances are feces and a variety of liquids, the latter of which include menstrual blood, respiratory secretions of the upper as well as the lower tracts, saliva, semen, tears, urine, and vaginal fluid. Sweat is another fluid that is naturally released from the body, however, it is not known to contain viruses. Viruses can also be found in blood released from wounds in the skin; blood acquired by blood-consuming parasitic insects, among which are the fleas, several groups of flies, ticks, and mosquitoes; and blood leaked from swollen or ruptured capillaries into mucosal tissues and pores of the skin during viral induced hemorrhages.

13.3.2 Route of Transmission between Hosts

Those natural routes by which viruses are transferred to and between humans are the same routes associated with all surface-dwelling terrestrial vertebrates. These routes are tightly associated with the portals of entry and exit which any particular virus family uses as it tries to survive and find its way from one host to the next. Viral transmission routes can be divided into two broad groups. The first of these groups is transmission by direct contact (also known as direct transfer) between two members of those species which host the virus. This first group includes both the possibility of transmission between two members of the principle host species as well as the possibility of transmission between a member of that principle host species and some alternate host species, the last of which may represent a vectoring species. The second group is transmission by indirect contact (also known as indirect transfer). These routes have been described in detail by Hurst and Murphy (1996) and are represented in [Figures 10 and 13 of chapter 1] of this book. As also explained in chapter 1 of this book, there are some routes of viral transmission which are termed as being unnatural vehicular routes; such routes represent the use of unnatural vehicles as a means to evade the host defenses associated with natural portals of entry. These unnatural routes involve invasive medical devices (such as syringes, endoscopes, and other surgical instruments) and transplanted tissues including transfused blood and blood products. The rest of this section describes the natural routes of viral transmission between hosts.

13.3.2.1 Direct Contact The direct contact approach offers to the virus one major advantage and also one major drawback. Those viruses transmitted by direct contact have an advantage in that they need not have evolved stability when exposed to the ambiental environments. The drawback which these viruses confront is that the number of new hosts to which they have potential access may be more limited than is the case for viruses transmitted by indirect contact. Viruses which are endogenous by their nature will survive for as long as the host survives. Although the endogenous viruses can only be transmitted to the host's progeny, these viruses neither have to adapt themselves to nor coevolve with any other hosting species. An example of this type of endogenous agent would be the endogenous retroviruses of the family Retroviridae. Those viruses which are venereal in nature, i.e. transmitted in semen and vaginal secretions during sexual activities, have a somewhat greater potential for contacting new hosts. Represented among the venereal viruses of humans are some species of the genera *Simplexvirus* (Herpesviridae family), *Deltaretrovirus* and *Lentivirus* (family Retroviridae), and papillomaviruses (Papillomaviridae family). Once a venereal virus infects a host, that virus and the hosting individual remain associated with one another for the rest of the host's lifetime in the form of a permanent infection. Thus, although the frequency with which the endogenous and venereal viruses can find a new host is limited, the viruses compensate for this to some degree by remaining with the host for a very long time. The next step upward on the scale of host access would be represented by those viruses transmitted via direct contact with insect vectors. These viruses have a greatly increased access to new hosts, and tend not to remain with their present host for the rest of the host's life. Those viruses which are transmitted by biting insects are commonly referred to as being arboviruses, a term which is an abbreviation of "arthropod-borne viruses". Included among those arboviruses which infect humans are members of the genera *Alphavirus* (family Togaviridae), *Orthobunyavirus*, *Nairovirus* and *Phlebovirus* (all three of these genera belong to the family Bunyaviridae), and *Flavivirus* (family Flaviviridae).

Viruses transmitted by way of saliva may be perceived as bridging the categories of direct contact and indirect contact. If any particular type of virus that is secreted into saliva either has no stability when exposed to the ambiental environments in oral secretions,or else has only a limited stability under those conditions, then that virus will have to be transmitted by saliva which is transferred during oral contact between hosting species. Conversely, if that particular virus type has a good stability when exposed to the ambiental environments in oral secretions, then that virus can be transmitted on shared food or in association with fomites. Some of the viruses transmitted in saliva do remain associated with the host as a permanent infection, and these often are the viruses which possess limited stability in the ambiental environments, such as the members of the family Herpesviridae. Many of those viruses which are secreted into saliva and can be transferred to a new host in association with fomites do not remain associated with the host as a permanent infection, such as members of the family Picornaviridae. It can likewise be noted that, in general, those viruses which are transmitted by indirect contact between hosting individuals tend to produce only transient infections of their individual hosts rather than to remain associated with the individual host as a permanent infection. These several latter points bring forth the suggestion that there may be some evolutionary relationship between either the ease of viral transmission to a new potential host individual or the frequency of opportunities for viral transmission, and the length of time that the virus must be capable of remaining with its present host in order to have a reasonable chance of eventually achieving transmission.

13.3.2.2 Indirect Contact (Vehicle Borne) The indirect contact approach likewise offers to the virus one major advantage and also one major drawback. Those viruses transmitted by indirect contact have an advantage in that they have potential access to a far greater number ot hosting individuals than is the case for viruses transmitted by direct contact. The drawback which these viruses confront is that they must have evolved stability when exposed to the ambiental environment. The vehicles which viruses may utilize for achieving transmission between hosting individuals by indirect contact are divided into the following four categories: foods, water, air (in actuality this is a reference to aerosols), and fomites. Transmission by any one of these four categories of vehicles usually will be associated with some specific physical activity on the part of the present host, and will always be associated with some physical activity on the part of the proximate (next) host.

Foods are, of course, items intentionally ingested for their caloric or nutritional value. Contamination of foods can occur by way of the food being a virally infected animal that is being consumed by the proximate host. A presumed example includes the human prion which caused the disease Kuru among the Fore people of Papua, New Guinea, and this also is know to be the route of transmission for the prion which causes both Bovine spongiform encephalopathy in cattle (BSE, Mad Cow disease) and the disease termed either "Variant" or "New variant" Creutzfeldt-Jakob disease in humans. In these cases, there has been no specific physical activity on the part of the present host (the one which is being eaten) which can be identified as having caused the proximate host to be ingesting contaminated food (indeed, perhaps it is a lack of physical activity on the part of the present host which is to blame!). Otherwise, viral contamination of foods can result from fecal material being transferred via contact with unwashed hands and via contaminated aerosols falling into the food. A particularly notable example of a virus of humans which is transmitted via foods is the hepatitis A virus of the genus *Hepatovirus* (family Picornaviridae).

Water usually serves as a vehicle after it has been contaminated with fecal material. The acquisition of a viral infection from water usually results from the proximate host ingesting contaminated water. Physical contact of the proximate hosts' skin with contaminated water, as may occur both during recreational activities and washing of the body, can also result in the acquisition of infection. Notable examples of viruses transmitted by these waterborne routes are those belonging to the viral families Astroviridae and Caliciviridae, and numerous members of the viral family Picornaviridae, most specifically the Human enteroviruses of the genus *Enterovirus* and the Hepatitis A virus of the genus *Hepatovirus*.

Viral contamination of air can occur by two principle mechanisms. The first, and most significant, of these mechanisms involves the release of aerosols that contain droplets of respiratory secretions (i.e., nasal, oral or pulmonary mucus). This type of transmission route is referred to as being the route of droplet aerosols. Notable examples of viruses transmitted by this route are those belonging to the viral families Coronaviridae, Orthomyxoviridae, Paramyxoviridae, and the Human rhinoviruses belonging to the genus *Enterovirus* of the family Picornaviridae. The second mechanism is that of particulate aerosols. This mechanism involves the generation of aerosols composed of soil particles coated with dried urine or dried feces. Notable examples of viruses transmitted by this route belong either to the genera *Arenavirus* (family Arenaviridae) or *Hantavirus* (family Bunyaviridae).

Viral contamination of fomites (defined as solid environmental surfaces which can serve in the transmission of infections) can occur in many ways. The variety of things which represent fomites include: items used for warmth (includes blankets and clothing), items used for eating (included are cups, dinner plates and utensils), changing tables used in diapering infants, doorknobs, medical devices, toilet seats, and toys. The ways by which these environmental items become contaminated include the projection of droplet aerosols onto environmental objects during either sneezing or coughing, the falling of aerosols onto objects, and the unintended contamination of surfaces (including childrens clothing, blankets and toys) with blood, feces, fluid from skin lesions (rashes), nasal secretions, saliva or urine. The task of achieving viral transmission via this route occurs when these objects subsequently are handled or used by a potential proximate host. Examples of viral genera whose members can be transmitted via fomites include *Orthopoxvirus* (family Poxviridae) and *Enterovirus* (family Picornaviridae).

13.4 SUMMARY OF VIRAL FAMILIES THAT AFFLICT HUMANS

Twenty five of the viral families contain members which are capable of infecting humans. Together, these cause a broad range of illnesses in humans. The terminology used in describing these illnesses is presented in Table 13.1. The rest of this section summarizes the ecology of those twenty five viral families and there are obvious differences in the extent to which these groups have received scientific study. Figure 13.1 helps to describe the manner in which the different aspects of a viral infection fit together. The reference sources used for compiling this information were: Hurst and Murphy, 1996; and the ICTV Master Species List of November 2009 (2009_5F00_v3) which is available on the website of the International Committee on Taxonomy of Viruses http://www.ictvdb.org/.

13.4.1 Viral Family Adenoviridae

Genus Affecting Humans *Mastadenovirus* (the 7 species infective for humans are designated alphabetically as Human adenovirus A through Human adenovirus G).

Familial Nature with Respect to Members Affecting Humans Viruses of humans.

Alternate Hosts Species affecting humans seem naturally limited to humans.

Types of Illnesses Induced in Humans Adenopathy (represents the origin of the name

TABLE 13.1 Terminology of Human Illnesses Induced by Viruses and Prions

Term	Definition
Acquired immunodeficiency syndrome	Syndrome of immunodepletion resulting in the establishment of frequent and chronic infections caused by opportunistic pathogens
Adenopathy	Physiological changes associated with the degeneration of adenoid tissues
Anemia	Low iron level in the blood
Arthralgia	Pain in skeletal joints
Arthritis	Syndrome associated with pain and swelling in the skeletal joints
Auditory	Relating to hearing
Biliary atresia	Blockage of the ducts which carry bile from the liver to the gall bladder
Biphasic behavioral disease	Episodic variations in excitability including hyperactivity, movement and postural disorders; also termed either Bipolar Disorder or Manic-Depressive Disorder
Broncheolitis (bronchitis)	Syndrome associated with swelling or the related dysfunction of bronchiole tissues
Carcinoma	Tumor of epithelial tissues
Cardiological	Relating to the heart
Carditis	Syndrome associated with swelling or the related dysfunction of heart tissues
Cellular displasia	Changes or abnormal development in either cells or tissues resulting in their having a modified appearance and sometimes indicative of a precancerous condition
Conjunctivitis	Swelling and reddening of the conjunctiva (the mucosal membranes which line the eyelids)
Coryza	Watery discharge from eyes and nose
Demyelination	Destruction of the myelin tissue which surrounds neurons
Diabetes	Syndrome associated with underproduction of insulin
Diabetic	A person who has diabetes, also can refer to some health characteristic related to diabetes
Diarrhea	A condition of producing at least three bowel movements per day which could be described as being either loose or liquid
Dysuria	Painful urination
Encephalitis	Syndrome associated with swelling or the related dysfunction of brain tissues, generally diffuse
Encephalomyelitis	Syndrome associated with swelling or the related dysfunction of brain and spinal cord tissues
Encephalomyocarditis	Combined syndrome of encephalitis and myocarditis
Encephalopathy	Physiological changes associated with the degeneration of brain tissues
Encephalopathy, demyelinating	Encephalopathy associated with the myelin tissue which surrounds nerve cells
Encephalopathy, spongiform	A condition in which tiny holes appear in the brain tissue representing areas where the neurons have died (prions cause a transmissible spongiform encephalopathy)
Edema	Swelling
Encephalitic	Relating to the brain
Enteritis	Syndrome associated with swelling or the related dysfunction of intestine tissues, frequently evidenced as diarrhea
Erythemia	Abnormal reddening of the skin
Exanthem	Focalized reddening of the skin (rash)
Facial	Relating to the face

(*continued*)

TABLE 13.1 (*Continued*)

Term	Definition
Fetal	Relating to the fetus
Fetal developmental abnormalities	Defects occurring during development of the fetus
Fetal loss	Spontaneous abortion
Fever	Abnormally elevated body temperature
Focal neurological deficits	Focal neurological signs or "focal CNS signs", perceptual or behavioral impairments caused by localized lesions in particular areas of the central nervous system, interpreted to mean that a disease process is focal as opposed to diffuse; diffuse neurological disease processes include encephalitis and meningitis
Gastritis	Syndrome associated with swelling or the related dysfunction of stomach tissues, frequently evidenced as vomiting
Gastroentestinal	Refers to the stomach and intestines, especially in association with gastroenteritis
Gastroenteritis	Combined syndrome of gastritis and enteritis, frequently evidenced as vomiting and diarrhea
Hematemesis	Vomiting of blood
Hematuria	Blood in the urine
Hemolytic uremia	Syndrome consisting of hemolytic anemia (anemia due to blood cell lysis), reduced level of thrombocytes and acute degeneration of kidney tissues
Hemorrhage	Abnormal presence of bleeding
Hemorrhagic conjunctivitis	Bleeding of the conjunctival tissues
Hemorrhagic cystitis	Diffuse inflammation of the bladder leading to dysuria, hematuria and hemorrhage
Hemorrhagic fever	Syndrome consisting of massive external, internal, and transdermal hemorrhage in combination with high fever
Hepatic	Relating to the liver
Hepatitis	Syndrome associated with swelling or the related dysfunction of liver tissues
Hepatomegaly	Enlargement of the liver
Herpangina	Lesions within the mouth which can include ulcers, generally accompanied by throat pain and fever
Immunodepletion	Reduced level of circulating immune cells
Immunosuppression	Suppressed functioning of the immune system
Intraparenteral	Delivered by injection through the skin
Intrapartum	During birth, refers to labor and delivery
Keratoconjunctivitis	Combined syndrome of swelling or the related dysfunction of both the cornea and conjunctiva tissues
Leukemia	Cancerous syndrome associated with an extremely high level of circulating white cells in the blood
Leukoencephalopathy	Encephalopathy affecting the white matter of the brain, which consists mostly of myelinated axons, this process may take the form of a progressive multifocal leukoencephalopathy (a progressive focalized demyelinating encephalopathy)
Lymphadenopathy	Broadly defined as disease of the lymph nodes, but almost synonymously used in reference to swelling and enlargement of the lymph nodes
Lymphadenitis	Syndrome characterized by inflammation of the lymph nodes
Lymphangitis	Syndrome characterized by inflammation of lymph channels

TABLE 13.1 *(Continued)*

Term	Definition
Lymphoma	Solid tumor of the immune system
Malaise	Syndrome characterized by an extremely low level of motivational energy
Malignancy	Spreading cancer
Melena	Bleeding into the lumen of the intestines evidenced by the voiding of tar-like fecal material
Meningitis	Syndrome associated with swelling or the related dysfunction of meninges tissues (membranes that enclose the brain and spinal cord), generally diffuse
Meningoencephalitis	Combined syndrome of encephalitis with meningitis
Mesothelioma	Cancer that develops from the mesothelial tissue which covers many of the body's internal organs
Mucosa	Tissues which secrete mucus
Myalgia	Tenderness or pain in the muscles
Myelopathy	Physiological changes associated with degeneration of the spinal cord
Myocarditis	Syndrome associated with swelling or the related dysfunction of heart muscle tissues
Myositis	Syndrome associated with swelling or the related dysfunction of muscle tissues
Nasopharyngitis	Combined syndrome of swelling or the related dysfunction of both nasal passage and pharynx tissues
Necrosis	Death of tissue cells
Necrotic lesion	Focal area of tissue cell death
Nephritis	Swelling or the related dysfunction of the kidneys
Nephropathy	Physiological changes associated with degeneration of the kidneys
Nerve deafness	Loss of hearing resulting from reduced functioning of nerve cells
Neuralgia	Severe sharp pain along the course of a nerve
Neurodegeneration	A general term describing the progressive loss of either structure or function of neurons including the death of neurons
Neuronal	Relating to the nerve cells
Neuronal degeneration	Degeneration of the nerve cells
Nodule	A small knot-like protuberance or swelling of tissue
Oncolytic	Capable of destructively attacking tumor cells
Orchitis	Syndrome associated with swelling or the related dysfunction of testicle tissues
Otitis media	Syndrome associated with swelling or the related dysfunction of middle ear tissues
Paralysis	Loss of mobility
Paraparesis	Partial paralysis of the lower limbs
Paresthesia	A sense of either tingling, prickling or numbness which, in the case of viral infections, indicates affects upon the peripheral neurons.
Parturitional	Refers to events associated with the act or process of giving birth to offspring, may refer to the acquisition of infections during the birth process
Pericarditis	Syndrome associated with swelling or the related dysfunction of pericardium tissues (fibrous membrane sack which surrounds the heart)
Perinatal	The time period 'around' birth, variably defined but especially refers to the period from 5 months before to 1 month following birth

(continued)

TABLE 13.1 (*Continued*)

Term	Definition
Pharyngitis	Syndrome associated with swelling or the related dysfunction of the pharynx tissues
Pharyngoconjunctival fever	Combined syndrome of conjunctivitis and pharyngitis with fever
Pleurodynia	Abrupt occurences of intense pain in either the chest or abdominal muscles
Pneumonia	Syndrome associated with swelling or the related dysfunction of lung tissues (by definition, this term indicates that the swelling was induced by an infection)
Pneumonia, hemorrhagic	Pneumonia accompanied by bleeding into the lungs
Pneumonitis	Syndrome associated with swelling or the related dysfunction of lung tissues (by definition, this term indicates that the swelling was induced by an unknown irritation)
Rash, hemorrhagic (petechial)	Bleeding within the skin evidenced by purple spots termed petechiae
Rash, macular	Discolored spots of various sizes and shapes on the skin (or on mucosa) that neither are elevated nor depressed
Rash, maculopapular	The presence of both macules and papules on the skin or mucosa
Rash, papular	Small, red circular elevated solid areas on the skin (or on mucosa) that may progess to become either vesicles (filled with clear liquid) or pustules (filled with pus), or may first fill with clear liquid and then with pus
Renal dysfunction	Dysfunction of the kidneys
Retinitis	Syndrome associated with swelling or the related dysfunction of the retina
Retro-ocular pain	Pain centered behind the eyeball
Salivary glands	Glands which produce saliva
Sarcoma	Cancer arising from connective tissue such as bone, and some soft tissues including muscles
Sinusitis	Syndrome associated with swelling or the related dysfunction of nasal sinus tissues
Splenomegaly	Enlargement of the spleen
Tonsilitis	Syndrome associated with swelling of the tonsils
Tracheobronchitis	Syndrome associated with swelling or the related dysfunction of both trachea and bronchiole tissues
Transdermal	Action which crosses through the barrier presented by the skin
Trimester	One third of a time period, usually refers to either the first, second or third three-month segment of the nine-month human fetal gestation period
Tumor	Solid abnormal growth of cells
Uremia	Renal failure (failure of the kidneys) resulting in urea and other waste products that normally would be excreted into the urine instead being retained in the blood
Viremia	Presence of infectious viral particles in the blood
Visual	Relating to vision

of this viral family), conjunctivitis, coryza, encephalitis, gastroenteritis, keratoconjunctivitis, pharyngitis, pharyngoconjunctival fever, pneumonia.

Familial Strategies

Infection course: is productive, variously presenting as either short term-initial or recurrent.

Viral replication: at the individual host level, the primary tissue and organ tropisms are towards the cervix, conjunctiva, pharynx, small intestine, and urethra the secondary tissue and organ tropisms are towards the brain, kidney, lungs, and lymph nodes; at the host population level, these viruses generally are endemic and initially acquired at a very early age with the infections very often asymptomatic in young children.

Evasion of host defenses: uncertain.

Predominant Routes of Transmission Between Hosts Direct contact via host to host plus indirect contact (vehicle borne) via fecally contaminated water, food, fomites, and fomites contaminated with respiratory secretions.

13.4.2 Viral Family Anelloviridae

Genus Affecting Humans *Alphatorquevirus* (the 29 species infective for humans are designated numerically as Torque teno virus 1 through Torque teno virus 29).

Familial Nature with Respect to Members Affecting Humans Viruses of humans.

Alternate Hosts Those species infecting humans presumably are naturally limited to humans.

Types of Illnesses Induced in Humans These viruses are very broadly infective for humans, with prevalence rates ranging from approximately 12 to 95 percent as a viremia even among healthy populations of individuals; there may be some increased level of viral presence in those persons who are immunosuppressed, including those who either are renal transplant recipients or have hepatic illness, and there have been associations made between this virus and acute respiratory diseases in children. Related viruses of livestock have been associated with generalized wasting disease in their corresponding host animals.

Familial Strategies

Infection course: is productive, persistent but inapparent.

Viral replication: at the individual host level, these viruses are produced in healthly individuals worldwide with replicating viruses found in the liver, bone marrow, lung tissues, lymph nodes, spleen and pancreas.

Evasion of host defenses: uncertain; presence of this virus in respiratory patients has been observed to correspond with depressed total T-cell and helper T-cell populations while not corresponding with B-cell populations, although it is thus far unclear whether the virus produces immune supression or if suppression of the immune system caused by some other factor then enables the virus to demonstrate pathogenicity.

Predominant Routes of Transmission Between Hosts By bodily secretions (blood transfusion and saliva) and possibly also fecal oral.

13.4.3 Viral Family Arenaviridae

Genus Affecting Humans *Arenavirus* (the species infective for humans are Chapare virus, Flexal virus, Guanarito virus, Junín virus, Lassa virus, Lymphocytic choriomeningitis virus, Machupo virus, Pichindé virus (Trapido and Sanmartín, 1971) which infects humans as determined serologically and possibly is a human pathogen, and Sabiá virus).

Familial Nature with Respect to Members Affecting Humans Generally presumed zoonotic.

Natural Hosts Rodents including commensal voles and mice, and also commercial colonies of hamsters and nude mice.

Types of Illnesses Induced in Humans Arthralgia, carditis, encephalomyelitis, encephalopathy, facial edema, fetal loss, focal necrosis of liver, gastritis, hemorrhagic fever (primarily Lassa virus), hepatitis, inhibition of platelet functioning (causes the fatal bleeding associated with this virus family), malaise, meningitis, myalgia, nerve deafness, pneumonia.

Familial Strategies

Infection course: is productive, short term-initial.

Viral replication: at the individual host level, the primary tissue and organ tropisms presumably are towards the liver and lungs; the secondary tissue and organ tropisms are towards the brain, fetus, heart, joints, and nerves; at the host population level, these viruses can be extremely devastating to individual hosts but are poorly transferred between humans, which usually represent dead-end hosts.

Evasion of host defenses: differentially interfere with type 1 interferon.

Predominant Routes of Transmission Between Hosts Generally presumed to be indirect contact (vehicle borne) via inhalation of particulate aerosols bearing dried rodent urine or acquisition of infectious materials through skin abrasion (a form of surface contact), although the discovery of Pichindé virus in mites and ticks suggests that direct contact via invertebrate vectors may also play a role in the transmission of arenaviruses (Trapido and Sanmartín, 1971).

13.4.4 Viral Family Astroviridae

Genus Affecting Humans *Mamastrovirus* (the species infective for humans is Human astrovirus).

Familial Nature with Respect to Members Affecting Humans Viruses of humans.

Alternate Hosts Species affecting humans seem naturally limited to humans.

Types of Illnesses Induced in Humans Enteritis, gastroenteritis.

Familial Strategies

Infection course: is productive, short term-initial.

Viral replication: at the individual host level, the primary tissue and organ tropisms are towards the small intestine the secondary tissue and organ tropisms presently are unknown; at the host population level, these viruses are endemic, principally causing a mild enteritis seen in young adults.

Evasion of host defenses: avoids host's non-immune defenses by resistance to proteolytic attack (their infectivity is actually increased by proteolytic attack).

Predominant Routes of Transmission Between Hosts Indirect contact (vehicle borne) via fecally contaminated water, food, and fomites.

13.4.5 Viral Family Bornaviridae

Genus Affecting Humans *Bornavirus* (the species infective for humans is Borna disease virus).

Familial Nature with Respect to Members Affecting Humans Zoonotic.

Alternate Hosts This viral species is naturally infective for a very broad range of avians and mammals, notably including wild and domestic cats, dogs, horses, sheep and cattle.

Types of Illnesses Induced in Humans Generally asymptomatic but possibly associated with neurobehavioral symptoms which mimic biphasic behavioral disease (bipolar disorder). The clinical symptoms result from the immune response of the host. Relevant knowledge from experimentally infected animals shows that the onset on symptoms coincides with the onset of inflammation in the brain which can cause neuronal degeneration. The result of natural infections of animals can be a noncytolytic progressive encephalomyelitis and those brain structures which show the most severe inflammation are the caudate nucleus, dentate gyrus and hippocampus.

Familial Strategies

Infection course: in humans, generally presumed to be persistent but inapparent; in animals such as horses and sheep the infection often increases to end stage.

Viral replication: at the cellular level, the virus establishes a persistent infection in the host cell nucleus and can at least partially integrate its genome into that of the host cell; initial tropism is towards the neurons located at the site of viral entry into the host's body, after which the virus moves intraaxonally towards the central nervous system and enters the brain where it shows a preferential tropism for the limbic system including the hippocampus, additionally infecting astrocytes, oligodendrocytes and Schwann along its path; the virus subsequently moves centrifugally, presumably intraaxonally, to the peripheral nerves. If virus delivery via the peripheral nerves is sustained, then eventually non-neural organs and tissues can become infected.

Evasion of host defenses: the possibility of evasion techniques remains uncertain, the clinical symptoms result from immunological response against infected cells.

Predominant Routes of Transmission Between Hosts Natural infection presumably occurs intranasally by contact with infectious bodily excretions and secretions. The possibility of blood-borne transmission is also of consideration.

13.4.6 Viral Family Bunyaviridae

Genera Affecting Humans *Hantavirus* (the species infective for humans are Andes virus, Bayou virus, Black Creek Canal virus, Caño Delgadito virus, Dobrava–Belgrade virus, Hantaan virus, Laguna Negra Virus, New York virus, Puumala virus, Saaremaa virus, Seoul virus, Sin Nombre virus, and Tula virus; *Nairovirus* (the species infective for humans are Crimean-Congo hemorrhagic fever virus, Dera Ghazi Khan virus, Dugbe virus, Hughes virus, Qalyub virus, Sakhalin virus and Thiafora virus); *Orthobunyavirus* (examples of species infective for humans include Garissa virus, La Crosse virus, Oropouche virus and Tahyna virus); and *Phlebovirus* (examples of species infective for humans are Rift Valley fever virus and Sandfly fever Naples virus).

Familial Nature with Respect to Members Affecting Humans Zoonotic.

Natural Hosts Largely rodents (especially *Hantavirus*), but also hares and rabbits, and some ungulates.

Types of Illnesses Induced in Humans Arthralgia, encephalitis, hematemesis, hematuria, hemolytic uremia, hemorrhagic fever, hemorrhagic pneumonia, hemorrhagic (petechial) skin rash, hepatitis, melena, myalgia, pneumonia, renal dysfunction, retinitis, retro-ocular pain. Members of the genus *Hantavirus* are causative agents for two different characterized zoonotic syndromes: "Hemorrhagic Fever with Renal Syndrome" (HFRS; associated with the viruses named Caño Delgadito, Dobrava–Belgrade, Hantaan, Puumala, Saaremaa, and Seoul) and "Hantavirus Pulmonary Syndrome" which also sometimes is called "Hantavirus Cardiopulmonary Syndrome" (HPS is associated with the viruses named Andes, Bayou, Black Creek, Canal, Laguna Negra, and New York; HCPS is associated specifically with the Sin Nombre virus). The Hantaviral species named Tula virus is associated with fever and rash. Members of the genus *Nairovirus* are asociated with the syndrome called "tick borne fever". The *Phlebovirus* species named Sandfly fever Naples virus causes the syndrome "Pappataci fever" formerly identified as "Sandfly fever".

Familial Strategies

Infection course: is productive, short term-initial.

Viral replication: at the individual host level, the primary tissue and organ tropisms are towards the kidneys, liver, and lungs; the secondary tissue and organ tropisms are towards the brain and eyes; at the host population level, these viruses are not well sustained within human populations and humans usually represent dead-end hosts.

Evasion of host defenses: uncertain, but may include avoiding host's immume defenses by infecting immune cells.

Predominant Routes of Transmission Between Hosts Direct contact via host to vector by gnats, midges, and mosquitoes (genus *Orthobunyavirus*), sandflies (genus *Phlebovirus*), ticks (genus *Nairovirus*), and either indirect contact (vehicle borne) via particulate aerosols containing dried rodent urine or contact with rodent excreta or contaminated fomites (genus *Hantavirus*).

13.4.7 Viral Family Caliciviridae

Genera Affecting Humans *Norovirus* (the species infective for humans is Norwalk virus) and *Sapovirus* (the species infective for humans is Sapporo virus).

Familial Nature with Respect to Members Affecting Humans Viruses of humans and zoonotic.

Natural or Alternate Hosts Fish, terrestrial as well as marine mammals.

Types of Illnesses Induced in Humans Gastroenteritis, myalgia.

Familial Strategies

Infection course: is productive, short term-initial.

Viral replication: at the individual host level, the primary tissue and organ tropisms are towards the small intestine; at the host population level, subsequent transmission of all caliciviruses within human populations occurs readily by fecally contaminated waste and thus these viruses are very widespread.

Evasion of host defenses: avoids host's non-immune defenses by resistance to proteolytic attack.

Predominant Routes of Transmission Between Hosts Indirect contact (vehicle borne) via fecally contaminated water, food including infected animals, and fomites.

13.4.8 Viral Family Coronaviridae

Genera Affecting Humans *Alphacoronavirus* (the species infective for humans are Human coronavirus 229E and Human coronavirus NL63); *Betacoronavirus* (the species infective for humans is Human coronavirus HKU1); and *Torovirus* (the species infective for humans is Human torovirus).

Familial Nature with Respect to Members Affecting Humans Viruses of humans.

Alternate Hosts Some terrestrial ungulates and carnivores.

Types of Illnesses Induced in Humans Coryza, gastroenteritis.

Familial Strategies

Infection course: is productive, short term-initial.

Viral replication: at the individual host level, the primary tissue and organ tropisms are towards the intestines, lungs (possibly), nasopharynx, and sinuses; at the host population level, these viruses are very widespread and essentially non-fatal.

Evasion of host defenses: avoids host's immune defenses by viral mutation and recombination.

Predominant Routes of Transmission Between Hosts Indirect contact (vehicle borne) via fecally contaminated water, food, and fomites.

13.4.9 Viral Family Filoviridae

Genera Affecting Humans *Ebolavirus* (the species infective for humans are Cote d'Ivoire ebolavirus, Reston ebolavirus, Sudan ebolavirus, and Zaire ebolavirus) and *Marburgvirus* (the species infective for humans is Lake Victoria marburgvirus).

Familial Nature with Respect to Members Affecting Humans Generally zoonotic.

Natural Hosts Unknown but may include bats and rodents, with primates serving as intermediary hosts leading to human exposure. The Genus *Marburgvirus* has been shown to come from contact with infected African green monkeys (*Cercopithecus aethiops*). Ingestion of "Bushmeat", meat from wild animals which may contain viruses pertinent to those animal species, might represent the initiation point for human epidemics of *Ebolavirus*.

Types of Illnesses Induced in Humans Conjunctivitis, hemorrhagic fever (frequently fatal, with death possibly resulting from extreme inflammatory response), hepatic necrosis, myalgia, pharyngitis.

Familial Strategies

Infection course: is productive, short term initial.

Viral replication: at the individual host level, the primary tissue and organ tropisms are towards the immune cells and possibly also the liver; the secondary tissue and organ tropisms are towards the adrenal glands, kidneys, liver and spleen; at the host population level, these viruses are transferred between humans but seem unable to be sustained in human populations and humans usually represent dead-end hosts.

Evasion of host defenses: avoids host's immune defenses by infecting immune cells.

Predominant Routes of Transmission Between Hosts Transmission between humans occurs by direct contact via exposure to contaminated tissues and bodily fluids, especially blood.

13.4.10 Viral Family Flaviviridae

Genera Affecting Humans *Flavivirus* (examples of species infective for humans are Dengue virus, Kyasanur Forest disease virus, Murray Valley encephalitis virus, Omsk hemorrhagic fever virus, St. Louis encephalitis virus, Tick-borne encephalitis virus, West Nile virus, and Yellow fever virus) and *Hepacivirus* (the species infective for humans is Hepatitis C virus).

Familial Nature with Respect to Members Affecting Humans Includes viruses of humans (Hepatitis C virus of the genus *Hepacivirus*, and Dengue virus of the genus *Flavivirus*), a virus which clearly has become one of humans in addition to retaining its zoonotic existance (Yellow fever virus of the genus *Flavivirus*), and all others are considered zoonotic.

Natural or Alternate Hosts Variously either zoonotic, human, or both.

Types of Illnesses Induced in Humans Arthritis with rash, encephalitis, hemorrhagic fever, hepatitis (chronic, which may lead to hepatocellular carcinoma).

Familial Strategies

Infection course: is short term-initial for the genus *Flavivirus*, increasing to end-stage for the Hepatitis C virus.

Viral replication: at the individual host level, the primary tissue and organ tropisms are towards the immune cells (principally monocytes and macrophages);

the secondary tissue and organ tropisms are towards the brain, with the liver various percieved as being either a primary or secondary tropism; at the host population level, most of these viruses are zoonotic with humans representing dead-end hosts, however, some can be sustained within human populations and those occasionally involve high lethality rates.

Evasion of host defenses: avoids host's immune defenses by infecting immune cells.

Predominant Routes of Transmission Between Hosts direct contact via host to vector for the flaviviruses, direct contact via host to host transfer of contaminated bodily fluids for the Hepatitis C virus.

13.4.11 Viral Family Hepadnaviridae (and genus *Deltavirus*)

Genera Affecting Humans *Orthohepadnavirus* (the species infective for humans is Hepatitis B virus). The Hepatitis delta virus (HDV), which is a member of the floating genus *Deltavirus*, is a defective satellite virus that can only infect humans in association with the Hepatitis B virus (HBV) because HDV encapsidates itself with proteins encoded by the genome of the co-infecting HBV.

Familial Nature with Respect to Members Affecting Humans viruses of humans.

Alternate Hosts Only one species of the viral family Hepadnaviridae (Hepatitis B virus) is known to infect humans and it seems naturally limited to humans.

Type of Illness Induced in Humans hepatitis which may become chronic in adults.

Familial Strategies

Infection course: is productive, variously presenting as either short term-initial or increasing to end-stage.

Viral replication: at the individual host level, the primary tissue and organ tropisms are towards the liver; the secondary tissue and organ tropisms are towards the bile duct epithelium, circulating immune cells, and pancreatic acinar cells; at the host population level, when acquired by adults and older children these viruses generally cause an acute but short-term illness that sometimes can be fulminant; when acquired by neonates or younger children the infection initially tends to be subclinical but can become chronic, and the tendency to be chronic can be racially associated (Chinese and possibly also Black African).

Evasion of host defenses: avoids host's immune defenses by infecting immune cells.

Predominant Routes of Transmission Between Hosts direct contact via host to host transfer of contaminated bodily fluids and perinatally from contaminated maternal blood.

13.4.12 Viral Family Hepeviridae

Genera Affecting Humans *Hepevirus* (the species infective for humans is Hepatitis E virus.

Familial Nature with Respect to Members Affecting Humans presumably zoonotic, but can sustain itself in a human population.

Alternate Hosts Cervines, wild and domesticated swine.

Types of Illnesses Induced in Humans Hepatitis, which is debilitating but usually self limiting and not progressive, albeit there is an extraordinarily high fatality rate (from 10 to greater than 25%) for women if this virus is contracted during the third trimester of pregnancy.

Familial Strategies

Infection course: is productive, usually short term-initial but can prove increasing to end-stage.

Viral replication: at the individual host level, primary tropisms presumably are towards the small intestine, lymph nodes and colon, with secondary tropism towards the liver; at the host population level, transmission occurs via ingestion of fecally contaminated material.

Evasion of host defenses: avoids host's non-immune defenses by resistance to proteolytic attack.

Predominant Routes of Transmission Between Hosts Ingestion of fecally contaminated food or water, ingestion of pork from virally infected wild animals, and ingestion of bivalve molluscs harvested from fecally contaminated water in geographic areas that are epidemic for this virus.

13.4.13 Viral Family Herpesviridae

Genera Affecting Humans *Cytomegalovirus* (the species infective for humans is Human herpesvirus 5); *Lymphocryptovirus* (the species infective for humans is Human herpesvirus 4); *Rhadinovirus* (the species infective for humans is Human herpesvirus 8); *Roseolovirus* (the species infective for humans are Human herpesvirus 6 and Human herpesvirus 7); *Simplexvirus* (the species infective for humans are Human herpesvirus 1 and Human herpesvirus 2); and *Varicellovirus* (the species infective for humans is Human herpesvirus 3).

Familial Nature with Respect to Members Affecting Humans Viruses of humans.

Alternate Hosts Species affecting humans seem naturally limited to humans, but may pass to primates.

Types of Illnesses Induced in Humans Carcinoma, carditis, chronic gastrointestinal infection, encephalitis, hepatomegaly, keratoconjunctivitis, lymphoma, myelitis, neuralgia, papular rash of skin and mucosa, paralysis, paresthesia, retinitis, splenomegaly.

Familial Strategies

Infection course: is productive, recurrent.

Viral replication: at the individual host level, the primary tissue and organ tropisms are viral specific and variously affect the genital and oral mucosa, pharynx, salivary glands, skin, and regional neurons; the secondary tissue and organ tropisms variously include the esophagus, eyes, kidneys, liver, lymph nodes, central nervous system including brain, and spleen; at the host population level, these viruses are ubiquitous, most tend to be acquired in childhood or early adulthood, and they seldom directly result in the death of the host with a general exception of encephalitis. The outcome of parturitional infection can be severe.

Evasion of host defenses: Avoids host's immune defenses by producing a protein which mimicks interleuken 10, downregulation of the major histocompatibility complex, inhibiting cytokine synthesis, infection of the immune cells, and by maintaining its long term replicative presence within the host cells as viral genomic nucleic acid (latent infection) while only sporadically producing infectious virions (lytic infection).

Predominant Routes of Transmission Between Hosts Generally by direct contact via host to host transfer of either fluid from viral-induced lesions of skin and mucosa or saliva contaminated by chronically infected salivary glands; infection can occur from exposure to contaminated fomites (genus *Roseolovirus*); transmission to the offspring can occur either transplacentally, intrapartum (during the birth process), or via breast milk.

13.4.14 Viral Family Orthomyxoviridae

Genera Affecting Humans *Influenzavirus A* (the species infective for humans is Influenza A virus); *Influenzavirus B* (the species infective for humans is Influenza B virus); *Influenzavirus C* (the species infective for humans is Influenza C virus); and *Thogotovirus* (the species infective for humans are Dhori virus and Thogoto virus).

Familial Nature with Respect to Members Affecting Humans Generally viruses of humans except for the genus *Thogotovirus* whose member species are zoonotic, normally infective of terrestrial mammals, and transmited by ticks and mosquitos.

Alternate Hosts Swine and possibly birds (genera *Influenzavirus A*, *-B*, and *-C*); terrestrial mammals including mongoose and rodents (genus *Thogotovirus*).

Types of Illnesses Induced in Humans Coryza, encephalitis (genus *Thogotovirus*) fever, malaise, myalgia, nasopharyngitis, pneumonia, retro-ocular pain, tracheobronchitis.

Familial Strategies

Infection course: is productive, short term-initial.

Viral replication: at the individual host level, the primary tissue and organ tropisms are towards the ciliated columnar epithelium of the respiratory tract with the exact tissue tropism being directly related to the viral HA (hemagglutin) serotype; at the host population level, these viruses constantly undergo antigenic drift and antigenic shift and cause wide-scale seasonal epidemics in humans; infection related fatality usually is limited to humans of age 65 or older (most notably age 75 or older) and individuals with impairment of lung functioning.

Evasion of host defenses: avoids host's immune defenses by antigenic mimicry and by rapid viral mutation.

Predominant Routes of Transmission Between Hosts Transmission of viral species belonging to the genera *Influenzavirus A*, *-B* and *- C* occurs through indirect contact (vehicle borne) via either droplet aerosols from sneezing and coughing or exposure to aerosol-contaminated fomites, transmission of viral species belonging to the genus *Thogotovirus* occurs via tick bites.

13.4.15 Viral Family Papillomaviridae

Genera Affecting Humans *Alphapapillomavirus* (the species infective for humans are designated numerically as Human papillomavirus 2, 6, 7, 10, 16, 18, 26, 32, 34, 53, 54, 61, 71, and candidate 90); *Betapapillomavirus* (the species infective for humans are designated numerically as Human papillomavirus 5, 9, 49, candidates 92 and 96); *Gammapapillomavirus* (the species infective for humans are designated numerically as Human papillomavirus 4, 48, 50, 60, and 88); *Mupapillomavirus* (the species infective for humans are Human papillomavirus 1 and Human papillomavirus 63); and *Nupapapillomavirus* (the species infective for humans is Human papillomavirus 41). Why are there numbers that are not accounted for in this listing? Some viral isolates were assigned numbers but later determined not to be distinct species.

Familial Nature with Respect to Members Affecting Humans Viruses of humans.

Alternate Hosts Species affecting humans seem naturally limited to humans.

Types of Illnesses Induced in Humans These viruses generally are associated with benign tumors of the skin and mucosa, a result which can in rare instances become disfiguring, obstructive and debilitating; although papillomaviral infections seldom directly result in the death of the host, they can result in squamous cell displasias leading to carcinomas of many epithelial tissues including those of the esophagus, fingers, lung, and oropharynx. Most notably, however, carcinomas induced by papillomaviruses are associated with cancers of the anal and genital tissues including the penis, uterine cervix, vagina and vulva. Vaginal carcinomas tend to stay localized but may spread to the liver and lungs. The outcome of parturitional infection can be especially severe.

Familial Strategies

Infection course: is productive, recurrent.

Viral replication: at the individual host level, the primary tissue and organ tropisms are specific for each viral species and largely either the skin or mucosal tissues of the genital and oral regions, although the list of primary tropisms can encompass the esophagus, larynx, pharynx, salivary glands, sinuses, tonsils and vocal cords; the secondary tissue and organ tropisms include the eyes, kidneys, liver, lymph nodes, lungs, spleen and urethra; at the host population level, these viruses are ubiquitous and most papillomaviral infections tend to be acquired in childhood or early adulthood.

Evasion of host defenses: avoids host's immune defenses by antigenic mimicry, and by maintaining its long term replicative presence within the host cells as viral genomic nucleic acid (termed "plasmid replication"), with the production of infectious virions (termed "vegetative replication") occuring only in terminally differentiated cells of the epidermis.

Predominant Routes of Transmission Between Hosts Presumably direct contact via host to host or indirect contact (vehicle borne) via fomites.

13.4.16 Viral Family Paramyxoviridae

Genera Affecting Humans *Henipavirus* (the species infective for humans are Hendra virus and Nipah virus), *Metapneumovirus* (the species infective for humans is Human metapneumovirus), *Morbillivirus* (the species infective for humans is Measles virus), *Pneumovirus* (the species infective for humans is Human respiratory syncytial virus); *Respirovirus* (the species infective for humans are Human parainfluenza virus 1 and Human parainfluenza virus 3); and *Rubulavirus* (the species infective for humans include three designated numerically as Human parainfluenza virus 2, 4 and 5, plus Mumps virus and Simian virus 41).

Familial Nature with Respect to Members Affecting Humans Viruses of humans except for the henipaviruses and Simian virus 41.

Alternate Hosts Species affecting humans seem naturally limited to humans except for Simian virus 41, which presumably is acquired as a vaccine contaminant, and the henipaviruses which naturally are harbored by the fruit bats known as "flying foxes" and are capable of also infecting both horses and pigs.

Types of Illnesses Induced in Humans Bronchiolitis, conjunctivitis, coryza, encephalitis, glandular enlargement (especially salivary glands), immunosuppression, macular rash, nerve deafness, orchitis, pneumonitis, and respiratory fever. Measles virus causes an immunosuppression which is temporary but arguably the most severe that is induced by a virus of humans and this suppression can result in a far greater risk of death associated with other co-infecting pathogens, such as enteric protozoans, that normally would not cause human fatality.

Familial Strategies

Infection course: is productive, short term-initial.

Viral replication: at the individual host level, the primary tissue and organ tropisms are towards the epidermis and mucosa (including conjunctival, oral and respiratory); the secondary tissue and organ tropisms are towards the brain, breasts, circulating immune cells, and testicles; at the host population level, these viruses tend to be acquired at a young age and almost never are directly responsible for death of the host, although severe sequela can result if acquired beyond early childhood.

Evasion of host defenses: avoids host's immune defenses by infecting immune cells and interferon antagonism associated with degradation of the Stat-1 protein.

Predominant Routes of Transmission Between Hosts indirect contact (vehicle borne) via aerosols.

13.4.17 Viral Family Parvoviridae

Genera Affecting Humans *Dependovirus* (the species infective for humans are designated numerically as Adeno-associated virus 1, 2, 3, 4, and 5); *Erythrovirus* (the species infective for humans is Human parvovirus B-19); and possibly also *Bocavirus* (the species presumably infective for humans is unofficially designated Human bocavirus).

Familial Nature with Respect to Members Affecting Humans Dependoviruses seem capable of replication within all vertebrates and in that regard they are only limited by the host range of the virus with which they must coinfect, also known as the helper virus, and for humans that generally is an adenovirus; Human parvovirus B-19 seems exclusively a virus of humans; the bocaviruses are zoonotic.

Alternate Hosts The genera *Dependovirus* and *Bocavirus* broadly infect terrestrial vertebrates including chiropterans (bats), equines, and swine.

Types of Illnesses Induced in Humans The dependoviruses are not known to cause illness although the Human adenoviruses which serve as their helper viruses do cause illness; Human parvovirus B-19 causes aplastic anemia, arthralgia, erythemia, and myalgia; bocaviral infections have been associated with lower respiratory illness and diarrhea.

Familial Strategies

Infection course: is productive, short term-initial.

Viral replication: at the individual host level, the primary tissue and organ tropisms variously are towards the intestines, lungs and throat; secondary tissue and organ tropisms include the circulatory system, erythrocyte precursor cells in bone marrow, possibly reticulocytes in blood, and skin; at the host population level, these viruses usually cause a disease of early childhood and they are capable of replicating only in mitotically active cells.

Evasion of host defenses: these viruses seem to be poorly antigenic, dependoviruses can persist by inserting their genome into human chromosome 19, and Human parvovirus B-19 may cause a reduction in white blood cells.

Predominant Routes of Transmission Between Hosts Uncertain, but potentially direct contact via host to host including transplacental and also indirect contact (vehicle borne) via aerosols of either bodily excretions or secretions, as well as fecally contaminated water, food, and fomites.

13.4.18 Viral family Picobirnaviridae

Genera Affecting Humans *Picobirnavirus* (the species infective for humans is Human picobirnavirus).

Familial Nature with Respect to Members Affecting Humans The species Human picobirnavirus presumably is exclusive to humans, but cross infectivity with related viruses of other mammals such as swine is possible.

Alternate Hosts Swine

Types of Illnesses Induced in Humans This virus is fecally excreted by a small percentage (about 2 percent) of asymptomatic individuals who are not immunologically suppressed and in those people its presence seems to follow a pattern which is persistent but inapparent. This virus is fecally excreted by a higher percentage (approximately 9–12 percent) of immunosuppressed patients with episodic diarrhea, and those people continue to excrete the virus even between diarrheic episodes suggesting that in these hosts the viral infection follows a persistent -episodic pattern. The episodic association is strongest in the case of individuals who are in the late stage of infection by Human immunodeficiency virus 1, and for that reason this viral species often is referred to as being opportunistic when considered as a pathogen.

Familial Strategies

Infection course: is productive, either persistent - episodic or persistent but inapparent.

Viral replication: there is an obvious tropism towards the intestines, little more is known.

Evasion of host defenses: uncertain but obviously effective, its episodic disease presence in humans occurs in association with T-cell suppression.

Predominant Routes of Transmission Between Hosts Presumably acquired either by contact with feces from an infected individual, fecally contaminate fomites, or ingestion of fecally contaminated food and water.

13.4.19 Viral Family Picornaviridae

Genera Affecting Humans *Cardiovirus* (the species infective for humans are Encephalomyocarditis virus and Theilovirus); *Enterovirus* (the 7 species infective for humans are designated alphabetically as Human enterovirus A through Human enterovirus D and Human rhinovirus A through Human rhinovirus C); *Hepatovirus* (the species infective for humans is Hepatitis A virus); *Kobuvirus* (the species infective for humans is Aichi virus); *Parechovirus* (the species infective for humans is Human parechovirus); and *Senecavirus* (the species infective for humans is Seneca Valley virus).

Familial Nature with Respect to Members Affecting Humans Viruses of humans except for Encephalomyocarditis virus, which is zoonotic.

Alternate Hosts Species affecting humans seem naturally limited to humans, but may pass to primates and canines.

Types of Illnesses Induced in Humans Encephalitis, encephalomyelitis, encephalomyocarditis, and myocarditis (genus *Cardiovirus*); acute hemorrhagic conjunctivitis, diabetes, encephalitis, herpangina, macular and maculopapular rashes of skin and mucosa, meningitis, myalgia, myocarditis, otitis media, paralysis of skeletal muscles occasionally including the diaphragm, pericarditis, pleurodynia, retro-ocular pain, and sinusitis (human enteroviruses, genus *Enterovirus*); coryza (human rhinoviruses, genus *Enterovirus*); non-progressive hepatitis (genus *Hepatovirus*); gastroenteritis (genus *Kobuvirus*); encephalitis, gastroentestinal and respiratory illness, myocarditis (genus *Parechovirus*). Seneca Valley virus (genus *Senecavirus*) has thus far not been associated with illness and may in fact be oncolytic.

Familial Strategies

Infection course: is productive, short term-initial.

Viral replication at the individual host level, the primary tissue and organ tropisms initially are towards the submucosal lymphatic tissues at the initial sites of infection including the nasopharynx and small intestine, followed sequentially by regional lymph nodes, often leading to a minor transient viremia after which the infection spreads to other lymph nodes and organs; the secondary tissue and organ tropisms are highly genus and species specific and include the beta cells of the pancreas, conjunctiva, liver, meninges, muscles (including the heart), neurons (including those of the central nervous system), and skin; at the host population level, infections caused by the enterovirus members of the genus *Enterovirus* can be severe but usually are non-fatal and both *Enterovirus* and *Hepatovirus* tend to result in asymptomatic infections if acquired in infancy albeit the likelihood of severe symptomatology increases with the age at acquisition, infections caused by the rhinovirus members of the genus *Enterovirus* generally are symptomatic but essentially are non-fatal regardless of the age of the host.

Evasion of host defenses: members of genus *Enterovirus* avoid host's non-immune defenses by resistance to low pH (resistant to stomach acid) and to moderate alkalinity.

Predominant Routes of Transmission Between Hosts Indirect contact (vehicle borne) via aerosols and fecally contaminated water, food, and fomites

13.4.20 Viral Family Polyomaviridae

Genera Affecting Humans Polyomavirus (the species infective for humans are BK polyomavirus, Human polyomavirus, JC polyomavirus, and Simian virus 40).

Familial Nature with Respect to Members Affecting Humans These are considered viruses of humans except for Simian virus 40, which is of monkeys and not naturally zoonotic but was acquired by the human population as a contaminant from cultured cells used in the production of intraparenteral poliomyelitis vaccine ("Salk vaccine").

Alternate Hosts Species affecting humans seem naturally limited to humans except for Simian virus 40 (SV40) which naturally infects monkeys.

Types of Illnesses Induced in Humans These viruses are ubiquitous and almost never are directly responsible for death of the host unless their relationship with the host is compounded by the presence of a coinfecting agent, most notably by human immunodeficiency viruses. Poylomaviruses generally are seen as producing infections of the respiratory system (both BK virus and JC virus), urinary tract infections variously involving the kidneys (both BK virus and JC virus) and bladder (BK virus), and infections of the brain (JC virus). Infection of the kidneys may present as either nephritis or nephropathy. Infection of the bladder often presents as hemorrhagic cystitis, frequently ocurring in children. Infection of the brain tends to take the form of a progressive demyelinating encephalopathy named "Progressive multifocal leukoencephalopathy", often abbreviated "PML", which involves destruction of the oligodendrocytes as well as astrocytes. PML is caused by a reawakening of the JC polyomavirus, which normally is present quiescently as a latent infection within the body and kept under control by the immune system. Development of PML occurs almost exclusively in people who are immunosuppresed either as the result of concurrent infection by another microorganism such as the Human immunodeficiency viruses, are receiving immunosuppresssive medications, or whose level of immune function has declined naturally because of age. There has been some reporting of polyomaviruses associated with human cancers including mesothelioma, Merkel cell carcinoma of the skin, and lymphoma. The effect of Simian virus 40 in humans remains an open question, the virus naturally replicates in the kidneys of monkeys without causing symptoms but does cause sarcomas when injected into hamsters.

Familial Strategies

Infection course: is productive, becoming either recurrent (both BK virus and JC virus) or persistent but inapparent (BK virus); recurrence of JC virus can result in the infection pattern changing into one that represents increasing to end-stage.

Viral replication: at the individual host level, the primary tissue and organ tropisms presumably are towards the upper respiratory tract, tonsils, and gastroentestinal tract; tropisms towards two additional sites may be considered primary in nature, if transmission is waterborne, and those are the tubular epithelial cells of the kidneys (both BK virus as well as JC virus) and the bladder (BK virus) where the viruses may continuously reproduce and resultingly shed virions into the urine; JC virus demonstrates a definate secondary tropism towards the brain marked by the fact that JC viral DNA can be detected in both PML-affected as well as PML-unaffected tissues; the BK virus can remain quiescent in a latent condition within lymphocytes and the urogenital tract, the JC virus can remain quiescent in a latent condition within the gastroentestinal tract and lymphocytes, and both BK virus as well as JC virus have been suggested to establish quiescences in a latent condition within the brain; at the host population level, polyomaviruses generally are acquired either in childhood or early adulthood with the initial infections mostly considered asymptomatic, both the BK virus and JC virus are widespread with perhaps 70 - 90 percent of the adult human population being serologically positive for these two viral species.

Evasion of host defenses: avoids host's immune defenses by antigenic mimicry and infection of the immune cells.

Predominant Routes of Transmission Between Hosts Presumably transmission occurs via indirect contact (vehicle borne); aerosols may be involved because these viruses can be found in respiratory secretions; BK virus additionally is shed in the urine of pregnant women and JC virus is found in wastewater, suggesting the possibility that waterborne routes also may play a role in the transmission of human polyomaviruses.

13.4.21 Viral Family Poxviridae

Genera Affecting Humans *Molluscipoxvirus* (the species infective for humans is Molluscum contagiosum virus); *Orthopoxvirus* (the species infective for humans are Camelpox virus, Cowpox virus, Monkeypox virus, Vaccinia virus, and Variola virus; *Parapoxvirus* (the species infective for humans are Bovine papular stomatitis virus, Orf virus, and Pseudcowpox virus; and *Yatapoxvirus* (the species infective for humans is Tanapox virus).

Familial Nature with Respect to Members Affecting Humans Some are considered viruses of humans (Molluscum contagiosum virus, Vaccinia virus, Variola virus) and the others either are know or presumed to be zoonotic.

Alternate Hosts The species which causes the disease known as smallpox (Variola virus) seems naturally limited to humans; monkeypox presumably always is acquired from monkeys; several of the other poxviral species that affect humans may cycle between humans and domesticated mammals including bovines and ovines, commonly appearing as lesions on the teats and udder of those domesticated animals.

Types of Illnesses Induced in Humans Necrotic lesions of abdominal organs and skin, nodules and tumors in skin, papular rash.

Familial Strategies

Infection course: is productive, short term-initial.

Viral replication: at the individual host level, the primary tissue and organ tropisms are towards the skin at the initial site of infection; a transient viremia subsequently can lead to expression of secondary tissue and organ tropisms directed towards the internal organs and lymph nodes; at the host population level, these viruses generally have very low transmissibility.

Evasion of host defenses: avoids host's immune defenses by antigenic mimicry.

Predominant Routes of Transmission Between Hosts Direct contact via host to host contact with skin lesions and indirect contact (vehicle borne) via lesion-contaminated fomites (very notably blankets and other bedding items). These viruses demonstrate extreme resistance to dessication and thus can have prolonged survivability on fomites.

13.4.22 Viral Family Reoviridae

Genera Affecting Humans *Coltivirus* (the species infective for humans is Colorado tick fever virus); *Orbivirus* (the species infective for humans are Changuinola virus and Orungo virus; *Orthoreovirus* (the species infective for humans is Mammalian orthoreovirus); and *Rotavirus* (the 3 species primarily infective for humans are designated alphabetically as Rotavirus A through Rotavirus C).

Familial Nature with Respect to Members Affecting Humans Those species affecting humans variously are considered viruses of humans (genera *Rotavirus*), zoonotic (genera *Coltivirus* and *Orbivirus*), or broadly infective of birds and mammals (genus *Orthoreovirus*).

Natural or Alternate Hosts Colorado tick fever virus (genus *Coltivirus*) has terrestrial mammals, most notably chipmunks, ground squirrels and mice, as its natural hosts; Changuinola virus (from Panama and South America, genus *Orbivirus*) and Orungo virus (notably from Nigeria and Uganda, genus *Orbivirus*) presumably have either terrestrial mammals or birds as their natural hosts; Mammalian orthoreovirus (genus Orthoreovirus) broadly cross-infects endothermic vertebrates.

Types of Illnesses Induced in Humans *Coltivirus* infection causes Colorado tick fever (CTF) which generally is survivable but can result in death from hemorrhagic complications; *Orbivirus* infections of humans are associated with confusion, diarrhea, encephalitis, fever, hemorrhagic fever, meningoencephalitis, myalgia, nausea, seizures, vomiting, and in rare instances, focal neurological deficits; *Orthoreovirus* infections of humans generally are asyptomatic although they can indeed be fatal, and there are associative linkages between orthoreoviral infections and upper respiratory illnesses such as broncheolitis and pneumonia which can be accompanied by enteritis in the form of mild diarrhea, orthoreoviral infections also can cause encephalitis, fever, hepatitis, maculopapular rash, meningitis, myocarditis, pharyngitis, pneumonia, pneumonitis, and tonsilitis, plus there is speculation that orthoreoviral infections may be invloved in biliary atresia and type 1 diabetes, the latter resulting from induction of anti-beta cell autoimmunity; *Rotavirus* infection produces a gastroenteritis which usually is nonfatal, except for its troubling nature in infants of the developing countries where the infection often is inadequately treated with a result that rotaviruses alone have been estimated to account for perhaps 5 - 10 million deaths worldwide per year which would represent 10 - 20 percent of total gastroenteritis deaths in those regions.

Familial Strategies

Infection course: as productive, short term-initial.

Viral replication: at the individual host level, the primary tissue and organ tropisms are highly genus-specific; *Coltivirus* has a broad range of primary tropisms including transiently infecting the bone marrow, and also targeting the erythroblasts, heart, liver, lymph nodes, reticulocytes, and spleen, plus human erythrocytes are known to carry the virus where its presence is presumed to result from viral replication in hematopoietic erythrocyte precursor cells after which the virus remains in those cells during their maturation; *Orbivirus* has a number of tropisms all of which are presumed to be primary, and these are the brains, intestines, and either skeletal muscles or perhaps more likely the innervations of those muscles; *Orthoreovirus* has primary tropisms asociated with its acquisition and transmission as an upper respiratory infection and those include the facial sinuses, lungs, pharynx, and tonsils, presumably its tropism towards the intestines represents a secondary tropism (assuming that fecal-oral transmission does not also represent a primary route of disease spread, in which case the intestines would have to be considered a primary tropism), other secondary tropisms for this virus genus include the blood, brain, heart, intestines, liver, kidney, meninges, myocardium, skin, spleen, and possibly the adrenal glands; *Rotavirus* demonstrates tropism towards the site which it initially infects, which is the vili of the small intestine where the virus mutiplies in the cytoplasm of the enterocytes; at the host population level, these viruses can be generalized as having high transmissibility especially among newborns for whom they usually produce asymptomatic infections and, while they likewise have a tendency to produce asymptomatic infections in older children and adults, these viruses can be particularly fatal in those children who are undernourished.

Evasion of host defenses: avoids host's immune defenses by infecting immune cells (genus *Coltivirus*); avoids host's non-immune defenses by resistance to heat, low pH, and proteolytic attack with their infectivity actually being increased by proteolytic attack (members of the genera *Orthoreovirus* and *Rotavirus*.

Predominant Routes of Transmission Between Hosts Indirect contact transmission associated with vehicles occurs via aerosols for the genus *Orthoreovirus* and via fecally contaminated water, food, and fomites for members of the genus *Rotavirus*; direct contact transmission associated with vectors variously occurs via ticks (Colorado tick fever virus, genus *Coltivirus*), sandflies (Changuinola virus, genus *Orbivirus*), and mosquitos (Orungo virus, genus *Orbivirus*).

13.4.23 Viral Family Retroviridae

Genera Affecting Humans *Deltaretrovirus* (the species infective for humans are Primate T-lymphotrophic virus 1, which includes Human T-lymphotrophic virus 1, and Primate T-lymphotrophic virus 2, which includes Human T-lymphotrophic virus 2); and *Lentivirus* (the species infective for humans are Human immunodeficiency virus 1 and Human immunodeficiency virus 2).

Familial Nature with Respect to Members Affecting Humans Viruses of humans.

Alternate Hosts Species affecting humans seem naturally limited to humans.

Types of Illnesses Induced in Humans Carcinoma, encephalitis, leukemia (adult T-cell), lymphoma (adult T-cell), paraparesis, progressive chronic immunosuppression and immunodepletion (including acquired immunodeficiency syndrome), progressive myelopathy, sarcoma.

Familial Strategies

Infection course: is productive, short term-initial, often followed by increasing to end-stage, also may seem non-productive in the case of some endogenous retroviruses.

Viral replication: at the individual host level, the primary tissue and organ tropisms are towards the immune cells (largely the T-cell populations); the secondary tissue and organ tropisms are towards the brain and intestine; at the host population level, those viruses which are considered transmissible (i.e., excludes the endogenous retroviruses) have a very low transmissibility rate, produce infections whose incubation times are very long (10–40 years), and may pass through breast milk; the endogenous retroviruses are endosymbiotic and permanently integrated into the human genome, they are passed genetically to all offspring.

Evasion of host defenses: avoids host's immune defenses by rapid viral mutation, by infecting immune cells, and by maintaining its long term presence within the host cell as viral genomic nucleic acid. Endogenous retroviruses produce infectious virions only sporadically if ever.

Predominant Routes of Transmission Between Hosts Direct contact via host to host transfer of virally contaminated bodily fluids.

13.4.24 Viral Family Rhabdoviridae

Genera Affecting Humans *Lyssavirus* (the species infective for humans are Australian bat lyssavirus, Duvenhage virus, and Rabies virus); and *Vesiculovirus* (the species infective for humans are Chandipura virus, Isfahan virus, and Vesicular stomatitis Indiana virus.

Familial Nature with Respect to Members Affecting Humans Zoonotic.

Natural Hosts The natural hosts for those lyssaviruses which affect humans include members of the order Chiroptera (bats, including vampire bats), and three families of the suborder Caniformia, specifically Canidae (foxes, dogs, wolves), Mephitidae (skunks), and Procyonidae (raccoons), it seems reasonable to speculate that acquisition of the *Lyssavirus* Rabies virus by these carnivore hosts results from their curious investigation of ill bats; the natural hosts for those vesiculoviruses which affect humans are members of the familes Bovidae (cattle), Equidae (horses), and Muridae (gerbils).

Types of Illnesses Induced in Humans Lyssaviral infections produce neuronal symptoms, which sometimes begin as paresthesia at the site of a bite wound, eventually leading to an encephalitis which appears invariably fatal although there have been a very few instances of survival when patients received extensive medical treatment; vesiculoviral infections are mostly asymptomatic but can result in arthralgia, blisters of oral mucosa, encephalitis, hemorrhagic fever, and myalgia.

Familial Strategies

Infection course: is productive, short term-initial.

Viral replication: at the individual host level, for *Lyssavirus* the primary tissue and organ tropisms are towards the neurons, including those in the spinal cord and limbic system of the brain, as well as the salivary glands, with the known secondary tropisms being towards the adrenal cortex and pancreas; for *Vesiculovirus* the known primary tropism is towards the oral mucosa while other tropisms are neuronal including the brain, along with the joints and muscles (or perhaps more likely the corresponding innervations); at the host population level, these viruses are considered essentially non-transmissible between humans.

Evasion of host defenses: avoids host's immune defenses by limited antigenic exposure within the host because the virus largely remains within neuronal cells until near the end-stage (genus *Lyssavirus*).

Predominant Routes of Transmission Between Hosts For the lyssaviruses, transmission is by direct contact via host to host in association with the deposition of contaminated saliva into a bite wound, is possibly also associated with contamination of skin or mucosal wounds by other types of bodily fluids containing the virus, has in some rare instances resulted from implantation of organs harvested from rabies-infected donors, and in very rare instances infection has been acquired by indirect contact via inhalation of infectious aerosols in bat-colonized caves; for the vesiculoviruses, specifically the Vesicular stomatitis Indiana virus, transmission is known to occur by two means, the first being direct contact via host to host occuring during milking operations as the result of mucosal contact with contaminated saliva containing vesicular fluids, and the second being indirect contact associated with fomites in the form of milking equipment which had become contaminated with saliva containing vesicular fluids; it is presumed that transmission is by direct contact via phlebotomous vectors (sandflies) for the other known vesiculoviruses of humans.

13.4.25 Viral Family Togaviridae

Genera Affecting Humans *Alphavirus* (examples of the species infective for humans are Barmah Forest virus, Chikungunya virus, Mayaro virus, O'nyong-nyong virus, and Ross River virus; *Rubivirus* (the species infective for humans is Rubella virus).

Familial Nature with Respect to Members Affecting Humans Viruses of humans and zoonotic.

Natural or Alternate Hosts Species of the genus *Alphavirus* cross-infect a wide variety of terrestrial vertebrates, mostly via mosquitoes and ticks; one species of the genus *Rubivirus* affects humans and it seems restricted to humans.

Types of Illnesses Induced in Humans Alphaviral infections are associated with arthralgia, arthritis, fever, malaise and myositis; the rubivirus which affects humans causes the disease syndrome variously identified as either rubella, German measles, or Three-day measles, for which the symptoms include most notably a transient exanthem (hence the name "Three-day") typically accompanied by arthralgia, conjunctivitis, fever, and lymphadenopathy; a serious consequence of fetal Rubella virus infection termed "Congenital rubella syndrome" occurs if the viral infection is contracted during the first trimester of pregnancy and the results include diabetes, encephalitis, and fetal developmental abnormalities (cardiological, diabetic, and neurological including auditory, encephalitic, and visual).

Familial Strategies

Infection course: is productive, short term-initial.

Viral replication: at the individual host level, the primary tissue and organ tropisms are towards the immune cells (specifically monocytes and macrophages in bone marrow, liver, lymph nodes, and spleen) and oropharynx, the secondary tissue and organ tropisms are towards the beta cells of the pancreas, muscles, neurons of the central nervous system including the brain, skin, and synovial cells of joints; at the host population level, most members of the genus *Alphavirus* seem poorly transmitted between humans and indeed humans probably represent a dead-end host, infection by the genus *Rubivirus* is widespread as a childhood disease and seldom fatal.

Evasion of host defenses: avoids host's immune defenses by infecting immune cells.

Predominant Routes of Transmission Between Hosts Direct contact via either host to vector usually involving mosquitos (genus *Alphavirus*), or host to host (genus *Rubivirus*); indirect contact (vehicle borne) via aerosols (genus *Rubivirus*).

13.5 SUMMARY OF PRIONS THAT AFFLICT HUMANS

Family (Not classified into Linnean taxonomy).

Prions Affecting Humans The prions affecting humans are the $HuPrP^{Sc}$ isoforms of the protein PrP, and these are variously known as either CJD prion (Creutzfeldt–Jakob disease); vCJD prion (Variant Creutzfeldt-Jakob disease) also referred to as nvCJD (causing New Variant Creutzfeldt-Jakob disease); FFI prion (Fatal familial insomnia); GSS prion (Gerstmann-Sträussler-Scheinker syndrome); or Kuru prion (Kuru).

Familial Nature with Respect to Members Affecting Humans Agents of humans and zoonotic.

Natural or Alternate Hosts Humans, cattle.

Types of Illnesses Induced in Humans Spongiform encephalopathy and transmissible spongiform encephalopathy.

Familial Strategies

Infection course: is increasing to end-stage.

Replication: Replication is within the brain and lymphatic system, these all represent variations in a normal human protein whose conformation becomes modified to a form which then accumulates in areas of the brain resulting in neurodegeneration.

Evasion of host defenses: low antigenicity attributable to the fact that the protein of which prions are made (PrP) is naturally produced and found in its normal configuration (PrP^{c}) throughout the body in healthy people, albeit the protein found in infectious material has an abnormal conformational structure (PrP^{Sc}); avoids hosts non-immune defenses by resistance to protease and low pH.

Predominant Routes of Transmission Between Hosts Indirect contact by several routes including ingestion of tissues and organs contaminated with prions (this route of transmission is facilitated by resistance of prions to protease and low pH), medically associated transfer of infected bodily tissues including blood, use of medical instruments contaminated with prions, and the suspected possibility for accidental contamination of wounds by infective bodily material.

13.6 CONCLUSIONS

There are numerous types of viruses which afflict humans and there also are human prions. We have managed to coevolve with some of these to lessen our misery. The struggle against viruses will continue as new ones appear and as the existing ones change their antigenicity and genomic capabilities by mutation. Our struggle against the prions will perhaps be even more difficult. In the end, the contest is a struggle of biology versus biology, and the basic biology of these infectious agents is the same as ours.

ACKNOWLEDGEMENT

I am thankful to my friend and former colleague Noreen J. Adcock for her assistance with the preparation of a much older version of this chapter. My hopes and best wishes go out to her.

REFERENCES

Hurst, C. J. and Murphy, P. A. (1996). The transmission and prevention of infectious disease. In *"Modeling Disease Transmission and Its Prevention by Disinfection"* (C. J. Hurst, ed.), pp. 3–54. Cambridge University Press, Cambridge.

Trapido, H. and Sanmartín, C. (1971). Pichindé virus a new virus of the tacaribe group from Colombia. *Amer. J. Trop. Med. Hyg.* 20(4), 631–641.

Villareal, L. P. (1997). On viruses, sex and motherhood. *J. Virol.* 71, 859–865.

CHAPTER 14

ECOLOGY OF AVIAN VIRUSES

JOSANNE H. VERHAGEN and RON A.M. FOUCHIER
Department of Virology, National Influenza Center, Erasmus Medical Center, Rotterdam, The Netherlands

VINCENT J. MUNSTER
Laboratory of Virology, Rocky Mountain Laboratories, Division of Intramural Research, National Institute of Allergy and Infectious Diseases, National Institutes of Health, Hamilton, MT

CONTENTS

14.1 INTRODUCTION

The spread of the highly pathogenic avian influenza (HPAI) H5N1 virus from Asia to Europe and Africa and the spread of a virulent form of West Nile virus (WNV) across North America have revealed links between migratory birds and animal and human health. The transmission of these viruses and their geographical spread is dependent on the tight connection between the ecology of the migrating host and the ecology of the pathogen. The ecology of avian viruses within the host is determined by viral characteristics such as host cell receptor use, tissue tropism, replication efficiency, and the capacity to evade the host's immune system; and by host characteristics, such as species, diversity and distribution of virus-specific receptors, host cell transcription and translation machinery, and the capacity of the immune system to recognize and fight the viral infection. In addition, the ecology of avian viruses depends largely on the behavior of the host species, such as diet and foraging

Studies in Viral Ecology: Animal Host Systems: Volume 2, First Edition. Edited by Christon J. Hurst.

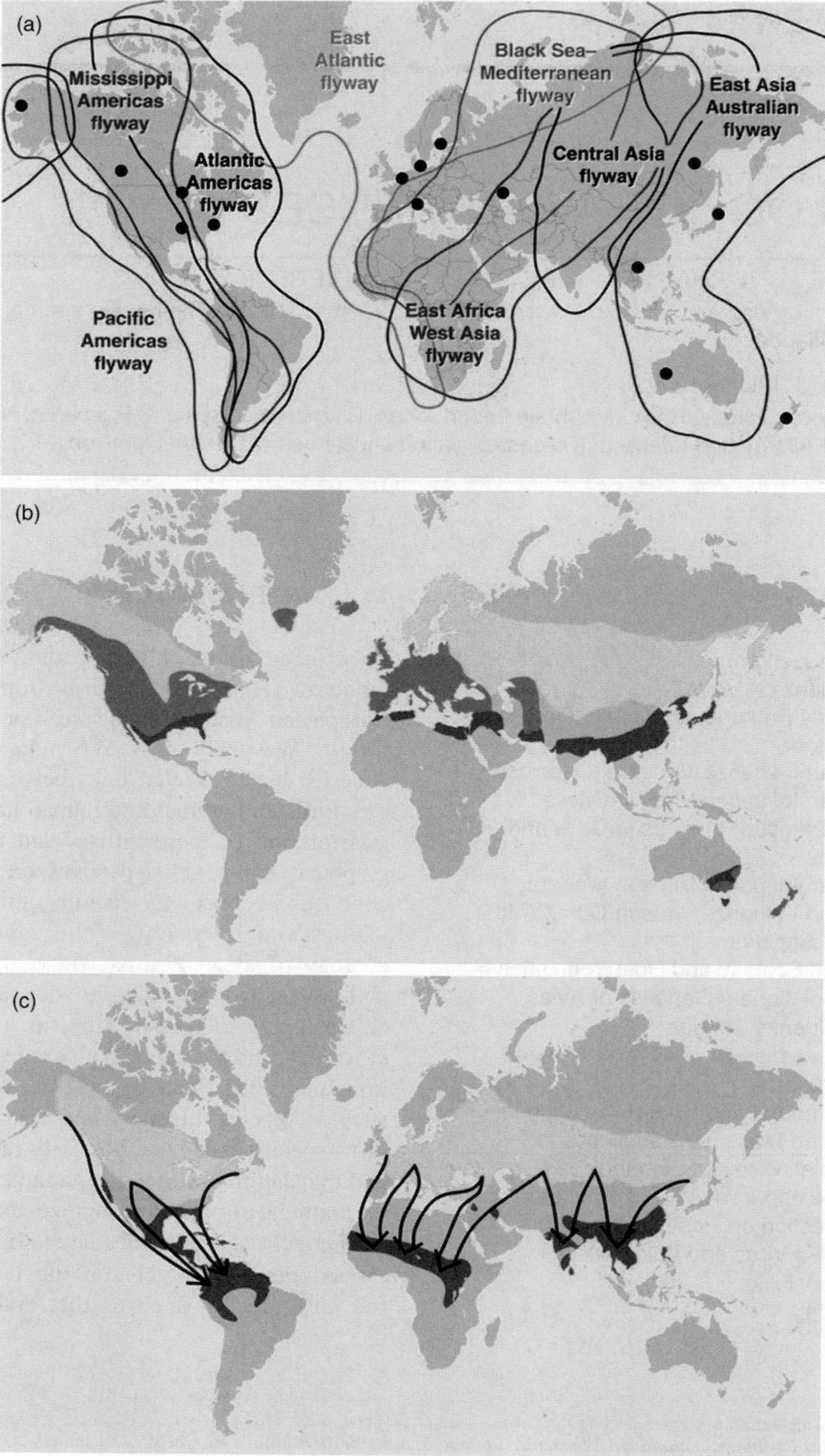
(a)
East Atlantic flyway
Black Sea– Mediterranean flyway
East Asia Australian flyway
Mississippi Americas flyway
Atlantic Americas flyway
Central Asia flyway
Pacific Americas flyway
East Africa West Asia flyway
(b)
(c)

behavior, habitat use, migratory patterns and behavior, population size and density, group size, and frequency of aggregation, and on biotic and abiotic factors outside the host such as viral environmental persistence.

The world's avifauna has a number of features that make wild birds unique as potential vectors for emerging infectious diseases, such as the large species diversity, long life span, seasonal aggregation, the spatial population structure, and the capacity for long-distance migration. Migratory birds can disperse pathogens, particularly those that do not significantly affect the birds' health status and consequently interfere with migration, either as biological or mechanical carriers. As a result, the distribution of birds and their associated pathogens is continuously changing—in regular seasonal patterns, and on local, regional, or global scales (Hubalek, 2004). For instance, migratory birds can carry ticks for several days and drop the ticks, including tick-borne viruses (e.g., tick-borne encephalitis viruses), in new geographical areas. Many bird species are known to perform regular long-distance migrations, thereby potentially distributing viruses between countries or even continents. Bird species breeding in one geographic region often follow similar migratory flyways, for example, the East Asia–Australian flyway from eastern Siberia south to eastern Asia and Australia. However, the major flyways are simplifications, and there are numerous exceptions where populations behave differently from the common patterns (Figure 14.1). Within the large continents and along the major flyways, migration connects many bird populations in time and space, either at common breeding areas, during migration, or at shared nonbreeding areas. As a result, one bird population may transmit their pathogens to another population that may subsequently carry these pathogens to new areas. It is important to realize that the transmission of the viruses and their geographical spread is dependent on the ecology of the migrating hosts. For instance, migrating birds rarely fly the full distance between breeding and nonbreeding areas without stopping over and "refueling" along the way. Rather, birds make frequent stopovers during migration, and spend more time foraging and preparing for migration than actively performing flights. Many species aggregate at favorable stopover or wintering sites, resulting in high local densities. Such sites may be important for transmission of viruses between wild and captive birds and between different species.

It is reasonable to assume that the viral diversity in avian species is in line with the host species diversity and with the viral diversity for instance observed in mammalian species. The number of viral pathogens associated with birds is probably far greater than presently known. So far, research into avian viruses has primarily focused on viruses with pathogenic properties either from a veterinary or a public health perspective. Influenza A virus is known to affect a wide range of host species, including

FIGURE 14.1 Migratory flyways of wild bird populations. A worldmap with the main general migratory flyways of wild bird populations is shown (adapted from information collected and analyzed by Wetlands International). (a) Black dots indicate the locations of historical influenza virus surveillance sites from which data have been used in this book chapter. These global migration flyways are simplifications, and there are situations where populations behave differently from the common patterns. Migration patterns of mallard (*A. platyrhynchos*) (b) and garganey (*Anas querquedula*) *in Eurasia and Africa and blue-winged teal* (*Anas discors*) *in the Americas* (c) (right and left parts of the map, respectively) are provided. Yellow color indicates breeding areas in which species are absent during winter, green indicates areas in which species are present around the year, and blue indicates areas in which species are only present in winter and do not breed. Arrows indicate the seasonal migration patterns (Reproduced from Olsen et al., 2006 with permission.) (*See the color version of this figure in Color Plates section.*)

nonbird species. Some avian viruses are also known to infect a broad range of bird species (e.g., avian reoviruses) while other viruses are known to only infect a small number of bird species (e.g., Buggy Creek virus, Crane herpesvirus). These infections with avian viruses can range from acute (e.g., West Nile virus, Newcastle disease virus) to chronic (e.g., avian pox virus) and latent or asymptomatic (e.g., St. Louis encephalitis virus, avian influenza virus). An overview of avian virus families and their main viruses is presented in Table 14.1.

The ecology of most avian viruses has been studied to a very limited extent, with the exception of classic poultry diseases, such as Newcastle disease virus and especially avian influenza virus that have been studied extensively in domestic and wild birds. The recent introductions of HPAI H5N1 virus, a highly pathogenic avian influenza variant that emerged from a low pathogenic H5N1 influenza virus in the late 1990s in China, in wild birds and its subsequent spread throughout Asia, the Middle East, Africa, and Europe has put a focus on the role of wild birds in the geographical spread of HPAI H5N1 virus. Large-scale surveillance programs are ongoing to determine a potential role of wild birds in H5N1 virus spread and to serve as sentinel systems for introductions into new geographical regions. The unprecedented scale and coverage of these surveillance programs has made avian influenza virus the best studied of all wildlife diseases in general. Therefore, avian influenza viruses are used as an example to illustrate the ecology of avian viruses throughout the remainder of this chapter.

14.2 INFLUENZA A VIRUS

Influenza A viruses are probably best known for their ability to cause pandemics and subsequent annual epidemics in humans, with the 1918 H1N1 Spanish influenza and the 2009 H1N1 Swine origin pandemic as prime examples. In addition, outbreaks of HPAI virus, such as the HPAI H5N1 outbreaks, recently gained a high profile in both the scientific community and the general public. Less well known is the fact that avian influenza A viruses circulating in wild birds are the progenitors, either directly or indirectly, of all pandemic and highly pathogenic influenza viruses (Webster et al., 1992). Besides being prevalent in humans, influenza A viruses have been isolated from many other species including pigs, horses, mink, dogs, cats, marine mammals, and a wide range of domestic birds. However, wild migratory birds are the original virus reservoir in nature (Figure 14.2).

Influenza A virus is an enveloped RNA virus, belonging to the family of Orthomyxoviridae (Wright and Webster, 2001). The influenza A virus particle is pleomorphic, with a diameter of approximately 120 nm (Figure 14.3). The viral envelope is derived from the host cell membrane. Influenza A viruses are classified on the basis of the viral surface glycoproteins hemagglutinin (HA) and neuraminidase (NA), which mediate cell entry and release of virus particles, respectively. In wild birds, influenza viruses representing 16 distinct types of HA and 9 of NA have been found, which can be found in numerous combinations (also called subtypes, e.g., H1N1, H5N1, H16N3) (Wright and Webster, 2001; Fouchier et al., 2005).

The influenza A virus genome consists of eight segments of negative-sense, single-stranded RNA. The 8 gene segments of influenza A virus encode 11 different proteins (Lamb and Krug, 2001). The segmented nature of the influenza virus genome enables evolution by a process known as genetic reassortment (Webster et al., 1992), that is, the mixing of gene segments from two or more influenza A viruses (see Section 14.10).

14.3 HIGHLY PATHOGENIC AVIAN INFLUENZA VIRUSES

The HA protein of influenza A viruses is initially synthesized as a single polypeptide

TABLE 14.1 Avian Virus Families[a] and Respective Viruses Isolated from Domestic and Wild Birds

Genome	Family	Genus	Virus	Primary Host Species	Distribution	Associated Avian Disease
dsDNA	*Adenoviridae*	*Aviadenovirus*	*Avian adenoviruses group 1*	Chicken (*Phasianidae*), geese, domestic duck (*Anatidae*)	Worldwide	Egg-drop syndrome
	Herpesviridae	*Iltovirus*	*Gallid herpes virus 1*	Chicken (*Phasianidae*)	Worldwide	Laryngotracheitis
	Herpesviridae		*Anatid herpesvirus*	Domestic duck (*Anatidae*)	Americas, Europe, Asia	Duck virus enteritis (Duck plague)
	Herpesviridae	*Mardivirus*	*Gallid herpes-virus 2 and 3 (Marek's Disease Virus serotype 1)*	Chicken (*Phasianidae*)	Europe, Americas, Asia	Marek's disease
	Polyomaviridae	*Polyomavirus*	*Goose hemorrhagic polyoma virus (GHPV)*	Goose (*Anatidae*)	Europe	Hemorrhagic nephritis enteritis
	Polyomaviridae	*Polyomavirus*	*Budgeriar fledging polyoma-virus*	Budgeriar (*Psittacidae*)	Worlwide	Acute fatal illness
	Poxviridae	*Avipoxvirus*	*Avian pox virus*	Numerous	Worldwide	Nodular lesions nonfeathered areas
dsDNA and ssDNA	*Hepadnaviridae*	*Avihepadnavirus*	*Duck hepatitis B virus*	Domestic duck, mallard (*Anatidae*)		

(*continued*)

TABLE 14.1 (*Continued*)

Genome	Family	Genus	Virus	Primary Host Species	Distribution	Associated Avian Disease
ssDNA	*Circoviridae*	*Gyrovirus*	*Chicken Infectious anemia virus*	Chicken (*Phasianidae*)	Worldwide	Anemia
	Circoviridae	*Circovirus*	*Psittacine beak and feather disease virus*	Psittacine birds (Psittacidae)	Worldwide	Feather dystrophy and loss
	Circoviridae	*Circovirus*	*Pigeon circovirus*	Racing pigeons (*Columbia livia*) Senegal doves (*Streptopelia senegalensis*)	USA, Africa, Europe, Canada	Diarrhea
	Parvoviridae	*Eiythiovirus*	*Goose parvovirus*	Geese (*Anatidae*)	Asia, Europe	Fatal disease
dsRNA	*Birnaviridae*	*Avibirnavirus*	*Infectious bursal disease virus*	Chicken, turkey (*Phasianidae*)	Worldwide	Infectious bursal disease (Gumboro disease)
	Reoviridae	*Orthoreovirus*	*Reovirus*	Chicken, turkey (*Phasianidae*) (symptomatic), numerous (asymptomatic)	Worldwide	Tenosynovitis, viral arthritis
ssRNA	*Astroviridae*		*Duck hepatitis virus (DHV) type 2*	Domestic duck (*Anatidae*)	Europe	Hepatitis
	Bunyaviridae	*Nairovirus*	*Crimean-Congo hemorrhagic fever*		Europe, Asia, Africa	
	Coronaviridae	*Coronavirus*	*Avian infectious bronchitis virus*	Chicken (*Phasianidae*)	Worldwide	Infectious bronchitis
	Coronaviridae	*Coronavirus*	*Turkey coronavirus enteritis*	Turkey (*Phasianidae*)		Turkey coronavirus enteritis
	Flaviviridae	*Flavivirus*	*West-Nile virus (WNV)*	Numerous	Africa, Europe, Asia, Americas	WNV encephalitis

Orthomyxoviridae	*Influenza A virus*	*Avian influenza*	Waterfowl (*Anatidae*), shorebirds (*Charadriiformes*)	Worldwide	Avian influenza
Paramyxoviridae	*Paramyxovirus*	*Newcastle Disease Virus (APMV-1)*	Numerous	Worldwide	Newcastle disease
Paramyxoviridae	*Metapneumovirus*	*Avian pneumovirus*	Turkey (*Phasianidae*)	Americas, Asia, Europe, Africa	Turkey rhinotracheitis
Picornaviridae	*Hepatovirus*	*Avian encephalomyelitis virus*	Chicken, Turkey (*Phasianidae*)	USA	Encephalitis
Picornaviridae		*Duck hepatitis virus (DHV) type 1 and 3*	Domestic duck (*Anatidae*)	Americas, Asia	Hepatitis
Retroviridae	*Alpharetrovirus*	*Avian leukosis sarcoma virus (ALSV)*		Europe	Leukotic disease
Togaviridae		*Eastern equine encephalitis (EEE)*	Songbirds (*Passeriformes*)	Americas	EEE

The classification of the avian viruses in this table is based on genome type.
[a]This table represents the currently best studied avian viruses and does not present a complete overview of all avian viruses currently known.

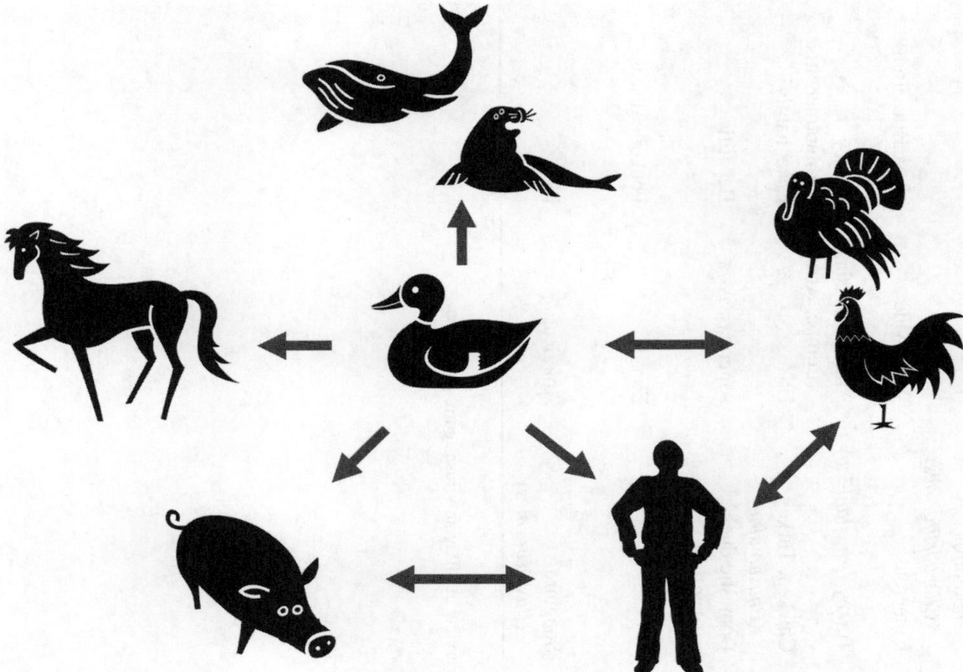

FIGURE 14.2 Influenza A virus reservoir. Wild birds are the main reservoir of influenza A viruses and can be transmitted from this reservoir to a wide variety of other host species.

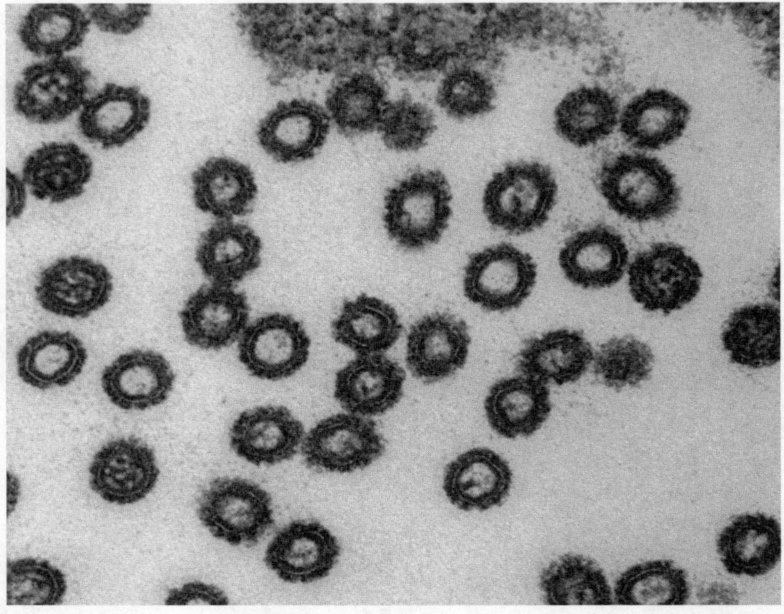

FIGURE 14.3 Electron micrograph image of an HPAI H7N7 influenza A virus responsible for causing the HPAI H7N7 outbreak in the Netherlands in 2003 (Fouchier et al., 2004).

precursor (HA_0) that is cleaved into HA_1 and HA_2 subunits by host cell proteases. Influenza A viruses of subtypes H5 and H7, but not of other HA subtypes, may become highly pathogenic after introduction into poultry and cause HPAI outbreaks. The switch from a low pathogenic avian influenza (LPAI) virus phenotype, commonly circulating in wild birds, to the HPAI virus phenotype is achieved by the introduction of basic amino acid residues into the HA_0 cleavage site (Table 14.2), which facilitates systemic virus replication and a mortality of up to 100% in poultry (Alexander, 2000; Webster and Rott, 1987). HPAI virus isolates have been obtained primarily from commercially raised poultry (agricultural birds such as chickens, turkeys, and quail). It is important to distinguish between the origin of LPAI (wild birds) and HPAI (predominantly poultry); the majority of influenza A viruses circulating in the wild bird reservoir is of the LPAI phenotype.

In the past two decades, HPAI outbreaks have occurred frequently, caused by influenza viruses of subtype H5N1 in Asia, Russia, the Middle East, Europe, and Africa (ongoing since 1997), H5N2 in Mexico (1994), Italy (1997), and Texas (2004), South Africa (2004), H7N1 in Italy (1999), H7N3 in Australia (1994), Pakistan (1994), Chile (2002), and Canada (2003), H7N4 in Australia (1997), H7N7 in the Netherlands (2003), North Korea (2005), and England (2008) (Alexander, 2000; Alexander and Brown, 2009). While most HPAI outbreaks have been controlled relatively quickly, HPAI H5N1 virus has been circulating in poultry continuously since 1997. These recent introductions of HPAI H5N1 virus in wild birds and the subsequent spread of the virus throughout Asia, the Middle East, Africa,

TABLE 14.2 The Putative HA_0 Cleavage Sites of All Avian Influenza A Virus HA Subtypes

HA Subtype	Putative Cleavage Site Sequence of HA_0																		
	HA_1									HA_2									
H1	P	S	I	Q	S	·	·	**R**	↓	G	L	F	A	I	A	G	F	I	E
H2	P	Q	I	E	S	·	·	**R**	↓	G	L	F	A	I	A	G	F	I	E
H3	P	E	K	Q	T	·	·	**R**	↓	G	L	F	A	I	A	G	F	I	E
H4	P	E	K	A	T	·	·	**R**	↓	G	L	F	A	I	A	G	F	I	E
H5	P	Q	R	E	T	·	·	**R**	↓	G	L	F	A	I	A	G	F	I	E
$H5_{HPAI}$[a]	P	Q	**R**	**R**	**R**	**K**	**K**	**R**	↓	G	L	F	A	I	A	G	F	I	E
H6	P	Q	I	E	T		·	**R**	↓	G	L	F	A	I	A	G	F	I	E
H7	E	I	P	K	G	·	·	**R**	↓	G	L	F	A	I	A	G	F	I	E
$H7_{HPAI}$[a]	E	I	P	**K**	**R**	**R**	**R**	**R**	↓	G	L	F	A	I	A	G	F	I	E
H8	P	S	I	E	P	·	·	**K**	↓	G	L	F	A	I	A	G	F	I	E
H9	P	A	A	S	D	·	·	**R**	↓	G	L	F	A	I	A	G	F	I	E
H10	E	V	V	Q	G	·	·	**R**	↓	G	L	F	A	I	A	G	F	I	E
H11	P	A	I	A	T	·	·	**R**	↓	G	L	F	A	I	A	G	F	I	E
H12	P	Q	A	Q	N	·	·	**R**	↓	G	L	F	A	I	A	G	F	I	E
H13	P	A	I	S	N	·	·	**R**	↓	G	L	F	A	I	A	G	F	I	E
H14	P	G	K	Q	A	·	·	**R**	↓	G	L	F	A	I	A	G	F	I	E
H15	E	K	I	R	T	·	·	**R**	↓	G	L	F	A	I	A	G	F	I	E
H16	P	S	I	G	E	·	·	**R**	↓	G	L	F	A	I	A	G	F	I	E

The occurrence of basic amino acid residues in the cleavage motifs are indicated in bold. Only the H5 and H7 subtype influenza A viruses with the multiple basic amino acid residues are HPAI viruses. The reason behind the subtype restriction in the emergence of HPAI viruses is currently unknown.

[a]Only one of multiple recognized MBCS motifs is shown for HPAI H5 and H7 viruses.

and Europe has put a focus on the role of wild birds in the geographical spread of HPAI H5N1 virus. Large-scale wild bird surveillance programs are currently implemented in many parts of the world to determine the role of wild birds in the spread of HPAI H5N1 virus and to serve as a sentinel system for the introduction of HPAI H5N1 virus into new geographical regions (Siembieda et al., 2010; Munster et al., 2006; Fereidouni et al., 2010; Gaidet et al., 2007; Sharshov et al., 2010; Chen et al., 2006b; Haynes et al., 2009; Kou et al., 2009; Dusek et al., 2009). Whereas the initial influenza A virus surveillance studies in the 1960s and 1970s relied on classical virological tools such as virus isolation in embryonated chicken eggs and were able to process, at most, a couple of thousand samples on a yearly basis, the implementation of high-throughput molecular diagnostic methods has allowed the intensity to increase to an astonishing 400,000 samples a year (Munster et al., 2009), making avian influenza viruses one of the most extensively studied wildlife diseases of our time.

14.4 AVIAN INFLUENZA VIRUS HOST SPECIES

Avian influenza viruses have been isolated from over 100 wild bird species belonging to more than 25 families (Olsen et al., 2006). Although many wild bird species may occasionally harbor avian influenza viruses, birds of wetlands and aquatic environments such as belonging to the orders of Anseriformes (ducks, geese, and swans) and Charadriiformes (gulls, terns, and waders) appear to function as the main reservoirs for avian influenza viruses (Olsen et al., 2006). The classical influenza surveillance studies of wild ducks and shorebirds in North America were the first to reveal the high avian influenza virus prevalence in species such as mallards and turnstones (Webster et al., 1992; Krauss et al., 2007; Kawaoka et al., 1988). An overview of the prevalence of influenza A virus in wild birds is presented in Table 14.3.

14.5 AVIAN INFLUENZA VIRUS IN DUCKS

Dabbling ducks of the *Anas* genus, with mallards (*Anas platyrhynchos*) by large the most extensively studied species (Wallensten et al., 2006, 2007; Munster et al., 2007; Krauss et al., 2004; Hanson et al., 2003; Ellstrom et al., 2008; Hinshaw et al., 1980a, 1980b; Parmley et al., 2008), have been found to be infected with avian influenza viruses more frequently than other duck species, including diving and seafaring ducks (Munster et al., 2007; Olsen et al., 2006). In addition, all avian influenza HA and NA subtypes, with the exception of H16, have been detected in wild ducks and the largest diversity of HA/NA subtype combinations has been detected in ducks (Figure 14.4). The differences in virus prevalence between ecological guilds of ducks are likely related in part to their behavior. Dabbling ducks feed mainly on the water surface allowing effective fecal–oral transmission, while diving ducks forage at deeper depths and more often in marine habitats. Dabbling ducks display a propensity for migration and the switching of breeding grounds between years, in part due to mate choice (Del hoyo et al., 1996). This behavior could provide an opportunity for influenza A viruses to be transmitted between different host subpopulations.

14.6 AVIAN INFLUENZA VIRUS IN GULLS AND TERNS

The first recorded isolation of influenza virus from wild birds was from an outbreak in common terns (*Sterna hirundo*) in 1961 in South Africa, where at least 1300 of these birds died (Becker, 1966). This outbreak remains unusual in that it is the only recorded case of an outbreak of an HPAI virus in wild birds with no direct evidence for association with poultry. The most frequently detected avian influenza virus subtypes in gulls are H13 and H16 (Munster et al., 2007; Hinshaw et al., 1982; Fouchier

TABLE 14.3 Prevalence of Influenza A Viruses in Wild Birds[a]

Family	Species	Sampled	Positive (*n*)	Positive (%)
Ducks	36 species	34,503	3,275	9.5
	Mallard (*Anas platyrhynchos*)	1,525	1,965	12.9
	Northern pintail (*Anas acuta*)	3,036	340	11.2
	Blue-winged teal (*Anas discors*)	1,914	220	11.5
	Common teal (*Anas crecca*)	1,314	52	4.0
	Eurasian wigeon (*Anas penelope*)	1,023	8	0.8
	Wood duck (*Aix sponsa*)	926	20	2.2
	Common shelduck (*Tadorna tadorna*)	881	57	6.5
	American black duck (*Anas rubripes*)	717	130	18.1
	Green-winged teal (*Anas carolinensis*)	707	28	4.0
	Gadwall (*Anas strepera*)	687	10	1.5
	Spot-billed duck (*Anas poecilorhyncha*)	574	21	3.7
Geese	8 species	4,806	47	1.0
	Canada goose (*Branta canadensis*)	2,273	19	0.8
	Greylag goose (*Anser anser*)	977	11	1.1
	White-fronted goose (*Anser albifrons*)	596	13	2.2
Swans	3 species	5,009	94	1.9
	Tundra swan (*Cygnus columbianus*)	2,137	60	2.0
	Mute swan (*Cygnus olor*)	1,597	20	1.3
	Whooping swan (*Cygnus cygnus*)	930	14	1.5
Gulls	9 species	14,505	199	1.4
	Ring-billed gull (*Larus delawarensis*)	6,966	136	2.0
	Black-tailed gull (*Larus crassirostris*)	1,726	17	1.0
	Black-headed gull (*Larus ridibundus*)	770	17	2.2
	Herring gull (*Larus argentatus*)	768	11	1.4
	Mew gull (*Larus canus*)	595	0	0.0
Terns	9 species	2,521	24	0.9
	Common tern (*Sterna hirundo*)	961	16	1.7
Waders	10 species	2,637	21	0.8
Rails	3 species	1,962	27	1.4
	Eurasian coot (*Fulica atra*)	1,861	23	1.2
Petrels	5 species	1,416	4	0.3
	Wedge-tailed shearwater (*Puffinus pacificus*)	794	4	0.5
Cormorants	1 species	45	18	0.4
	Great cormorant (*Phalacrocorax carbo*)	45	18	0.4

Reproduced from Olsen et al. (2006) with permission.
[a]Influenza virus prevalence in specific species is given only if tests on <500 birds have been reported Of the 36 species of ducks, 28,955 were dabbling ducks and 1011 were diving ducks, with influenza virus prevalence of 10.1 and 1.6%, respectively.

et al., 2005; Kawaoka et al., 1988). Both subtypes are rarely found in other bird species. Influenza viruses can be detected in a small proportion of gulls, with the highest virus prevalence reported in late summer and early fall. Most gull species breed in dense colonies, with adults and juveniles crowded in a small space, potentially creating good opportunities for virus spread. Breeding in dense colonies contrasts with dabbling ducks that do not breed

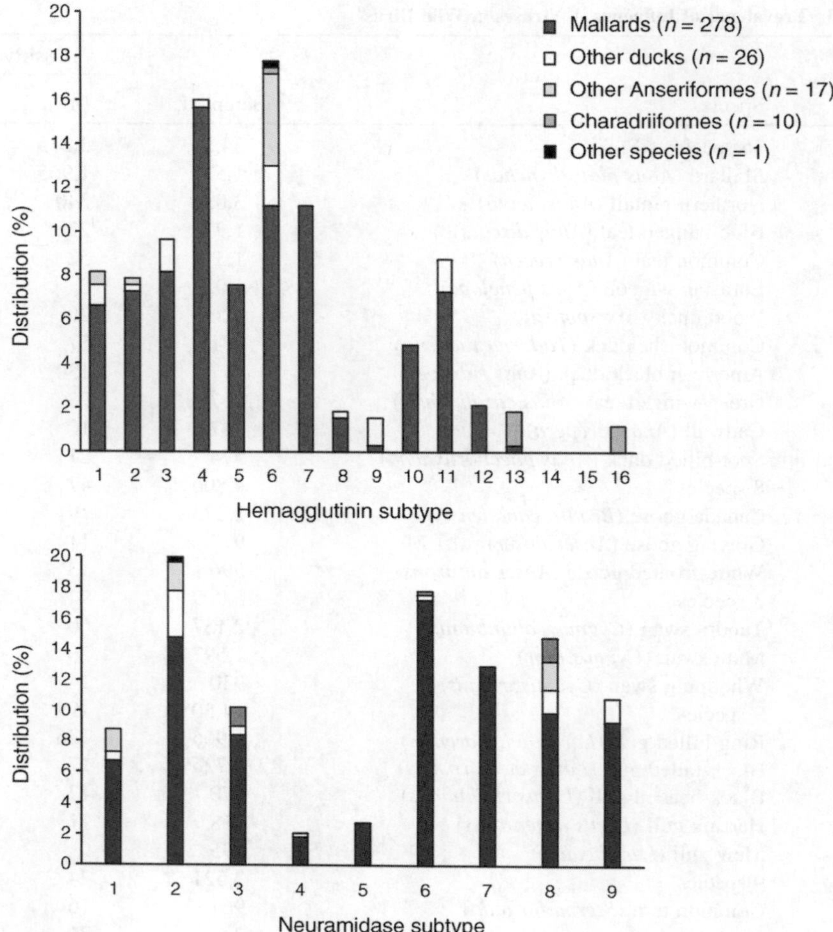

FIGURE 14.4 Distribution of HA and NA subtypes in influenza A virus isolates obtained from wild birds. Almost all HA and NA subtypes circulate in the dabbling duck populations (e.g., mallards). H13 and H16 influenza A viruses circulate almost exclusively in gull populations. (Reproduced from Munster et al., 2007 with permission.)

in dense colonies (Del hoyo et al., 1996), and epizootics are likely to be more easily initiated when birds congregate in large numbers during molt, migration, or wintering.

14.7 AVIAN INFLUENZA VIRUS IN WADERS

The wader species in the Charadriidae and Scolopacidae families are adapted to marine or freshwater wetland areas and often live side by side with ducks. Data from North America suggests a distinct role of these birds in the ecology of avian influenza viruses (Krauss et al., 2004; Hanson et al., 2008; Munster et al., 2007), but long-term avian influenza virus surveillance studies at different geographical locations are needed to clarify this. Influenza viruses of subtypes H1–H12 have been isolated in birds migrating through the eastern USA, with a high prevalence of certain HA subtypes (H1, H2, H5, H7, H9–H12) in combination with a large variety of NA subtypes (Krauss et al., 2004; Kawaoka et al., 1988). Although the bulk of data on the circulation of avian influenza A viruses in waders originates from North America, avian influenza viruses have been detected in waders in Africa and Australia as well, but not to the same extent (Gaidet et al., 2007; Hurt et al., 2006).

14.8 AVIAN INFLUENZA VIRUS IN OTHER WILD BIRD SPECIES

Avian influenza viruses have been found in numerous other bird species (Olsen et al., 2006), but it is unclear whether avian influenza virus is endemic in these species or whether the virus is a transient pathogen. Species in which avian influenza viruses are endemic share the same habitat at least part of the year with other species in which influenza viruses are frequently detected including geese, swans, rails, petrels, cormorants, and, to a lesser extent, passerine species. In these and other bird species, influenza A virus prevalence seems to be lower than in dabbling ducks, but studies that sample during the full annual cycle are limited, and it is possible that peak prevalence has been missed because of its seasonal nature or location. In addition, avian influenza virus surveillance has typically shown considerable bias toward species that are easily caught or are present in accessible areas at high concentrations. Therefore, the current status of our knowledge may only partly reflect the true ecology of avian influenza viruses with respect to host reservoir species.

14.9 VIRUS ECOLOGY AND HOST POPULATIONS

Large-scale surveillance studies have identified a predominant role for dabbling ducks (Webster et al., 1992; Munster et al., 2005; Parmley et al., 2008; Krauss et al., 2004, 2007). Factors contributing to this role of dabbling duck populations as influenza A virus host species include population size and mode of transmission (see Section 14.11). The importance of population size, age structure, and herd-immunity on the epidemiology of infectious diseases has been investigated in detail for human pathogens such as measles (Rohani et al., 1999). Large populations are probably more capable of sustaining a large variety of different influenza A virus subtypes, as observed in dabbling ducks. The dabbling duck population is estimated to consist of 10,000,000 birds in Europe alone, with the mallard being the most abundant species (~5,000,000 in Europe and ~27,000,000 worldwide) (Scott and Rose, 1996). The estimated yearly turnover rate of mallards in Northern Europe is roughly 1/3 (Bentz, 1985). A large part of the population is therefore rejuvenated every year, potentially allowing simultaneous cocirculation of multiple genetic lineages and subtypes within one metapopulation of potential hosts for influenza A virus. In contrast, the population estimates for the different goose species in Europe are significantly lower compared to the dabbling ducks with a total population size of ~1.2 million geese (Scott and Rose, 1996). Smaller population sizes would likely limit the perpetuation and maintenance of multiple influenza A virus subtypes and allow only a limited number of influenza A virus subtypes to cocirculate within these populations. The predominant avian influenza virus detected within geese in the Netherlands over the last decade was of the H6 subtype, with around 60% of all viruses isolated from geese populations of this subtype (Munster et al., 2007; Kleijn et al., 2010). The relative abundance of the detection of the H6 subtype within the geese populations does not correlate with the predominant subtypes detected within mallards (Kleijn et al., 2010). The global populations of Laridae (gulls and terns) species appears to be large enough to allow cocirculation of two distinct influenza A virus lineages of H13 and H16 influenza A viruses, although other avian influenza virus subtypes are also occasionally detected in terns and gulls (Munster et al., 2007; Kawaoka et al., 1988; Krauss et al., 2004).

14.10 EVOLUTIONARY GENETICS OF AVIAN INFLUENZA VIRUSES

The segmented nature of the influenza virus genome enables evolution by a process known as genetic reassortment, that is, the mixing of genes from two or more influenza viruses (Webster et al., 1992). Reassortment is one of the driving forces behind the variability of influenza viruses and contributes greatly to the phenotypic variability among these viruses. The three most recent human influenza pandemic viruses and the multitude of viral genotypes associated with the outbreaks of HPAI H5N1 viruses were the result of reassortment of gene segments. Few details of the capacity for reassortment of different lineages of influenza A viruses, the exact rate of reassortment in nature or the effects of reassortment on the virus population are currently known (Macken et al., 2006). A recent study of influenza A viruses obtained from ducks in Canada indicates that genetic "sublineages" do not persist, but frequently reassort with other viruses (Hatchette et al., 2004). In addition, analysis of the genome constellation (the set of eight gene segments as a whole) of five H4N6 influenza A viruses isolated from mallards at the same day and location revealed four different genome constellations (at least one of the eight gene segments within the virus has a different evolutionary origin compared to the other four viruses) with only one pair of viruses sharing the same genome constellation (Dugan et al., 2008). Influenza viruses of a particular subtype therefore do not necessarily have the same genetic makeup, even within a single day, location, or host species. Combined with the high prevalence of influenza virus in some wild bird species, and the detection of concomitant infections in single birds, this supports the notion that reassortment occurs at a relatively high rate in nature (Macken et al., 2006; Dugan et al., 2008). The continuous cocirculation of several influenza A virus sub- and genotypes in a staging population of hosts, together with the replacement of viruses in the individual hosts, sets a scene where recombination of coinfecting viruses is very likely to occur at a high rate. This indicates that influenza A viruses do not circulate as "fixed" genome constellations but rather the continuous reassortment leads to "transient" genome constellations.

The increasing number of available complete avian influenza virus genomes has enabled the detailed analysis of the patterns of genetic diversity in relation to host species and geographical location (Obenauer et al., 2006). The evolutionary genetics of influenza A viruses are for instance studied phylogenetically, where the gene segments of a viral genome can be associated with distinct clades of viruses in segment-specific phylogenies.

Avian influenza viruses can be divided into two main phylogenetic lineages: the Eurasian and American lineage (Olsen et al., 2006; Webster et al., 1992; Widjaja et al., 2004; Munster et al., 2005) (Figure 14.5). The major geographic segregation is observed between viruses isolated from bird species that utilize the migratory flyways of the America's and Eurasia/Africa/Australia, respectively. Apparently, this led to a long-term ecological and geographical separation of these bird populations and hence the viruses circulating within these hosts. This allopatric separation has resulted in a major phylogenetic split between the Eurasian and American genetic lineages of influenza A viruses. Despite this phylogenetic split, the separation of these virus populations is not absolute. The avifauna of North America and Eurasia are not completely separated; some ducks and shorebirds cross the Bering Strait during migration or have breeding ranges that include both the Russian Far East and North-Western America (Del hoyo et al., 1996). The majority of tundra shorebirds from the Russian Far East winter in Southeast Asia and Australia, but some species winter along the West coast of the Americas. The overlap in distribution of ducks is not as profound as that of shorebirds, but a few species (e.g., Northern pintail, *Anas acuta*) are common in both North America and Eurasia and could also provide an

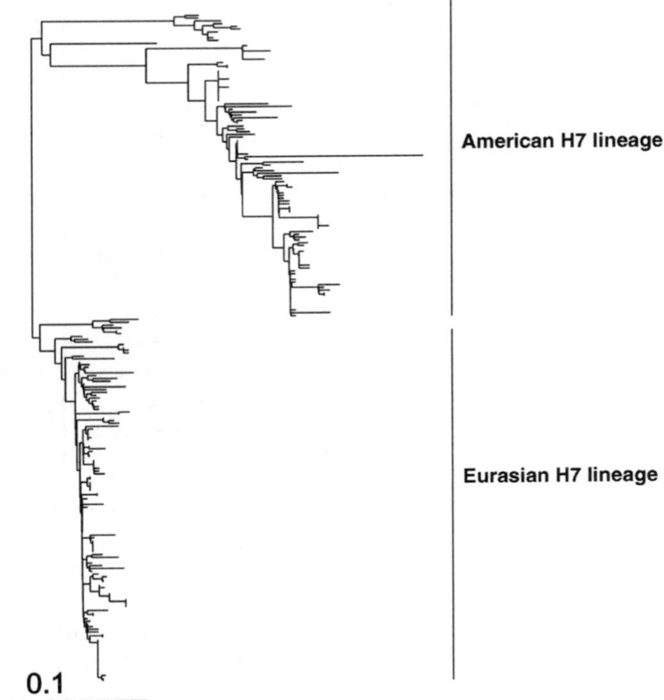

FIGURE 14.5 Phylogeny of the HA of American and Eurasian H7 avian influenza viruses. The tree displays the allopatric evolution of H7 viruses. The scale bar represents ≈10% of amino acid changes between close relatives. (Reproduced from Munster et al., 2005 with permission.)

intercontinental bridge for influenza A virus. Indeed, influenza viruses carrying a mix of genes from the American and Eurasian lineages have been isolated, indicating that allopatric speciation is only partial and that frequent exchange of gene segments occurs between the two virus populations (Wallensten et al., 2005; Wahlgren et al., 2008; Makarova et al., 1999; Koehler et al., 2008; Krauss et al., 2007). Whole-genome analyses of influenza isolates obtained from northern pintails in Alaska, a species that migrates between north America and Asia, suggests intercontinental virus exchange at a relatively high frequency (Pearce et al., 2009). Of the analyzed influenza viruses, 44% had at least one gene segment originating from the Eurasian lineage, whereas all other gene segments were of the American lineage. In addition, analyses of H6 avian influenza viruses suggest the introduction of Eurasian H6 gene segment in North America on several occasions (zu Dohna et al., 2009). However, so far there has been no evidence for cross-hemisphere circulation of entire virus genomes but only introduction of single gene segments that reassorted with other segments found in the new hemisphere. The partial geographic isolation of influenza virus hosts therefore seems sufficient to facilitate divergent evolution and continue the existence of separate gene pools.

Besides the influence of geographical separation on the evolutionary genetics of avian influenza A viruses, differences in host species affinity have also resulted in clearly distinguishable virus populations. A good example of this genetic separation are the influenza A viruses of the H13 and H16 subtypes that are predominantly isolated from gulls and terns (Fouchier et al., 2005; Hinshaw et al., 1982). These viruses belong to a group of distinct

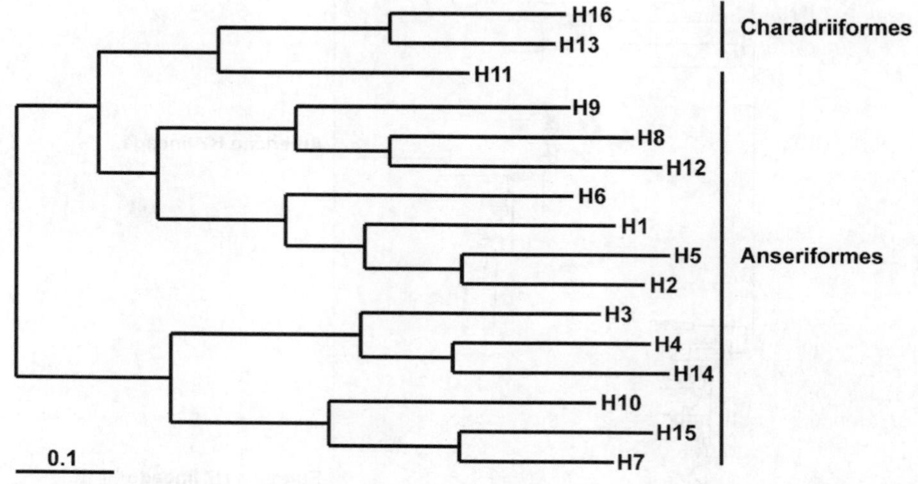

FIGURE 14.6 Phylogeny of 16 influenza virus HA subtypes. The scale bars represent approximately 10% nucleotide changes between close relatives.

influenza viruses based on genetic, functional, and ecological properties and have evolved into separate genetic lineages from influenza A viruses isolated predominantly from Anseriformes and Charadriiformes (H1–H12 subtypes) (Figure 14.6). Gene segments of gull viruses are genetically distinct from those circulating in other wild birds, suggesting that they have been separated for a sufficient amount of time to allow genetic differentiation by sympatric speciation (Fouchier et al., 2005). Gull influenza viruses do not readily infect ducks upon experimental inoculation (Hinshaw et al., 1982), providing a biological explanation for the limited detection of these viruses in other avian influenza host species, although a limited number of gull viruses has been isolated from ducks and vice versa (Munster et al., 2007; Kawaoka et al., 1988; Krauss et al., 2004).

14.11 TRANSMISSION OF AVIAN INFLUENZA VIRUSES

The maintenance and circulation of avian influenza viruses within the wild bird host populations relies on the effective transmission of the virus between susceptible hosts and susceptible host populations. Avian influenza A viruses generally infect cells lining the intestinal tract and are transmitted via the fecal–oral route (Webster et al., 1992). Avian influenza viruses can stay infectious for prolonged periods of time in surface water dependent on pH, salinity, and temperature (Stallknecht et al., 1990; Brown et al., 2009), potentially allowing temporal and spatial connectivity of different host subpopulations by their respective virus populations.

The observed differences in avian influenza virus prevalence in various bird species and families can be partially explained by differences in the effectiveness of bird-to-bird transmission. A broad comparative analysis of available influenza wild bird surveillance data indicated that occupation of aquatic environment alone was not an indicator of avian influenza prevalence, but feeding in surface waters was (Garamszegi and Moller, 2007). In addition, research using the data obtained from influenza virus surveillance in the Camargue area (France) highlighted waterborne transmission as the main determinant of the avian influenza dynamics and prevalence observed in that region (Roche et al., 2009). Therefore, differences in feeding behavior and diet could account for the differences in virus prevalence between bird families and species. Whereas dabbling ducks feed and defecate on the surface water, thereby allowing effective indirect fecal–oral transmission, geese and certain swan species graze in pastures and agricultural fields (Munster and Fouchier, 2009). This could potentially lead to a less efficient fecal–oral transmission and thereby explain the lower avian influenza virus prevalence in geese and swan species as observed in avian influenza virus surveillance studies. The lower overall virus prevalence in geese and the limited number of HA subtypes identified in influenza A viruses isolated from geese is in agreement with this hypothesis (Munster et al., 2007). However, influenza is a respiratory disease in mammals, where transmission primarily occurs via aerosols. In wild birds, avian influenza viruses are not exclusively a gastrointestinal infection; besides replicating in the intestinal tract, the viruses also replicate in the respiratory tract of wild birds. For mallards, the frequency of avian influenza detection in cloacal samples was twice as high as in oropharyngeal samples (Munster et al., 2009; Ellstrom et al., 2008). The majority of mallards with positive oropharyngeal swabs also had positive cloacal swabs. The amount of avian influenza virus present was higher for the samples obtained from the cloaca than for those obtained from the respiratory tract (Munster et al., 2009). This suggests that the respiratory tract plays a limited role in the replication and transmission cycle of avian influenza viruses in mallards. However, in white-fronted geese (*Anser albifrons albifrons*), the detection frequency was 2.4 times higher in oropharyngeal samples than in cloacal samples and a comparable amount of virus was shed via both routes (Kleijn et al., 2010). This indicates that direct infection from respiratory tract to respiratory tract (as in humans and other mammals) may be a more effective route of transmission in this species. The site of replication and route of transmission may therefore be species specific and transmission via the respiratory route may be relevant for bird species in which fecal–oral transmission would prove difficult. Thus, differences in behavior may select for viruses that can switch from fecal–oral to respiratory transmission in these species.

14.12 TEMPORAL AND SPATIAL VARIATION IN AVIAN INFLUENZA PREVALENCE IN RELATION TO HOST ECOLOGY

The prevalence of avian influenza A viruses within their natural hosts depends on the combination of geographical location, season, and species. The influenza A virus prevalence can vary greatly between years, locations, and species, and although certain gross patterns have been observed, large variations occur within these patterns (Figure 14.7). The prevalence in the most studied reservoir species, the mallard, varies in a seasonally predictable way, from low prevalence (<1%) during spring and summer to high prevalence (up to 30%) during autumn migration and early winter (Webster et al., 1992; Munster et al., 2007; Wallensten et al., 2007; Halvorson et al., 1985). These temporal patterns have been observed in North America as well as in Europe, and seem to be a general feature of influenza A virus ecology in this host species. Surveillance of mallards at two geographical locations within the same flyway in northern Europe showed that influenza A virus prevalence during fall migration (roughly a north–south migration) varied between the two locations at the same time point. The prevalence in mallards at the northern sampling site was approximately threefold higher as compared to the prevalence in mallards at the wintering grounds, suggesting that timing relative to migration is an important determinant of influenza A virus prevalence (Munster et al., 2007) (Figure 14.8). High virus prevalence early in fall migration likely gradually declines as the migration proceeds, thus forming a north–south gradient of influenza A virus prevalence (Munster et al., 2007). This explains geographical differences in influenza

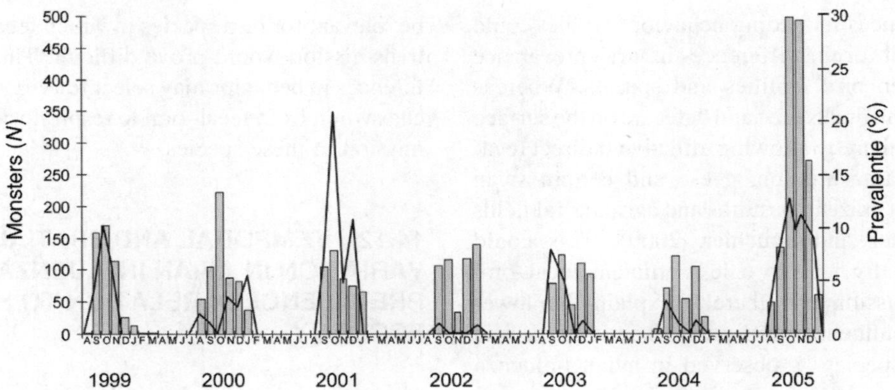

FIGURE 14.7 Annual influenza A virus prevalence in mallards during fall migration in The Netherlands from 1999 to 2005. Bars indicate the number of samples collected per month (left *y*-axis), and the red line indicates the number of samples positive for influenza A virus by RT-PCR (right *y*-axis). Clear seasonal patterns are observed, with peak prevalence occurring in September and October. Large variation in avian influenza prevalence between different years is observed. (Reproduced from Munster et al., 2007 with permission.) (*See the color version of this figure in Color Plates section.*)

A virus prevalence between Northern and more Southern latitudes observed in different surveillance studies. The peak in prevalence during fall migration is believed to be related to the large numbers of young, immunologically naive birds of that breeding season that aggregate prior to and during their southbound migration (Webster et al., 1992). The difference in prevalence of influenza A viruses with respect to age was determined for mallards and Eurasian wigeons in an European study. In this study, the prevalence differed between

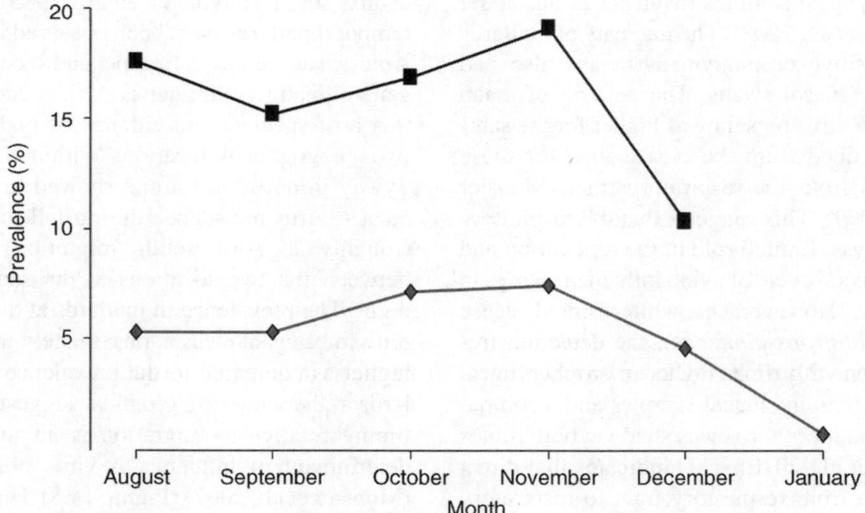

FIGURE 14.8 Trend lines for influenza A virus prevalence in Mallards caught in Sweden and The Netherlands during fall migration. The blue line and filled squares (■) represent the proportion (%) of influenza A virus positive mallards caught and sampled in Sweden between 2002 and 2005 at Ottenby bird Observatory and the red line and filled diamonds (◆) represent mallards caught at various locations in The Netherlands. (Reproduced from Munster et al., 2007 with permission.) (*See the color version of this figure in Color Plates section.*)

juveniles (first year) and adults (consecutive years). Influenza A virus prevalence was 6.8% for juvenile ducks and 2.8% for adults (Munster et al., 2007). In addition, consecutive or simultaneous infections with different subtypes of influenza A viruses are common in dabbling ducks, suggesting that only partial homo- and heterosubtypic immunity is induced by infection of the birds with an influenza A virus (Latorre-Margalef et al., 2009b). Different experimental infection studies were performed in mallards to assess the influence of prior infection with influenza A virus on the reinfection with another influenza A virus either of the same or a different subtype combination. It was shown that although the mallards could be reinfected with another subtype of influenza A virus, the duration of the infection and the virus shedding was markedly reduced. This reduction was even more pronounced when the subsequent reinfection was performed with influenza A virus of the same HA subtype (Jourdain et al., 2010; Fereidouni et al., 2009). The duration of influenza A virus shedding has been investigated in field studies as well as in experimental infections. The duration of infection, when using 18 wild caught mallards where the infection could be followed over the full infection cycle, was estimated at approximately 10 days (Latorre-Margalef et al., 2009b). This estimate is in agreement with duration of shedding between 7 and 17 days, obtained from experimental infections (Jourdain et al., 2010; Kida et al., 1980; Fereidouni et al., 2009). The transient infection in combination with the relatively short shedding time implies that the spatial dynamics of influenza A viruses are mainly explained by circulation within bird flocks or by relay transmission between staging areas where the birds congregate. Given the maximum duration of shedding observed, single birds are not likely to carry the virus on a continental scale. The yearly replenishment of the susceptible host pool (Bentz, 1985), the ability of reinfection of the host and large host populations sizes likely result in a critical community size of the host species large enough to allow endemicity and persistence of the genetically and antigenically diverse group of avian influenza viruses. Thus, although it has been speculated that influenza A viruses may persist in abiotic reservoirs such as arctic lakes, the continuous prevalence in dabbling ducks in combination with the abundance of these species (Wallensten et al., 2006; Scott and Rose, 1996), appears to be sufficient for year-round perpetuation of the virus in these species without a need for environmental persistence.

As opposed to the endemicity of avian influenza viruses in dabbling ducks, the avian influenza virus prevalence in other Anseriformes species suggests that avian influenza virus infections behave epidemically in those species (Kleijn et al., 2010; Munster et al., 2007). In white-fronted geese in the Netherlands, the absence of avian influenza virus prevalence upon arrival on their wintering grounds is explained by the introduction of the virus after their arrival on their wintering grounds likely through spillover from other reservoir species, such as the ubiquitous mallards. This suggests that frequent spillover from the endemic host (able to sustain long-term continuous circulation of the avian influenza virus) occurs to more transient hosts (able to sustain circulation of the influenza virus only for a limited period of time).

The complexity of the ecology of avian influenza viruses is reflected by the temporal, spatial, and species variation observed in the circulation of these viruses in wader species. Waders of the Charadriidae and Scolopacidae families, including species such as turnstones, sandpipers, knots, redshanks, and bar-tailed godwits, are adapted to marine or freshwater habitats. In general, the annual cycle of wader species consists of breeding in the arctic regions, performing long-distance migrations during late summer–early fall to their wintering grounds and returning back to their breeding grounds in spring and early summer (Del hoyo et al., 1996; van de Kam et al., 2004). Migratory pathways of these wader species include the Pacific flyway from western North America into western South America, the Atlantic

America flyway from eastern North America into South America, the East Asia–Australasian flyway from the eastern Siberian tundra and taiga regions south to eastern Asia and Australia and the East Atlantic Flyway from Greenland and the western Siberian tundra and taiga regions south to western Africa (van de Kam et al., 2004).

Within the population of waders migrating along the Atlantic America flyway, generally following the Atlantic Coast of North America, the influenza A virus dynamics appear to be reversed as compared with the previously discussed dabbling ducks. Peak prevalence of influenza virus (~14%) in waders (especially ruddy turnstones, *Arenaria interpres*) was observed during spring migration in Delaware Bay (Krauss et al., 2004). Surveillance activities performed at other geographical locations, such as Europe, Alaska, and Australia, only identified very limited circulation of influenza A viruses in the respective wader populations (Munster et al., 2007; Winker et al., 2008; Haynes et al., 2009). Influenza A virus circulation in waders therefore appears to be localized, species specific, and highly variable with respect to virus prevalence. In addition to being a geographical location with high peak prevalence of influenza A viruses in waders, Delaware Bay also stands out as the location hosting one of the largest seasonal aggregations of waders in the world. During spring migration, over a million waders gather in Delaware Bay when horseshoe crabs (*Linudus polyphemus*) are spawning (Figure 14.9). During this stopover, the birds refuel on the horseshoe crab eggs to finish migration to their breeding grounds in the Arctic. This high wader density potentially facilitates efficient circulation and transmission of influenza A viruses between various birds and bird species present during spring migration stopover at Delaware Bay. The reversion of the temporal dynamics in influenza A virus prevalence in the Delaware Bay shorebird population, as compared to those observed within dabbling duck populations, suggest that the influenza A viruses are introduced from ducks or gulls annually at this site. This is in agreement with the limited detection of influenza A viruses in shorebirds from other locations and the fact that recent genetic analyses have not revealed striking differences between influenza viruses isolated from ducks or waders in the Americas (Dugan et al., 2008), suggesting

FIGURE 14.9 Waders foraging at horseshoe crabs in Delaware Bay during their spring migration stopover. Mating horseshoe crabs are visible in the middle and back of the picture.

that these viral gene pools are not separated. Thus, the Delaware Bay area, where the initial shorebird studies were performed, combines a unique set of ecological factors that meet the requirements for efficient influenza A virus circulation and transmission. These ecological factors are probably a combination of host- and virus-specific ecological factors such as foraging behavior, migratory behavior, habitat preference, susceptibility, local bird density and species composition, aggregation of host species, environmental persistence of the virus, receptor specificity, tissue tropism, and replication and transmission kinetics. Locations comparable to Delaware Bay with respect to virus–host ecology have so far not been identified elsewhere in the world.

14.13 THE IMPACT OF AVIAN INFLUENZA VIRUS INFECTION ON HOST ECOLOGY

Little is known about the effect of avian influenza virus infection on the individual wild bird, the wild bird population as a whole, and the ecosystems occupied.

The clinical signs of disease observed following infection of domestic birds with low pathogenic avian influenza viruses vary with species, age, immune system development, presence of other microorganisms, and environmental factors. In general, symptoms range from no noticeable clinical signs to mild respiratory disease; depression and decreased egg production. The HPAI viruses can cause severe disease and high mortality in domestic poultry and will be discussed in detail in the next chapter.

In contrast to commercial poultry, wild waterfowl represent a broad genetic heterogeneity, have relatively long life cycles, live under variable animal densities and have well-developed immune systems (Caron et al., 2009). In addition, wild birds are exposed to variable ecological environments during different stages in the annual cycle. Extensive avian influenza virus surveillance studies have shown high virus prevalence in cloacal swabs of asymptomatic wild ducks. Upon experimental infection of ducks, avian influenza viruses replicate in the epithelial cells of the intestine of birds and virus may be shed in high concentrations in the feces, without inducing apparent signs of disease (Kida et al., 1980; Jourdain et al., 2010; Fereidouni et al., 2009). However, it is hard to extrapolate these data directly to the situation in wild birds, where mild or subclinical infections may have significant ecological fitness consequences. Wild birds can experience physiological stress as a result of limited nutritional resources and variable energy expenditure during the year, which could have an effect on the course of disease within the host and therefore on the host population.

An extensive longitudinal study in migratory mallards in Sweden showed that avian influenza virus infection had a negative impact on the body mass of the infected mallards (Latorre-Margalef et al., 2009b). However, the overall weight loss was limited and no effects on migration speed or distance were observed, suggesting that the impact on the infected individuals was relatively limited. In addition, it has been discussed whether the lowered body mass of the avian influenza virus infected mallards is a direct effect of the infection or whether birds in poor physical condition are more susceptible to acquiring an infection (Flint and Franson, 2009; Latorre-Margalef et al., 2009a). Although several studies have shown that dabbling ducks did not reveal clinical signs of disease upon infection with avian influenza virus (Webster et al., 1992; Kida et al., 1980; Jourdain et al., 2010), these infections still may have consequences for the ecological fitness of the host species.

A study on the effects of avian influenza on free-living white-fronted geese naturally infected with LPAI virus did not show differences in dispersal pattern of the birds within 12 days after sampling (van Gils et al., 2007). A study on the effect of natural avian influenza virus infection on free-living Bewick's swans (*Cygnus columbianus bewickii*) showed that infected Bewick's swans fueled and fed at reduced rates, displayed delayed migration, and traveled shorter distances in comparison with uninfected Bewick's swans (van Gils et al., 2007) (Figure 14.10). However, the number of infected Bewick's swans in this study was low ($n = 2$) and it is unknown whether the change in feeding and migrating pattern is a cause or consequence of the infection as with the mallard study.

Late arrival on the breeding grounds has a negative effect on the reproduction success as a result of the occupation of the best breeding sites, the decreased quality of nutrition and a higher pathogen pressure. The observations that infection of Bewick's swans with avian influenza virus may result in altered feeding and migration patterns, indicates that avian influenza virus infections in wild birds may well have a higher clinical, epidemiological, and ecological impact than previously recognized. However, the impact of avian influenza virus infections should not be generalized. Infection with avian influenza viruses will likely have less impact in wild bird species that are regularly exposed to avian influenza viruses than in wild bird species that are less frequently exposed. Avian influenza virus infections could therefore have a larger behavioral impact on transiently infected species, such as swan species, and limited impact on endemically infected species such as dabbling ducks.

Due to the scarceness of studies linking virus ecology to host ecology, it is currently not known how avian influenza virus infections affect the various wild bird species during their annual stages and consequently affect the reproduction success and survival of these wild bird species.

14.14 HPAI VIRUS AND WILD BIRDS

HPAI viruses can emerge when LPAI viruses of the H5 or H7 subtype are introduced from wild birds into poultry, through a change in the HA cleavage site. These HPAI viruses have a devastating impact on chickens and turkeys, with mortality rates of ~100% (Alexander, 2000).

FIGURE 14.10 Bewick's Swan fitted with GPS-collar. Satellite telemetry allows the monitoring of survival and migration patterns of wild birds. (Reproduced from van Gils et al., 2007 with permission. Photo by W. Tijsen.)

Outbreaks of HPAI virus in poultry have occurred on a relatively regular basis during the last decades and with the exception of HPAI H5N1 virus, these have all been contained by preventive measures focused on eradication of the causative agent, such as "stamping out" procedures aimed at infected poultry flocks and preemptive culling aimed at preventing the spread of the virus. Before the unprecedented spread of HPAI H5N1 viruses, there was only one report on the outbreak of an HPAI virus in a colony of common terns (Becker, 1966) with no direct evidence for association with poultry.

Compared to all other HPAI virus outbreaks, the current epizootic of HPAI H5N1 virus is highly unusual in many regards, such as the spread of HPAI H5N1 virus throughout Asia and into Europe and Africa, the large number of countries affected, the loss of hundreds of millions of poultry, the zoonotic transmission to humans and other mammals, the continuously changing genotypes and the spill-back of the virus into wild birds, leading to outbreaks and circulation of HPAI H5N1virus in those birds. The ancestral HPAI H5N1 virus likely originated from a virus circulating in domestic geese in Guandong province (Chen et al., 2004), China in 1996 and was introduced in Hong Kong poultry markets in 1997. The 1997 outbreak in poultry in Hong Kong led to another paradigm shift, as direct transmission of a purely avian influenza virus from poultry caused respiratory disease and death in humans, something that was previously deemed impossible (de Jong et al., 1997). After the containment of the HPAI H5N1 virus outbreak in Hong Kong, the virus reappeared in 2002 when it caused an outbreak in waterfowl and various other bird species in two waterfowl parks in Hong Kong (Ellis et al., 2004; Sturm-Ramirez et al., 2004; Sims et al., 2003). In 2003, the HPAI H5N1 virus was again transmitted to humans, leading to at least one fatal case. There is little information on the circulation of HPAI H5N1 virus from 1997 to 2002, but it is believed that the virus continued to circulate in China during that period. HPAI H5N1 virus resurfaced again in 2004 to spread across a large part of Southeast Asia, including Cambodia, China, Hong Kong, Indonesia, Japan, Laos, Malaysia, South Korea, Thailand, and Vietnam. Until 2005, HPAI H5N1 viruses had been isolated only sporadically from wild birds. In 2005, the first reported outbreak in wild migratory birds occurred in April–June at Lake Qinghai, China. This HPAI H5N1 virus outbreak in wild birds affected large numbers of birds such as bar-headed geese (*Anser indicus*), brown-headed gulls (*Larus brunnicephallus*), great black-headed gulls (*Larus ichthyaetus*), and great cormorants (*Phalacrocorax carbo*) (Liu et al., 2005; Chen et al., 2005). After the HPAI H5N1 virus outbreak in wild birds, the virus rapidly spread westwards across Asia, Europe, Middle East, and Africa. Affected wild birds have been reported in several countries, predominantly in mute swans (*Cygnus olor*), whooper swans (*Cygnus cygnus*), and tufted ducks (*Aythya fuligula*), although small numbers of cases in other species have been reported as well (raptors, gulls, and herons) (Olsen et al., 2006; Hesterberg et al., 2009).

While in the years prior to 2005 the transmission of the HPAI H5N1 virus is thought to have primarily occurred via movement of poultry and poultry products (Alexander and Brown, 2000; Gilbert et al., 2006; Sims et al., 2003), dispersal via wild migratory birds seems to be a likely route for several of the other reported outbreaks in addition to the poultry-based transmission routes. It has been much debated whether wild birds have played an active role in the geographic spread of the HPAI H5N1 virus. Some have argued that infected birds would be too severely affected to continue migration and would thus be unlikely to spread the HPAI H5N1 virus (Feare and Yasue, 2006). However, it has been shown that the pathogenesis of the HPAI H5N1 virus infection and the susceptibility of wild bird species to this infection may vary considerably, depending on bird species and previous exposure to viruses of the same or other avian influenza virus subtypes. Recent experimental infections suggest that preexposure to LPAI viruses of homologous or heterologous subtypes may result in partial immunity to HPAI H5N1 virus infection (Fereidouni et al., 2009). Such preexisting immunity might protect birds from developing severe disease upon infection but may still allow replication and thus shedding and spreading of the virus. Upon experimental HPAI H5N1 virus infection, some duck species proved to develop minor, if any, disease signs while still excreting the virus, predominantly from the respiratory tract, whereas other species developed a largely fatal infection that would not allow them to spread the virus efficiently over a considerable distance (Keawcharoen et al., 2008; Brown et al., 2006, 2008; Kalthoff et al., 2008). The outcome of HPAI H5N1 virus infections in wild bird species generally ranges from high morbidity and mortality (geese, swan, and certain duck species) to minimal morbidity without mortality (dabbling duck species). The present situation in Europe, where infected wild birds have been found in several countries that have not reported outbreaks in poultry (Hesterberg et al., 2009; Globig et al., 2009), suggests that wild birds can indeed carry the virus to previously unaffected areas. In addition, analysis of the spread of HPAI H5N1 virus indicated that it was likely that most of the introductions into several parts of Asia were likely poultry related, whereas in Europe most introductions were probably caused by migratory birds (Starick et al., 2008; Si et al., 2009). Although swan deaths have been the first indicator for the presence of the HPAI H5N1 virus in several European countries, this does not necessarily imply a role as predominant vectors; they could merely have functioned as sentinel birds infected via other migrating bird species.

Despite intensive surveillance programs, HPAI H5N1 virus has predominantly been found in dead wild birds (Artois et al., 2009; Hesterberg et al., 2009). Only in limited cases was HPAI H5N1 virus detected in apparently healthy birds (Chen et al., 2006a, 2006b). Many national surveillance programs aimed at the early detection of HPAI H5N1 virus have therefore focused on collecting samples from birds exhibiting morbidity or mortality. The intrinsic problem associated with establishing a clear idea of the prevalence of HPAI H5N1 virus in wild bird populations is the number of birds that have to be caught and sampled for this purpose. The more prevalent a virus is in the respective bird population, the fewer individuals need to be sampled to actually detect the virus. However, the number of birds that would need to be caught and sampled to detect viruses with a very low prevalence with a 95% probability of detection will rapidly become unfeasible, as may currently be the case with the lack of detection of HPAI H5N1 virus in wild bird populations (Artois et al., 2009) (Table 14.4).

TABLE 14.4 The Number of Birds Within a Given Population that Would Need to be Sampled to have a 95% Chance of Detecting Avian Influenza

	Number of Birds to be Sampled	
Prevalence (%)	95% CI[a]	99% CI
0.1	3000	4600
1	300	460
2	150	230
5	60	90
10	30	45

Adapted from Artois et al. (2009) with permission.

[a]95% CI is the confidence intervals of prevalence (of avian influenza) in an independent wild bird population for a given number of samples and determines the level of reliability. Increasing the desired confidence (e.g., from 95% to 99% CI) will result in a larger number of birds needed to be sampled.

This raises the question whether these infections have indeed become endemic in wild bird populations, or whether HPAI H5N1 virus is being reintroduced repeatedly by poultry or human activities. A recent study from China reported high prevalence of HPAI H5N1, suggesting that HPAI H5N1 circulates endemically in China (Kou et al., 2009). Whether HPAI H5N1 viruses will eventually also cross the Atlantic or the Pacific Oceans to reach the Americas, remains a matter of speculation.

14.15 CONCLUSION

Undoubtedly, the ecology of avian viruses is intrinsically related to the ecology of their host, as it is for all viruses. Birds form a highly distinct class of vertebrates with features that distinguish them from other groups of virus host organisms, most notably by their ability to fly. Few virus hosts are as mobile as birds. This mobility enables them to access a wide variety of environments and habitats. Birds are found throughout the world and on all continents, including remote places such as the world's oceans, the arctic tundra's, and Antarctica. One of the most important features of birds as potential vectors for emerging infectious diseases is the seasonal migration performed by many bird species, the regular movement between breeding and wintering grounds and the number of birds involved in these movements, which potentially allow very effective dispersal of pathogens over vast geographical areas and even between continents. The outbreaks of HPAI H5N1 virus and the westward progression of the virus caused an increased awareness of the role of wild birds in the spread and transmission of pathogens. As for many other species, birds have been shown to be infected with a wide range of different viruses. However, it is clear that only a very limited amount of the viral diversity within the avifauna has been identified to date. Unfortunately, the ecology of most avian viruses has only been studied to a limited extent, if studied at all.

To increase our understanding of the complex relationship between avian viruses and their hosts, it is crucial to integrate virus and host ecology within the currently performed long-term surveillance studies, not only for avian influenza viruses but also for other viruses. With the various large data sets available from the numerous ornithology networks around the globe and the incorporation of behavioral ecology in virus surveillance networks this will enable an interdisciplinary and integrated approach to study virus and avian host ecology. The incorporation of tools such as GPS-collars allows continuous tracking of birds, and subsequently the monitoring of infection status of individual wild birds in real time, allowing detailed investigations into virus prevalence in individual birds in relation to their social status, migration pattern, condition, habitat choice, and aggregation intensity. Furthermore, it will lead to more detailed information on the potential pathological effects of the infection, the effect of previous infections on reinfection, duration of virus shedding, and differences between avian viruses. The ecology of avian viruses is a fascinating and challenging field of research and the recent progress in understanding the relationship between avian influenza viruses and wild birds will hopefully have a stimulating impact on the study of the ecology of avian viruses in general.

REFERENCES

Alexander, D. J. (2000). A review of avian influenza in different bird species. *Vet. Microbiol.* 74, 3–13.

Alexander, D. J. and Brown, I. H. (2000). Recent zoonoses caused by influenza A viruses. *Rev. Sci. Tech.* 19, 197–225.

Alexander, D. J. and Brown, I. H. (2009). History of highly pathogenic avian influenza. *Rev. Sci. Tech.* 28, 19–38.

Artois, M., Bicout, D., Doctrinal, D., Fouchier, R., Gavier-Widen, D., Globig, A., Hagemeijer, W., Mundkur, T., Munster, V., and Olsen, B. (2009). Outbreaks of highly pathogenic avian influenza in Europe: the risks associated with wild birds. *Rev. Sci. Tech.-Offic. Int. Epizooties* 28, 69–92.

Becker, W. B. (1966). The isolation and classification of Tern virus: influenza A-Tern South Africa—1961. *J. Hyg. (Lond.)* 64, 309–320.

Bentz, P.-G. (1985). Studies on some urban Mallard Anas platyrhynchos populations in Scandinavia. Part I: causes of death, mortality and longevity among Malmö Mallards as shown by ringing recoveries. *Fauna Norv. Ser. C* 8, 44–56.

Brown, J. D., Goekjian, G., Poulson, R., Valeika, S., and Stallknecht, D. E. (2009). Avian influenza virus in water: infectivity is dependent on pH, salinity and temperature. *Vet. Microbiol.* 136, 20–26.

Brown, J. D., Stallknecht, D. E., Beck, J. R., Suarez, D. L., and Swayne, D. E. (2006). Susceptibility of North American ducks and gulls to H5N1 highly pathogenic avian influenza viruses. *Emerg. Infect. Dis.* 12, 1663–1670.

Brown, J. D., Stallknecht, D. E., and Swayne, D. E. (2008). Experimental infection of swans and geese with highly pathogenic avian influenza virus (H5N1) of Asian lineage. *Emerg. Infect. Dis.* 14, 136–142.

Caron, A., Gaidet, N., de Garine-Wichatitsky, M., Morand, S., and Cameron, E. Z. (2009). Evolutionary biology, community ecology and avian influenza research. *Infect. Genet. Evol.* 9, 298–303.

Chen, H., Deng, G., Li, Z., Tian, G., Li, Y., Jiao, P., Zhang, L., Liu, Z., Webster, R. G., and Yu, K. (2004). The evolution of H5N1 influenza viruses in ducks in southern China. *Proc. Natl. Acad. Sci. U. S. A.* 101, 10452–10457.

Chen, H., Li, Y., Li, Z., Shi, J., Shinya, K., Deng, G., Qi, Q., Tian, G., Fan, S., Zhao, H., Sun, Y., and Kawaoka, Y. (2006a). Properties and dissemination of H5N1 viruses isolated during an influenza outbreak in migratory waterfowl in western China. *J. Virol.* 80, 5976–5983.

Chen, H. X., Shen, H. G., Li, X. L., Zhou, J. Y., Hou, Y. Q., Guo, J. Q., and Hu, J. Q. (2006b). Seroprevalance and identification of influenza A virus infection from migratory wild waterfowl in China (2004–2005). *J. Vet. Med. B Infect. Dis. Vet. Public Health* 53, 166–170.

Chen, H., Smith, G. J., Zhang, S. Y., Qin, K., Wang, J., Li, K. S., Webster, R. G., Peiris, J. S., and Guan, Y. (2005). Avian flu: H5N1 virus outbreak in migratory waterfowl. *Nature* 436, 191–192.

de Jong, J. C., Claas, E. C., Osterhaus, A. D., Webster, R. G., and Lim, W. L. (1997). A pandemic warning? *Nature* 389, 554.

Del hoyo, J., Elliot, A.,and Sargatal, J. (eds.) (1996). *Handbook of the Birds of the World.* Lynx Edicions, Barcelona.

Dugan, V. G., Chen, R., Spiro, D. J., Sengamalay, N., Zaborsky, J., Ghedin, E., Nolting, J., Swayne, D. E., Runstadler, J. A., Happ, G. M., Senne, D. A., Wang, R., Slemons, R. D., Holmes, E. C., and Taubenberger, J. K. (2008). The evolutionary genetics and emergence of avian influenza viruses in wild birds. *PLoS Pathog.* 4, e1000076.

Dusek, R. J., Bortner, J. B., DeLiberto, T. J., Hoskins, J., Franson, J. C., Bales, B. D., Yparraguirre, D., Swafford, S. R., and Ip, H. S. (2009). Surveillance for high pathogenicity avian influenza virus in wild birds in the Pacific Flyway of the United States, 2006–2007. *Avian Dis.* 53, 222–230.

Ellis, T. M., Bousfield, R. B., Bissett, L. A., Dyrting, K. C., Luk, G. S., Tsim, S. T., Sturm-Ramirez, K., Webster, R. G., Guan, Y., and MalikPeiris, J. S. (2004). Investigation of outbreaks of highly pathogenic H5N1 avian influenza in waterfowl and wild birds in Hong Kong in late 2002. *Avian Pathol.* 33, 492–505.

Ellstrom, P., Latorre-Margalef, N., Griekspoor, P., Waldenstrom, J., Olofsson, J., Wahlgren, J., and Olsen, B. (2008). Sampling for low-pathogenic avian influenza A virus in wild Mallard ducks: oropharyngeal versus cloacal swabbing. *Vaccine* 26, 4414–4416.

Feare, C. J. and Yasue, M. (2006). Asymptomatic infection with highly pathogenic avian influenza H5N1 in wild birds: how sound is the evidence? *Virol. J.* 3, 96.

Fereidouni, S. R., Starick, E., Beer, M., Wilking, H., Kalthoff, D., Grund, C., Hauslaigner, R., Breithaupt, A., Lange, E. & Harder, T. C. 2009. Highly pathogenic avian influenza virus infection of mallards with homo- and heterosubtypic immunity induced by low pathogenic avian influenza viruses. *PLoS ONE*, 4(8), e6706.

Fereidouni, S. R., Werner, O., Starick, E., Beer, M., Harder, T. C., Aghakhan, M., Modirrousta, H., Amini, H., Moghaddam, M. K., Bozorghmehrifard, M. H., Akhavizadegan, M. A., Gaidet, N., Newman, S. H., Hammoumi, S., Cattoli, G., Globig, A., Hoffmann, B., Sehati, M. E., Masoodi, S., Dodman, T., Hagemeijer, W., Mousakhani, S., and Mettenleiter, T. C. (2010). Avian influenza virus monitoring in wintering waterbirds in Iran, 2003–2007. *Virol. J.* 7, 43.

Flint, P. L. and Franson, J. C. (2009). Does influenza A affect body condition of wild mallard ducks, or vice versa? *Proc. R. Soc. B-Biol. Sci.* 276, 2345–2346.

Fouchier, R. A., Munster, V., Wallensten, A., Bestebroer, T. M., Herfst, S., Smith, D.,

Rimmelzwaan, G. F., Olsen, B., and Osterhaus, A. D. (2005). Characterization of a novel influenza a virus hemagglutinin subtype (H16) obtained from black-headed gulls. *J. Virol.* 79, 2814–2822.

Fouchier, R. A., Schneeberger, P. M., Rozendaal, F. W., Broekman, J. M., Kemink, S. A., Munster, V., Kuiken, T., Rimmelzwaan, G. F., Schutten, M., Van Doornum, G. J., Koch, G., Bosman, A., Koopmans, M., and Osterhaus, A. D. (2004). Avian influenza A virus (H7N7) associated with human conjunctivitis and a fatal case of acute respiratory distress syndrome. *Proc. Natl. Acad. Sci. U.S.A.* 101, 1356–1361.

Gaidet, N., Dodman, T., Caron, A., Balanca, G., Desvaux, S., Goutard, F., Cattoli, G., Lamarque, F., Hagemeijer, W., and Monicat, F. (2007). Avian influenza viruses in water birds, *Africa. Emerg. Infect. Dis.* 13, 626–629.

Garamszegi, L. Z., and Moller, A. P. (2007). Prevalence of avian influenza and host ecology. *Proc. Biol. Sci.* 274, 2003–2012.

Gilbert, M., Chaitaweesub, P., Parakamawongsa, T., Premashthira, S., Tiensin, T., Kalpravidh, W., Wagner, H., and Slingenbergh, J. (2006). Free-grazing ducks and highly pathogenic avian influenza, *Thailand. Emerg. Infect. Dis.* 12, 227–234.

Globig, A., Staubach, C., Beer, M., Koppen, U., Fiedler, W., Nieburg, M., Wilking, H., Starick, E., Teifke, J. P., Werner, O., Unger, F., Grund, C., Wolf, C., Roost, H., Feldhusen, F., Conraths, F. J., Mettenleiter, T. C., and Harder, T. C. (2009). Epidemiological and ornithological aspects of outbreaks of highly pathogenic avian influenza virus H5N1 of Asian lineage in wild birds in Germany, 2006 and 2007. *Transbound. Emerg. Dis.* 56, 57–72.

Halvorson, D. A., Kelleher, C. J., and Senne, D. A. (1985). Epizootiology of avian influenza: effect of season on incidence in sentinel ducks and domestic turkeys in Minnesota. *Appl. Environ. Microbiol.* 49, 914–919.

Hanson, B. A., Luttrell, M. P., Goekjian, V. H., Niles, L., Swayne, D. E., Senne, D. A., and Stallknecht, D. E. (2008). Is the occurrence of avian influenza virus in Charadriiformes species and location dependent? *J. Wildlife Dis.* 44, 351–361.

Hanson, B. A., Stallknecht, D. E., Swayne, D. E., Lewis, L. A., and Senne, D. A. (2003). Avian influenza viruses in Minnesota ducks during 1998–2000. *Avian Dis.* 47, 867–871.

Hatchette, T. F., Walker, D., Johnson, C., Baker, A., Pryor, S. P., and Webster, R. G. (2004). Influenza A viruses in feral Canadian ducks: extensive reassortment in nature. *J. Gen. Virol.* 85, 2327–2337.

Haynes, L., Arzey, E., Bell, C., Buchanan, N., Burgess, G., Cronan, V., Dickason, C., Field, H., Gibbs, S., Hansbro, P., Hollingsworth, T., Hurt, A., Kirkland, P., McCracken, H., O'Connor, J., Tracey, J., Wallner, J., Warner, S., Woods, R., and Bunn, C. (2009). Australian surveillance for avian influenza viruses in wild birds between July 2005 and June 2007. *Aust. Vet. J.* 87, 266–272.

Hesterberg, U., Harris, K., Stroud, D., Guberti, V., Busani, L., Pittman, M., Piazza, V., Cook, A., and Brown, I. (2009). Avian influenza surveillance in wild birds in the European Union in 2006. *Influenza Other Respi. Viruses* 3, 1–14.

Hinshaw, V. S., Air, G. M., Gibbs, A. J., Graves, L., Prescott, B., and Karunakaran, D. (1982). Antigenic and genetic characterization of a novel hemagglutinin subtype of influenza A viruses from gulls. *J. Virol.* 42, 865–872.

Hinshaw, V. S., Webster, R. G., Bean, W. J., and Sriram, G. (1980a). The ecology of influenza viruses in ducks and analysis of influenza viruses with monoclonal antibodies. *Comp. Immunol. Microbiol. Infect. Dis.* 3, 155–164.

Hinshaw, V. S., Webster, R. G., and Turner, B. (1980b). The perpetuation of orthomyxoviruses and paramyxoviruses in Canadian waterfowl. *Can. J. Microbiol.* 26, 622–629.

Hubalek, Z. (2004). An annotated checklist of pathogenic microorganisms associated with migratory birds. *J. Wildlife Dis.* 40, 639–659.

Hurt, A. C., Hansbro, P. M., Selleck, P., Olsen, B., Minton, C., Hampson, A. W., and Barr, I. G. (2006). Isolation of avian influenza viruses from two different transhemispheric migratory shorebird species in Australia. *Arch. Virol.* 151, 2301–2309.

Jourdain, E., Gunnarsson, G., Wahlgren, J., Latorre-Margalef, N., Brojer, C., Sahlin, S., Svensson, L., Waldenstrom, J., Lundkvist, A., and Olsen, B. (2010). Influenza virus in a natural host, the mallard: experimental infection data. *PLoS ONE* 5, e8935.

Kalthoff, D., Breithaupt, A., Teifke, J. P., Globig, A., Harder, T., Mettenleiter, T. C., and Beer, M. (2008). Highly pathogenic avian influenza virus

(H5N1) in experimentally infected adult mute swans. *Emerg. Infect. Dis.* 14, 1267–1270.

Kawaoka, Y., Chambers, T. M., Sladen, W. L., and Webster, R. G. (1988). Is the gene pool of influenza viruses in shorebirds and gulls different from that in wild ducks? *Virology* 163, 247–250.

Keawcharoen, J., van Riel, D., van Amerongen, G., Bestebroer, T., Beyer, W. E., van Lavieren, R., Osterhaus, A. D., Fouchier, R. A., and Kuiken, T. (2008). Wild ducks as long-distance vectors of highly pathogenic avian influenza virus (H5N1). *Emerg. Infect. Dis.* 14, 600–607.

Kida, H., Yanagawa, R., and Matsuoka, Y. (1980). Duck influenza lacking evidence of disease signs and immune response. *Infect. Immun.* 30, 547–553.

Kleijn, D., Munster, V. J., Ebbinge, B. S., Jonkers, D. A., Muskens, G. J., Van Randen, Y., and Fouchier, R. A. (2010). Dynamics and ecological consequences of avian influenza virus infection in greater white-fronted geese in their winter staging areas. *Proc. Biol. Sci.* 277(1690), 2041–2048.

Koehler, A. V., Pearce, J. M., Flint, P. L., Franson, J. C., and Ip, H. S. (2008). Genetic evidence of intercontinental movement of avian influenza in a migratory bird: the northern pintail (*Anas acuta*). *Mol. Ecol.* 17, 4754–4762.

Kou, Z., Li, Y. D., Yin, Z. H., Guo, S., Wang, M. L., Gao, X. B., Li, P., Tang, L. J., Jiang, P., Luo, Z., Xin, Z., Ding, C., He, Y. B., Ren, Z. Y., Cui, P., Zhao, H. F., Zhang, Z., Tang, S. A., Yan, B. P., Lei, F. M., and Li, T. X. (2009). The survey of H5N1 flu virus in wild birds in 14 provinces of China from 2004 to 2007. *PLoS ONE* 4(9), e6926.

Krauss, S., Obert, C. A., Franks, J., Walker, D., Jones, K., Seiler, P., Niles, L., Pryor, S. P., Obenauer, J. C., Naeve, C. W., Widjaja, L., Webby, R. J., and Webster, R. G. (2007). Influenza in migratory birds and evidence of limited intercontinental virus exchange. *PLoS Pathog*. 3, e167.

Krauss, S., Walker, D., Pryor, S. P., Niles, L., Chenghong, L., Hinshaw, V. S., and Webster, R. G. (2004). Influenza A viruses of migrating wild aquatic birds in North America. *Vector Borne Zoonotic Dis.* 4, 177–189.

Lamb, R. A. and Krug, R. M. (2001). Orthomyxoviridae: the viruses and their replication. In: Knipe, D. M.& Howley, P. M. (eds.), *Fields Virology,* 4th edition. Lippincott Williams & Wilkins, Philadelphia, PA, pp. 1487–1531.

Latorre-Margalef, N., Gunnarsson, G., Munster, V. J., Fouchier, R. A. M., Osterhaus, A. D. M. E., Elmberg, J., Olsen, B., Wallensten, A., Fransson, T., Brudin, L., and Waldenstrom, J. (2009a). Does influenza A affect body condition of wild mallard ducks, or vice versa? A reply to Flint and Franson. *Proc. R. Soc. B-Biol. Sci.* 276, 2347–2349.

Latorre-Margalef, N., Gunnarsson, G., Munster, V. J., Fouchier, R. A., Osterhaus, A. D., Elmberg, J., Olsen, B., Wallensten, A., Haemig, P. D., Fransson, T., Brudin, L., and Waldenstrom, J. (2009b). Effects of influenza A virus infection on migrating mallard ducks. *Proc. Biol. Sci.* 276, 1029– 1036.

Liu, J., Xiao, H., Lei, F., Zhu, Q., Qin, K., Zhang, X. W., Zhang, X. L., Zhao, D., Wang, G., Feng, Y., Ma, J., Liu, W., Wang, J., and Gao, G. F. (2005). Highly pathogenic H5N1 influenza virus infection in migratory birds. *Science* 309, 1206.

Macken, C. A., Webby, R. J., and Bruno, W. J. (2006). Genotype turnover by reassortment of replication complex genes from avian influenza A virus. *J. Gen. Virol.* 87, 2803–2815.

Makarova, N. V., Kaverin, N. V., Krauss, S., Senne, D., and Webster, R. G. (1999). Transmission of Eurasian avian H2 influenza virus to shorebirds in North America. *J. Gen. Virol.* 80(Pt. 12), 3167–3171.

Munster, V. J., Baas, C., Lexmond, P., Bestebroer, T. M., Guldemeester, J., Beyer, W. E. P., de Wit, E., Schutten, M., Rimmelzwaan, G. F., Osterhaus, A. D. M. E., and Fouchier, R. A. M. (2009). Practical considerations for high-throughput influenza A virus surveillance studies of wild birds by use of molecular diagnostic tests. *J. Clin. Microbiol.* 47, 666–673.

Munster, V. J., Baas, C., Lexmond, P., Waldenstrom, J., Wallensten, A., Fransson, T., Rimmelzwaan, G. F., Beyer, W. E., Schutten, M., Olsen, B., Osterhaus, A. D., and Fouchier, R. A. (2007). Spatial, temporal, and species variation in prevalence of influenza A viruses in wild migratory birds. *PLoS Pathog*. 3, e61.

Munster, V. J. and Fouchier, R. A. (2009). Avian influenza virus: of virus and bird ecology. *Vaccine* 27, 6340–6344.

Munster, V. J., Veen, J., Olsen, B., Vogel, R., Osterhaus, A. D., and Fouchier, R. A. (2006). Towards improved influenza A virus surveillance in migrating birds. *Vaccine* 24(44–46) 6729–6733.

Munster, V. J., Wallensten, A., Baas, C., Rimmelzwaan, G. F., Schutten, M., Olsen, B., Osterhaus, A. D., and Fouchier, R. A. (2005). Mallards and highly pathogenic avian influenza ancestral viruses, northern Europe. *Emerg. Infect. Dis.* 11, 1545–1551.

Obenauer, J. C., Denson, J., Mehta, P. K., Su, X., Mukatira, S., Finkelstein, D. B., Xu, X., Wang, J., Ma, J., Fan, Y., Rakestraw, K. M., Webster, R. G., Hoffmann, E., Krauss, S., Zheng, J., Zhang, Z., and Naeve, C. W. (2006). Large-scale sequence analysis of avian influenza isolates. *Science* 311, 1576–1580.

Olsen, B., Munster, V. J., Wallensten, A., Waldenstrom, J., Osterhaus, A. D., and Fouchier, R. A. (2006). Global patterns of influenza a virus in wild birds. *Science* 312, 384–388.

Parmley, E. J., Bastien, N., Booth, T. F., Bowes, V., Buck, P. A., Breault, A., Caswell, D., Daoust, P. Y., Davies, J. C., Elahi, S. M., Fortin, M., Kibenge, F., King, R., Li, Y., North, N., Ojkic, D., Pasick, J., Pryor, S. P., Robinson, J., Rodrigue, J., Whitney, H., Zimmer, P., and Leighton, F. A. (2008). Wild bird influenza survey, Canada, 2005. *Emerg. Infect. Dis.* 14, 84–87.

Pearce, J. M., Ramey, A. M., Flint, P. L., Koehler, A. V., Fleskes, J. P., Franson, J. C., Hall, J. S., Derksen, D. V., and Ip, H. S. (2009). Avian influenza at both ends of a migratory flyway: characterizing viral genomic diversity to optimize surveillance plans for North America. *Evol. Appl.* 2, 457–468.

Roche, B., Lebarbenchon, C., Gauthier-Clerc, M., Chang, C. M., Thomas, F., Renaud, F., van der Werf, S., and Guegan, J. F. (2009). Water-borne transmission drives avian influenza dynamics in wild birds: the case of the 2005–2006 epidemics in the Camargue area. *Infect. Genet. Evol.* 9, 800–805.

Rohani, P., Earn, D. J., and Grenfell, B. T. (1999). Opposite patterns of synchrony in sympatric disease metapopulations. *Science* 286, 968–971.

Scott, D. A. and Rose, P. M. (1996). *Atlas of Anatidae Populations in Africa and Western Eurasia*. Wetlands International, Wageningen, Netherlands.

Sharshov, K., Silko, N., Sousloparov, I., Zaykovskaya, A., Shestopalov, A., and Drozdov, I. (2010). Avian influenza (H5N1) outbreak among wild birds, Russia, 2009. *Emerg. Infect. Dis.* 16, 349–351.

Si, Y. L., Skidmore, A. K., Wang, T. J., de Boer, W. F., Debba, P., Toxopeus, A. G., Li, L., and Prins, H. H. T. (2009). Spatio-temporal dynamics of global H5N1 outbreaks match bird migration patterns. *Geospatial Health* 4, 65–78.

Siembieda, J. L., Johnson, C. K., Cardona, C., Anchell, N., Dao, N., Reisen, W., and Boyce, W. (2010). Influenza A viruses in wild birds of the Pacific flyway, 2005–2008. *Vector Borne Zoonotic Dis.* 10(8), 793–800.

Sims, L. D., Ellis, T. M., Liu, K. K., Dyrting, K., Wong, H., Peiris, M., Guan, Y., and Shortridge, K. F. (2003). Avian influenza in Hong Kong 1997–2002. *Avian Dis.* 47, 832–838.

Stallknecht, D. E., Kearney, M. T., Shane, S. M., and Zwank, P. J. (1990). Effects of pH, temperature, and salinity on persistence of avian influenza viruses in water. *Avian Dis.* 34, 412–418.

Starick, E., Beer, M., Hoffmann, B., Staubach, C., Werner, O., Globig, A., Strebelow, G., Grund, C., Durban, M., Conraths, F. J., Mettenleiter, T., and Harder, T. (2008). Phylogenetic analyses of highly pathogenic avian influenza virus isolates from Germany in 2006 and 2007 suggest at least three separate introductions of H5N1 virus. *Vet. Microbiol.* 128, 243–252.

Sturm-Ramirez, K. M., Ellis, T., Bousfield, B., Bissett, L., Dyrting, K., Rehg, J. E., Poon, L., Guan, Y., Peiris, M., and Webster, R. G. (2004). Reemerging H5N1 influenza viruses in Hong Kong in 2002 are highly pathogenic to ducks. *J. Virol.* 78, 4892–4901.

van de Kam, J., Ens, B., Piersma, T., and Zwarts, L. (2004). *Shorebirds: An Illustrated Behavioural Ecology*. KNNV Publishers, Utrecht.

van Gils, J. A., Munster, V. J., Radersma, R., Liefhebber, D., Fouchier, R. A., and Klaassen, M. (2007). Hampered foraging and migratory performance in swans infected with low-pathogenic avian influenza A virus. *PLoS ONE* 2, e184.

Wahlgren, J., Waldenstrom, J., Sahlin, S., Haemig, P. D., Fouchier, R. A. M., Osterhaus, A. D. M. E., Pinhassi, J., Bonnedahl, J., Pisareva, M., Grudinin, M., Kiselev, O., Hernandez, J., Falk, K. I., Lundkvist, A., and Olsen, B. (2008) Gene segment reassortment between American and Asian lineages of avian influenza virus from waterfowl in the Beringia area. *Vector Borne Zoonotic Dis.* 8, 783–790.

Wallensten, A., Munster, V. J., Elmberg, J., Osterhaus, A. D., Fouchier, R. A., and Olsen,

B. (2005). Multiple gene segment reassortment between Eurasian and American lineages of influenza A virus (H6N2) in Guillemot (*Uria aalge*). *Arch. Virol.* 150, 1685–1692.

Wallensten, A., Munster, V. J., Karlsson, M., Lundkvist, A., Brytting, M., Stervander, M., Osterhaus, A. D., Fouchier, R. A., and Olsen, B. (2006). High prevalence of influenza A virus in ducks caught during spring migration through Sweden. *Vaccine* 24(44–46) 6734–6735.

Wallensten, A., Munster, V. J., Latorre-Margalef, N., Brytting, M., Elmberg, J., Fouchier, R. A., Fransson, T., Haemig, P. D., Karlsson, M., Lundkvist, A., Osterhaus, A. D., Stervander, M., Waldenstrom, J., and Bjorn, O. (2007). Surveillance of influenza A virus in migratory waterfowl in northern Europe. *Emerg. Infect. Dis.* 13, 404–411.

Webster, R. G., Bean, W. J., Gorman, O. T., Chambers, T. M., and Kawaoka, Y. (1992). Evolution and ecology of influenza A viruses. *Microbiol. Rev.* 56, 152–179.

Webster, R. G. and Rott, R. (1987). Influenza virus A pathogenicity: the pivotal role of hemagglutinin. *Cell* 50, 665–666.

Widjaja, L., Krauss, S. L., Webby, R. J., Xie, T., and Webster, R. G. (2004). Matrix gene of influenza a viruses isolated from wild aquatic birds: ecology and emergence of influenza a viruses. *J. Virol.* 78, 8771–8779.

Winker, K., Spackman, E., and Swayne, D. E. (2008). Rarity of influenza A virus in spring shorebirds, southern Alaska. *Emerg. Infect. Dis.* 14, 1314–1316.

Wright, P. F. and Webster, R. G. (2001). Orthomyxoviruses. In: Knipe, D. M. and Howley, P. M. (eds.), *Fields Virology*, 4th edition. Lippincott Williams & Wilkins, Philadelphia, PA, pp. 1533–1579.

zu Dohna, H., Li, J., Cardona, C. J., Miller, J., and Carpenter, T. E. (2009). Invasions by Eurasian avian influenza virus H6 genes and replacement of the virus' North American clade. *Emerg. Infect. Dis.* 15, 1040–1045.

INDEX

Studies in Viral Ecology: Animal Host Systems: Volume 2, First Edition. Edited by Christon J. Hurst.